Gerd Herold

HEROLD's INTERNAL MEDICINE

First English Edition

Volume Two

ISBN: 978-1-4467-6368-1

This book is available at Lulu.com (ID 10003821)

Author/Publisher:
Dr. med. Gerd Herold
Bernhard-Falk-Str. 27
D - 50737 Köln
Germany

www.herold-internal-medicine.com

Dedicated to my parents

The translation into English was done by:

Ralf Brücker (physician):
Endocrinology, Haematology

CLS Communication, Basel (translation service):
Parts of: Cardiology, Metabolic System Disorders, Rheumatology

Hedwig Möller (certified translator):
Gastroenterology, Pulmonology

Sylvia Möller (physician):
Angiology, Biochemistry and Haematology References, Infectious Diseases + Appendix, Nephrology, Salt and Water Homeostasis

Bettina Mues (physician):
Parts of: Cardiology, Metabolic System Disorders, Rheumatology

The following medical doctors did a proofreading of the translation: Talib Abubacker, Ben Braithwaite, Annabel Buxton, Saima Chaudary, Una Duffy, Tim Ladbrooke, Alastair Mitchell, Elisabeth Pearson, Connie Smith, Preema Vig

Project coordination: Björn H. Gemein

Legal Disclaimer

All diagnostic and therapeutic procedures in the field of science and medicine are continuously evolving:
Therefore the herein presented medical evidence is state of the technical and scientific art at the time of the editorial deadline for the respective edition of the book.

All details provided herein related to a particular therapeutic mode of administration and dosing of drugs are screened employing utmost measures of accuracy and precision.
Unless explicitly specified otherwise, all drug administration and dosing regimens presented herein are for healthy adults with normal renal and hepatic function.
Liability cannot be assumed for any type of dosing and administration regimen presented herein.
Each reader is advised to carefully consult the respective market authorization holders directions for use of the recommended drugs, medical devices and other means of therapy and diagnosis.

This applies in particular also to market authorization holders summaries of product characteristics for pharmaceutical products. Despite all care, there may be translation errors. The reader is advised to carefully check information related to the indication for use, contraindications, dosing recommendations, side effects and interactions with other medications!
All modes of medication use and administration are at the consumers risk. The author and his team cannot be held responsible for any damage caused by wrong therapy or diagnosis.

Please refer to, i.e: *www.leitlinien.de or www.guideline.gov*

The trade name of a trade name registered product does not provide the legal privilege to employ the trade name as a free trade mark even if it is not specifically marked as such.

Drugs which are sold as generics are referred to throughout the book by their generic name and not necessarily by a particular brand name.

Remark: ICD 10 code version 1.3 has been employed for the index. Since in the current edition ICD 10 code version 2.0 was used, different code numbers may have been assigned to indexed items!

No part of the book – neither in part or in toto – may be reproduced in any format (print, copy, microfilm, electronic storage, use and/or distribution by any electronic format including the internet) in the absence of an explicit written permission by the editor.

Be careful about reading health books. You may die of a misprint. (Mark Twain)

7

Abbreviations

AAA	Abdominal aortic aneurism	FUO	Fever of unknown origin
AB	Antibodies	GBM	Glomerular basement membrane
ACE	Angiotensin converting enzyme	GFR	Glomerular filtration rate
ADB	Anti-Desoxyribonucleotidase B	GI	Gastrointestinal
ADH	Antidiuretic hormone	GN	Glomerulonephritis
AET	Aetiology	HBV	Hepatitis B Virus
AF	Atrial fibrillation	HCT	Haematocrit
AG	Antigen	HI	Histology
AIDS	Acquired Immunodeficiency Syndrome	HIT	Heparin induced thrombocytopenia
AIHA	Autoimmune haemolytic anaemia	HIV	Human Immunodeficiency Virus
AN	Autonomic neuropathy	HLT	Half life time
ANA	Anti nuclear antibodies	HPV	Human Papilloma Virus
ANCA	Anti neutrophile cytoplasmatic antibodies	HSV	Herpes Simplex Virus
ANP	Atrial natriuretic peptide	HUS	Haemolytic uraemic syndrome
A.O.	And others	HX	History
APC	Activated Protein C	IA	Interaction
APS	Antiphospholipid syndrome	IBS	Irritable bowel syndrome
APPROX	Approximately	ICA	Internal carotid artery
ARDS	Adult respiratory distress syndrome	ICD	Implantable Cardioverter - Defibrillator
ARF	Acute renal failure	ICF	Intracellular fluid
ASAP	As soon as possible	ICP	Intracranial pressure
ASD	Atrial septum defect	ICU	Intensive care unit
ASL	Antistreptolysin	IFAT	Indirect immunofluorescence antigen test
AT	Antithrombin	IG	Immunoglobulin
AT	Antithrombin	IHA	Indirect Haemagglutinin test
ATP	Adenosine Triphosphate	I.M.	Intramuscular
AXR	Abdominal X-ray	INC	Incidence
BMI	Body Mass Index	IND	Indication
BMT	Bone marrow transplant	INR	International normalised ratio
BNP	brain natriuretic peptide	ISF	Interstitial fluid
BP	Blood pressure	ITP	Idiopathic thrombocytopenia
BU	Bread unit	IU	International units
BW	Body weight	I.V.	Intravenous
C	Celsius	IVF	Intravascular fluid
CA	Carcinoma	LAB	Laboratory tests
CA.	Circa	LAS	Lymphadenopathy syndrome
CDC	Centres for disease control	LDH	Lactate Dehydrogenase
CH	Carbohydrate	LDL	Low density lipoprotein
CHD	Coronary heart disease	LDV	Lymphocyte doubling time
CI	Contraindications	LMWH	Low molecular weight Heparin
CK	Creatine Kinase	LOC	Localisation
CL	Clinical picture	M	Male
CMV	Cytomegaly Virus	MDS	Myelodysplastic syndrome
CNP	Type C natriuretic peptide	MI	Mentzer index
CNS	Central Nervous System	MI	Myocardial infarction
CO	Complications	MIO	Million
CON	Contagiousness	MM	Multiple myeloma
COPD	Chronic obstructive pulmonary disease	MOA	Mode of action
COX	Cyclooxygenase	MRI	Magnet resonance imaging
CRP	C-reactive protein	MW	Molecular weight
CSE	Cholesterol Synthesis Enzyme	NK (cells)	Natural killer (cells)
CSF	Cerebrospinal Fluid	NSAID	Non steroidal anti-inflammatory drug
CT	Computer tomography	OAC	Oral anticoagulation
CU	Carbohydrate unit	OAD	Occlusive atherosclerotic disease
CVI	Chronic venous insufficiency	OCC	Occasionally
CVP	Central venous pressure	OCC	Occurrence
CXR	Chest X-ray	PA	Pulmonary artery
DD	Differential diagnosis	PAT	Pathogen
DEF	Definition	PAT	Pathology
DHD	Dengue haemorrhagic fever	PCR	Polymerase chain reaction
DHS	Dengue haemorrhagic shock	PE	Pulmonary embolism
DI	Diagnosis	PEP	Post exposure prophylaxis
DIC	Disseminated intravascular coagulation	PET	Positron emission tomography
DNA	Deoxyribonucleic acid	PFO	Persistent foramen ovale
DSA	Digital subtraction angiography	PG	Pathogenesis
DVT	Deep vein thrombosis	PI	Protease inhibitor
EBV	Epstein Barr Virus	PID	Pelvic inflammatory disease
ECF	Extracellular fluid	PMC	Pseudomembranous Enterocolitis
ECG	Electrocardiogram	PN	Parenteral nutrition
EEG	Electroencephalogram	PNH	Paroxysmal nocturnal haemoglobinuria
EF	Effects	POSS	Possibly
EHEC	Enterohaemorrhagic E. coli	PPC	Phenprocoumon
ELISA	Enzyme linked immunosorbent assay	PPH	Pathophysiology
EN	Enteral nutrition	PPSB	Prothrombin proconvertin Stuart-Power factor
ENT	Ear Nose & Throat		antihaemophilic factor B
EP	Epidemiology	PRG	Prognosis
ESP	Especially	PRO	Prophylaxis
ESR	Erythrocyte Sedimentation Rate	PTCA	Percutaneous transluminal coronary angioplasty
ET	Aetiology	PTH	Parathyroid hormone
EU	European Union	PTS	Post Thrombotic Syndrome
F	Female	PTT	Prothrombin time
FFP	Fresh Frozen Plasma	RA	Rheumatoid arthritis

RES	Reticular Endothelial Syncythium		TH	Therapy
RF	Rheumatoid factor		THR	Total hip replacement
RNA	Ribonucleic acid		TIA	Transitory ischaemic attack
RS	Raynaud's Syndrome		TOA	Thrombangiitis obliterans
RV	right ventricular		TPHA	Treponema Pallidum Haemagglutinin
SC	Subcutaneous		TPN	Total parenteral nutrition
SE	Side effects		TSH	Thyroid stimulation hormone
SIADH	Syndrome of inadequate ADH secretion		TURP	Transurethral resection of the prostate
SK	Streptokinase		UFH	Unfractioned Heparin
SLE	Systemic lupus erythematosus		URTI	Upper Respiratory Tract Infection
ST	Stage		US	Ultrasound
STD	Sexually transmitted disease		UTI	Urinary tract infection
SU	Sulfonyl urea		UV	Ultraviolet
SYM	Symptoms		VUR	Vesico-ureteric-renal reflux
SYN	Synonym		VV	Varicose veins
TAA	Thoracic aortic aneurism		VZV	Varicella Zoster Virus
TEA	Thrombendarteriectomy		WHO	World health organisation
TEE	Transoesophageal echocardiography		Y	Year

Find more medical abbreviations at   *www.medizinische-abkuerzungen.de*

Find more general abbreviations at   *www.acronymdb.com*
www.chemie.fu-berlin.de/cgi-bin/acronym

NOTE OF THANKS

My thanks for their support gos to:

Prof. Dr. med. E. Erdmann (Köln)   **Prof. Dr. med. W. Krone (Köln)**
Prof. Dr. med. R. Gross (Köln)   **Prof. Dr. med. H. Schicha (Köln)**

I'd like to thank these colleagues for their work on several chapters:

Dr. med. Schahin ALIANI
Arzt für Kinderheilkunde, Saarlouis
(Immunodeficiencies)

Dr. med. Christopher AMBERGER
Arzt für Innere Medizin/Rheumatologie, Grafschaft
(Rheumatology)

Prof. Dr. med. Helmut BAUMGARTNER
EMAH-Zentrum / Kardiologie, Universitätsklinikum Münster
(Acquired heart disease in adulthood)

Dr. med. Heinz BECKERS
Arbeitsmedizinisches Zentrum DEUTZ® AG, Köln
(several chapters, especially Infectious diseases)

Prof. Dr. med. Dr. h.c. Helmut BORBERG
Hämapherese-Zentrum, Köln
(Haemapheresis)

Dr. med. Dennis BÖSCH
Arzt für Innere Medizin, Bremen
(COPD)

Dr. med. Dipl. oec. med. Jürgen BRUNNER
Universitätsklinik Innsbruck, Kinder- und Jugendheilkunde
(Rheumatology, Hereditary fever syndromes)

Dr. med. Georg BÜHLER
Universitätsklinik Ulm
(Psychosomatic disorders, Eating disorders)

Dr. med. Ulrich DEUSS
Arzt für Innere Medizin / Endokrinologie, Köln
(Endocrinology)

Prof. Dr. med. Manfred O. DOSS
Konsultation Porphyrie, Marburg an der Lahn
(Porphyria)

Prof. Dr. med. Hans DREXLER
Arbeitsmedizinisches Institut der Universität Erlangen
(Occupational diseases)

PD Dr. med. Lothar FABER
Herz- und Diabeteszentrum NRW, Bad Oeynhausen
(Cardiomyopathies)

Prof. Dr. med. Meinrad GAWAZ
Universitätsklinikum Tübingen)
(Endocarditis)

Prof. Dr. med. Ulrich GERMING
Heinrich-Heine-Universität Düsseldorf
(Hematology)

Prof. Dr. med. Hartmut GÖBEL
Schmerzklinik Kiel
(Pain therapy)

Dr. med. Pontus HARTEN
Arzt für Innere Medizin; Strande
(Antiphospholipid syndrome, Rheumatology)

Dr. Barbara HAUER, MPH
Deutsches Zentralkomitee zur Bekämpfung der Tuberkulose (DZK), Berlin
(Tuberculosis and nontuberculous mycobacteriosis)

Dr. med. Joachim HEBE
Klinikum Links der Weser / Elektrophysiologie und Kardiologie
(Heart rhythm disturbances)

PD Dr. med. Jan Heidemann
Universitätsklinikum Münster
(Gastroenterology)

PD Dr. med. Tobias HEINTGES
Städt. Kliniken Neuß Lukaskrankenhaus
(Gastroenterology)

Dr. med. Björn HOFFMANN
Heinrich-Heine-Universität Düsseldorf
(Lysosomal storage diseases)

Dr. med. Guido HOLLSTEIN
Fachsanitätszentrum Kiel
(several chapters)

Dr. med. Alfred JANSSEN
Arzt für Innere Medizin / Angiologie / Phlebologie, Köln
(Angiology)

Prof. Dr. Dr. med. Harald KAEMMERER
Deutsches Herzzentrum München
(Congenital and acquired heart disease in adulthood, Marfan syndrome)

Prof. Dr. med. Joachim KINDLER
Medizinisches Zentrum Kreis Aachen, Würselen
(Hypertension, Nephrology)

Dr. med. Stefan KINTRUP
Arzt für Innere Medizin, Nephrologie, Dülmen
(Organ transplantation)

Dr. med. Peter KREBS
Arzt für Innere Medizin, Köln
(Eating disorders, Expertises)

Dr. med. Dirk LÖHR
Heidelberg
(several chapters)

Dr. med. Klaus MAGDORF
Deutsches Zentralkomitee zur Bekämpfung der Tuberkulose (DZK), Berlin
(Tuberculosis and nontuberculous mycobacteriosis)

Dr. med. Achim MELLINGHOFF
Arzt für Innere Medizin, Lindau-Bodensee
(several chapters)

Dr. med. Klaus-Peter MELLWIG
Herz- und Diabeteszentrum NRW, Bad Oeynhausen
(Coronary heart disease, Heart attack)

Dr. med. Guido MICHELS
Herzzentrum Köln
(several chapters)

Prof. Dr. med. Gynter MÖDDER
Arzt für Radiologie und Nuklearmedizin, Köln
(Thyroid)

Dr. med. Michael Montemurro
Centre Hôpitalier Universitaire Vaudoise (Lausanne)
(Oncological topics)

Prof. Dr. med. Kurt OETTE
Köln
(Lipid metabolism, Clinical chemistry, Laboratory values)

Dr. med. Mark OETTE
Heinrich-Heine-Universität Düsseldorf
(HIV/AIDS)

Prof. Dr. med. Hans-Georg PREDEL
Deutsche Sporthochschule Köln
(Physical Activity and Health)

Prof. Dr. med. Winfried RANDERATH
Krankenhaus Bethanien, Solingen
(Pneumology)

Dr. med. Alexander RÖTH
Universitätsklinik Essen
(Hematology)

Dr. med. Wolfgang SAUER
Arzt für Innere Medizin, Gastroenterologie, Angiologie, Bonn
(Angiology, Gastroenterology)

Dr. med. Henning Karl SCHMIDT
Herz- und Diabeteszentrum NRW, Bad Oeynhausen
(Coronary heart disease, Heart attack)

Dipl.-Psych. Josef SCHWICKERATH
Klinik Berus, Überherrn-Berus
(Mobbing)

Prof. Dr. med. Jörg SPITZ
Gesellschaft für medizinische Information und Prävention, Schlangenbach
(Smoking, Vitamin D)

PD Dr. med. Heinz ZOLLER
Universitätsklinik Innsbruck
(Siderosis/Haemochromatosis)

Thank you to the following colleagues for their hints:

Viktor Bäuerle (Augsburg)
Dr. med. J. Bargfrede (Köln)
Dr. med. D. Bastian (Ingolstadt)
Prof. Dr. med. K. Bauch (Chemnitz)
Dr. med. J. Beier (Köln)
Dr. med. J. Beller (Stuttgart)
Felix Bermpohl (Berlin)
Dr. med. R. Bergert (Berlin)
Gudrun Binder (München)
Dr. med. S. Binder (Bergisch-Gladbach)
Dr. med. H. Binsfeld (Drensteinfurt)
Prof. Dr. med. M. Blüher (Köln)
Dr. med. U. Böck (Dülmen)
Dr. med. B. Böll (Köln)
Prof. Dr. med. R. Braun (Genf)
Prof. Dr.H.-P. Brezinschek (Graz)
Patrick Brunner (Wien)
Angelika Bublak (Berlin)
Jan Bucerius (Köln)
Dr. med. Th. Butz (Bad Oeynhausen)
Dr. med. Dipl.-Psych. W. Carls (Überherrn-Berus)
Prof. Dr. med. Ch. Chaussy (München)
Dr. med. P. Dahl (Kassel)
Dr. med. A. Derstroff (Wiesbaden)
Prof. Dr. med. H.J. Deutsch (Frechen)
Dr. med. C. Dworeck (Berlin)
Knut Ehlen (Düsseldorf)
Michael Ehren (Aachen)
Manfred Eidt (Karlsruhe)
Oliver Eisen (Köln)
Dr. med. Th. Eisenbach (Leverkusen)
Knut Ehlen (Düsseldorf)
PD Dr. med. S. Fetscher (Lübeck)
Dr. med. F. Fortenbacher (Jettingen)
Dr. med. M. Friebe (Mönchengladbach)
Dr. med. J. Fuchs (Köln)
Patrick Gerner (Mainz)
Dr. med. T. Giesler (Erlangen)
Dr. med. B. Göhlen (Köln)
Dr. med. E.M. Göllmann (Dülmen)
Michael Göner (Münster)
Dr. med. K. Götz (Freiburg)
Dr. med. S. Götze (Berlin)
Dr. med. S. Gromer (Bad Schönborn)
Dr. med. F. Gundling (München)
Dr. med. S. Haack (Frankfurt a.M.)
Peter Häussermann (Bochum)
PD Dr. Dr. T. Haferlach (München)
Dr. med. H. Hagenström (Lübeck)
Anton Hahnefeld (München)
Andreas Hammer (Rauenberg)
Fabian Hammer, MD (Birmingham, UK)
Dr. med. G. Hansmann (Freiburg)
Carmen Heilmann (Jena)
Ursula Hein (Hannover)
Dr. med. W. Hein (Kassel)
Dr. med. U. Heinrich (Adendorf)
PD Dr. med. H. Herfarth (Regensburg)
Dr. med. B. Heßlinger (Freiburg)
Drs. med. D. und M. Hestermann (Bonn)
Ulrike Höcherl (Fürstenfeldbruck)

Arnd Hönig (Vellmar)
Dr. med. J. Hohlfeld (Hannover)
Dr. med. H. Hohn (Koblenz)
Dr. med. G. Hollstein (Kronshagen)
Dr. med. D. Holtermann (Moers)
Dr. med. Th. Holtmeier (Neustadt a.d.Waldnaab)
Dr. med. G. Hübner (Otterberg)
Alexander von Hugo (Hamburg)
Jan Humrich (Würzburg)
Ralf Husain (Berlin)
Dr. med. S. Jäckle (Villingen)
Dr. med. Ph. Jansen (Frankfurt a.M.)
Dr. med. M. Jost (Hamburg)
Dr. med. J. Jordan (Berlin)
Dr. med. C. Jürgensen (Heide)
PD Dr. med. W. Jung (Bonn)
Bernhard Kaess (München)
Dr. med. P. Kalin (Kiel)
Dr. med. J. Kavan (Dortmunde)
Dr. med. I. Kaya (Düsseldorf)
Dr. med. K. Kenn (Schönau/Königsee)
Akhtar Khawari (Leipzig)
Dr. med. P. Kirchhof (Münster)
G. Klausrick (Greifswald)
Dr. med. T. Klever (Bremen)
Dr. med. J. Klünemann (Regensburg)
Dr. med. G. Klug (Würzburg)
Dr. med. S. Klumpe (Münster)
Dr. med. M. Knechtelsdorfer (Wien)
Dr. med. T. Koch (Kassel)
Dr. med. M. Köhler (VS-Villingen)
Dr. med. M. Körner (Bad Oeynhausen)
Gabriele Komesker (Köln)
Prof. Dr. med. T. Kraus (Aachen)
Clemens Krauss (Graz)
Patrick Kreisberger (München)
Dr. med. H. Kriatselis (Nürnberg)
Andreas Krier (Mannheim)
Florian Krötz (München)
Dr. med. C. Krüger (Schriesheim)
Dr. med. D. Kügler (Halle/Saale)
Dr. med. M. Kunze (Villingen-Schwenningen)
Dr. med. M. Kupfer (Freising)
Dr. med. O. Laakmann (Mainz)
Dr. med. A.C. Lambrecht (Coesfeld)
Dr. med. M. Lange (Osnabrück)
Dr. med. J. Leidel (Köln)
Prof. Dr. med. A. Lechleuthner (Köln)
Dr. med. G. Lennartz (Recklinghausen)
Dr. med. J. Letzel (Niesky)
Jin Li (Heidelberg)
Drs. med. H.-J. + T. Lindner (Euskirchen)
Prof. Dr. med. R. Loddenkemper (Berlin)
Dr. med. M. Ludwig (Bonn)
Jan Dirks Lünemann (Berlin)
Dr. med. Th. Lüthy (Berlin)
PD Dr. med. L.S. Maier (Göttingen)
Dr. med. J. Maiß (Erlangen)
Nadja Makansi (Berlin)
Gerrit Matthes (Bochum)
Jan Matthes (Köln)

Dr. med. M.E. Meis (Vietnam)
Dr. med. D. Menche (Bremen)
Dr. med. U.J. Mey (Bonn)
Dr. med. F. Michold (Erlangen)
Dr. med. W. Mönch (Recklinghausen)
Dr. med. S. Moll (Chapel Hill, North Carolina)
Jens Mommsen (Bonn)
Dr. med. H. Montag (Wittlich)
Dr. med. F. Moos (Herdecke)
Dr. med. F. Moosig (Kiel)
Prof. Dr. V. Mühlberger (Innsbruck)
Dr. med. F. Müller (Stralsund)
L. Müller-Lobeck (Niedernhausen)
Dr. med. B. Mues (Köln)
Dr. med. A. Nacke (Wolfenbüttel)
Dr. med. Ch. Ndawula (Köln)
Dr. med. M. Neugebauer (Krefeld)
Dr. med. J. Neuss (Basel)
Dr. med. M. A. Neusser (Nürnberg)
Christian Nickel (Freiburg)
Thorsten Nickel (Kiel)
Dr. med. M. Opel (USA)
PD Dr. med. B. Otto (München)
Dr. med. V. Pabst (Aachen)
Dr. med. M. Parpart (Nigeria)
Dr. med. Graf A. von Perponcher (Tegernsee)
Dr. med. Th. Pfab (Berlin)
PD Dr. med. U. Platzbecker (Dresden)
Dr. med. Th. Poehlke (Münster)
Dr. med. J. Rachl (Graz)
Dr. med. J. Radke (Dresden)
Till Reckert (Tübingen)
Prof. Dr. med M. Reincke (München)
Dr. med. S. Reiter (Bonn)
Ltd. RMD Dr. med. H.-D. Reitz (Köln)
Andreas Reuland (Dossenheim)
Dr. med. S. Reuter (Ulm)
Dr. med. A. Ricke (Köln)
Dr. med. F. Rieder (Regensburg)
Dr. med. E. Ritter (Nürnberg)
Prof. Dr. med. I. Rockstroh (Bonn)
Lars Rommel (Eschweg)
Prof. Dr. med. P. Sawicki (Köln)
Dr. med. M. Schiffer (Hannover)
Dr. med. R. Schimpf (Mannheim)
Caroline Schirpenbach (Freiburg)
Dr. med. A. Schlesinger (München)
Dr. med. C. Schlüter (Nürnberg)
PD Dr. med. A. Schmidt-Matthiesen (Frankfurt a.M.)
Dr. med. M. Schneider (Offenbach)
Dr. med. N. Schönfeld (Berlin)
Dr. med. A. Schönian (Hage)
Dr. med. M. Schopen (Köln)

Dr. med. R. Schorn (Zug / Schweiz)
Martin Schünemann (Nörten-Hardenberg)
Dr. med. E. Schumacher (Köln)
Dr. med. J. M. Schwab (Tübingen)
Dr. med. S. Schwartz (Berlin)
Dr. med. W. Sicken (Mülheim a.d. Ruhr)
Christiane Siefker (Würzburg)
Dr. med. B. Siegmund (München)
Dr. med. A. Skarlos (Mannheim)
Dr. med. B. M. Stadler (Stuttgart)
Daniela Stennke (Berlin)
Dr. med. C. Sticherling (Berlin)
Dr. med. B. Stoschus (Bonn)
Dr. med. Switkowski (Berlin)
Dr. med. D. Tamm (Koblenz)
Andreas Theilig (Aachen)
Thore Thiesler (Gießen)
Prof. Dr. med. G. Trabert (Nürnberg)
Dr. med. F. Treusch (Villingen)
PD Dr. med. J. Truckenbrodt (Zeitz)
Gert Tuinmann (Göttingen)
Dr. med. M. Uffelmann (Gemünden)
Roland Ullrich (Köln)
Christian Vatter (Essen)
Prof. Dr. med. F. Vogel (Hofheim)
Dr. med. M. Vogel (Bonn)
Prof. Dr. med. A. Vogt (Köln)
Annett Wagner (Oerlinghausen)
PD Dr. med. A. A. Weber (Düsseldorf)
Dr. med. C. Weber (Berlin)
Hermann Weber (Fürstenfeldbruck)
Dr. med. Maria Weber (Bad Soden)
Dr. med. M. Weidenhiller (Erlangen)
Andreas Weimann (Erftstadt)
Gerrit Weimann (Linden)
Dr. med. D. Werner (Erlangen)
Dr. med. D. Werner-Füchtenbusch (Regensburg)
Dr. med. E. Wessinghage (Fulda)
Dr. med. T. Wetzel (Witten)
Dr. med. J. Wiechelt (Mainz)
Dr. med. H.-C. Wilken-Tergau (Celle)
Prof. Dr. med. U. J. Winter (Essen)
Dr. med. A. Wolff (Nürnberg)
Dr. med. T. Wollersheim (Köln)
Dr. med. S. Wüsten (Düsseldorf)
Özgür Yaldizli (Düsseldorf)
Dr. med. Ö. Yildiz (Uelzen)
Dr. med. G. Zachow (Berlin)
PD Dr. med. R. Zankovich (Köln)
Dr. med. L. Zell (Homburg/Saar)
Dr. med. R. Zell (München)
Dr. med. D. Zielske (Kiel)
Dr. med. M. Zimmermann (Philippsburg)

I'd like to thank

Prof. Dr. H.-P. Brezinschek (Medizinische Universitätsklinik Graz) for his cooperation on the chapter Osteoarthritis.

Dr. med. Björn Gemein, Frankfurt am Main, for creating my website and e-book and for managing the translation project

Dr. med. Robert Zell (München) for the „mini-compendium" on my website

Dr. med. Oliver Adolph, Ulm, for the new design of the graphics and for the pictures on the cover page

Dr. med. Angelika Demel (Günzburg) for several suggestions and for her lyrics and paintings on my website

Angelika Karger, Köln, for her help on typing

Dr. med. Heinz Beckers for his faithful editorial accompaniment

Gerd Herold

I. GASTROENTEROLOGY

Internet info: *www.dgvs.de* Deutsche Gesellschaft für Verdauungs- und Stoffwechselerkrankungen
(German Society for digestive and metabolic diseases

PANCREAS

1. Exocrine function – 2. Endocrine function
While diseases of the endocrine part of the pancreas (diabetes) do not show any exocrine dysfunction, in advanced chronic pancreatitis exocrine functional impairments and occ. also symptoms of an endocrine reduction of function can occur: insulinopenic diabetes.

Exocrine function:
About 1,5 l alkaline pancreatic secretion is produced, this consists of:
1. Water and ions (esp. HCO_3^- und Cl^-)

  HCO_3^- und Cl^- are secreted in the opposite ratio: with increase of the secrete amount the HCO_3^- concentration rises while the Cl^- concentration declines.

2. Digestive enzymes:
   - Proteolytic enzymes are produced for the protection of the pancreatic tissue as inactive cymogenics: trypsin, chymotrypsin, elastase, carboxypeptidase. The inactive cymogenics are activated only in the duodenum by the enterokinases produced there. This also applies to the phospholipase A.
   - Protease inhibitors: These deactivate prematurely activated proteases.
   - Amylase, lipase, nuclease (which are not able to attack own tissue) are secreted in active forms.
   Symptoms of a maldigestion occur if 90 % of the exocrine pancreas function has failed.

Regulation of the exocrine function:
1. Nerval: vagus nerve preferably stimulates the enzyme secretion.
2. Hormonal: The stimulus of the duodenal mucous membrane by hydrochloric acid, bile acids and foods leads to the secretion of hormones of the duodenal mucous membrane:
   - Secretin: stimulates the pancreas to producing water and HCO_3^-
   - Pancreocymin (= cholecystokinin): stimulates the pancreas to producing enzymes

Pancreas diagnostics:
- Imaging diagnostics:
  (Endo-) sonography, (angio-)CT, MRCP, ERCP, pancreaticoscopy
- Laboratory diagnostics:
  - Inflammation parameters: lipase, pancreas-elastase 1, pancreoisoamylase i.s.
  - Tumor marker: CA 19-9 (significant only for tumour aftercare examinations)
- Pancreas function testing:
  - Direct: secretin-pancreocymin test (most sensitive)
  - Indirect: fluorescein-dilaurat test; chymotrypsin and elastase 1 in the stool
- Bacteriologic, cytologic diagnostics: ultrasound-monitored fine needle puncture

ACUTE PANCREATITIS [K85]

Aet.:   1. Bile duct diseases = acute biliary pancreatitis (about 55 %): choledochal duct stones, stenosis of the ampulla of Vater
2. Alcohol abuse (about 35 %)
3. Other causes (10 %):
   - Drugs (2%): diuretics, beta-blockers, ACE inhibitors, methyldopa, oestrogens, glucocorticosteroids, antibiotics (erythromycin, rifampicin, tetracyclines), virostatics (Didanosin, Zalcitabin), anticonvulsants (valproic acid, carbamazepine), NSAIDs, Mesalazin, Sulfasalazin, gold, cyclosporin A, cytostatic drugs (Azathioprine, Mercaptopurin)
   - Hereditary pancreatitis: Rare, autosomal-dominant inheritance:
     - Mutation in the cationic trypsinogen = PRSS1 gene. Most frequent mutation R122H ~ 50% and N29 ~ 20%
     - Mutations of the serin-protease inhibitor Kazal type 1 = SPINK1-gene (N34S most frequent)
   Both groups of gene mutation can also lead to chronic pancreatitis
5. Other, less frequent causes
   - Abdominal trauma, after abdominal operations, after ERCP
   - Viral infections (e.g. mumps, AIDS, viral hepatitis)
   - Duodenal diverticulum (parapapillar), penetrating duodenal/gastric ulcer
   - Pure hypertriglyceridemia

- hypercalcemia (primary hyperparathyroidism)
- Ascarids in bile ducts
- pancreas divisum
- Pancreas transplantation → 2 forms of transplant pancreatitis: early form (= post-ischaemic transplant pancreatitis) - late form by rejection, drainage disturbance or cytomegalic infection.
- Idiopathic (exclusion diagnostics)

Severity degrees	Frequency	Lethality
I. Acute interstitial (oedematous) pancreatitis	80 - 85 %	0 %
II. Acute necrotic pancreatitis	15 - 20 %	
▶ with partial necrosis		about 15 %
III. ▶ with total necrosis		> 50 %

Progress:

Phase 1: Pancreatic oedema or necrosis: increase of pancreatic enzymes, CRP, leucocytes
Phase 2: Healing
Phase 3: Facultative in necrotic pancreatitis: infection of the necroses, sepsis, abscess: renewed increase of CRP and leucocytes

Pg

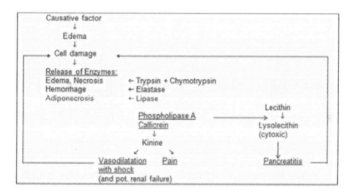

Cl.: Chief complaints: upper abdominal pain + increase of pancreatic enzymes i.s. + i.u.
Acute beginning with severe abdominal pain (90 %) which can radiate to all sides (also into the thorax → DD: cardiac infarction). Often, the pain stretches like a belt around the abdomen.

Further symptoms:	Frequency (%)
- Nausea, vomiting	85
- Meteorism, paralytic (sub)ileus	80
- Ascites	75
- Fever	60
- Shock signs, hypotonia	50
- ECG-changes (ST-distance)	30 (!)
- Pleural effusion left (right)	25
- Jaundice	20
- Face redness	

- Rarely bluish patches periumbilically (Cullen's sign) or in the flank area (Grey-Turner's sign) → prognostically unfavourable signs

Co.: • Bacterial infection of necroses with complication caused by sepsis
• Circulatory shock
• Consumption coagulopathy
• ARDS, acute renal failure
• Erosion of vessels with massive gastrointestinal haemorrhage, erosion of small or large bowel with development of intestinal fistulae
• Splenic and portal vein thrombosis
• Pancreatic abscess
• Post-acute pancreatic pseudocysts (5 %) [K86.3]

Di.:   - Fever, leucocytosis
           - Feeling of pressure in the epigastrium
           - Palpable resistance
           - Ultrasound, CT, MRI

Lab.:   ▶ Pancreatic enzyme diagnostics:
- Lipase and elastase 1 i.S. are pancreas-specific . For pancreatitis diagnose, determining the lipase is sufficient
- Amylase i.s. is not pancreas-specific, as it consists of iso amylases of which the salivary amylase constitutes the greatest part (60 %). In about 2% of the population, a slightly increased amylase is found physiologically. Increased values of total amylase are also found in extrapancreatic diseases (parotitis, acute abdomen, diabetic coma, alcololic intoxication).

Enzyme-increase < the 3fold of the upper norm does not indicate pancreatitis (this also applies to lipase)

Other causes of false-positive enzyme increases:
- Macroamylasemia: (about 0.5% of all people) → Cause: Complex formation of amylase with proteins, polysaccharides or hydroxyethylstarch (HES); as these complexes can not be eliminated renally, the urine amylase values are normal, as are Lipase/Elastase 1 i.S. ; harmless findings.
  Di.: low urine amylase, normal lipase and elastase 1 i.s.
- Familial idiopathic hyperamylasemy: Autosomal-dominant inheritance, mostly of no pathologic value.
- In renal insufficiency elevated amylase and lipase values are found (up to 3 times normal), since both enzymes are eliminated renally.

▶ Indicators of a necrotic pancreatitis with close progress monitoring: Persistent or renewed increasing values of CRP (> 15 mg/dl), LDH.
Other laboratory parameters are hardly more convincing than CRP: PMN-Elastase, phospholipase A_2, trypsin-activated peptide (TAP) and pancreatitis associated protein (PAP).

▶ In obstruction of the choledochal duct:
Increase of the cholestasis indicating enzymes (γGT, LAP, AP) and the conjugated bilirubin

▶ Prognostically unfavourable Laboratory parameters which may indicate necrotising pancreatitis:
- Leukocytosis > 16,000/μl
- Age > 55 years
- BMI > 30
- Serum-calcium concentration < 2 mmol/l
- Lactate dehydrogenase > 350 u/l
- Hyperglycemia, haematocrit increase
- Hypoxemia
- Creatinine increase
These parameters are also found in prognostic scores (e.g. Ranson-criteria).

Note: The level of the amylase and lipase does not give any indication of the severity and prognosis of the disease, however a decrease of the calcium i.s. The cause of the hypocalcemia is not exactly known (saponification of fatty acids with calcium bonding can play a part).

Imaging methods:
- Endo-/sonography: enlarged, distortion of the pancreatic contour, necroses, abscesses, pseudocysts; diagnosis of a peritonal-/pleural effusion, proof of biliary stones and an extrahepatic cholestasis in acute biliary pancreatitis.
  Remark.: The missing diagnosis of biliary stones does not exclude biliary genesis, since it might be that stones have already been excreted and microliths have escaped the imaging.
- Plain abdominal radiography: pancreas calcification? (chronic pancreatitis), biliary stone shadows .
  Air filled gastric intestinal segments are frequent in pancreatitis, esp. in the left upper-/middle abdomen.
- Angio-CT: Visualisation of necroses fails during the first days.
                - slight pancreatitis: interstitial oedematous pancreatitis
                - Severe pancreatitis: pancreatic necroses, pot. abscess streets
- Thorax-radiography: plate atelectases, pleural effusions, complicating basal pneumonia
- MRCP and ERC (No imaging of the pancreatic ducts with ERC → aggravation of the pancreatitis)
  Ind: With suspicion of obstruction of the choledochal duct
- Ultrasound controlled skinny needle puncture of necrosis-suspicious pancreatic areas with cytology (proof of necrotic cells) and bacteriology (proof of infected necroses; most frequently, bacterias of the intestinal flora, like Escherichia coli, enterococcus, klebsiella, pseudomonas aeruginosa are found.
  Ind: necrotising pancreatitis

DD.:   Often very difficult ! Unfortunately , frequent misdiagnoses !
      ▶ Acute abdomen:
           Leading symptoms of acute abdomen are:
           - severe pain of the abdomen (localized or diffuse)
           - peritoneal symptomatic (guarding)
           - disturbance of intestinal peristalsis (meteorism, nausea, vomiting)
           - bad general condition, circulatory trouble

- Ureter-/renal colic (microscopic haematuria, visceral – difficult to localize- convulsive pain, stone history, CT)
- Biliary colic (often simultaneous pancreatitis). Visceral pain difficult to localize; ultrasound: three-shifting of the gallbladder wall in acute cholecystitis, gallbladder hydrops, stone detection.
- Perforation of stomach/duodenum (ulcer), bowel (e.g. sigmoid diverticulitis), gallbladder. After initially severe pain possible temporary easing pain (" illusion"-stage), afterwards renewed pain increase. Abdomen in pancreatitis not hard as a board ("rubber abdomen")
Detection of free air in the abdomen: Air quantities starting with 1 ml can be detected by sonography ventrally to the liver . Larger volumes of air can also be identified in plain abdominal radiography with picture of the diaphragm: subphrenic air sickle? Radiography in left side position is more sensitive than in standing position. Free air can also be detected by CT.
- Mechanical ileus:
  1. Small bowel ileus (Adhesive strangulation of intestines in 50% hernias, malignant tumours, Crohn's disease)
  2. Large bowel ileus (carcinomas in 50%, diverticulitis, hernias, volvolus) In every acute abdomen there is more or less a paralytic ileus. Therefore, a mechanical ileus must be excluded:
  - Hyperperistalsy in a localized position with sonorous sound; sonography: Great enlargement of the small bowel in mechanical small bowel ileus with dilated fluid-filled loops of the small bowel preceding the stenose, small bowel with keyboard signs, pendular peristalsy; standing radiography:fluid surface formation ; search for hernial canals, old abdomen scars (adhesive strangulation ileus)
  - feculent stomach contents (probe) and/or vomiting of stool (= copremesis) is a late symptom of distal obstruction.
- Acute appendicitis:
  - Somatic (well locatable) pain
  - Pain in McBurney, Lanz (initially, often epigastric or paraomphalic)
  - Pain on percussion, (contralateral) rebound tenderness as a symptom of peritoneal irritation
  - Rovsing's pain in retrograde colon compression
  - Psoas pain (pain in the right hypogastrium with elevation of the stretched leg)
  - Douglas`s pain (rectal examination)
  - Temperature difference (rectal - axillary ≥ 1°C)
  - Leukocytes, neutrophils, CRP↑
  - Sonography (target sign, tubular structure, abscess)
- Strangulated abdominal wall hernia (search for hernias, examination of the hernial canal; positive Carnett's test: The pain is the same or worsens on tension of the abdominal wall.
- Mesenteric ischaemia: 2 forms: Oclusive form (OMI in mesenteric infarction and non-oclusive form (NOMI: history: visceral angina with postprandial abdominal pain? Bloody diarrhoea? → angiography))
- Gynaecologic diseases : Extrauterine pregnancy (absent period, positive pregnancy test, ultrasound), torsional or ruptured ovarian cyst/tumour (ultrasound), acute salpingitis, tubo-ovarian abscess.
▶ Acute attack of chronic pancreatitis (history)
▶ Myocardial infarction, esp. posterior myocardial infarction: difficult DD:
Both the infarction and the pancreatitis can show similar pain, collapse and ECG-changes: On one hand, the pancreatitis can show the typical picture of an outside layer damage with terminally negative T, on the other hand during the fresh infarction the ECG can be negative within the first 24 h. After all, even the transaminases can be slightly elevated during acute pancreatitis ! Decisive indications by pancreas enzymes, CK, CK-MB, troponin I and T.
As long as in differential diagnosis myocardial infarction is not excluded, no i.m.-injections (because of CK-increase and pot. lysis therapy)
▶ Pulmonary embolism with infradiaphragmatic symptomatic complex (history, echocardiography, D-dimers, Troponin I/T, BNP)
▶ Aneurysma dissecans (Echo CG, transesophageal)
▶ DD of unclear collapse:
Only about 5 % of all cases of acute pancreatitis do not show any pain, so that these patients might have only collapses (possible slight redness of the skin instead of paleness).
▶ Peudoperitonitis: precoma diabeticum / porphyria / Addisonian crisis / vasooclusive crisis with sickle cell anaemia

Procedure in acute abdomen:
1. History: Ulcer? Biliary stones? Last period, pregnancy testing (with women)?
2. Development of the pain (peracute: perforation)
3. Character of the pain: - constant (like in perionitis)
   - Rhythmic (colic)
4. Radiation (e.g. with ureter stone into the external genital organ)
5. Guarding (board-hard with generalized peritonitis)
6. Hernial canals (inguens,umbilicus)
7. Auscultation ("deathlike silence" or increasing rhythmic, ringing murmurs ?)
8. Rectal, gynaecologic examination
9. Temperature (rectal, axillary)

10. Laboratory, esp.:- CK-MB, troponin I/T   : myocardial infarction
  - Pancreatic enzymes   : pancreatitis
  - Blood glucose   : diabetic coma
  - Hoesch's test   : porphyria
  - Urine status   : microscopic haematuria with ureter stone
  - Blood count, creatinine, electrolytes and others

11. ECG, Echo CG
12. Ultrasound, thorax + general abdomen x-ray, spiral CT

Di.: • Of acute pancreatitis: history/clinical picture + lipase + ultrasound
  • Of biliary pancreatitis: Increase of transaminases, cholestase parameters : γGT, AP, bilirubin and stone detection by (endo)sonography, MRCP and ERCP
  • Of necrotising pancreatitis: fine-meshed progress monitoring of the clinical picture (pain, fever) and inflammation parameters (CRP; leukocytes, among other things) → in case of deterioration : angio-CT-scan or MRI
  • Of infected necrosis: Angio-CT: pot. gas inclusion in necrotic foci; fine needle aspiration (sonography- or CT-controlled) with bacteriologic examination

Th.: a) **Conservative**
  1. Close monitoring of the patient on the Intensive Care Unit:
    - Abdomen findings:pain? palpation, auscultation (peristalsis?) + ultrasound
    - Supplementary examinations: x-ray overview of the abdomen, chest x-ray, poss. angio-CT
    - Cardiovascular /volume status (blood pressure, pulse, CVP)
    - Balance of fluids, kidney function, electrolytes
    - Pancreatitis-relevant laboratory parameters: lipase, CRP, calcium, glucose, blood count + haematocrit, serum creatinine, blood gas analysis, coagulation state
  2. Fasting until resolution of pain. If reintroducing oral intake under therapy does not seem recommendable within 48 hours, early tube feeding. With enteral feeding, infectious complications are less frequent than in the parenteral one. When the patient is free of symptoms careful constitution of a light, lipid-restricted diet.
  A stomach probe usually is not necessary; indication: emesis, subileus
  3. Parenteral volume-, electrolyte- and glucose substitution: Since there often exists a considerable hypovolemia, 3 to 4 l/24 h are at least necessary (often more) ! Volume supply under CVP-control (target value: 6-10 cm of H_2O).
  4. Analgesics as required:
    - Light pain: e.g. tramadol i.v.
    - Severe pain: e.g. pethidine (Dolantin®) i.v.
      Morphine derivates are contraindicated (except for pethidin) because they produce papilla spasm.
  5. Prophylaxis of thromboembolism (low dose heparin, support stockings)
  6. Prophylaxis of stress ulcer with severe course (proton pump inhibitors)
  7. Indication for antibiotics with necrotising or biliary course, infected pseudocysts, abscess. Range of antibiotics: e.g. Carbapenemes (Imipenem or Meropenem) or Ciprofloxacin, combined with metronidazole, respectively . Duration: about 10 days
  8. Therapy of complications: Continuous venovenous or haemodialysis in acute renal failure, monitored ventilation in ARDS, appropriate antibiotics in case of sepsis

  b) **MINIMAL INVASIVE THERAPY:**
  1. Choledocholiths (mostly pre-papillar congestion): endoscopic papillotomy (EPT) + stone extraction
  2. Pancreatic pseudocysts can withdraw spontaneously in about 50 % of cases.Asymptomatic pseudocysts do not need to be treated. Symptomatic pseudocysts > 5 cm Ø are drained interventionally (percutaneous catheter drainage, endosonographic cysto gastroscopy or -duodenostomy)
  Co.: Bleeding, infection. The drainage occurs 6 weeks after development of the pseudocyst (after development of a cyst wall) at the earliest
  3. Pancreatic abscess: puncture drainage + irrigation

  c) **SURGICAL THERAPY:**
  Ind: infected pancreatic necrosis, failure of the minimal-invasive therapy
  Method: Careful digital necrectomy + lavage procedure (In open or closed technique)
  Hospital lethality: 15 %

Prg.: The progress of an acute pancreatitis is difficult to forecast. Various prognostic scores have been developed. (e.g. Ranson-/Imrie score→ see internet). A close intensive-care monitoring is decisive, in order to recognize necrotising pancreatitis with its complications early enough and treat it consistently. The mortality depends on the severity degree of the disease (see above). Infected necroses have a worse prognosis than aseptic necroses. Septic complications are most frequently the cause of death within the setting of necrotising pancreatitis.

Pro.: Clearance of possible causes: e.g. rescue of the bile ducts, avoiding of alcohol, treatment of hyperlipidemia, hyperparathyroidism, omission of pancreas-toxic drugs

Classification (Marseille 1984):

Chronic pancreatitis
a) With focal necrosis
b) With segmental or diffuse fibrosis
c) Calcifying
Variety: obstructive chronic pancreatitis (pancreas atrophy due to obstruction in the duct system)

Aet.: 1. Chronic alcohol abuse: about 80 %
2. Idiopathic (no detectable cause): 15 %
3. Other causes (5%):
- drugs (see acute pancreatitis)
- hypertriglyceridemia
- hyperparathyroidism
- hereditary pancreatitis: At present, mutations in the following 3 molecules are known of: Cationic trypsinogen (PRSS1), serinprotease-inhibitor Kazal type 1 (SPINK1) and chloride transporters (CFTR) = cystic fibrosis
- Autoimmune pancreatitis (AIP): In about 30% association with other autoimmune diseases; identification of auto-antibodies (ANA, RF) and IgG/IgG4 ↑
Hi.: Cellular lymphoplasmatic infiltrations (CD4-/CD8-positive)
Th.: Prednisolone

Cl.: Often, the clinical picture is oligosymptomatic.

1. Chief symptom is the relapsing pain which is not colicky (DD: biliary colic) and can last for hours up to days. Pain is present in over 90 % of cases. The pain is located deep in the epigastrium (palpation) and can radiate to both sides into the back (belt-like); occ. as a late pain after eating. The late stage of chronic pancreatitis is often pain-free again.

2. Food intolerance (fat): induction of dyspeptic discomforts, nausea, vomiting and pain

3. Maldigestion: weight loss, fatty stools, meteorism, diarrhoea. Symptoms of a maldigestion only appear when the exocrine pancreatic function is reduced to around 10 % of the norm.

4. Insulinopenic diabetes (about 1/3 of the patients with advanced disease)

5. Relapsing jaundice

Co.: • Pancreatic pseudocysts (with bleeding and haemobilia), abscess
• Spleen and portal vein thrombosis with portal hypertension
• Stenoses of the pancreatic duct system, formation of multiple intraductal concrements (pancreatolithiasis); pancreatic duct fistulae
• Stenosis of the distal choledochal duct with (relapsing) jaundice; duodenal stenosis
• Pancreatic carcinoma as late complication (esp. in hereditary pancreatitis)

DD.: • Acute relapsing pancreatitis
• Other epigastrium diseases, e.g. ulcer disease, gastric carcinoma, cholelithiasis
• Pancreatic carcinoma (internal ultrasound, MRCP/ERCP; HRCT, PET)

Di.: A) Proof of a pancreatic attack: Increase of pancreatic enzymes i.s.: lipase, elastase 1 (specific), amylase (less specific). Normal pancreatic enzymes do not exclude chronic pancreatitis!

B) Aetiologic clarification (in patients < 20 years with idiopathic pancreatitis exclusion of hereditary pancreatitis

C) Proof of an exocrine pancreatic insufficiency:

• Pancreocymin-secretin test: Most sensitive, but for the practice too extensive a direct pancreatic function test. First , the water and bicarbonate secretion is stimulated with secretine (i.v.): in the process, duodenal liquid is fractionally obtained by means of a probe and the HCO_3-concentration is measured. Afterwards, the pancreas is stimulated by pancreocymin in order to determine the amount of enzyme in the duodenal secretion liquid (amylase, lipase, trypsin, chymotrypsin).

• Indirect pancreatic function tests:
Due to low sensitivity the indirect pancreas function tests are not suitable for the early diagnosis of chronic pancreatitis:
- Fluorescein-Dilaurat-Testing (= pancreolauryl-test):
Orally administered Fluorescein-Dilaurat is split by pancreas-specific arylesterases. The split fluorescein is absorbed, conjugated in the liver and passed renally and measured in the urine. The elimination of fluorescein via urine correlates with the arylesterase secretion of the pancreas. False pathologic test results are relatively frequent (10 %): e.g. in malabsorption, liver cirrhosis, renal insufficiency
- Determination of chymotrypsin and elastase 1 in the stool:
In pancreatic insufficiency the stool concentrations are lowered, the intake of pancreatic enzyme

preparations must be stopped 5 days before chymotrypsin measurment. This does not apply to the determination of the elastase 1 which is also more sensitive than the chymotrypsin measurment.
Lowered values with normal pancreatic function are also found in diarrhoea, malabsorption and condition after Billroth II-operation.

C) Imaging methods:
- • Diagnosis of morphologic modifications of the pancreas:
  - - Pancreas calcifications: (endo-/sonography, plain film-X-ray of the upper abdomen, CT, MRI) indicate chronic pancreatitis and are found most frequently in alcohol-toxic chronic pancreatitis.
  - - Pancreatic duct stones and calibre irregularities of the pancreatic ducts (MRCP, ERCP, pancreaticoscopy): short-distance of stenoses and dilatation of the pancreatic duct (string -like)
- • Detection of complications:
  - - Choledochal stenosis (MRCP, ERCP)
  - - Duodenal stenosis (MDP)
  - - Pseudocysts (endo-/sonography, CT)
  - - Pseudocysts with bleeding (colour duplex)

  For diagnosis safeguarding, Score systems are used (e.g. Mayo-clinic score or the Lüneburg-score) → for details see internet

Th.: A) Causal: e.g. avoiding of alcohol
B) Symptomatic:

▶ Conservative:
1. Treatment of inflammatory attacks (as in acute pancreatitis)
2. Treatment of excretory pancreatic insufficiency:
   - • Carbohydrate-rich nutrition with frequent (5 - 7) small meals, ban on alcohol. In steatorrhoea increase of the lipase dose and pot. fat-reduced diet and supply of medium-chain fatty acids (MCT-fats) which can also be absorbed without splitting.
   - • Pancreatic enzyme substitution in case of exocrine pancreatic insufficiency. Preparations must have high enzyme activity, they have to be protected against the inactivating influence of the gastric juice (gastric juice resistant micropellets) and quickly be released at the place of effect. The dose must be adapted to the meals (3 x 1 dose/d is always wrong).
     An effective lipase substitution is important. 60,000 FIP units (Fédération Internationale Pharmaceutique) lipase are necessary to digest the fat content of an average meal → dos.: 25,000 – 50,000 u lipase/meal. The enzyme preparations are taken with the meals.
     It is controversial whether pancreatic enzyme substitution can reduce the pain in part of the patients.
   - • Supplementation of fat-soluble vitamins (ADEK)

3. Treatment of endocrine pancreatic insufficiency:
   In pancreatogenic diabetes small insulin doses + sufficient enzyme substitution (otherwise danger of hypoglycemia).

4. Treatment of pancreatic pain:
   - - Removal of outflow obstructions in the pancreatic duct system (protein precipitates, stones, strictures) reduces the pain in 50 % of cases, because pain correlates with pre-stenosal pressure.
   - - Avoidance of analgesics if possible (danger of addiction and analgesic nephropathy); morphine derivates are contraindicated .

▶ Endoscopic therapy:
- • Endoscopic treatment in case of pancreatic duct stones: endoscopic papillotomy + extracorporal shock wave lithotripsy (ESWL). Remaining fragments are removed endoscopically, e.g. with catch bag or extraction balloon; possibly laser lithotripsy.
- • Endoscopic treatment of pancreatic duct stenoses: balloon dilatation, possibly with subsequent support of plastic endoprostheses (stents). Various complications can arise from stents (haemorrhage, pancreatitis, stent occlusion, stent migration) so that changing a prosthesis is often necessary.
- • Endoscopic treatment of pancreatic pseudocysts and abscesses: Clarification by ERCP, whether the cyst or the abscess are connected with the pancreatic duct system or if a ductal stenosis is the cause. Depending on the finding, a transpapillar or transmural drainage is effectuated (cystogastric, cystoduodenal). A wait-and-see attitude can be taken towards asymptomatic pseudocysts .

▶ Surgery:
- • Drainage operations → Ind.:
  - - Isolated obstruction of the pancreatic duct → pancreatico-jejunostomy
  - - Isolated choledochal stenosis → choledocho-jejunostomy
  - - Large pseudocyst (with unsuccessful internal drainage) → cysto-jejunostomy
- • Partial resection of the pancreas → Ind.:
  Chronic pain symptoms, stenosis complications (pancreatic duct, choledochal duct, duodenum), portal and splenic vein thrombosis, fistulation, suspected carcinoma

Standard procedure:
Duodenum maintaining resection of the pancreatic head
The left-sided pancreas resection leads only in 50 % of cases to absence of pain after 2 yrs.

MUCOVISCIDOSIS [E84.9]

Syn: Cystic fibrosis (CF)

Def.: Autosomal-recessive hereditary disease in which the epithelial cell membranes have <u>defective chloride channels.</u> The <u>CFTR-gene</u> is on the long arm of chromosome 7 (7q31-32). Of about > 1,000 CFTR-mutations the mutation delta-F-508 is the most frequent in Western Europe (70 %). Consequence is a <u>pathological gene product,</u> the cystic fibrosis transmembrane regulator- <u>(CFTR-)protein.</u> This membrane protein represents chloride channels which cause the formation of <u>viscous</u> mucus secretions in all exocrine glands: pancreas, small bowel, bronchial system, bile ducts, gonads, sweat glands.

Remark.: The mutation of the CFTR-gene can lead to 4 different types of function defects:
1. Production of incomplete protein
2. Defective protein processing: CFTR does not reach its destination in the cell membrane (e.g. delta-F-508-mutation)
3. Defective regulation: mutation of the nucleotide bonding point
4. Defective Cl⁻-conductivity

Occ.: Most frequent congenital metabolism disease of the white population of Europe and of the USA.
Disease frequency (homozygous frequency) 1 : 2,500 births; gene carriers frequency (heterozygous frequency) about 4 % of the population.

Cl.: Differently marked progress according to the CFTR mutation.

Chief complaints:
- <u>Bowel:</u> Meconium ileus with the birth (10 %). In about 20% of older children and adolescents , occurrence of <u>d</u>istal intestinal <u>o</u>bstruction <u>s</u>yndromes (DIOS) = meconium ileus equivalents
- <u>Airways:</u> The extent of the pulmonary manifestation varies from patient to patient: chronically pertussive cough, relapsing bronchial infections (staphylococcus aureus, pseudomonas aeruginosa, Burkholderia cepacia and other gram-negative problematic germs), bronchiectases, obstructive emphysema
  Co.: Pulmonary hypertension and respiratory insufficiency, pneumothorax in about 10 % of cases, haemoptyses; allergic bronchopneumonic aspergillosis.
- <u>Pancreas:</u> Exocrine pancreatic insufficiency with chronic diarrhea and maldigestion syndrome; pancreatogenic diabetes
- <u>Liver and bile ducts:</u> Adult patients develop a biliary cirrhosis in 10 %; cholelithiasis
- Growth disorder and inadequate weight gain of the child
- Reduced fertility in women, infertility in men (bilateral vas deferens-obliteration

Di.: <u>Pilocarpine-iontophorese sweat test:</u> Cl content of the sweat > 60 mmol/l (in newborns > 90 mmol/l); determination of the CFTR-gene (also as a possible prenatal diagnostic).

<u>Newborn-screening</u> (still no routine test): Quantitative determination of trypsinogen in the blood (↑)

Th.:
- <u>Consultation by mucoviscidosis-ambulance.</u>
- <u>Symptomatic:</u>
  - Sufficient supply of NaCl
  - Mucolysis (including DNAse-inhalation therapy) and autogenic drainage of the viscous bronchial secretion (drainage positioning, tapping massage); tobramycin-inhalation for prophylaxis/therapy of pseudomonas infections; systemically targeted antibiotic therapy of bronchial infections; in case of spasms bronchospasmolytics
  - Substitution of pancreatic enzymes +supplementation dose of fat-soluble vitamins (ADEK).
  - Ursodeoxycholic acid in biliary cirrhosis.
  - In case of intestinal obstruction hyperosmolar enemas and oral application of intestine cleaning agents.
  - In case of increasing respiratory insufficiency long-time O₂ treatment and lung transplantation.
- <u>Somatic gene therapy:</u> transfer of healthy CFTR-genes (in clinical testing).

Prg.: Median life expectancy about 32 years (m>f)

PANCREATIC CARCINOMA [C25.9]

Internet Info: *www.med.uni-marburg.de/fapaca*

Occ.: Incidence: 10/100,000 inhabitants yearly; after colon and stomach carcinoma third-frequent tumour of the digestive tract. Average age of disease m: 67 years – f: 75 years; m > f

Aet.: Unknown, genetic predisposition is of some importance; risk factors: cigarette smoking (3fold increased risk) and alcohol abuse ; ASA possibly has a protective effect.

Hereditary syndromes with increased risk of pancreatic cancer

Syndrome of tumour predisposition	Gene	Rel. risk of pancr. carcinoma
Peutz-Jeghers-syndrome	STK 11	> 100 (1)
Hereditary pancreatitis	PRSS1	About 85 (1)
Familial pancreatic carcinoma	?	About 40 (2)
FAMMM- and pancreatic carcinoma-melanoma syndrome	CDKN2A	About 20 (1)
Familial breast- and ovarian carcinoma	BRCA2	About 5 (1)

(1) For carriers of the corresponding gene resp. germ line mutation
(2) For first degree-relatives of a person with pancreatic carcinoma

Definition of familial pancreatic carcinoma:
≥ 2 first degree reliaths with histologic diagnosis of pancreatic carcinoma or
≥ 3 second degree relatives with histologic diagnosis of pancreatic carcinoma, of whom
≥ 1 person with disease < 50 years

Prg.: Tumour progression model of ductal adenocarcinoma (DAC): The tumour progression from the normal tissue via preneoplastic duct lesions to DAC is caused by an accumulation of miscellaneous gene mutations: activation of the oncogene K-ras (100 %) and inactivation of the tumour suppressor genes: p53, p16, DPC4.

Preneoplastic duct lesions : Mucinous duct cell hypertrophy (PanIN1A) →ductal papillary hyperplasia = DPH (panIN1B) → DPH with moderate intraepithelial neoplasia = IEN (PanIN2) → severe ductal IEN (PanIN3) →DAC

Pat.: Most of the pancreatic carcinomas are adenocarcinomas which most frequently affect the pancreatic head (70 % of cases). The epithelium of the small pancreatic ducts is the starting point in 90% (ductal cancer), in 10 % the acinus epithelium (acinal carcinoma), early lymphogenicand haematogenic metastasis.
Papilla-(ampulla-)carcinomas are defined as an independent tumour group.
Pancreatic carcinoma is charatarised by:
1. difficult diagnosis, 2. difficult therapy, 3. bad prognosis

Cl.: The absence of early symptoms is a great diagnostic problem
  1. Symptoms as with chronic pancreatitis (difficult DD):
     • Loss of appetite, unspecific epigastrium pain, nausea, weight loss
     • Accompanying pancreatitis (lipase increase)
     Gnawing back pains are mostly a sign of inoperability.
  2. Jaundice: Can be an early symptom of pancreatic head carcinoma (25 %), in case of papillary carcinoma the jaundice can occur intermittently; in the late stage jaundice is mostly present (90 %).
     Courvoisier's sign (= chabby-elastic palpable painless gallbladder + jaundice) is a consequence of tumour-caused occlusion of the choledochal duct.
  3. Less frequent symptoms:
     - Thromboses, thrombophlebitis, thrombophlebitis migrans

     Note: Also think of carcinomas of the pancreas, stomach and the prostate in case of inexplicable relapsing thromboses

     - Pathologic glucose tolerance or diabetes
     - In the extremely rare "metastatic enzyme producing pancreatic adenoma", the triad: 1. purple, necrotic fat tissue nodes, 2. polyarthritis, 3. eosinophilia, is found.

Staging: TNM classification; simplified UICC-stage classification:
  I: Tumour limited to the pancreas
  II: Pancreas + neighbouring tissues affected
  III: Involvement of the regional lymph nodes
  IV: Distant metastases

DD: Chronic pancreatitis, jaundice of other origins, among other things

Di.:
- Sonography and endosonography = application of the ultrasound probe to the posterior wall of stomach/duodenum (most sensitive diagnostic)
- "One-stop-shop"-MRI = MRI with MRCP and MR-angiography (3D-MRA): Detection of pancreatic tumour, duct modifications (pancreatic duct break-off, prestenosal duct dilatation, choledochal stenosis, vessel break-off

  Note: Endosonography and "One-stop-shop"-MRI achieve the greatest score ratio (90 %). 10 % of cases can only be clarified intraoperatively. Tumours with a diameter under 1 cm can only be identified preoperatively by endosonography.

- Spiral-CT and ERCP: More extensive alternative to "One-stop-shop MRI"
- PET with FDG (Fluorodeoxyglucose): Early tumour detection; disadvantage: Limited availability, considerable expense
- Supplementary examinations:
  - Endoscopy of the pancreatic duct (pancreaticoscopy) with targeted biopsy of suspicious duct changes.
  - GIP (radiography imaging of the gastro-intestinal passage) + endoscopy: diagnosis of delayed complications: gastric outlet or duodenal stenosis, expansion of the duodenal C-form.
  - Determination of the tumour marker CA 19-9 and CA 50: These do not enable early diagnosis, but help monitoring the absence of relapses postoperatively. Elevated CA 19-9-values are also found in cholestasis and inflammatory diseases of the gastrointestinal tract and the liver.
  - Cytology of the pancreatic secretion
  - Celiacography and spleno-portography for the clarification of operability
  - In case of suspected familial pancreatic carcinoma molecular-genetic diagnostics

  Note: With fine needle biopsy danger of puncture channel metastasis (seeding) ! If all findings indicate curatively resectable pancreatic carcinoma, renunciation of fine needle biopsy and directly clarifying laparotomy are recommended.

Th.:   1. Radical operation with lymphadenectomy:
- Pancreatic head carcinoma on UICC-stages I-III:
Since the results from partial duodenopancreatectomy by Kausch-Whipple are similar to those of the less radical pylorus-maintaining partial duodenopancreatectomy, the latter is applied preferably
- Pancreatic body-/tail carcinoma
Left pancreatic resection with splenectomy

2. Palliative therapy: e.g.

  - Systemic chemotherapy with gemcitabine
  - Regional chemotherapy in specialized centres
  - In jaundice: endoscopic transpapillary installation of stents for keeping open the choledochal duct or installation of biliodigestive anastomosis
- In gastric outlet stenosis: Introduction of a duodenal stent or installation of a gastroenterostomy
- With tumour pain: Palliative therapy according to the WHO-staging and, should this not be enough, blockage of the coeliacal ganglion or analgesia procedure close to the spinal cord
3. New therapeutic approaches in clinical testing
- Adjuvant radio-chemotherapy post-operative (on UICC-stages I-III)
- Thyrosinkinase inhibitors , monoclonal antibodies

Prg.:   Pancreatic carcinoma can be resected only in 10 - 20 % of the patients.
5 year survival rates:
0 % with palliative therapy
15 % after resection with curative intention (about 30% after postoperative radiochemotherapy)
Up to 40% with resection on stage $T_1N_0M_0$ (small pancreatic carcinomas up to 2 cm $\varnothing$ without regional lymph node diseases and without distant methastases)

Pro.:   Abstention from smoking; in hereditary syndromes (see above) regular checkup + familial examination

---

PAPILLARY CARCINOMA

Occ.:   Infrequent

Cl.:   Early cholestatic jaundice

Di: MRCP, ERCP, CT-scan

Th.:   Whipple's procedure

Prg.:   5 year survival rate of the radically operated persons 30 %

NEUROENDOCRINE TUMOURS (NET) OF THE GASTRO-ENTERO-PANCREATIC SYSTEM (GEP)

Def.: During the last few decades, a number of names were given to this kind of tumours (APUDoma, carcinoids or neuroendocrinoma). Depending on the corresponding secretion products, the functionally active NET are referred to as gastrinoma, insulinoma, etc. The combination synaptophysine and chromogranine A are applied for immunocytochemical diagnostic.

Pat.:
- 1a Highly differentiated endocrine tumour: benign or questionable valency
- 1b Highly differentiated endocrine carcinoma: low malignant behaviour
- 2 Low differentiated neuroendocrine carcinoma: highly malignant behaviour

Loc.: 50% of the NET are found in the appendix, 20% in the anterior intestine (see below),15% in the last fragment of the illeus and 15% in the posterior intestine (see below)

Cl.: The clinical picture of the NET depends on their secretion pattern. Accordingly, there are functional and non-functional tumours.

Secretion products of the NET and associated symptoms		
Hormone/Neuro-transmitter	**Tumour**	**Symptom/Syndrome**
Tumours of the anterior intestine (pancreas, stomach, Duodenum)		
Insulin	Insulinoma	Hypoglycaemia
Gastrin	Gastrinoma	Peptic ulcers, diarrhoea
Glucagon	Glucagonoma	Diabetes mellitus, exanthema
Somatostatin	Somatostatinoma	Diabetes mellitus, gallstones
VIP = Vasoactive in-testinal Polypeptides	VIPoma	Watery diarrhoea
Midgut tumours (jejunum, ileum, colon ascendens)		Carcinoid syndrome
Serotonin Neurotensin B	NET with liver methastasis	
Tumours of the rectum (colon transversum, colon descendens, sigmoid, rectum)		
Chromogranin A		Non-functional

NET of the stomach

4 different types; type 1, with 75% the most frequent one, develops on the basis of an autoimmune, chronic-athrophic corpus gastritis (Type A) and does not metastize. Type 2 develops in association with MEN 1.

NET of the duodenum and the proximal jejunum

5 different types of duodenal NET. In the duodenum, gastrinoma are found in 65% of cases.
Gastrinoma: see there

NET of the ileum and the appendix

Def.: Epithelial tumour, coming from the enterochromaffin cells (EC cells) of the DNES (diffuse neuroendocrine system) with production of serotonin, callicrein, tachykinines and prostaglandins. In 25 % multiple localisation in the ileum.

Occ.: Incidence: 1/100,000 inhabitants/year. Frequency peak between age 40 - 70 (exception: the carcinoid of the appendix occurs as an incidental finding in 0,3 % of all appendectomies, often in younger patients).

Loc.: 1. Intestine (about 90 %): most frequently appendix (50 %) and distal 60 cm of the ileum (15 %),
2. Extraintestinal (10 %), mostly bronchus carcinoids

The solitary carcinoid of the appendix is mostly a benign incidental finding in every 300th appendectomy. The residual carcinoids metastize like a carcinoma (regional LN→ liver). Except for some bronchial carcinoids they cause only symptoms (carcinoid - syndrome) by their liver metastases. As long as no liver metastases are present, serotonin is catabolized by monoaminooxidases of the liver. The metastasis is dependent on localisation and size of the tumours: between 1 - 2 cm ∅ in 10 % metastasizing, > 2 cm ∅ in 80 % metastasizing.

PPh.: Serotonin → diarrhoea and endocardial fibrosis
Callicrein → change from kininogen to bradykinin; bradykinin causes flush syndrome and activates the prostaglandin synthesis → prostaglandin F: asthmatic attacks.

Cl.: Non-functional NET of the small bowel usually shows stenosis symptoms as first symptoms .
Functional NET are sign of metastasis with carcinoid-syndrome: The carcinoid - syndrome, consisting of the triad flush, diarrhoea and cardiac symptoms, can be provoked from time to time by stress, alcohol consumption and nutrition.
- Flush (70 %): paroxysmal heat attacks, redness of the face and neck that turns into cyanosis, tachycardia, sweating
- Intermittent painful subileus (50 %)
- Diarrhoea (70 %), weight loss
- Cardiac manifestation of the carcinoid - syndrome (Hedinger`s syndrome): endocardiac fibrosis, mainly of the right heart, leading to tricuspidal insufficiency, pulmonary stenosis
- asthmatic attacks
- teleangiectasias
- pellagra-like skin changes (caused by lack of tryptophan that is transformed into serotonin by the tumour cells)
- Cushing syndrome caused by ectopic ACTH-production
- palpable liver tumour

DD.: Systemic mastocytosis: flush attacks with pruritus, headache, fever and collapse condition, vomiting-diarrhoea, bone marrow cytology: mast cell infiltrates

Di.: 1. 5-Hydroxyindolacetic acid (= catabolic product of serotonin) in the 24 h-urine↑ (omission of serotonin-
  rich foods before measurment: bananas, egg plants, avocadoes, melons, tomatoes, walnuts, pineapples; if possible no antihistaminics, antihypertensive drugs, neuroleptic drugs
  2. Serotonin and chromogranine A i.s. pot.↑
  3. Proof of the primary tumour: endosonography, "one-stop-shop"-MRI (including MRCP and MRI-angiography); alternative: Spiral-CT
     Somatostatin- (octreotid-) receptor scintigraphy
     Angiography
  In case of suspected carcinoid of the bronchial tree bronchoscopy.
  4. Search liver for metastases: ultrasound, CT

Th.: 1. Surgical removal of the primary tumour and the regional lymph nodes with octreotid-protection
  2. In case of inoperability or metastases conservative therapy:
     - Octreotid (somatostatin analogues) inhibits the hormone secretion, in higher dosage cytostatic effects are also observed.
     - α-interferon (also in combination with cotreotid)
     - Radionuclide therapy (application of a beta radiator) in somatostatine-receptor-secreting NET
     - Symptomatically serotonin antagonists (methysergid, cyproheptadin)
     - Local procedures to destroy liver metastases
     - With rapid progressive tumours: palliative chemotherapy (e.g. 5-fluorouracil + streptozocin)

Prg.: 5 year survival rate:
  - Appendiceal carcinoid: 99 %
  - Localized small bowel carcinoid: 75 %
  - All small bowel carcinoids: 55 %

NET of the colon/rectum

Very rare tumours, already metastasized when identified

Pancreatic NET

55% are functionally/hormonally active. Depending on the predominant secretion, these tumours are referred to as insulinoma, gastrinoma, VIPoma, glucagonoma (see there). Insulinoma are mostly benign, the remaining pancreatic NET are frequently malignant.

INSULINOMA [D13.7]

Def.: Most frequent endocrine pancreatic- (B-cell-)tumour - f : m = 2 : 1, mostly benign (> 90 %), mostly solitary (90 %) and frequently small (< 2 cm). In about 10 % multiple adenomas, in 4 % within a multiple endocrine neoplasia type I (MEN I). Insulinomas produce in 50 % of cases only insulin, in the residual cases also other gastrointestinal hormones.

Cl.: Whipple's triad:
1. Spontaneous hypoglycemia< 45 mg/dl (< 2,5 mmol/l) provoked by fasting.
2. - <u>Autonomous symptoms:</u> sweating, sensation of heat, palpitations, tachycardias, tremor, weakness, anxiety, excessive hunger, nausea
   - <u>Neuroglucopenic symptoms:</u> visual difficulties, vertigo, headache, confusion, changes of behaviour (concentration impairment, aggression), paraesthesias, hemiplegia, aphasia, convulsions, coma, death (misdiagnosis: neuropsychiatric disease)
3. Prompt improvement after oral or i.v. glucose supply.
   The excessive hunger caused by hypoglycemia often leads to weight gain.

DD.: Hypoglycemia of other genesis (for details see there).

Di.: • <u>Fasting test:</u> The easiest and most reliable test is provoking hypoglycemia by <u>fasting</u> (in-patient over a maximum of 72 h) with close blood glucose controls as well as determination of insulin and C - peptide. Termination with symptomatic hypoglycemia.
   Typical of the insulinoma is the absence of physiologic insulin suppression with the drop of blood glucose during the <u>hunger test</u>. The insulin-/glucose ratio (µU/ml)/(mg/dl) drops in healthy persons and increases in insulinoma patients > 0,3.
   In <u>factitious hypoglycemia</u> due to insulin injections , a high insulin and a low C – peptide is found.

   • Proinsulin ↑

   • Localisation diagnostic:
   - <u>Preoperative localisation diagnostics:</u> with small tumours < 1 cm ∅ unsure (30 % of cases):
     Endosonography, "one-stop-shop"-MRI (including MRCP and MRI- angiography); alternative: Spiral-CT, somatostatin receptor scintigraphy, <u>p</u>ercutane <u>t</u>ranshepatic <u>p</u>ortal catheterism (PTP) with selective insulin measurment
   - <u>Intraoperative localization diagnostics:</u> palpation, ultrasound, poss. selective insulin measurment in the portal vein

Th.: <u>The surgical removal of an adenoma is the method of 1st choice.</u> Preoperative and in case of inoperability medicamentous inhibition of the insulin secretion by diazoxid (Proglicem®), octreotid (Sandostatin®), lanreotid. These preparations only have an effect on insulinomas with typical secretory granules – but <u>not</u> on agranular tumours (= 50 % of cases).

   Options in case of liver metastases:
   - locally applied methods to destroy metastasis
   - Chemotherapy (streptozotocine + 5-fluorouracil)
   - radionuclide therapy in somatostatine-receptor-secreting NET

GASTRINOMA [D37.7]

Syn: Zollinger-Ellison syndrome [E16.4]

Def. : - Localized in the <u>pancreas</u> (80 %) or duodenum, antrum, hepatoduodenal ligament, mostly <u>malignant</u> tumour (60 - 70 %), which in 50 % is already metastasized when establishing the diagnose

   - Gastric acid hypersecretion and <u>multiple ulcerations</u> in the upper gastrointestinal tract
   - <u>Formation of gastrin</u> and often also of other gastrointestinal hormones
   - In 75%, gastrinoma occur sporadically, in 25% within a MEN-I-syndrome. Age of manifestation mostly between 20-50 years

Cl.: - Therapy-resistant, <u>relapsing, often atypically localized ulcers (95%)</u> in the stomach, duodenum or even jejunum.
   - <u>Diarrhoea</u> (about 50 % of cases), occ. steatorrhoea (as hydrochloric acid inactivates the lipases)

DD.: Other causes of hypergastrinemia(<500 ng/l)
   - Therapy with H_2-blockers, proton pump inhibitors
   - Chronical-atrophic type A-gastritis, HP- gastritis
   - Retained antrum in patients after partial gastric resection; gastric outlet stenosis
   - Renal insufficiency

28

Di.:
- Increased basal (= fasting) gastrin level (values > 1,000 ng/l are almost diagnostic)
- Secretin test: Increase of the gastrin level by > 100% after challenge testing with secretin (unlike hypergastrinemias of other genesis)
- Localisation diagnostics:
  - Of ulcers: endoscopy
  - Of the tumour/metastasis: Imaging methods (see chapter Insulinoma)

Th.:
- Resection of the tumour with curative target is possible only with absence of metastases (about 30 %).
- Medicinal acid blockage with proton pump inhibitors
- Therapy options in case of metastasis/inoperability: see insulinomas

VERNER-MORRISON SYNDROME [D37.7]

Syn: Vipomas, WDHH-syndrome

Def.: Very rare, mostly malignant tumour of the pancreas with increased production of VIP ("vasoactive intestinal polypeptide") and other pancreatic polypeptides.

Cl.: WDHH-syndrome : watery diarrhoea, hypokalemia, hypochlorhydria or achlorhydria (like cholera toxin, VIP activates the intestinal and pancreatic adenylcyclase, which leads to an intensive pancreas-/small bowel secretion). Diabetes, weight loss, dehydration, abdominal spasms, confusion

Di.: Measurment of VIP (and other pancreatic peptide hormones), imaging diagnostics (see chapter insulinomas)

DD.: Ganglioneuroblastomas (esp. in children), other gastro-enteropancreatic tumours, abuse of laxatives

Th.: Surgical tumour removal is rarely possible. Options on therapy: see insulinomas

GLUCAGONOMA [D13.7]

Extremely rare, mostly malignant islet cell tumour of the A-cells with increased secretion of glucagon

Cl.: Facial and acral erythema necrolyticum migrans, diabetes

Di.: Clinical picture, glucagon i.s.↑ , imaging diagnostics (see above)

Th.: Options on therapy: see insulinomas

MULTIPLE ENDOCRINE NEOPLASIA (MEN) [D44.8]

The multiple endocrine neoplasia can occur in miscellaneous organs and is divided into 3 subgroups: the MEN-syndrome are passed on autosomal-dominant. MEN I is caused by mutations in the menin-gene (11q13), a tumour-supressor gene. Mutations of the ret-proto-oncogene on chromosome 10q11.2 are the genetic cause of the MEN-II-syndrome. Sporadic cases of MEN IIb are caused by new mutations.

Occ.: About 1 : 50,000 (in each case for MEN I and MEN II)

MEN 1: Wermer`s syndrome:
Combination pattern variable in the individual generations. Primary hyperparathyroidism (95 %)
Main tumour: tumours of the pancreas: gastrinoma, insulinoma, rarely other (50 %)
Tumors of the hypophysis (about 30 %)
Familiy members of MEN I patients should be offered genotype diagnostic with a geneticist. In order to detect the aforementioned tumours the earliest possible, regular checkups are performed in gene carriers (for further information see: www.uni-marburg.de/gastro/MEN I).

MEN 2: MEN 2a: Sipple`s syndrome (70 % of the MEN II cases)
Chief-tumour: medullar (C-cell-) thyroid carcinoma (100 %)
Pheochromocytoma (50 %)
Primary hyperparathyroidism (20 %)

MEN 2b: Gorlin`s syndrome (10 % of all MEN 2 cases) – like MEN 2a, additionally:
Ganglioneuromatosis (tongue, intestine) + marfanoid habitus (leptosomal, slim habitus, long limbs, arachnodactyly, hyperextendibility of the joints)
(for further details on MEN 2: see chap. thyroid gland tumours)

FMTC-only = non-MEN (20 % of the MEN 2 cases): only familiar medullary thyroid carcinoma (FMTC)

Note: Always think of the possibility of a MEN-syndrome in case of medullary thyroid carcinomas, endocrine pancreatic tumours (gastrinomas, insulinomas), pheochromocytomas and primary hyperparathyroidism (esp. with positive family history of the same diseases), and offer genetic testing in case of reasonable suspicion.

LIVER

Examination course

1. History:
 Familial and genetic history (Liver diseases?), questions about blood transfusions, trips abroad, alcohol consumption, drug intake, dealing with liver-toxic agents
2. Medical examination
 • Inspection:
Skin characteristics linked to liver problems (Spider naevi, palmar erythema, glazed lips, beefy red tongue, leukonychia), icterus and signs of scratching due to itching, collateral veins in the area of the abdominal wall, ascites, gynaecomastia, feminine hairiness type, foetor hepaticus
 • Examination of the liver:
 ► Determination of the size during the physical examination:
The information about of the liver size in cm below the right costal arch is inaccurate, since this value is dependent on the diaphragm state (e.g. phrenoptosis in pulmonary emphysema).The distance between the border of lung and liver and the lower liver edge is measured in the centre of the medioclavicular line (MCL): up to 12 cm is normal in adults.
 The lung-liver border and the lower liver edge are determined by:
- Percussion in the right MCL (Determination of the lung-liver border and the transition from the abdominal cavity's tympania to hepatic dullness).
 Additionally, the lower liver edge is determined by:
- Palpation (with deep inhalation the liver is forced against the palpating hand): Consistency, surface structure, tenderness to palpation
- Scratch-auscultation: Put stethoscope in the right MCL to a location with hepatic dullness and scratch from below towards the stethoscope until the sound suddenly becomes louder.
 ► Determination of size by sonography: Craniocaudal distance in the MCL: max. 14 cm
3. Biochemical laboratory parameters:
 a) Enzyme tests in order to assess the integrity of the hepatocyte. The following enzymes are indicators of a hepatocytic damage:
 • Glutamate-pyruvate-transaminase (GPT)
 = alanine-aminotransferase = (ALT))
 • Glutamate-Oxalacetat-transaminase (GOT)
 = Aspartates aminotransferase = (AST)
 • Glutamate-Dehydrogenase (GLDH))
 • Gamma-Glutamyl-Transferase (γGT)
The γGT is the most sensitive indicator showing disorders of the liver and the biliary tree. The highest values are found in case of cholestasis and alcohol-toxic hepatitis (most sensitive parameter of an alcohol-toxic liver change). In case of elevated osteoblast activity the γGT is normal, which allows a differentiation to an osteogenically caused elevation of the alkaline phosphatase.

Enzym	Localisation		Liver specific
	Cell plasma	Mitochondria	
GOT = AST = ASAT	+	+	No DD: card. Infarction Muscle trauma
GPT =ALT	+		yes
GLDH		+	yes
γGT	Membrane-bound		yes

The amount of the enzyme increase correlates with the extent of the hepatocytedamage.
De Ritis-ratio : AST/ALT or GOT/GPT.
Slight liver-cell damages lead to an increase in membrane-bound γGT and in cytoplasmatic enzymes (GPT and partially GOT). In this case, the de Ritis-ratio GOT/GPT < 1
Severe liver-cell damages additionally lead to an increase of mitochondrial enzymes (GLDH and partially GOT). In this case, the de Ritis-ratio shifts in favour of the GOT (> 1).
b) Synthesis performanceof the liver
- Cholinesterase (CHE)

CHE is produced in the liver and is reduced in case of necrotising hepatitis, chronic-active hepatitis and liver cirrhosis. Lowered values are found also in difficult clinical pictures with catabolism , cachexia and intoxication with organic orthophosphoric acid esters (in heavy intoxication, for example with diethyl-p-nitrophenyl thiophosphate,the CHE is no longer measurable).

- Vitamin K- dependent blood clotting factors

The majority of all factors of the coagulation and fibrinolysis system are formed in the liver, whereas the synthesis of the following factors is vitamin K-dependent:

- Factors II, VII, IX and X (so-called prothrombin complex)
- Protein C and S Protein

Vitamin K is a fat-soluble vitamin, either taken in by nutrition or produced by the intestinal flora. In the case of vitamin K deficiency the liver forms dysfunctional precursors of the coagulation factors which lack the γ carboxylation of the Glutamyl - side chains.

Causes for a reduction of the Vitamin K blood clotting factors

1. Synthesis disorder of the liver: Liver damage
2. Vitamin K deficiency:
- Newborn (→ oral vitamin K-prophylaxis)
- Malabsorption syndrome
- Faulty intestinal flora due to antibiotics
- Obstructive jaundice with fat malabsorption due to bile deficiency

3. Therapy or intoxication with vitamin K antagonist (Cumarine: Phenprocoumon = Marcumar®, Falithrom®; Warfarin (Coumadin®)

With the Koller test the cause of a reduced prothrombin complex can be determined, after applying vitamin K1 and measuring the PT (after 24 h). This is of special importance for the differential diagnosis of icterus.

In obstructive jaundice the PTnormalizes after applying a dose of vitamin K1, with a parenchymal damageno normalisation occurs . Vitamin K1= phytomenadione (Konakion®) in rare cases can lead to allergic shock after i.v. - injection; therefore inject very slowly and keep resuscitation drugs at hand

- Other blood clotting factors

In case of severe liver function disorders the factors V, XI, XII, XIII, fibrinogen and antithrombin also drop. Differentiation between slighter (Factor V normal) and severe synthesis disorder of the liver (Faktor V lowered).

- Albumin: Is produced in the liver. In case of liver cirrhosis - with increasing functional impairmentof the liver - the albumin level in the serum also drops .

c) Ammonia:

In advanced hepatic failure- in particular in hepatic encephalopathy- an increase of ammonia is found in the blood as an expression of a reduced detoxification performanceof the liver.

d) Enzymes indicating cholestasis :

In intra-and extrahepatic cholestasis there is an increase of the following enzymes:

- Alkaline phosphatase (ALP):

The activity of the ALP is the sum of miscellaneous isoenzymes:

Liver-ALP, small bowel-ALP, bone-ALP

The placenta-ALP physiologically passes over to the plasma from the 12th week of pregnancy on. The gamete ALP is found in some tumours (Seminoma, ovarian carcinoma, hypophysis and thymus tumours).

Causes of elevated ALP:

• Physiological:
  - In children/youths due to bone growth (bone-ALP)
  - In the last trimester of a pregnancy (placenta-ALP)
• Pathologic:
 • Osteogenic (bone-ALP):

Increased osteoblastic activity: Rachitis, osteomalacia, fracture healing, fluoride therapy, hyperparathyroidism, Paget's disease, bone tumours, osteoblastic metastases.

 • Hepatic/biliary:
- Slightly elevated values in hepatitis (liver-AP)
- High values in cholestasis syndrome (biliary duct-AP)
- Leucin-aminopeptidase (LAP) = Leucin-arylamidase

LAP is found particularly in the epithelium of the bile duct in high concentrations and is also elevated in cholestasis; however, it is of no greater expressiveness than the AP and therefore practically does not play any part.

- Gamma-Glutamyl-Transferase (γGT) see above.

e) Virus serology and immunologic diagnosis (refer to chapter hepatitis)

f) Tumour marker: α1-Fetoprotein (refer to chapter liver cell carcinoma)

4. Imaging procedures:

• Ultrasound, 3D ultrasound, endosonography, contrast medium sonography:
- Position, form, size of the liver, blood vessels
- Echo structure (for example diffuse reflex densification in fatty liver)
- Detection of defined liver changes. Differentiation between solid and cystic (anechoic) changes; the most sensitive way to detect liver metastases is by contrast medium sonography.

- Assessment of intra-and extrahepatic biliary ducts and gallbladder
- Assessment of flow conditions in the liver, portal and lienal vein (Duplex-sonography), for example in thromboses or Budd-Chiari syndrome
- Detection of extrahepatic characteristics of a hepato-portal hypertension (Ascites, splenomegaly)
- Proof of fibrosis and cirrhosis (elastometry by means of "fibroscan")
- Fine needle puncture under sonographic control
• ERCP (s. Chap. "Diseases of the biliary tract")
• Spiral CT with and without contrast media
• "One-stop-shop"-MRI including MRCP and MR-angiography
• Hepato-biliary sequenced scintigraphy: Diagnosis of a focally nodular hyperplasia
• Blood-pool sequenced scintigraphy with marked erythrocytes; typical finding for liver hemangiomas
5. Invasive diagnostic
• Angiography:
 - Hepaticography: Imaging of the liver arteries, e.g. for tumour diagnostic
- Splenomesentericoportography: Direct method with splenic puncture (risky); indirect method: measuring of the venous phase after the application of a contrast medium into the A. mesenterica superior. Differentiation between pre-, intra-, posthepatic block in a case of hepatoportal hypertension
- Presentation of the liver veins via V. cava
- Ultrasound controlled fine-needle biopsy (Needle diameter < 1 mm) with histology
Co.: Haemorrhage, bile peritonitis, pneumothorax, haematothorax
Danger of lethal complication: approx. 1 : 100,000
Cl.: Haemorrhagic diathesis, ascites
• (Mini-)laparoscopy with targeted biopsy + histology assessment of a liver cirrhosis
Co.: Haemorrhage, bile peritonitis (after puncture), skin emphysema, air embolism, liver or intestine injury
Danger of lethal complication: approx. 0,04 ‰

ICTERUS - Jaundice

Def.: • Icterus
Xanthochromia of skin/mucosa and sclerae due to deposition of bilirubin in the tissue.
Icterus is identified by the conjunctives, when the total bilirubin i.s. > 2 mg/dl (> 34 µmol/l).
DD: "pseudoicterus" due to colour depositions (for example intense carrot intake; after fluorescent-angiography)
• Cholestasis:
"bile stasis" with icterus, itch and elevated so-called cholestasis enzymes (AP, LAP, γGT)
PPh: About 85% of the bilirubin is a degradation product of haemoglobin. Approx. 300 mg of bilirubin are produced daily and transported to the liver bound to albumin. With the help of the UDP-glucuronyltransferase bilirubin and glucuronic acid are conjugated to the hydrosoluble form which is excreted through the biliary tract.
In the intestine, the bilirubin is reduced to urobilinogen, 80 % of which are excreted in feces and 20 % reach the liver after re-absorption via the enterohepatic cycle; a part is excreted renally.
The physiologic neonatal icterus has its cause in a reduced activity of key enzymes, the UDP-Glucuronyltransferase as well as a shortened life spanof fetal erythrocytes.

Classification and causes of icterus:
1. Haemolytic icterus (prehepatic ilcterus):
   Haemolytic anaemia, ineffective erythropoiesis (details: See there)
2. Hepatocellular icterus (hepatic or hepatocellular jaundice):
- Familial hyperbilirubinemia syndromes (see below)
- Infectious hepatitis (viruses, bacterias, malaria)
- Chronic hepatitis and liver cirrhosis
- Toxic hepatitis (alcohol, carbon tetrachloride, amanita)
- Hepatitis caused by medication hepatitis)
- Congestive hepatopathy
3. Cholestatic (obstructive-) Icterus:
This is the result of a disturbed passage of bilefor which the cause can be localized at any location, from the liver up to the papilla Vateri. In cholestasis, all gall-dependent agents can pass into the blood (bilirubin, bile acids, cholesterol, gall enzymes).
Cl.: Icterus, bright (acholic) feces, beer-brown urine, pruritus

Lab • Increase of cholestasis-indicating enzymes (AP, LAP and γGT)

• Increase of the direct conjugated bilirubin

• Absorption disorder of fat-soluble vitamins (particularly Vitamin K → reduced synthesis of the coagulation factors of the prothrombin complex with possible lowered quick value → normalisation after i.v.dose of vitamin K - see Kollerassay

• Serum iron : serum copper in obstructive jaundice< 0,8 (iron dischargefrom hepatocytes in hepatitis, copper elimination with bile)

3.1. Intrahepatic cholestasis: disorder of bile secretion in the liver → causes:

a) Hepatitis, liver cirrhosis
- Viruses (A, B, C, D, E, cytomegaly, Epstein-Barr)
- Bacteria (e.g. leptospiroses)
- Autoimmune hepatitis
- Haemochromatosis, Wilson's disease
- Alcohol-toxic liver damage
- Drugs: drug-induced jaundice: Phenothiazines, sexual hormones, antithyroid drugs
- "Fat overloading syndrome" with parenteral nutrition

b) Progressive destruction or hypoplasia of the biliary tract:
- primary biliary cirrhosis (PBC) and primary sclerosing cholangitis (PSC)
- "Vanishing bile duct syndrome" after liver transplantation
- Idiopathic ductopenia in adults (exclusion diagnose)
- Congenital dysfunctions with cholestasis already in infants: Byler's disease, Alagille syndrome, biliary atresia

c) Vascular diseases:
- Ischaemic cholangitis after infusion of 5-FU or after liver transplantation
- Budd-Chiari-syndrome (see there)

d) Idiopathic functional cholestasis
- Pregnancy cholestasis (see there)
- Summerskill-Tygstrup-syndrome (see there)
- Idiopathic postoperative jaundice (spontaneous remission after 2-3 weeks)

e) Cholestasis due to missing transporters in the canalicular membrane:
   Manifestation of the disease after birth
- Cholestasis in mucoviscidosis (dysfunction of the CFT-regulator)
- Dubin-Johnson-syndrome (MOAT –multi organic anion transporter- deficiency leads to the disturbance of bilirubin) secretion in the gall bladder
- Byler's syndrome with elevated γGT (missing MDR-transporter)

f) Bile acid synthesis disturbance:
   Congenital enzym defects
   e.g. Zellweger's syndrome with damaging of the peroxisomes

3.2 Extrahepatic cholestasis: Draining disorders of the bile into the extrahepatic large biliary ducts
   - Intracanalicular occlusion (Choledochal calculipapillary stenosis, cholangitis, tumour, stricture, parasites: Askarides, bilharziosis, fasciola hepatica, chlonorchis sinensis
   - Extracanalicular duct compression (Pericholecystitis, pancreatitis, pancreatic carcinoma and other tumours, pancreatic pseudocysts, liver abscess)

Familial hyperbilirubinemia syndromes:

A) With elevated indirect unconjugated bilirubin:

• Icterus intermittens juvenilis (Meulengracht's disease or Gilbert's disease) [E80.4]
   Aet.: Reduced UDP-Glucuronyltransferase, conjugation disorder with disturbed bilirubin absorption in the liver cell
   Various mutations of the UDPGT, mostly in the exon 1A1 (TATA5 or TATA 7 instead of the normal TATA6); autosomally–dominant inheritance
   Ep.: Most frequent familial hyperbilirubin syndrome: About 9% of the population; predominantly in males ; age of manifestation: usually around the 20th year
   Cl.: uncharacteristic: Headache, tiredness, depressive mood, symptoms of dyspepsia.
   Lab.: Elevated indirect bilirubin < 6 mg/dl. The residual values are normal, no signs of haemolysis. Fasting or Niacin test (elevation of the indirect bilirubin after fasting or after the application of niacin);
   Detection of the mutation (UDPGT – TATA –analysis)
   Histology normal, reduced UDP-Glucuronyltransferase-activity in the punch biopsy cylinder
   Th.: None
   Prg.: good (harmless anomaly)

Note: reduced decomposition of some drugs (ketoconazol, amitriptylin, ketoprofen, irenotecan)

- Crigler-Najjar-syndrome
- Type I:
Absence of UDP-bilirubin glucuronosyltransferase, autosomal-recessive inheritance, biliary encephalopathy directly after birth, UV light therapy accelerates the bilirubin decomposition; without liver transplantationor gene therapy lethal outcome.
- Type II: (= Arias-syndrome)
Strong reduction of the UDP-bilirubin glucuronosyltransferase, autosomal-dominant inheritance, manifestation of the jaundice in the 1st year.
Th.: usually no therapy
Prg.: favourable
B) With elevated direct conjugated bilirubin (very rare):
- Dubin-Johnson syndrome: [E80.6] autosomal-recessive elimination disorder for bilirubin (conjugation normal): Direct bilirubin increased; cause: Mutation of the Multidrug Resistance Protein MRP 2, a canalicular membrane transporter of the ABC-trans-porter-super-families(ABCC2), with disorder of the hepatobiliary bilirubin excretion and conjugated hyperbilirubinemia.
f > m; Manifestation occ. only with pregnancy; protoporphyrin I increased in urine
Diagnosis by liver biopsy: Brownish- black centriacinar pigment
Th.: None
Prg.: good
- Rotor syndrome [E80.6] (without pigment): Elimination disorder with elevated conjugated bilirubin, increased coproporphyrin III found in urine.
Th.: None; good prognosis.
- Idiopathic recurrent cholestasis (Summerskill-Tygstrup) [K83.1]: Rare autosomal-recessive heritabledisorder with intermittently occurring intrahepatic obstructive jaundice in children and young adults; good prognosis

DD:

	Haemolitic jaundice	Obstructive jaundice (cholestasis)	Hepatic jaundice
Serum			
- indirect bilirubin	+ +	--	+
- direct bilirubin	--	+ +	+
Urine			
- Bilirubin	--	+ +	+
- Urobilinogen	+ +	--	+
Stool	Dark	Light	
Additional examinations	Haptoglobin ↓ Reticulocystosis	γGT, AP, LAP ↑	GPT↑ GOT↑

Note: Indirect = unconjugated bilirubin; direct = conjugated bilirubin

DD cholestasis:
1. Intrahepatic cholestasis:
- Cholestatic progress of a virus hepatitis, liver cirrhosis
  Signs indicating virus hepatitis and excluding mechanical occlusion:
  - Positive virus serology
  - Sonography/ERCP: No dilated biliary ducts
  - Typical liver histology
- Drug-induced jaundice (Drug history)
- Primary biliary cirrhosis: antimitochondrial antibodies
- Primary sclerosing cholangitis: ERCP
2. Extrahepatic cholestasis (mechanical obstructive jaundice)
- Ultrasound, 3D-ultrasound, endosonography: Congested biliary ducts in extrahepatic cholestasis, obstruction high or deep (→ chart) Stones ? Pancreatic tumour? lymphomas in the liver hilus?

Intrahepatic cholestasis

Extrahepatic cholestasis

Deep obstruction   High obstruction

- "one stop-shop" MRI including MRCP and Angio-MR
- ERCP: Diagnostic and therapeutic procedure

Note: In extrahepatic cholestasis the most important diagnostic methods are ERCP + sonography .

LIVER DISEASES DURING PREGNANCY

Liver diseases during pregnancy

A) Pregnancy-independent liver diseases:

Acute viral hepatitis is the most frequent cause of jaundice in pregnancy.

B) Pregnancy-specific liver diseases:

1. Idiopathic icterus gravidarum [O26.6]

Syn: Benign recurrent pregnancy cholestasis, pregnant women's cholestasis

Incidence: 1 / 2,000 f – 8,000 births; second commonest cause of jaundice in pregnancy

In the last trimester of pregnancy, with given familial disposition, light intrahepatic cholestasis with itch and icterus with bilirubin up to 5 mg/dl can occur ; reversible benign disorder that can recur in subsequent pregnancies.

Disease not dangerous for the mother; however, there exists an elevated perinatal mortality rate (10 %) for the child and an elevated rate of immature births (20 %)→ administration of ursodeoxychol-acid and pre-termdelivery.

2. Icterus in hyperemesis gravidarum: [O21.1]

50 % of all pregnant women complain about nausea, in 0,3 % of the cases this goes up to severe vomiting: Hyperemesis gravidarum. Bilirubin + aminopherases can increase, a fattening of the liver with acinus-central lobule necroses is found histologicallyThe prognosis is good, therapy is unnecessary in most of the cases..

3. Icterus in pregnancy-induced hypertension (PIH) and (Pre-)eclampsia

Classification of the PIH:

I.   Isolated PIH = gestation hypertension

II.   PIH with albuminuria and possibly oedema = preeclampsia (formerly EPH gestosis, eclamptic toxemia)

Co.:   •HELLP-syndrome (haemolysis, elevated liver enzymes, low platelet count); clinical chief complaint: Upper abdominal pain(tenderness in liver capsule)

      • Eclampsia with neurological complaints (flickering, hyperreflexia, convulsion)

Note: The extent of the hypertonia determines the perinatal mortality of mother + child .

In.: 10 % of all pregnant develop a pregnancy induced hypertonia (PIH); 1 % of all pregnant women develop a pre-eclampsia; 0.5 % a HELLP-syndrome; 0.1 % an eclampsia. In 20 % of HELPP-syndrome cases hypertoniaand albuminuria can be missing (HELLP-syndrome sine preeclampsia).

Hi.: Thrombi in the portal vein branches, haemorrhagic hepatic necroses(in addition kidney damages and brain oedema with eclampsia).

DD: See chart

Th.: Immediate Caesarean section, supportive therapy

4. Acute gestational fatty liver [O26.6] :

Very rare disease (1 : 1 million pregnancies) after the 30th pregnancy week with high mortality (30 - 70 %): Unclear aetiology (defect in fatty acid-metabolizing enzyme system?)

Hi.: Rapid development of microvesical hepatocyte adipositas, particularly in the central lobule, single-cell necroses, round cell infiltrations

Memo: In most cases liver biopsy is contraindicated because of haemorrhagic disease

Cl.: Fulminant liver failure with icterus, vomiting, somnolence; transaminases normal or slightly elevated (DD: Acute viral hepatitis → high transaminases)

Co.: Disseminated intravascular coagulation (DIC), shock, renal failure, liver failure

Th.: Immediate caesarean section supportive therapy, in fulminant liver failure possibly liver transplantation

Criteria	HELLP	Acute gestational fatty liver	Viral-hepatitis	Intrahep. gestational cholestasis	TTP	HUS
Haemolysis (Haptoglobin)	++	(+)	-	-	+++	+++
Aminopherases	++	++	+++	+	(+)	(+)
Thrombozytope never	++	+	-	-	+++	+++
Hypertonia	90 %of cases	40 % of cases			(+)	(+)
Albuminuria	+++	--			+	+
HypercytosisRenal insufficiency	-	+++	++		(+)	(+)
	+ → +++	+			+	+++
Neurological complaints						
Icterus	+ → +++ (+)	++ +	- +++	++	+++ ++	+ variable ++
Other	DIC	hypoglycemia DIC → bleedings	Bilirubin ↑ Serology	Pruritus Cholestasis	Great von-Willebrand-multimeres	may also oc-cur in puer-perium

DIC =disseminated intravascular coagulation	
TTP =thrombotic-thrombocytopenic purpura	(+) = irregular
HUS =Haemolytic-uremic syndrome	+ to +++ = state of manifestation

---

ACUTE VIRAL HEPATITIS
(GENERAL INFORMATION)

Suspicion, disease and death notifiable!

Internet info: *www.kompetenznetz-hepatitis.de*

Def.: Diffuse (non-purulent) hepatitis, caused by miscellaneous viruses. There is no cross-immunity between the individual hepatitis forms.

PPh.: 5 viruses which are indicated with capital letters from A to E cause approx. 95 % of all virus hepatitis. Unknown hepatitis viruses are probably responsible for the rest. Details follow after the general presentation of the acute viral hepatitis.

Ep.: Frequency in % of world population:
1) HA: most frequent acute viral hepatitis - 2) HB: 6% virus carriers - 3) HC: 3% virus carriers- 4) HD: 5% ofHB-virus carriers – 5) HE: special at risk- areas

Inf.

Hepatitis	A	B and D	C	E
fecal-oral	+	-	-	+
Blood/-products	(-)	+	+	(-)
Sexual	(-)	+	(+)	(-)
Perinatal	-	+	+	-

The sexual transmission of the HB is frequent, of HC it is rare
(-) HA and HE are only exceptionally transferable by blood or sexually during the short phase of viraemia.

Inc:

Hepatitis	A	B and D	C	E
Days	15 - 50	30 - 180	15 - 180	15 - 60

Remark.: The incubation period varies a little with different authors.

PPh.: Histologic characteristics of acute viral hepatitis:
1. Proliferation of the Kupffer' star cells
2. Single-cell necroses and Councilman bodies (= necrotic cell residue)
3. Ballooned hepatocytes
4. Mild, inflammatory co-reaction of the Glisson' areas (Lymph cells macrophages)
5. Accumulation of ceroid pigment and iron in phagocytes in the recovery phase of the hepatitis

Cl.: There is basically no differing symptomatic of the individual virus hepatitis. The majority of the infections runs asymptomatically (2/3 of the cases), particularly in childhood

1. Prodromal stage
   Duration: Approx. 2 - 7 days (in B hepatitis longer than in A hepatitis)
   • Flu-like symptoms: Subfebrile temperatures, exhaustion (wrong diagnosis: "influenza").
   • Gastrointestinal symptoms: Loss of appetite, nausea, pain on palpationin the right epigastrium (Hepatomegalia) with capsule tenderness, diarrhoea

• Possibly arthralgias and lightexanthema (rash); for B hepatitis this is thought to be caused by immune complexing of HBs-Ag and anti- HBS

2. Period of hepatic manifestation:

　　Duration: Approx. 4 - 8 weeks (shortest in hepatitis A)

　• Anicteric progression (70% of adults, > 90% of children)

　• Ictericprogression (30% of adults, < 10 % of children.)
　　- Dark colour of the urine + decolouration of the feces
　　- Icterus (first with the sclerae, then on the skin)
　　- Itch (caused by increase of the bile acids i.s.)
　　With the onset of icterus the patient mostly feels better

　• Frequently hepatomegalia (with growth in consistency and tenderness to palpation)

　• Pot. slightly enlarged spleen and lymph node swelling (10 - 20 % of the cases, respectively)

Co.: • **Cholestatic progression form** (approx. 5 %)

= hepatitis with intrahepatic occlusion syndrome

Strong increase of bilirubin and cholestasis-indicating enzymes (AP, LAP, γGT)

- Drug-induced Intrahepatic cholestasis (Drug history)

- Extrahepatic cholestasis (mechanical obstructive jaundicedue to stones, tumours)

Di.: Sonography, MRCP, ERCP inconspicuous; typical virus serology and liver histology

The prognosis of cholestatic virus hepatitis is good in most cases.

• **Protracting and recurrent hepatitis:**

Transaminase elevation (constant or relapsing) > 3 months

DD: 1.Development of chronic hepatitis
　　2. Second viral infection or noxious substances (for example alcohol, drugs)

• **Fulminant hepatitis :**

In.: 　HA (0.2%), HB (1%), HC (very rare); HD: (> 2 %); HE: (up to 3%; in pregnant women up to 20 %). HB-carriers falling ill with HA have a mortality of approx. 10 %.

PPh.: Bridge-forming and multilobular necroses (uncomplicated virus hepatitis: Single-cell necroses). Due to the necroses the liver becomes small and flabby

Cl.: 　Triad: icterus, coagulation disorder, disorder of consciousness
　　　For details see chap. Acute Liver Failure

• **Extrahepatic manifestations:**

- Immune complex syndrome with arthralgias and exanthema(5 - 10 % of all HB-Patients)

- Other extrahepatic manifestations in acute viral hepatitis are very rare, for example aplastic anaemia.

　Extrahepatic manifestations in chronic B and C hepatitis: See there.

• **Virus persistence (virus carrier):** HB, HC, HD

3 forms of the virus carrier's physical status:

- Asymptomatic (healthy) virus carrier in HBV-infection with mostly good prognosis

- Chronic hepatitis with small inflammatory activity: Prognosis favourable in most of the cases.

- Chronic hepatitis with high inflammatory activity: Danger of development of liver cirrhosis and primary liver cell carcinoma

Occ. - HB: Healthy adults: 5 - 10 % - newborns: > 90 %; immunosuppressed: 50%
　　　- HC: Symptomatic icterus patients with acute HC: about 50% (with IFN-therapy < 5%)
　　　　　Asymptomatic HC-infections usually turn chronic
　　　- HD-superinfections of a HBs-Ag-carrier: > 90 %
　　　- HD/HB-simultaneous infection as in HB

•**Primary liver cell carcinoma:** Risk in chronic HB > HC. The risk of development of a primary liver cell carcinoma is also determined by supplementary factors:

- Genetic predisposition (in Asians and Inuit increased risk)

- Ageat the time of infection (greatest danger withperinatal infection)

- In chronic hepatitis and liver cirrhosis increased risk

- Co-carcinogens (Alcohol, cigarette smoking, alfatoxin)

DD: 　A) Accompanying hepatitis in other infectious diseases:

　　　1. Viral infections

　　　　　• Herpes viruses: Epstein-Barr virus, cytomegaly virus (CMV; in immunosuppressed individuals also herpes simplex-virus and varicella-zoster virus

　　　　　• Coxsackie virus

　　　　　• "Exotic" viruses: Arbor viruses (Yellow fever, Dengue-fever, Rift Valley fever), Arenaviridae (Lassa-fever, south-american hemmorrhagic fever), Marburg virus, Ebola virus → history of travel, consultation of an institute for tropical medicine

2. Bacterial infections: for example.
- Brucelloses, Q fever
- Leptospiral diseases: Weil's disease is characterized by the combination: Hepatitis + nephritis (Icterus, haematuria, albuminuria; conjunctivitis, joint / calf pain).

3. Parasitical infections: Malaria, amebiasis, echinococcosis, bilharziosis (= schistosomiasis, liver fluke and other tropical infections

B) Drug - induced hepatitis and alcoholic hepatitis (see below)

C) Acute attack of a chronic hepatitis
(Progression, biopsy, laparoscopy)

D) Other liver diseases (for example autoimmune hepatitis, primary biliary cirrhosis hereditary metabolic diseases, tumours)

Lab.: - Increase of the transaminases (500 – 3,000 U/l), while GPT is stronger elevated than GOT (de Ritis-ratio GOT/GPT< 1).
- with icteric progression: Bilirubin i.S. ↑ , urobilinogen and bilirubin in urine ↑

-.Slight elevation of the γGT and alkaline phosphatase (stronger elevation with cholestatic progression)
- Further: Increase of the serum iron; Electrophoresis: increase of the globulins; blood count: lymphocythosis ESR and CRP ↑
- In acute (and fulminant) progression with development of hepatic insufficiency there are signs of reduced synthesis capacity of the liver: Cholinesterase↓ Quick-value↓, albumin i.s..↓
- Serology: IgM antibodies against the individual virus marker indicate a new infection, IgG of the antibody immune status after previous infections (Exception: In B hepatitis, IgM anti-HBc - except for an acute hepatitis -are also found in a chronic B hepatitis attack).

Di.: History + examination+ laboratory (possibly histology)

	H A	H B	H C	HD Superinf.	HD Simultan. inf.	HE
anti-HAV-IgM	+	-	-	-	-	-
anti-HBc-IgM	-	+	-	-	+	-
HCV-RNA	-	-	+	-	-	-
anti-HDV-IgM	-	-	-	+	+	-
anti-HEV-IgM	-	-	-	-	-	+

Part of the screening-markers in acute viral hepatitis are the HBs-Ag, which in 90 % is positive withhepatitis B, and always with D hepatitis. In early stage of HC, the HCV-RNA is positive and anti-HCV still negative. Anti-HCV only becomes positive 1 - 5 months after infection , and due to this gap it is unsuited for reliably excluding acute C-hepatitis → HCV-RNA-determination

Th.: A) General measures:
- Bed rest in the acute disease period (Circulation of the liver is better while lying than while standing). Too early rising can lead to renewed transaminase increase.
- Ban on alcohol and omission of all drugs which are not absolutely necessary; all hepatotoxic drugs, also oestrogens, are prohibited.
Corticosteroids - formerly used in case of severe progression from time to time - are contraindicated ("transaminase cosmetics"); they prevent the endogenic virus elimination, favour the transition to chronic hepatitis and can trigger a difficult hepatitis attack when the patient stops taking them.

B) Antiviral therapy:
In 95% of cases, acute hepatitis C can be cured by an early therapy with Interferon over a period of 24 weeks

Isolation: Only necessary in A hepatitis in infants and in fecal incontinence.

Prg.: for acute viral hepatitis in adults:
Curing rates: HA: almost 100 % (in HBV-carriers:10 % fulminantprogression); in patients > 50 y., mortality rate is about 3%)
HB: ≈ 95 % (5 % virus persistence)
HC: Symptomatic icteric patients with acute HC have a 50% chance of spontaneous virus elimination. Asymptomatic infections usually take a chronic course. With interferon, acute HC is cured in 95% of cases.

HD-simultaneous infection with HBV: Similar to HB
HD-superinfection of a HBsAg-carrier: Small curing chance
HS: 98 % (in pregnant women 20 % fulminantprogression)

Pro.:   General hygienic measures:
- Disinfecting and disposal measures in medical institutions; careful dealing with blood/-products + body protection: disposable gloves, safety drain tubes, safety lancets and other injury-safe instruments→ observance of technical recommendations for biologic working materials (In Germany :TRBA 250)
- HA/HE: Food/water hygiene, hand disinfection
- HB/HC/HD:
  - Blood donor screening for virus marker + transaminases
  - Separation of hepatitis virus carriers in dialysis wards
  - Avoidance of promiscuity, usage of condoms
  - No "needle sharing" with i.v.-drug addicts (education)
  - Occupational medicine check-up and active immunisation of occupationally at risk patients against HB
  - Screening for HBsAg after the 32nd pregnancy week
  - Hepatitis B-simultaneous prophylaxis in babies born to HB-infected mothers

▶ Active immunisation:
  1. **HA:** Inoculation with formalin-inactivated vaccine
     Ind: • HAV-endangered employees (public health service, laboratories, children's homes, psychiatric facilities, sewage works, etc.
     • Other at-risk persons (patients with chronic liver diseases, patients in psychiatric facilities, homosexually active men)
     • Travelers to HA- endemic areas
     SE: See chart in the appendix; very rarely neurologic disorders
     Last-minute-travelers can receive the first vaccination even shortly before the departure, since the protection is sufficient given the long incubation period.Testing anti- HAV in advance only pays off in populations with an increased epidemicity degree (in western Europe cohorts before 1950)
     CI: Allergy against components of the vaccine, febrile infections and other.
     Dos: for example Havrix 1440®, HAV pur® 2 doses i.m. (M. deltoideus) at the age of 0 and 6 months. Approx. 10 year protection, seroconversion rate > 99 %.
     Children from 1st year up to the 12th year receive half of the inoculating-dosage (for example Havrix 720®).

  2. **HB and HD:** gene-technologically set up vaccine from the cell-surface antigen (HBs-Ag). Successful inoculation anti-HBS10 IU/l protects against HBV-infection and HDV simultaneous infection and reduces the incidence of hepatocarcinoma (Taiwan study).
     Ind: 1. Pre-exposure:
        - Infants/children:
             General inoculation without pre- or after-testing (WHO vaccination-plan)
        - Adults: For financial reasons unfortunately still restricted to risk groups (see specific part) and patients with chronic liver diseases who are HBs-Ag-negative.
             1.1 Pre-testing is necessary for risk-groups only: if anti-HBc is negative, vaccination is indicated. If anti-HBc is positive: No vaccination - additional examination for anti-HBS and HBs-Ag.
             1.2 After-testing for inoculating-effect (anti-HBS titer) is necessary for risk groups, e.g.:
                - Persons endangered occupationally by blood contact (dentists /doctors, nursing personnel, persons in emergency service)
                - Patients who frequently receive blood products; dialysis-, haemophilia patients
             . • Patients with chronic liver diseases or liver-involvement in chronic HBs-Ag – negative diseases.
                - Persons at risk due to closecontact with virus carriers (high exposure risk)
                - Persons with immunodeficiency
                (For complete list, see recommendations of the STIKO, the Vaccination Commission of the Robert Koch- Institute, Germany
        2. Postexposure (always as active-passive immunisation): See below

     SE: See inoculating-chart in the appendix; very seldom: neurological disorders
     CI: Same as with HA inoculation
     Dos: for example HB-Vax Pro®, Engerix B®, Gen-HB-Vax® 3 x 1 dosage i.m. (for better action in deltoid) at the age of 0, 1, 6 months.
     HB- response to vaccination after basic immunisation: Normal response ≥ 100IU/l, low response 11-99 IU/l, non-response ≤ 10 IU/l. 96% of the vaccinated persons show normal response Reasons for low-/non-response: HIV-infection, renal insufficiency and other conditions of immunodeficiency.

Options for low- or nonresponders (anti-HBS titer < 10 IU/l):
1. Further boosting-vaccinations, possibly with higher serum dosage (for example Gen-HB-Vax® D)
2. Aftervaccinations withHA/HB-serum (for example Twinrix®)
Boosters after basic immunisation are not necessary for immunocompetent persons. For risk groups, the STIKO recommends further procedures depending on the anti-HBS titer measured 1 - 2 months after the completion of the 3rd inoculation:
Anti HBsAg titer < 100 IU/l: Immediate revaccination and check after 4-8 weeks
Anti- HBsAg titer≥ 100 IU/l: Revaccination after 10 years (In case of needle stick injury revaccination after 5 years)

Combined HA- und HB-vaccine (Twinrix®): Dos.: 3 x of 1 dosagei.m. (M. deltoideus) at 0, 1, 6 months
Remark.: Therapeutic HB-vaccinations with MPL (monophosphorlipid-vaccines) which are supposed to speed up seroconversion in HB-carriers are in clinical testing.

3. **HE:** Recombinant hepatitis E-vaccine has been tested successfully. 3 doses at 0, 1, 6 months give an approx. immunisation of 90%.

▶ Passive immunisation:
1. **HA:** Normal immunoglobulins (NIg) = standardimmunoglobulin = immunoglobulin (human) = gamaglobulin
Offers a relative protection for 3 months.
Ind: Only simultaneously with active immunisation for persons at risk with acute exposure. Post-exposure prophylaxis within 10 days after close contact with HA –patients prevents from infection in 80 %.
Dos: Adults: 5 ml NIg i.m. offers a relative protection for 3 months.

2. **HB and HD:** B-immunoglobulin hepatitis (HBIg)
The postexposure prophylaxis always occurs as active-passive immunisation (HBIg-administration + active inoculation) and only makes sense within 48h (best within 6 h) after infection from unprotected (anti-HBs < 10UE/l) or non-vaccinated persons

Ind: • Babies born to HBs-Ag-positive mothers (screening of all pregnant womenfor HBs-Ag after the 32nd week of pregnancy)
• Unprotected/non-inoculated medical employees after injury with HBV-containing material (anti-HBs-titer unknown or < 10 UE/l
Report injuries as occupational accidents to the trade co-operative association
Dos: For adults: 0.06 ml/kg bw i.m.

ACUTE VIRAL HEPATITIS (SPECIFIC PART)

Internet info: *www.kompetenznetz-hepatitis.de*

HEPATITIS A (HA)

[B15.9]
PPh.: Hepatitis A-virus (HAV), an RNA-enterovirus from the picornavirus family
The HAV is very resistant. It can survive in extreme cold; in seawater, it survives for about three months and in dryness for about 1 month. The virus cannot be inactivated by normal soaps.

Ep.: Most HAV-infections in industrialized countries concern persons returning from vacation from southern countries with bad hygienic conditions. High number of unrecorded cases due to anictericprogression.
• Endemic incidence: Countries with low hygienic state. Most infections there run oligo- or asymptomatic (unrecognized) in childhood and adolescence. The epidemic infiltration rate in Europe shows a southern-northern-difference (greatest frequency in the Mediterranean area) and is age-dependent: In Germany, approx. 10 % of 20-year-olds are anti-HAV-positive, compared to 40% in the 5th decade of life
• Epidemic incidence: Last great epidemic in Shanghai 1987 with 300,000 illpersons; small epidemics in community institutions (kindergardens, homes for mentally ill, barracks)
• Sporadic incidence: Holidaymakers who travel to endemic areas are especially at risk; further: (medical employees in children's clinics, kindergardens, medical laboratories, sewerage workers, homosexuals, drug addicts

Inf.: Mostly fecal-oral (contaminated water, foods, raw seafoods, vegetables manured with feces, salads); very seldom parenteral (i.v.-drug addicts) or by anal-oral contacts

Inc: 15 - 50 days

    - Serology: anti-HAV- <u>IgM new</u> infection
                    Anti HAV-IgG: Former infection (remains positive for life)

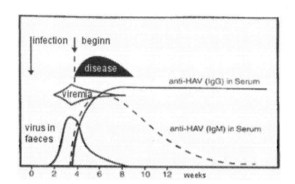

Infectivity: Corresponds to the duration of the HAV elimination with feces (2 weeks before until 2 weeks after the disease's beginning and/or 1 week after occurrence of a possible jaundice).

Progression: - Almost regular consolidation
                  - Icteric progression: in children < 6 y.: < 10%
                                    in children 6-10 y.: about 45%
                                    in adults: about 75%
           - Fulminant progression relatively rare (0.2 %), in HBV-carriers up to 10%
           - No virus carriers
           - No chronic hepatitis
           - Life-long immunity

HEPATITIS B (HB)
[B16.9]

PPh.:   The B virus Hepatitis (HBV) belongs to the group of the <u>Hepatitis-DNA-</u>(Hepadna)-viruses. Electron-microscopically the HBV corresponds to the so-called <u>Dane-Particle</u> . The HBV consists of a <u>Surface</u>, the <u>core,</u> the DNA and DNA-polymerase.

    - **In routine diagnostics the following virus components can be found:**
       • HBV-DNA
       • Surface-antigen (HBs-Ag)
       • Core antigen (HBc-Ag)
       • Envelope-antigen (HBe-Ag): A protein that is encoded by
         the precore/core-factor = secretory form of the HBc-Ag
       <u>Corresponding antibodies are called:</u>
       anti-HBS, anti-HBc, anti-HBe

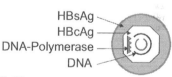

HBsAg
HBcAg
DNA-Polymerase
DNA

    - <u>Virus components not defined in routine check:</u>
       The HBs-Ag consists of the small (SHBs), middle (MHBs)
       and large HBs-Protein (LHBs) with the domains S, PreS1 and PreS2.
       The <u>nucleocapsid</u> is formed by the HBc-Ag, apart from HBV-DNA it contains the HBV-specific polymerase and a host- encoded protein kinase.
       8 <u>genotypes</u> of HBV (A - H). A and D are the most frequent ones in Western Europe.The HBs-Ag shows the <u>subtype determinants</u> in the s domain d or y as well as w1-4 or recombinant (only important for epidemiologic examinations).
    - <u>HBV-DNA – detection in the serum</u> is an expression for a continuing virus replication and <u>means infectiousness</u>
    - HBc-Ag can be determined only histologically in hepatocyte cores, while all the other components can be determined by serology and immune-histologically from liver puncture (HBsAg in the cell plasma of the hepatocytes, HBe-Ag in the hepatocyte cores).

- HBV-mutants: for example.
  - Pre-S/S-Gene-Mutants ("Immune escape"-mutant): Miscellaneous amino acid exchanges within the so-called a-determinant of the HBV-S-Region lead to a missing recognition of the virus by the anti-HBS antibodies → in spite of formation of anti-HBS after inoculation, this mutant can escape the immune reaction and lead to a HBV-infection.
    Deletions in the PreS1 factor can prevent the efficient coverage of the virus → direct damage through accumulation of virus particles in the endoplasmatic reticulum.
  - Polymerase-gene-mutants:
    Individual amino acid exchanges in the so-called YMDD-motive or the b-domain of the polymerase-gene cause resistance of the virus to therapy with nucleoside analogues and the transition to the highly viremic phase.
  - Pre-core stop codon mutant (HBe-minus-mutant):
    The formation of the HBe-Ag is prevented by a stop codon in the Pre-core-region of the HBV-genome in spite of virus replication → Hbe-Ag-negative chronic HB (in Germany, up to 50% of chronic HBV-infections).
    - "Diagnostic escape"-mutant with negative result of the HB-Ag-test (rare)

Ep.: Globally 6% HBV-carriers and > 1 million deaths/year. The frequency of B hepatitis is determined by the prevalence of virus carriers in a population group: 3 zones: ≥ 8 % (Central Africa, China); 2 - 7 % (middle East, northern Africa, eastern/southern Europe); < 2 % residual areas (Germany 0.6%).
- Endemic incidence: In areas with high number of virus carriers (see above)
- Sporadic incidence: with particularly endangered groups: i.v.-drug addicts, sex tourists and promiscuous hetero- or homosexuals, tattooed persons, individuals receiving blood/-products, dialysis patients. HB is an important occupational infectious disease: employees in the medical field, in the emergency service (including cleaning personnel), mentally ill persons in homes, persons with close contact to HBs-Ag-carriers, travelers to HB-endemic areas with close contact to the indigenous population, newborns with HBs-Ag-positive mothers.

Inf.: • Parenteral: Directly through blood/-products, indirectly by contaminated implements; 20 % of all cases are transmitted through needles shared by i.v.-drug addicts
Memo: Average risk of infection after needle stick injury with virus-containing blood: about 30%
- Sexually transmitted (65 % of the hepatitis B infections are sexually transmitted)
- Perinatal (= vertical): In countries with a high incidence of virus carriers (for example Africa, southeast Asia) the HBV transmission frequently occurs perinatally from mother to child. → screening of all pregnant women for HBs-Ag
  after the 32nd week of pregnancy

Inc: 30 - 180 days (history up to half a year)

Di.: History (halfyear) clinic / laboratory
Serology: In acute B hepatitis IgM anti-HBc is always positive , while HBs-Ag is in 90 % of the cases.
Acute HBV-infection with curing

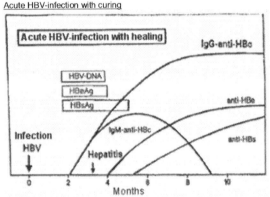

The function of the cellularimmune reaction determines the course of a HBV-Infection. The HBV itself is not cytopathogenic. Cytotoxic T-cells and α- interferon play an essential role in the eliminationof the HBV .
Hbs-Ag becomes detectable even before the beginning of clinical complaints and in 90 % of the cases is positive with the onset of disease (in 10 % HBs-Ag is not detectable). The HBV-DNA is detectable 2 - 4 weeks before the HBs-Ag.
Anti-HBS only turns positive when HBs-Ag has disappeared and signalizes a consolidation of the B hepatitis (10 % of patients do not develop anti- HBSAg). In cases for which HBs-Ag is not detectable at all as well as in the period between the disappearance of the HBs-Ag and the detection of anti-HBs ("diagnostic fenestra"), the formation of anti-HBcIgM is the single proof of acute B hepatitis.

Infectivity: No isolation obligation. As long as HBs-Ag is positive, potential infectiousness exists; the danger of infection can be clarified by determination of the HBV-DNA:
Infectious: Proof of HBV-DNA
Not infectious: Missing test of HBV-DNA (most sensitive assay: PCR)

Possible progression in HBV-infection:
1. Asymptomatic infection in adults approx. 65 %   } cure 90 %
2. Acute hepatitis with cure and virus elimination in adults approx. 25 %   }
3. Death with fulminant hepatitis (up to 1% of hospitalized patients)
4. Virus persistence → HBs-Ag-carrier: (HBV-carriers)
   • Immunocompetent adults: 5 % (m : f = 2 : 1)
   • Drug addicts: up to 20 %
   • Haemodialysis patients: up to 30 %
   • Immuno-suppressed persons after kidney transplantation: up to 50 %
   • Newborns with HBV-infected mothers: > 90 %
   • Babies: 70 %
   • infants: 35 %

HBV-carriers[Z22.5] can be clinically healthy (70 %) or develop chronic hepatitis (30 %). 20 % of the patients with chronic hepatitis develop liver cirrhosis within 10 years and of these , 15 % develop primary liver cell carcinoma within 5 years (interferon- and lamivudin-therapy diminish the HCC-risk)

Chronic hepatitis according to definition is present, if the acute hepatitis has not been healed after 6 months and is seen in a persistence of the cell-surface antigens (HBs-Ag) and a persistence of the marker of the virus replication (HBe-Ag, HBV-DNA). Anti-HBe and anti-HBS are not detectable (missing seroconversion).
If the HBV-DNA persists in the progression of an acute B hepatitis > 8 weeks, the possibility of a chronic progression must be taken into consideration.

Progression possibility of the HBV-infection in healthy adults:

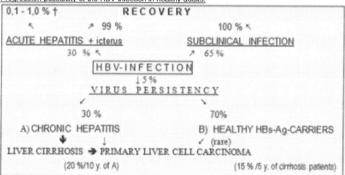

HEPATITIS D (HD)

[B18.0]

PPh.: Delta virus hepatitis (HDV), an incomplete ("naked") RNA virus (Viroid)
which needs the surface (HBs-Ag) of the HBV for its replication.
3 genotypes: I (Western world, Taiwan, Lebanon), II (Eastern Asia), III (South America)
In HDV the following componentsmay be found:

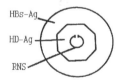

	Corresponding anti-bodies
• Surfacewith HBs-Ag	anti-HBs
• Core with HD-Ag	anti-HDV
• HDV-RNA	

43

Ep.: The spreadof the HDV is bound to the existence of the HBV. Globally approx. 5 % of the HB-virus-carriers are also carrying HDV
- Endemic incidence: for example Mediterranean area (in southern Italy > 50 % of the HBs-Ag-carriers are contaminated), Romania, middle East, some countries of Africa, region of the Amazon river.
- Sporadic incidence: In HBV-risk-groups (see above); HBs-Ag-carrier are especially at riskfor a HDV-infection in endemic areas.

Inf.: • Parenteral (like HBV)
- Sexually transmitted
- Perinatal (= vertical)

Inc: of the acute HDV-HBV- co-infection : 3-7 weeks

Di.: History (endemic areas, groups) - examination - serology

The following distinction can be made according to the chronologic connection of HBV and HDV-infection:
  ► Superinfection of a HBs-Ag-carriers with HDV (most frequent):
    In this case, there is a frequent adaptation of the HBV-infection from a replicant to a non-replicant form with the loss of HBe-Ag and occurrence of anti-HBe.
    Di.: - anti HDV IgMand HDV-RNA positive
        - anti-HBc IgM negative, HBs-Ag persistently positive
    Simultaneous infection (= co-infection) with HBV + HDV (less frequent):
    In this case the hepatitis often presents with 2 transaminaseapexes, the first of them being induced by HBV, the second by HDV.
    Di.: - anti HDV IgM and HDV-RNA positive
        - anti-HBc IgM positive, HBs-Ag at first positive, after consolidation negative

Prg.: Simultaneous infections (HDV + HBV): Severe progression of acute hepatitis 95 % cure
Superinfections of a HBs-Ag-carrier with HDV: More frequent fulminant course of disease. Most of the cases with chronic progression, often with transition to cirrhosis.

HEPATITIS C

[B17.1]

Internet info: *www.hepatitis-c.de*

PPh.: C virus hepatitis (HCV), a RNA virus (Flavi-Virus) with 6 genotypes + approx. 100 subtypes. Multiple infections with miscellaneous subtypes are possible →one-time HCV-infection does not protect against reinfection

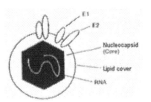

Worldwide spread are 1 A (60 %), 1 b, 2 and 3 a.
In Germany the following 3 subtypes are found most frequently: HCV-1b (50 %), HCV-1a and HCV-3a (with approx. 20 % each).

Ep.: About 3% of the global population have a chronic HCV- infection. Prevalence 0.2 - 2 % in Europe and USA, in the Third World up to 5 % (in some regions up to 20%). The frequency of virus carriers in Europe increases from north to south: Scandinavia 0.2 %, Germany 0.4 %, Mediterranean area 1 - 5 %. Worldwide, Russia is one of the countries with the highest HCV prevalence. In Europe and North America, HC causes 70 % of chronic virus hepatitis, 40% of cirrhosis and 60% of primary liver cell carcinomas.

Risk groups:
- i.v.-drug addicts (80% are HCV-positive)
- Unsterile piercing , tattooing, acupuncture
- Patients receiving blood/-products (for example patients after multiple transfusions, haemodialysis patients, haemophilia patients)
- transplant recipients
- Medical employees (needle stick injuries, blood splash into the eyes)
- Sexual partners of HC-virus-carriers: Dangerin HCV lower than in HBV

Inf.: - Parenteral (50 %)
Memo: Average risk of infection after needle stick injury with virus-containing blood: about 3%.
- Sexually transmitted (danger of infection in stable partner relationships with approx. 2,5 % rel. small.)
- Perinatal (less frequent than in HBV: Vertical transmission occurs only in viremic mothers and constitutes about 4% (in case of HIV-co-infection > 5%)
- Sporadic infection: Path of infection often unknown (45 %)

Inc: 15 - 180 days

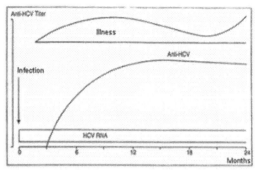

Prg.: • Acute HC: in 85 % asymptomatic and in 15 % symptomatic

Asymptomatic HCV-infections mostly take a chronic course

Spontaneous healing can occur in up to 50% of symptomatic icteric HCV-infections

• Chronic HC: About 75% of HCV-infections in adults show a chronic progression. 20 % of the patients with chronic HC develop a liver cirrhosis within 20 years. 3- 4 % of cirrhosis patients have a decompensation and 1-2% develop primary liver Some co-factors play a part in this process (alcohol drinking, fatty liver, infection with HBV-/HCV. In HIV- and HCV double infections (drug addicts, recipients of blood/products) the progression is often rapidly progredient and more frequently cholestatic.

Childhood-HCV-infections rarely lead to chronic hepatitis and liver cirrhosis.

Extrahepatic complications:

Essential mixed cryoglobulinemia, membranoproliferative glomerulonephritis, auto-immune thyroiditis, Sjögren-syndrome, cutaneous porphyria tarda.

Di.: - History taking (risk groups - history every ½ year) / examination

- Laboratory:

• Proof of an acute hepatitis:

Initially, detection of HCV-RNA. HCV-RNA proves viremia and therefore infectivity of the patient. In case of chronic progression HCV-RNA persists.

The quantitative determination of the HCV-RNA is important for the assessment of the therapeutic effect

• Anti-HCV: Anti HCV -positive patients in approx. 80 % of cases also show HCV-RNA in the blood and therefore are to be considered infectious. A positive anti-HCV assay must be verified by an acknowledgement assay (for example Immunoblot) since false positive findings are noticed. Anti-HCV becomes positive 1 - 5 months after the disease onset (diagnostic gap) and is therefore not suited for the exclusion of an acute HC . In case of chronic progression anti HCV-antibodies persist . Anti HCV-IgM correlates with the activity of the HC.

• Determination of the HCV-genotype

Th.: By therapy with PEG- INF-alpha over 24 weeks, consolidation of acute HCV occurs in 95% of cases (= normal aminopherases and HCV-RNA negative). In chronic HCV, PEG-IFN-alpha is combined with ribavirin.(see there)

HEPATITIS E

[B17.2]

PPh.: hepatitis E-virus (HEV), a RNA virus (Calicivirus), 2 serotypes; naturalreservoir in animals: for example sheep, pigs, monkeys, rats, mice

Ep.: Sporadic + epidemic incidence (for example Asia, middle East, Pakistan, North and Central Africa, Mexico). Disease under the age of 20 are rare.

Inf.: Fecal- oral transmission (as in A hepatitis); in the viremic phase, as an exception also parenterally

Inc: 15 - 60 days

Di.: - History / examination + laboratory

- Serology: Detection of anti-HEV-IgM and IgG. Detection of HEV-RNA in the feces and frequently also in blood is possible with the onset of disease.

<u>Course.:</u>1. Curing rates: 98 % (Pregnant womenonly 80 %)
2. Fulminant progression: Up to 3% (up to 20 % in pregnants)
Chronic progression is not known.

<u>Th.:</u>   Symptomatic
<u>Prg.:</u>   Hygienic rules as with A hepatitis
                Active immunisation (in preparation)

```
CHRONIC HEPATITIS (CH)
(GENERAL INFORMATION)
```

Internet info: *www.kompetenznetz-hepatitis.de*
<u>Def.:</u>   Hepatitis not healed after 6 months.

<u>Aet.:</u>   1. Virus-induced CH (HBV, HCV, HDV): 60 % of the cases.
           2. Auto-immune CH
           3. Diseases that can run under the terms of a chronic hepatitis (see DD)

<u>PPh:</u>   Classification of chronic hepatitis:
          Today pathologists no longer use old terms like chronically persisting   hepatitis (CPH) and chronically active
          hepatitis (CAH). The diagnosis of tissue examination should contain 3 declarations:
          1. Aetiology
          2. Degree of inflammatory activity (grading):
             • Minimal = exclusively small portal inflammatory infiltration
             • Mild = portal inflammatory infiltration with individual piece-meal-necroses and small inflammatory activity in the
               lobules
             • Moderate = numerous piece-meal necroses and necro inflammatory activity, lobule with individual group
               necroses
             • Severe: pronounced piece-meal necroses   and bridge necroses in the lobule
          3. Stage according to the extent of the fibrosis (Staging):
             • Minimal = Slight portal connective tissue growth
             • Mild = Increased portal connective tissue growth with slight connective tissue extraction
             • Moderate = portal connective tissue growth with formation of particular incomplete and also complete septa
             • Severe = development of numerous complete septa with transition to cirrhosis
             Remark: ground glass hepatocytes = HBV-containing liver cells with modified cytoplasm aspect due to
             hyperplasia of the smooth endoplasmatic reticulum with HbsAg overproduction are typical of chronic hepatitis B

<u>Cl.:</u>   **for chronic hepatitis**
          ▶ In case of minimum and mild inflammatory activity:
             • Mostly absence of symptoms and normal liver size
             • Possible deterioration in performance, tiredness uncharacteristic epigastric complaints
          ▶ In case of moderate and severe inflammatory activity:
             • Declining performance, tiredness, growing irritability

          Memo: Tiredness is the most frequent symptom in liver diseases

             • Loss of appetite
             • Pain on palpation in the liver area
             • Possibly arthralgias
             • During inflammatory attack icterus with dark urine
             • Liver mostly enlarged
             • Spleen enlarged in 1/3 of the cases. light leuko-/thrombocytopenia (Hypersplenia
             • Liver skin characteristic, for example:
               - Clean, beefy red tongue, glazed lips
               - Palmar and sole erythema
               - Spider naevi
               - Prurigo simplex with often strong itching and scratches
               - Skin atrophy with teleangiectases
               - White nails, Dupuytren's contraction
             • In women often menstrual disorders and secondary amenorrhoea
             • In men hypotrichosis of the body hair, orchiatrophy and possible gynaecomastia
               Causes:1. Hormonal disorder: Testosterone ↓/ oestrogen ↑
                       2. Iatrogenic: Gynaecomasty as side effect of a spironolactone therapy

<u>Co.:</u> 1. Liver cirrhosis with corresponding complications   } degree of risk.: See prognose
2. Primary liver cell carcinoma   }
3. Extrahepatic manifestations in HB and HC : See chronic hepatitis special part

<u>DD:</u> 1. Liver damages due to alcohol , drugs or chemicals (alcoholic, drug-,occupational, hobby history).
2. primary biliary cirrhosis (high AMA-titer, strong elevation of the γGT + AP, pruritus), sprue.
3. Hereditary metabolic activity diseases:
    - Haemochromatosis (Iron i.s + ferritin ↑, liver histology)
    - Wilson's disease (total copper i.s. + ceruloplasmin ↓)
    - Alpha1-antitrypsin deficiency

<u>Di.:</u> - History + clinic
- Laboratory with virus markers + autoantibodies
- Liver morphology (sonography, possibly CT, MRI, abdominoscopy)
- Liver histology

<u>Th.:</u> General measures:
- Avoidance of all potential liver toxins (alcohol, drugs)
- during the inflammatory attack period of rest, bed rest.
- Vaccinate all patients with chronic hepatitis against HA/HB (provided that they were not contaminated with these viruses before), because additional infestions with hepatitis lead to severe progression with increased mortality.

Specific measures:

• Virus-induced chronic B, C and D hepatitis:
Alpha-Interferon IFN α– 2a (Roferon A®) , IFN- α–2b (Intron A®), Peginterferon (Peg)-alfa-2a (Pegasys®) Peg-INFB-alpha-2b (Pegintron ®), IFNalfacon 1 (Inferax ®)
Ind: Chronic viral hepatitis with inflammatory activity (Transaminases elevated double the norm or more)
HB: IFNα is injected s.c. 3x /week , Peginterferon only 1x/week; as the dosages change, follow the guidelines / product information.
Duration: 16 weeks (in case of Hbe-negative HB up to 48 weeks)
HC: e.g. Peginterferon alfa 2a (180μg s.c. once a week) + ribavirin (Rebetol®) 800-1,200 mg/d (body weight-dependent) and under leukocyte check (dosage reduction with leukopenia)
Duration: HCV- genotypes 2 and 3: 24 weeks, HCV-genotypes 1 and 4: 48 weeks
If the viral load decreases by less than 2 log-stages within the first 12 weeks, the procedure can be terminated, since no effect can be expected in this case. If the viral load drops fast (rapid responder) or slowly (low responder), the duration of therapy should be adjusted accordingly.
SE: Frequent: Local reactions at the site of injection; grippal complaints with fever, myalgias and headache, gastrointestinal SE, thrombo- /leukocytopenia
    Infrequent Neurotoxic SE: depressions, loss of taste, confusion, amnesia, vertigo, paraesthesias, polyneuropathy; exacerbation of an autoimmune disease, auto-immune thyroiditis, elevation of the γGT and the bilirubin
CI: Absolute: Auto-immune hepatitis + depression, pregnancy, decompensated liver cirrhosis (aggravation under IFN therapy), immunosuppression, manifest endogenous psychosis, current drug- or alcohol-addiction, pregnancy, thrombocytopenia (< 50,000/μl), neutropenia (< 1,000/μl), bilirubin > 2.5 mg/dl
    Relative: Other autoimmune diseases
Rates of success and supplementary therapy options
HB: In about 40 % termination of viral replication with normalization of transaminases and seroconversion from HB to anti-Hbe (healthy HBs-Ag-carriers)
    For HB which does not respond to Interferon, there are additional therapy options:
    - Lamivudin (=Zeffix ®) – Dos.: 100mg/d orally ; duration : 1 year or longer ; success rate about 40% (end of viral replication). SE: Appearance of up to 40 %/4 years of viral mutants (esp. YMDD) . Patients with HBe-Ag-negative chronic hepatitis have a poor reaction to IFN and should be treated primarily with Lamivudin.
    - Adefovirdipivoxil (Hepsera®): Dos.:10 mg/d ; normalisation of transaminases /improvement of liver histology in about 40%
    SE: Nephrotoxicity → monitoring of the renal function
    - Entecavir (Baraclude®): Dos. 0.5 or 1.0 mg/d , if there are resistancies against Lamivudin
    - Tenofovir (Viread®): Dos. 300mg/d; SE: Nephrotoxicity (already approved for HIV)
    - Telbivudin (Sebivo®): Dos. 600 mg/d
    HC: Cure with the combination Peginterferon + ribavirin in 50% of genotype 1-infections and in about 80% of genotype 2- or 3-infections
HD: Rarely cure, but partial improvement of transaminases and histology

• Auto-immune CAH: see there

<u>Prg.:</u> - Chronic HB: Healthy virus carriers have a good prognosis. HBV-DNA-positive patients with chronically-active hepatitis have a more unfavourable prognosis; there is an imminent danger of liver cirrhosis (20% /10 years) and primary liver cell carcinoma (15%/ 5years in patients with liver cirrhosis).

- Chronic HC: Development of liver cirrhosis in 20% /20 years. 1-2% / year of cirrhosis patients develop HCC. - Chronic HD: More unfavourable prognosis than in chronic HB.
- Auto-immune CAH: More unfavourable prognosis than in chronic HB.

CHRONIC HEPATITIS (CH)
(SPECIFIC PART)

Aet.: ▶ VIRUS-INDUCED CHRONIC HEPATITIS [B18.9]

Def.: Virus hepatitis which has not healed after 6 months. In this case persistent virus replication and an absent virus elimination are found.

Internet info:*www.kompetenznetz-hepatitis.de*

• Chronic B hepatitis [B18.1]

Ep.: Prevalence correlates with the number of HBs-Ag-carriers (see above); m : f = 5 : 1.

Pg.:The functioning of the cellular immune reaction determines the progression of a HBV-Infection. The HBV itself is not cytopathogenic. Cytotoxic T-cells and IN-α play an essential role in the elimination of HBV.

5 - 10 % (in perinatal infection 90 %) of the HBV-infected can not eliminate the virus and become HBs-Ag-carrier who either stay healthy or develop chronic hepatitis. In the normal progression of HBs-Ag-carriers three phases can be distinguished:

	highly replicative phase	low replicative phase
Hepatitis	active	inactive
Transaminases	increased	mostly normal
HBs-Ag	+	+
HBe-Ag	+	-
anti-HBe	-	„Seroconversion" +
HBV-DNS: - in serum	+	- / + (PCR)
(HBV-DNA) - in hepatocytes	Episomal = extrachromosomal	Integrated in the host-DNA = intrachromosomal
Infectivity	high	with negative HBV-DNA no infectivity, with little viremia small risk of infectivity

1. Early phase of the virus replication (highly replicative phase):
   In this case complete HB-viruses are produced
   Characteristics:
   • Biochemical (Aminopherases) + histologic signs of inflammation
   • Proof of replication markers in the serum: The HBV-DNA is the most important marker (also for the assessment of infectivity); a second marker is the HBe-Ag
   • High infectivity

2. Late low replicative phase:
   In this case, almost only wrapping-particles (HBs-Ag) are produced. Approx. 5 % of the patients a year show a spontaneous transition from high- to low replicative phase (50 % under IFN-alpha therapy). In this case a transient inflammational attack can be observed. Subsequent normalisation of transaminases and disappearence of Hbe-Ag from the serum + formation of anti- Hbe → inactive and asymptomatic HBs-Ag-carrier

3. A third phase of definitive healing with loss of anti-HBs and disappearance of HBV-DNA in the PCR test is rarely observed in spontaneous progression of chronic hepatitis B, however in 10% under IFN-α therapy.

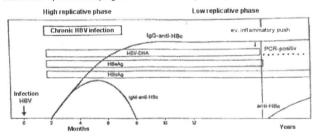

Progression forms of chronic HBV-infection:
- ▶ Inactive or asymptomatic (healthy) HBs-Ag-carrier:
  - serologic constellation:
    Positive : HBs-Ag, anti-HBc, anti-HBe
    Negative : HBe-Ag, anti-HBs
  - Normal liver enzymes and synthesis performance (Quick-value, albumin)
  - Liver histology in 80 % of the cases normal, hepatocytes containing HBs-Ag appear as so-called ground-glass hepatocytes (= HBV-containing liver cells with hyperplasia of the even endoplasmatic reticulum with overproduction of HBs-Ag)
  - HBV-DNA-negative persons are not infectious. With low viremia ($< 10^5$) the infectivity is relatively small. Favourable prognosis.
- ▶ Immunotolerant HBV-carrier
  - High viral load
  - Transaminases, however, normal or slightly↑
  - Typical progression after perinatal infection
  - After 10-30 years, possible conversion to immunoactive hepatitis
- ▶ Immunoactive replicative B hepatitis
  1. HBe-Ag-positive progression:
     - Transaminases ↑ (> 2fold normal value)
     - HBV-DNA ↑ (10^5 copies/ml)
     - Distinct histologic activity
     - Imminent danger of liver cirrhosis (8-20% / 5 years) and HCC (60-fold increased risk) in highly replicative immunoreactive HB.
  2. HBe-Ag-negative progression:
     - Antiviral therapy can lead to a HBe-Ag-seroconversion from positive to negative
     - Transaminases ↑ (> 2fold normal value)
     - HBV-DNA ↑ (lower by the factor 10 than in HBe-Ag-positive progression)
     - Histology + prognosis similar to HBe-Ag-positive progression
- ▶ Extrahepatic complications of chronic B hepatitis
  Panarteriitis nodosa, membranoproliferative glomerulonephritis
Co.: 1. Liver cirrhosis: 20% of patients with chronically active B hepatitis develop liver cirrhosis after 10 years.
2. Hepatocellular carcinoma (HCC): The relative risk for a healthy HbeAg-negative HBV-carrier is small. In chronic B hepatitis with positive HbeAg , the relative risk has increased by the factor 60 (compared with the healthy person). Of the patients with liver cirrhosis,15%/5 years fall ill with HCC. (Lamivudin can reduce this risk).)
Th./Prg.: The success rate of α-Interferon-therapy is about 40% (better in case of HBV-genotypes A and B, worse in case of the HBV-genotypes C and D)
Therapy options for IFN-non-responders
  1) Lamivudin (Zeffix®)
  2) Adefovirdipivoxil (Hepsera®)   } see chap. HC/ general part
  3) Entecavir (Baraclude®)   }
  4) Tenofovir (Viread®)
  5) Telbivudin (Sebivo®)
The prognosis for the healthy HBs-Ag-carrier is favourable while the prognosis for chronically active HB is not. (see above)

- **Chronic B and C hepatitis** [B18.1]
  While the simultaneous infection (HDV + HBV) has a progression similar to the pure HBV-infection (apart from the fact that fulminant progression is more frequent), the superinfection of HBs-Ag-carriers with HDV often (in 90%) leads to chronic B + D hepatitis.
  Di.: Clinic + hepatitis serology + (Proof of anti-HDV, HDV-RNA and HBs-Ag) liver biopsy + (with proof of HBV and HDV in hepatocytes).
  Th./Prg.: The chronic hepatitis B + D has a triple lethality compared with a simple chronic hepatitis. α-Interferon therapy is mostly ineffective.

- **Chronic C hepatitis** [B18.2]
  Internet info:*www.hepatitis-c.de*
  70 % of all chronic virus hepatitis are caused by a HCV-infection.
  Up to 50% of the acute symptomatic /icteric HCV infections heal spontaneously (virus elimination).
  Asymptomatic acute HCV-infections mostly have a chronic progression (virus persistence).
  20 % of the patients with chronic HC develop a liver cirrhosis within 20 years. 3-4% of the cirrhosis patients decompensate and 1-2 develop a primary liver cell carcinoma.

  Extrahepatic manifestations of the chronic hepatitis C: Cutaneous porphyria tarda, idiopathic cryoglobulinemia, membranoproliferative glomerulonephritis; Sjögren- syndrome, Hashimoto thyroiditis.

Di.: Clinic + hepatitis serology:
- Detection of anti-HCV (searching and confirmation-testing)
- Proof of HCV-RNA shows virus replication + infectivity
  The quantitative determination of the HCV-RNA is important for the assessment of the therapy effect
- Transaminases, γ GT(in 50 % ↑)
- Liver biopsy with determination of the inflammatory activity (grading) and the fibrosis build-up (staging)

> **Consider:** In case of chronic hepatitis C autoantibodies can be found: ANA (20 %), also anti-LKM (20 %) (→ mis-diagnosis: auto-immune hepatitis).
> Only the histology allows a statement on the condition of the liver in chronic hepatitis C Normal transaminases do not exclude chronic hepatitis C

Th./Prg.: Success rate (cure) with IFN-therapy (PEG-IFN-α2b) + ribavirin 50% (genotype 1), 80% (genotype 2 or 3). The results are most unfavourable with high viremia (> 3 million viruses/ml blood) and in the subtype HCV-1b (most frequent in Germany , with 50 %). The genotypes 2 and 3 can be eradicated in most cases.
Under therapy with ribavirin the Hb can decline by 3 - 4 g/dl. Consider SE + CI → blood count monitoring with leukocytes + thrombocytes

AUTOIMMUNE HEPATITIS (AIH)

Syn: Auto-immune chronic-active hepatitis [K75.4]

Occ.:   80 % of the cases affect women.
  In 50 % the disease begins before age 30; familial disposition (frequent incidence of HLA-B8, -DR3 or -DR4).

Characteristics:
- Features of chronic liver disease with considerable pains, frequent incidence of extrahepatic autoimmune diseases (for example autoimmune thyroiditis, rheumatoid arthritis, vasculitis, chronic inflammatory intestinal diseases, vitiligo)
- Continuously elevated transaminases with apexes during inflammatory attacks
- Early reduction of the synthesis achievementof the liver (Quick value, albumin).
- Total protein and gamma globuline (IgG) increased
- Histologic picture of the CAH
- virus marker negative, detection of typical autoantibodies

In > 90 % of cases typical autoantibodies are found:

		ANA, SMA, p-ANCA	LKM1, LC
Type 1	classic (lupoide) autoimmune CAH (80 %)	+	
Type 2	LKM1-positive autoimmune CAH (esp. in children)		+

Legend:
ANA	=	Antinuclear antibody
LKM	=	liver kidney microsome-Ab (Target antigen: Components of the P450 system)
SMA	=	Antibody against smooth musculature (Factin)
LC	=	anti-liver cytosol

Note: The variation with SLA formerly called type 3 (ab against soluble cytoplasmatic hepatocellular antigen) nowadays is assigned to type 1

Occasionally, overlap-syndromes can be observed, e.g.

- AIH / PBC-overlap syndrome. Histologic criteria of PBC + laboratory constellation of AIH (syn.: AMA-negative PBC)
- AIH /PSC-overlap syndrome: Association with ulcerative colitis
- AIH/C-hepatitis overlap syndrome
DD: LKM1-positive chronic C-hepatitis is the most important DD of LKM1-positive CAH. Only the HCV-negative variation is treated with immunosuppressives.

Di.:   Exclusion of other causes of chronic hepatitis + laboratory

Th.:   Immunosupressive treatment with corticosteroids + azathioprine (for details see *www.leitlinien.de*)
          Osteoporosis prophylaxis with calcium and vitamin D
          Therapy of overlap syndromes in hepatologic centres:
          In AIH +PBC or PSC: Therapy of AIH + application of UDC

Prg.: unfavourable for the auto-immune CAH without therapy, under immunodepressive therapy rel. favourable (10 year survival rate approx. 90 %).

PRIMARY CHOLESTATIC LIVER DISEASES

Primary biliary Cholangitis (PBC)

Syn: Primary biliary cirrhosis
Primary biliary cholangitis [K74.3]

Def.: Late cirrhotic stage of a chronic non-puriform destructive cholangitis of unknown cause

Ep.: Incidence approx. 5/100,000/year; approx. 1 % of all cirrhosis cases mostly (> 90 %) women > 40 years., sometimes familial clustering of PBC-cases (association with HLA-DR 8)

Aet.: Unknown (Infection with retroviruses ?), increased risk for patients with celiac disease

Pg.: Autoimmune disease? (immunodepressive therapy is ineffective)

PPh.: 4 histologic stages
St. I: lympho-plasmacellular infiltration of the portal area with destruction of the biliary duct epithels
St. II: Biliary duct proliferation with pseudo biliary ducts
St. III: Obliteration and cicatrisation of the portal areas piece-meal-necroses + loss of small biliary ducts (ductopenia)
St. IV: Cirrhosis (mostly micronodular), macroscopically dark green liver

Cl.: In the early stage asymptomatic (incidental laboratory findings); later:
• Pruritus: An excruciating itch long before the occurrence of a cholestatic icterus is an early symptom
• Fatigue, loss of energy
• Hepatomegaly (70 %), splenomegaly (20 %)
• Maldigestion as a result of reduced bile acid excretion → steatorrhoea
Further: sometimes xanthelasma/xanthoma dark skin tone (melanin)
• Extrahepatic diseases which are frequently associated with PBC: Autoimmune Hashimoto thyroiditis (20%), Sjögren- syndrome (about 70%), rheumatoid arthritis
• Overlap syndromes: In approx. 10 % the PBC is associated with auto-immune hepatitis, in 10 % with CREST syndrome

Co.: Complications of portal hypertension (Ascites, variceal bleeding) and of liver cirrhosis, malabsorption syndrome, osteoporosis

Lab.: • Antimitochondrial antibodies (AMA): > 95 %
of the 4 AMA-subtypes (Anti-M2, Anti-M4, Anti-M8, Anti-M9) Anti-M2 is specific for PBC (Acryltransferases of the interior mitochondrial membrane are the target antigen)
The main PBC-antigen is PDH-E2, against which auto-antibodies are found in 95% of all PBC patients. Autoantibodies against OADC-E2 and BCKD-E2 are found in about 50% of cases
• ANA (50%)
• Strong IgM-elevation
• Cholestasis-symptom: Elevation of alkaline phosphatase, LAP, γGT, bilirubin,
• Hypercholesterolemia

DD: • Intrahepatic cholestasis due to
- Cholestatic progressive form of virus hepatitis
- Cholestatic drug-induced jaundice
• Extrahepatic cholestasis due to biliary duct stones, tumours, roundworms, liver fluke; biliary duct stenosis after laparoscopic cholecystectomy (Sonography, MRCP, ERCP) → possible development of secondary biliary cirrhosis with chronic cholestasis and infection of the biliary ducts (Hi.: Suppurating cholangitis
• DD itch (without initially visible skin diseases) (Selection of important causes):
- Allergic skin reactions   - Renal insufficiency
- Intestinal parasites   - Polycythaemia vera
- Cholestasis   - Iron deficiency
- PBC and PSC   - Senile pruritus, dry skin
- Diabetes mellitus   - Psychogenic pruritus
- Lymphomas

- Zieve syndrome: Triad
  - Fatty liver due to alcohol abuse or alcoholic hepatitis (with or without icterus)
  - Hyperlipidemia (milky plasma due to triglyceride increase)
  - Haemolytic anaemia
- Auto-immune cholangitis = AMA-negative PBC
- Overlap syndrome PBC / auto-immune hepatitis (AIH) in 10%

Di.: Clinic(pruritus) + laboratory (Cholestasis enzymes + IgM ↑, AMA-PDH-E2) - exclusion of extrahepatic cholestasis (Ultrasound: normal biliary ducts) liver histology (Laparoscopy)

Th.: A causal therapy is unknown.
Ursodeoxycholacid (UDCA) improves the elimination of biliary acids.There has been proof of a more favourable prognosis. Dos: 15 mg/kg BW /d
Combination of UDCA + corticosteroids (e.g. budesonide) still in clinical testing

Symptomatic therapy:
- Pruritus: colestyramine absorbs the bile acids in the intestine and lowers the cholesterol level, substitution of the fat-soluble vitamins (A, D, E, K) in the process. Intake of colestyramine 3 h after UDCA
- Maldigestion syndrome: Low-fat diet, dose of medium-chain triglycerideslipase dose with meals
- Osteoporosis prophylaxis (Details: See there)
- Liver transplantation in terminal liver cirrhosis

Prg.: 5-year survival of symptom-free patients about 90%
5-year-survival of symptomatic patients about 50% on average
The best prognosis parameter is the development of the serum bilirubin . With exceeding of a bilirubin level of 6 mg/dl life expectancy mostly < 2y.

Primary sclerosing cholangitis (PSC) [K83.0]

Ep.: Incidence of approx. 1 - 5/100,000/ y; m : f = approx. 3 : 1; Disease most commonly found between age 30 -50. Up to 80 % of the patients with PSC have ulcerative colitis (up to 5 % of the patients with ulcerative colitis have PSC), mainly HLA-B8 and DR3.

Aet.: Unknown

Cl.: In the early stage asymptomatic (incidental laboratory findings); later: Icterus, itching, diffuse upper abdominal pains, weight loss

Co.: Biliary cirrhosis with all complications, in 8 % cholangiocellular carcinoma (CCC), also increased risk of colourectal carcinoma

Lab.: γGT and AP ↑, proof of antineutrophile cytoplasmatic antibody (ANCA) with perinuclear (pANCA) or atypical (x-ANCA) fluorescence pattern in 80 % of cases

Di.: Clinic (itching, ulcerative colitis) + laboratory (cholestasis enzymes↑) + ERC or MRC: beaded duct irregularities , dentations
Liver biopsy/-histology: Periductal fibrosis, inflammatory infiltratesand biliary duct proliferation
Remark: Apart from classical PSC, "small-duct-PSC" is diagnosed occasionally = typical liver histology (as in PSC), but normal ERC-findings. The prognose for these cases is more favourable than for classic PSC with pathologic ERC-findings.

DD:
- Intra- or extrahepatic cholestasis (see there)
- Pruritus of different origins
- Overlap syndrome PSC / auto-immune hepatitis (AIH) in 6%

Th.: As in PBC with UDCA (see above). UDCA also reduces the risk for CCC
- In biliary duct infection: Antibiotics
- In biliary duct stenosis: Endoscopic balloon dilatation and pot. stents
- Terminal stage: liver transplantation

Prg.: Average survival time (without liver transplantation): 10 - 20 years.

Non-alcoholic fatty liver disease (NAFLD) [K76.0]

Def.: 3 stages:

1. Pure fatty liver (Steatosis hepatis) – histologic grading
   Degree 1 : Mild fatty liver : fat deposit in < 1/3 of the hepatocytes
   Degree 2: moderate fatty liver: fat deposit in < 2/3 of the hepatocytes
   Degree 3: Severe fatty liver: fat deposit in > 2/3 of the hepatocytes
2. Non-alcoholic steatohepatitis (NASH))
   Liver cell damages (fat deposits,enlarging,necrosis) + inflammatory cell infiltrates (neutrophilic granulocytes > mononuclear cells) ± fibrosis
   The lesions in NASH are similar to or broadly equal those in alcoholic steatohepatitis (ASH), even without alcohol consumption of > 20g /d
3. Micronodular liver cirrhosis ("fatty cirrhosis")

Ep.: About 20% of adult population in industrial nations. In 90% , the reason is the methabolic syndrome and type 2 diabetes mellitus

Aet. : ▶ **Metabolic syndrome** (definition see chap. diabetes)
   ▶ **Type 2 diabetes mellitus**
   ▶ **Drugs**
   • Amiodaron (leads to NASH in about 25%)
   • Glucocorticoids
   • Nifedipine, diltiazem
   • Tamoxifen, synthetic oestrogen
   • Highly active antiretroviral therapy (HAART)
   ▶ **Rare causes**
   • Gastrointestinal surgery: Jejunoileal bypass, extended small bowel resection, pancreatico-duodenectomy, gastroplasty
   • Total parenteral nutrition
   • Wilson's disease

Cl.: No symptoms in fatty liver, in fatty liver hepatitis unspecific symptoms in 50 %

Lab.: in fatty liver often γ GT↑, in fatty hepatitis additionally transaminases ↑, de Ritis-ratio (GOT/GPT or AST/ALT) in NASH often <1, in ASH >1

Sono: - Wide range of variations in findings in the often enlarged liver:
   - In diffuse fatty liver :homogeneously condensed echo pattern ("bright" liver)
   - single echoes often enlarged
   - rounding up of the lower liver edge
   - In strong markedness reduced distal echo
   - Different fat content of left and right hepatic lobe
   - Fat - infiltration or non-infiltration (polycyclically limited areas, typical localisation often in the area of the portal bifurcation and the fossa of the gall bladder; no impairment of the vessels) - DD: tumour
   - Seldom: inhomogeneous fat infiltration: landscape echo-rich areas without impairment of the vessels - DD: metastases

DD: Alcohol-induced liver damages: see below

Di.: History, laboratory./ultrasound, possibly liver histology

Th.: Causal therapy (see above)

Prg.: The prognose is determined by the causing disease (metabolic syndrome, type 2-diabetes). The prognose for bland fatty liver is favourable. 5% develop liver cirrhosis in 10 years (which is then often misinterpreted as cryptogenic liver cirrhosis

Alcoholic fatty liver diseases (AFLD)

Def.:

3 stages 1. Pure fatty liver (fatty degeneration hepatitis) without inflammatory reaction
   Histologic grading
   Degree 1 : Mild fatty liver : fat deposit in < 1/3 of the hepatpcytes
   Degree 2: moderate fatty liver: fat deposit in < 2/3 of the hepatocytes
   Degree 3: Severe fatty liver: fat deposit in > 2/3 of the hepatocytes

2. Alcoholic fatty hepatitis = alcoholic steatohepatitis (ASH):
Fatty liver with inflammatory reaction

    Hi.:   Fatty liver hepatitis:
- Fatty liver
- honeycomb hepatocytes
- endocellular alcoholic hyaline (= Mallory-bodies)
- Granulocytes surround poured out and necrotic hepatocytes
- Wire netting fibrosis
- Inflammatorily infiltrated portal areas

3. Micronodular cirrhosis ("fatty cirrhosis")

Ep.: Prevalence 5-10% of population in Europe. In Germany, 1/3 of all liver diseases are caused by alcohol consumption

Aet.: Alcohol consumption
Toxicity limit of alcohol for the liver differs individually, depending on former diseases gender ADH capacity in women is much smaller than in men), mal- and wrong nutrition: Toxicity limit for men approx. 40 g ethyl alcohol for women only approx. 20 g/d . The harmless maximum drinking amount for alcoholic drinks per day is for example up to 3/4 l of beer or 3/8 l of wine for healthy men, for healthy women 1/2 l of beer or 2/8 l of wine. With chronic alcohol consumption exceeding this limit, 30 % of the persons develop a fatty hepatitis and the danger of liver cirrhosis is 6-fold higher.

Memo: Moderate alcoholl consumption (up to 15 g/d for women and to 30 g/d for men) can reduce the danger of myocardial infarction and ischaemic stroke

Alcohol quantity (g): Vol% x Volume of drinks (ml) x 0.8
                                             100

Pg.: Chronic alcohol consumption leads to the induction of the Cytochrom-P450-dependent microsomal ethanol - oxidation system (MEOS). The alcohol decomposition therefore not only takes place via ADH but increasingly with MEOS. The increased O_2-consumption due to MEOS leads to central lobule hypoxemia. The acetaldehyde arising from alcohol decomposition hasliver-toxic effects. A reduced oxidation of fatty acids leads to the development of fatty liver.

Cl.: Alcohol-induced fatty liver: Discrepancy between palpable hepatomegaliaand patients being mostly asymptomatic
Alcohol-induced fatty liver hepatitis:
Half of the patients are asymptomatic in the initial phase
- Hepatomegaly (90 %), splenomegaly (30 %)
- Poor appetite, nausea, weight loss
- Pain in the right epigastrium
- Icterus (50 %)
- Fever (45 %)

Co.: - Zieve syndrome (alcohol-toxic liver damage + haemolytic anaemia + hyperlipidaemia)
- Liver cirrhosis with hepatic failure, portal hypertension and its consequences
- Rarely: fulminant hepatitis
- Hypoglycemia (inhibition of the gluconeogenesis by alcohol)
- Extrahepatic damages caused by alcohol (see chap. alcohol disease)

Lab.: - CDT (Carbohydrate-Deficient-Transferrin)-increase is regarded as a marker for chronic alcohol abuse (specific in > 90 %, sensitivity good with men, inaccurate with women). Elevated values for CDT are also found in biliary cirrhosis and auto-immune hepatitis.
- in pure fatty liver: γGT and IgA
- in case of fatty liver hepatitis: in addition transaminases ↑ (while the de Ritis-ratio AST/ALT isoften > 1).
- Synthesis achievement of the liver (ChE, albumin, Quick-value ↓) reduced in hepatic insufficiency
- MCV frequent ↑ (unspecific)

Sono: Findings in fatty liver and liver cirrhosis: See there

CT: (diagnostically not necessary) quantification of the fat content of the liver.

DD: - Non-alcoholic fatty liver and non-alcoholic steatohepatitis (NASH): With NASH the histologiccriteria of the alcohol damage can be ascertained without alcohol playing a causal role. Main reasons:Metabolic syndrome and type 2 Diabetes mellitus (90% of cases)
- Acute and chronic hepatitis of other genesis
- Icterus with fever: biliary duct occlusion, cholangitis

Di.: - (Alcoholic) history + clinic
- typical sonography findings
- liver histology
- in suspected fatty liver cirrhosis additionally abdominoscopy

Th.: There is no drug therapy for alcohol-toxic liver damage. The only effective therapy is alcohol abstinence.
A possible lack of folic acid is substituted; in case of alcoholl-toxic CNS damages administration of thianine. In severe alcoholic hepatitis possibly short-term therapy with glucocorticoids.

Prg.: In stages 1 and 2 rather favourable (reversible) if the noxious substance (alcohol) is avoided. In stage 3 of fatty cirrhosis danger of complications like hepatic failure and portal vein hypertension

REYE SYNDROME [G93.7]

In.: Children up to age 15

Aet.: Unclear, frequently after respiratory infections and NSAID

Pg.: Diffuse mitochondrial damage

Cl.: • Intense vomiting, hypoglycemia
• Hepatic encephalopathy with brain oedema and possible convulsions
• Fatty hepatitis (diffuse small-drop fat infiltration

Th.: Symptomatic

Prg.: mortality up to 50 %, in 30 % neurologic damages

DRUG-INDUCED AND TOXIC LIVER DAMAGES [K71.9]

liver-toxic agents are divided into 2 groups:
A) obligatory hepatotoxine : liver damage occurs after brief latency dosage-dependent and therefore is foreseeable . The period between exposure and manifestation of the liver damage is short.
B) facultative hepatotoxine (majority): liver damage occurs after varied latency periods independent from the dosage in a small number of patients and therefore is not foreseeable.
1. metabolic idiosyncrasy in genetically conditioned enzyme defects
2. immunologically conditioned idiosyncrasy due to hypersensitivityreactions

Ep.: In the USA, hepatotoxic effects of drugs are considered to be the most common reason for acute liver failure. The most frequent drug in this context is paracetamol. Basically, most of the drugs, including naturopathic medicine (e.g. kava), have a liver-damaging potential. Hepatotoxic UDI (undesirable drug interactions) are the most frequent reason for the FDA for withdrawing a drug already licensed. As fatal hepatotoxic UDI occur relatively rarely (incidence 1 : 10,000 to 1 : 100,000), they mostly remain undiscovered during clinical trials conducted to obtain drug registration. It is only in broad usage that severe liver function disorders become apparent. Detecting drug-induced liver damages is simple if the doctors responsible for the treatment remember to monitor liver function tests.

Aet.: Examples of liver-toxic drugs and their predominant damaging pattern:
• Hepatocellular damage (ALT-increase)
paracetamol, allopurinol, amiodarone, antiretroviral agents, kava, isoniazid, ketoconazol, lisinopril, losartan, methotrexate, NSAID, omeprazole, pyrazinamide, rifampicin, statins, tetracyclines, valproic acid
• Cholestatic damage (AP- and bilirubin increase)
Amoxicilline-clavulanic acid, anabolic steroids, chlorpromazine, clopidogrel, oral contraceptives, erythromycin, oestrogens, phenothiazines, tricyclical antidepressant agents
• Mixed damages (ALT- and AP – increase)
Amitriptyline, azathioprine, captopril, carbamazepine, clindamycin, enalapril, nitrofurantoin, phenytoin, sulfonamides, cotrimoxazol, verapamil

PPh: Biotransformation of drugs and chemicals takes place in 2 steps:
1. oxidation supported by the monooxygenasesystem Cytochrom P 450; this can be stimulated by some agents (enzyme induction, for example by phenobarbital or alcohol), the enzyme system can be inhibited by other agents
2. conjugation of the agents' metabolites, for example with the help of glucuronyltransferase to glucuronic acid.
For some agents the biodegradation can also lead to toxic intermediates (for example with carbon tetrachloride the liver-toxic radical •CCl3 develops). In caseof hypersensitivity reactions it is presumed that the drug or its metabolite binds as hapten with the liver cell membrane and thus forms a neoantigen triggering an antibody formation.

Cl.:   The entire spectrum of possible liver damages can appear. In single cases the morphologic picture does not allow any conclusion about the triggering noxious substances.

- acute hepatitis: for example through isoniazid (INH), methyldopa
- fulminant hepatitis: for example Halothan (risk = 1 : 30,000), paracetamol intoxication, carbon tetrachloride
- chronic-active hepatitis (CAH): for example by methyldopa, isoniazid (INH)
- fatty liver: for example ethyl alcohol organic solvents, tetracyclines
- intrahepatic cholestasis: for example by chlorpromazine, thyreostatics, Ajmalin, C-17-alkylated steroids (risk in case of oestrogen-containing contraceptive-intake: 1 : 10,000)
  DD: cholestatic progression of a viral hepatitis, extrahepatic cholestasis
- mixed type of hepatitis and cholestasis: for example sulfonamides, PAS
- auto-immune hepatitis: for example minocycline, IFN alpha
- liver tumours: for example.
  - adenomas due to oestrogen-containing contraceptives
  - focally nodular hyperplasia (FNH) due to oestrogen-containing contraceptives
  - liver cell carcinoma after long-term therapy with androgen or oral contraceptives
  - angiosarcomas due to chloroethylene, arsenic, thorotrast

Note: In case of allergy-related drug damages of the liver extrahepatic hypersensitivity complaints (exanthema, arthralgias, fever, acidophilia) are found occasionally

Di.:   - Drug history - clinic - histology
- exclusion of other causes of liver disease.
- improved findings after omission of the presumed noxious substances

Remark: The assessment is difficult if several liver-toxic agentsplay a role (for example alcohol + drugs or chemicals).

Th.:   Withdrawal of the suspicious drugs. For allergic drug damages with extrahepatic complaints temporary dose of glucocorticosteroids.

Prg.:   The majority of the chronic liver damages improve after the patient has stopped taking the noxious agent; fulminant hepatitis and liver tumours have a poor prognosis

HEREDITARY CATABOLIC DISEASES OF THE LIVER

1. SIDEROSES (Iron storage diseases)   [J63.4] [J63.4]

In a healthy person, the need for iron regulates the iron absorption in the small intestine. The more accentuated the iron deficiency is, the more iron is absorbed from the intestine; vice-versa, if the iron reservoirs are saturated, iron absorption sinks. Therefore, an iron overload is the consequence of a faulty regulation of this protective mechanism or of parenterally administered iron (e.g. via transfusions).

In haemochromatosis, a genetic defect leads to an increased iron absorption from the duodenum. In anaemias with iron overload, (e.g. haemolytic anaemia or myelodysplastic syndrome) iron absorption is typically increased as well.

If the normal iron content of the body – 3.5 g (m) and 2.2 g (f) - is exceeded by the 5fold (or more), it leads to organ manifestation. The pattern of the organ manifestations indicates the origins (see chart)

The stored quantities of iron in the secondary forms do not reach the extent of the primary form. In advanced haemochromatosis, the body contains up to the 10fold of the normal iron concentration.

Def.:   • Pathologic-anatomic definition of haemochromatosis: Iron deposition with damages of the tissue

• Pathologic-anatomic definition of the haemosiderosis: Iron deposition without damages of the tissue

1. Primary form: HEREDITARY HAEMOCHROMATOSIS [E83.1]

Classification and aetiology:

Type	Gene locus	Affected gene	Typical mani-festation age	Frequency	Organ manifestation (organs according to frequency
1	Chr 6p21.3	HFE	30-50 y	1 : 1,000	Liver → cirrhosis Pancreas → diabetes **Heart → insufficiency** Joints → arthralgia Hypophysis → impotence
2a	Chr1q21	JH (HFE2)	10-20 y	~1:1 million	Heart → insufficiency Hypophysis → Hypogonadism Liver → cirrhosis
2b	Chr 19q13	HAMP1	5-15 y	Rarity	As type 2a
3	Chr 7q22	TFR2 (HFE3)	10-50 y	Rarity	
4	Chr 2q32	SLC1A2 (HFE4)	10-50 y	~1:1 million	Liver → cirrhosis Bone marrow → anaemia (Spleen → iron deposition)

1. 1 <u>Classical (adult) haemochromatosis:</u> In Europe, the prevalence of clinically manifest haemochromatosis is about 1:1,000. The most frequent organ manifestation of adult form haemochromatosis is liver cirrhosis, followed by diabetes and arthralgias as well as cardiac insufficiency. The typical presentation age lies between 30-50 years. Men are affected about 10 times more frequently than women. Over 90% of haemochromatosis patients are homozygotic for the C282Y-mutation (=Cys282Tyr) in the HFE-gene (vice-versa, the penetrance of C282Y homozygosis is under 25% - this means that only 25% of homozygotics develop a manifest haemochromatosis). About 5% of haemochromatosis cases are heterozygotic for C282Y-mutation and carry the H63D-mutation on the other allele (= compound -heterozygote). H63D-mutation alone (His63Asp-heterozygotic or homozygotic) does not lead to haemochromatosis. Heterozygotic gene carriers of theC282Y-mutation without supplementary H63D (frequency 1:10) do not fall ill with haemochromatosis; however, a moderate iron accumulation can occur, which, combined with other liver noxa/ diseases (e.g. alcohol consumption or hepatitis C) canlead to particularly severe liver damages (intensified liver damages caused by iron), m : f = 10 : 1; age of manifestation: starting with the 4th decade of life.

Pg.: The faulty HFE-gene product leads to an increased iron absorption from the small intestine , independent from the organism's iron demand (disregulated iron resorption) by diminished formation of Hepcidine in the liver, which inhibits the intestinal iron-absorption. In a healthy person 1-2 mg of iron are absorbed daily, in haemochromatosis about 3-4 mg.

1.2. <u>Juvenile haemochromatosis:</u> Rarely, age of manifestation: By definition, a clinically manifest iron overload before turning 30 corresponds with juvenile haemochromatosis. The most frequent organ manifestations are cardiac insufficiency and hypogonadism; liver cirrhosis is less frequent.

1.3 <u>Neonatal haemochromatosis:</u> Rarity. Already intrauterine liver cirrhosis – most frequent indication for liver transplantation within the first three months of life (without liver transplantation fatal)

2 Secondary sideroses (secondary iron storage diseases) :

2.1 Anaemias with iron overload : thalassemia and other haemolytic anaemias, myelodysplastic syndrome. On one hand , the iron absorption from the intestine is uncontrollably increased due to anaemia , on the other ,hand repeated transfusions lead to uncontrollable parenteral iron supply (250 mg of iron/ 500ml of blood). In these cases, a therapy with desferoxamine (Desferal) may be indicated. Typical organ manifestation: cardiac insufficiency and hypogonadism.

2.2 Alcoholic siderosis: In alcoholic disease (probably due to increased iron absorption

2.3 Sideroses within the setting of chronic liver diseases: Almost every liver disease in final stage can be accompanied by a secondary iron deposition/ ferritinine increase in the liver

Cl.:
- <u>liver cirrhosis</u> (75 % of cases), hepatomegalia (90 %), enlarged spleen (15 %)
- Hepatocellular carcinoma as complication of cirrhosis, but also in non-cirrhotic liver
- <u>diabetes mellitus (70</u> %) ("bronze diabetes" because of dark skin pigmentation)
- <u>dark skin pigmentation</u> especially in the axillas – therefore missing armpit hair (75 %)
- <u>secondary cardiomyopathy</u> due to iron deposition, possibly with arrhythmias and "digitalis -refractory" cardiac insufficiency
- <u>endocrine disorders:</u> damage of the hypophysis , pituitary hypogonadism in juvenile haemochromatosis, impotence in adults , damage of the adrenal cortex
- painfularthropathy : typical of the small joints of both hands (30 %)

2 stages of haemochromatosis:
1. Latent, pre-cirrhotic stage
2. Apparent, cirrhoticstage

DD: 1. Distinctionof secondary sideroses
2. Elevated ferritin values i.s. of other causes: liver diseases, inflammations, tumours

Di.: • History, clinic
• Laboratory: plasma ferritin ↑(> 300 µg/l) and transferritin saturation (f>45%, m>50%)
Remark: Transferrin saturation (= Serum iron : total iron binding capacity x 100)
Ferritin as an acute-phase-protein is also unspecifically increased in inflammatory diseases and tumours (esp. lymphomas)
• CT/MRI: Semi-quantitative assessment of the iron content of the liver (no early diagnosis possible)
• Non-invasive iron determination in the liver tissue by means of biomagnetometry (poor availability of equipment)
• Liver biopsy with histology and determination of iron concentration: elevation of the liver iron index (= iron concentration of the liver : Age) > 1,9
• Genetics: homozygosis for the C282Y mutation of the HFE-Gene makes the diagnose of haemochromatosis probable, but only 25% of the homozygotics develop clinically manifest haemochromatosis. A compound-heterozygosis (C282Y/ H63D) in the HFE-gene leads to haemochromatosis in > 10% ; on the other hand, a normal HFE-genotype does not exclude the diagnose of haemochromatosis (Non-HFE- haemochromatosis), i.e., gene findings can be assessed only in overview of clinic and serum-iron parameters.
• check-ups for the early detection of primary liver cell carcinoma (α1-Fetoprotein + sonography)
• Familyexamination in case of primary haemochromatosis (screening for HFE-Mutation)

Th.: An iron-poor diet is the basis: black tea, taken with meals, reduces the iron absorption. Ban on alcohol. Phlebotomy is the means of choice in primary haemochromatosis By erythroapherese (taking of packed blood cells) protein losses can be prevented. With 500 ml of blood 250 mg iron are ablated. The iron deposits are emptied once a week by bloodletting until the onset of (microcytary) anaemia The bloodlettings have to be continued until the serumferritin concentration drops < 50 µg/l. Afterwards, the frequency of bloodlettings is reduced to 4 x a year. The target of therapy is a ferritin-value of < 50 µg/l. If a bloodletting-therapy is contraindicated (e.g. anaemia, cardiac insufficiency), application of iron-chelation: Deferoxamin (Desferal ®) for parenteral therapy or deferasirox (Exjade ®), an oral preparation . Iron-chalation therapy is also applied in juvenile haemochromatosis and in transfusion-related secondary sideroses.
SE: Deferoxamin: Neurotoxic SE : parial deafness of the inner ear, tinnitus, sightdefects due to retina damages.
Deferasirox: Increase of creatinin, transaminases, gastro-intestinal discomforts, skin

Prg.: Patients who start therapy in the pre-cirrhotic phase have a normal life expectancy

2. WILSON'S DISEASE [E83.0]

Syn: Hepato -lenticular degeneration

Internet-informations: see *www.morbus-wilson.de*

Def.: Autosomal-recessive disorder caused by mutation in the Wilson-gene (ATPase 7B-geneon chromosome 13q14,3). > 250 variations of mutations complicate the genetic diagnosis. The most frequent mid-european mutation (40%): His 1069Gln. The Wilson-factor encodes the Wilson-protein, a P-type ATPase with copper transport function. The function loss of the Wison protein leads to a reduced biliary precipitation of copper and pathologic copper storage in liver and body ganglions. The disease presents as a liver disease at age 6 at the earliest and additionally as a neurologic disorder after age 12..

Ep.: Prevalence of M. Wilson about 1: 30,000, high number of unreported cases

PPh: The gene defect causes a reduced biliary copper excretion. In spite of renal copper excretion, copper accumulates in the body. The enteral copper absorption of a healthy adult totals 0,5 - 2,0 mg/24 h. Ceruloplasmin , which usually absorbs 95 % of the serum copper, is strongly reduced in Wilson's disease; free copper is cytotoxic and rapidly transfers from the bloodstream into the tissue. While an adult has 50-150 mg of total copper in the body, absorbs 4 mg daily and stores 20-50 µg /g of dry weight in the liver, in Wilson's disease the liver stores up to > 250 µg/g.

Cl.: • Hepatic manifestation (100 %): Manifestations range from asymptomatic transaminase increase over fatty liver to fulminant progression; final phase liver cirrhosis with all its complications.

Note: In unclear liver diseases at the age of < 35 years always exclude Wilson's disease.

• Neurologic-psychiatric manifestation (45 %) after age 10: Parkinson-like syndrome with rigidity, tremor, dysarthria, psychiatric disorders → MRI

- **Eye symptoms (ophthalmologic examination):** The Kayser-Fleischer' corneal ring is typical (gold brown-green discolouration of the corneal edge due to copper deposition:), which is always present in neurologic manifestation, chalcosis.
- Coombs-negative haemolytic anaemia (particularly in acute liver failure), acute haemolytic crises
- Less frequent: Kidney malfunction, cardiomyopathy with arrhythmias

Di.:
- Slitlamp examination (Kayser-Fleischer ring)
- Ceruloplasmin i.s. < 15 mg/dl
- Total copper i.s. < 70 µg/dl
- Free copper i.s. > 10 µg/dl
- Copper i.u. > 250 µg/day
- Copper content of the liver > 250 µg/g dry weight

Only in unclear diagnostic supplementatry examinations
- Penicillamin-load test: After administration of penicillamin distinct increase in copper excretion in 24-h-urine
- Radiocopper-test: After oral dose of ^{64}Cu a twin peak increase of radioactivity is usually registered in the serum. The 2nd peak marking the intake of copper in ceruloplasmin is missing in Wilson's disease.
- Proof of a mutation of the Wilson-gene (no routine test because of numerous mutations
- Familyexamination for possible further cases

Th.:
- Copper-poor diet (soft water that runs through copper pipes may contain copper: water analysis) Administration of the copper chelator D-Penicillamine ; regular urine checks: stop therapy in case of albuminuria (toxic nephrosis). Due to relatively frequent SE (skin rash, fever, leuko-/thrombocytopenia, nephrotic syndrome, Goodpasture syndrome, SLE, myasthenia), trientine (= trien or triethylentetramin) is mostly applied nowadays as means of first choice, which is also very effective and well tolerated. Additional dose of vitamine B6 (prophylaxis against optic nerve neuropathy)
- In fulminant hepatitis or terminal liver cirrhosis: liver transplant which repairs the gene defect
- Future prospects: somatic gene therapy

Prg.: With early initiated therapy: good, untreated: high mortality.

3. ALPHA₁- PROTEASE INHIBITOR DEFICIENCY [E88.0]

Syn: Alpha1 antitrypsin deficiency (AAT-deficiency)

Def.: Autosomal-recessive hereditary disease with AAT-deficiency and pulmonary and hepatic manifestation

In.:
- Homozygous severe form: Phenotype PIZZ
  1 : 10,000 in the population
  α1–PI- concentration 50 - 250 mg/dl
- Heterozygous less severe form: Phenotype PIMZ or PIMS
  α1–PI concentration 50 - 250 mg/dl

Pg.: Alpha1 anti trypsin (α1 At = AAT) = Alpha1 protease inhibitor (α1-PI): With 90% most important protease inhibitor in the serum, is formed in the liver and constitutes 85 % of the α1 globulins. Inactivation of serum proteases (neutrophilic elastase, (chymo)trypsin, collagenase, etc.) The PI gene is on the chromsome body locus14q32.1.
α1-PI is an acute phase protein (predominantly) produced in the hepatocytes → Also determine CRP while determining α1-PI (in case of inflammation pseudonormal α1-PI-values). While pulmonary emphysema is caused by α1-PI-deficiency, liver damage is the result of a secretion disorder in the hepatocytes for the changed α1-PI-molecule. The retained α1-PI is stored as a PAS-positive materialin the liver tissue. This secretion defect is observed particularly in the homozygous PIZZ-phenotype.

Cl.: for the severehomozygote form:
- prolonged icterus in the newborn (with direct hyperbilirubinemia)
- Emphysema development (see chap. Pulmonary emphysema)
- Chronic hepatitis and liver cirrhosis (withPIZZ type in 15 % at the age > 50 y.); co.: hepatocellular carcinoma

Di.:
- Reduction of the alpha1-peak in electrophoresis
- α1-PI-concentration i.s. ↑
- Liver biopsy with histology: detection of α1-PI-depositions
- Phenotypification

Th.: α1-PI-substitution in severe α1-PI-deficiency (contraindicated in liver cirrhosis)
Symptomatic therapy of liver cirrhosis and of pulmonary emphysema; avoidance of nicotine

Last resort: liver transplantation
Future prospects:: somatic gene therapy
See also chap. Pulmonary emphysema!

4. MUCOVISCIDOSIS → see chap. Pancreas

LIVER CIRRHOSIS [K74.6]

Def.: Destruction of the lobuleand vessel structure of the liver with inflammatory fibrosis, development of connective tissue bridges (septa) between adjacent portal areas (porto-portal) and between portal areas and central veins (porto-central) as well as development of regenerative nodes.
Functional consequences are:
- Liver insufficiency
- Portal hypertension (reduced total vessel cross-section of the liver)
- Formation of intrahepatic porto-systemic shunts between portal vessels and hepatic veins with minor perfusion of the liver

Ep.: Incidence in Europe and USA: approx. 250/100,000/year; m : f = 2 : 1

PPh.: 1. Micronodular liver cirrhosis   Regenerate nodule up to 3 mm⌀
2. Macronodular liver cirrhosis: Regenerate nodule 3 mm - 3 cm⌀
3. mixed-nodose liver cirrhosis: mixed picture of 1 + 2

Aet.: Liver cirrhosis is the late result of miscellaneous liver diseases. The pathologic classification usually does not give clues to the aetiology:
1. Alcohol abuse (in industrial nations approx. 50 %)
2. Viral hepatitis B, C, D (in industrial nations approx. 45 %)
3. Other causes (approx. 5 %):
    - Auto immune hepatitis
    - Primary biliary cirrhosis (PBC) and primary sclerosing cholangitis (PSC)
    - Drug-induced liver damages (for example methotrexate)
    - Chemical damages (for example carbon tetrachloride, arsenic)
    - Metabolism diseases: haemochromatosis, Wilson's disease, lack of α1-antitrypsin, mucoviscidosis.
    - Cardiac cirrhosis: Chronically congested liver with "armored heart" or chronic right-heart insufficiency
    - Budd-Chiari syndrome (occlusion of the hepatic veins)
    - Tropical diseases (Bilharziosis, liver fluke)

Cl.: 1. General symptoms:
    - Fatigue, exhaustion, decline of performance (70 %)
    - Feeling of pressure or bloated feeling with pain in the upper abdomen, bloating (60%)
    - Possible nausea, weight loss
2. Liver skin sign , for example.
    - Spider naevi (naevi aranei), esp. in the upper part of the body or in the face is not only found in cirrhosis (for example also in pregnancy)
    - Palmar and plantar erythema,
    "Glazed lips, beefy red tongue", angular cheilitis
    - Prurigo simplex (Pruritus) with possible scratches
    - Skin atrophy ("Banknote skin") with teleangiectasias
    - White nails, Dupuytren's contracture (unspecific)

    Consider: In about 50 % of all pregnancies discrete liver skin signs may occur which mostly disappear postpartally (palmar erythema, spider-naevi).

3. Hormonal disorders:
    - in men often loss of the male secondary hair (abdominal baldness), potency impairment, testicular atro-phy (Causes: Testosterone↓, oestrogen ↑), gynaecomastia (hormonally conditioned or as SE of a spironolactone therapy)
    - in women menstruation disorders, secondary amenorrhoea
4. Aetiology-specific symptoms: for example dark skin colouring with haemochromatosis; neurological symptoms with Wilson's disease.

5. Signs of decompensation = complications:
  - Icterus
  - Haemorrhagic diathesis
  - Malnutrition, cachexia
  - Portal hypertension and its consequences (variceal bleeding, ascites, oedemas, hypersplenia)
  - Hepatic encephalopathy and liver failure coma
  - Primary liver cell carcinoma as a late result

Remark.: Due to their great number and variety the complications of the liver cirrhosis will be presented separately following this exposition .

Palpation: - enlarged or reduced liver, indurate, possibly with bumpy surface;
  - Spleen: Splenomegaly (75 %)
  - Abdomen: bloating, ascites

Lab.: • Indicators of reduced synthesis achievement of the liver:
  - Vitamin K-dependent blood clotting factors of the prothrombin complex (Factors II, VII, IX, X) ↓, measurable in a decrease of the Quick-value that does not normalize on vitamin K dose (Koller test) .
  - Antithrombin (AT)↓
  - Albumin i.s..↓
  - Bilirubin ↑
  - Cholinesterase (CHE)↓
  • Hypergammaglobulinemia (approx. 80 %)
    DD: 1. unspecific symptom in liver cirrhosis
         2. typical symptom in auto-immune hepatitis
  • Thrombocytopenia in hypersplenia and in reduced hepatic thrombopoetin formation
  • In case of hepatic encephalopathy:
    Ammonia, ↑ respiratory alkalosis. with hypokalemia
  • In inflammatory attacks increase of the enzymes that indicate a hepatocyte damage:
    Transaminases (ALT, AST), GLDH, γGT
  • In PBC, PSC and cholestatic hepatitis attack increase of the cholestasis-indicating enzymes (AP, LAP, γGT) and possibly bilirubin.

Sono: - Irregular undulant liver surface (DD: metastasis liver)
  - Inhomogeneous liver parenchyma with regenerative nodes (DD: primary liver cell carcinoma)
  - liver vein rarefaction
  - reduced deformation ability
  - rounded liver edge
  - In portal hypertension reduction of the maximum flow speed in the central stem of the portal vein < 12 cm/s., even retrograde flow or pendular flow
  - Indirect indications of portal hypertension in the colour duplex: visible collaterals, ascites, splenomegaly, wide portal vein
  - Detection of advanced fibrosis and cirrhosis by means of elastometry ("fibrosis scan")

Transient elastography: The stiffness of the liver tissue is determined by means of a vibration sender and an ultrasound probe. It correlates with the extent of liver fibrosis.

  Child-Pugh-criteria for the classification of the severity degree of a cirrhosis:

	1 point	2 points	3 points
Albumin i.s.. (g/dl)	> 3.5	2.8 – 3.5	< 2.8
Bilirubin i.s.. (mg/dl)	< 2.0	2.0 – 3.0	> 3.0
Bilirubin µmol/l)	< 35	35 - 50	> 50
Bilirubin in PBC and PSC (mg/dl)	< 4	4 - 10	> 10
Bilirubin in PBC and PSC (µmol/l)	< 70	70 - 170	> 170
Quick test (%)	> 70	40 - 70	< 40
Ascites (Sonography)	0	mild	moderate
Encephalopathy	0	I - II	III - IV

Addition of the points: Child A = 5 - 6
                                    Child B = 7 - 9
                                    Child C = 10 - 15			

DD: of liver cirrhosis, for example.
  • Hepatomegaly of other genesis, for example in liver metastasis, primary liver cell carcinoma (as a delayed complication)
  • Splenomegaly of other genesis (see chap. spleen)
  • Ascites of other genesis (see chap. portal hypertension)
  • Encephalopathy of other genesis

- Icterus of other genesis (see chap. Icterus)
- in bleeding esophageal varices → other causes of upper gastrointestinal haemorrhage (see there)

Di.: ▶ of liver cirrhosis:
- History
- Clinical picture
- Laboratory
- Liver morphology (Sonography, CT, laparoscopy)
- Liver puncture (sonography guided) with histologyy

▶ of portal hypertension:
- History, clinical picture, colour duplex
- Proof of esophageal varices and hypertensive gastropathy (endoscopy)
- Detection of collaterals and blocking obstacles: Colour duplex, MRI-angiography, spiral-CT

▶ of hepatic encephalopathy:
- History
- Clinical picture
- Laboratory (Ammonia)
- Flickering test

▶ Aetiologic diagnostic: See chap. viral hepatitis, auto-immune hepatitis, PBC, PSC, metabolism diseases

Th.: of liver cirrhosis:

A) General measures:
- Ban on alcohol, omission of all potential liver-toxic drugs
- Adequate protein and calorie supply; in case of hepatic encephalopathy reduction of protein supply. Vitamin substitution on demand: In alcoholism substitution of folic acid and aneurin = vitamin B1. In biliary cirrhosis substitution of fat-soluble vitamins (A, D, E, K).

B) Therapy of the underlying disease, for example:
- Omission of the causal toxins in alcoholism, drug induced or toxic liver damages
- Immunosuppressive therapy in auto-immune hepatitis
- Attempt of a virus elimination in chronic viral hepatitis (see there)
- Iron elimination in haemochromatosis (phlebotomy)
- Copper elimination in Wilson's disease(D-penicillamine)

C) Treatment of complications (variceal bleeding, ascites, hepatic encephalopathy): See below, where the complications are presented separately.

D) Regular diagnostic for early detection of primary liver cell carcinoma (Sonography/ alpha1-Fetoprotein every 6 months)

E) Liver transplantation

Prg.: Dependent on:
- Aetiology of the liver cirrhosis and causal therapy possibilities (for example relatively favourable prognosis of alcohol-toxic liver cirrhosis with consistent alcohol abstinence)
- Complications: variceal bleeding (30 % of patients), liver failure, primary liver cell carcinoma
- Stage of liver cirrhosis:

1-year survival rates:	Child A:	Almost 100 %
	Child B:	85 %
	Child C:	35 %

Most frequent causes of death: Liver failure and/or variceal bleeding; further: liver cell carcinoma

PORTAL HYPERTENSION [K76.6]

Def.: Pressure enhancement in the portal vein > 13 mm Hg (reference range 3 - 13 mm Hg)

Aet.: Classification and causes:

1. Prehepatic block
Portal vein thrombosis (Strictly speaking, thromboses of the splenic vein do not lead to portal hypertension, but they can also indicate varices)
Aet.: - Thrombosis in patients with risk factors (for example polycythemia vera, intake oof estrogen-containing contraceptives)
- septic thrombosis due to navel string infection of the newborn
- Portal vein compression (tumours, pancreatic cysts, lymph nodes)
- Injuries, peritonitis
Cl.: Hypersplenia syndrome in normal liver function

2. Intrahepatic block (> 90 % of the cases.)
    a) Pre-sinusoidal (wedged hepatic vein pressure mostly normal)
        Aet.: Bilharziosis (Schistosomiasis - frequent cause in tropical zones), myeloproliferative diseases, liver metastases
    b) Sinusoidal
        Aet.: Liver cirrhosis (80 % of the cases with portal hypertension)
    c) Post sinusoidal (increased wedged hepatic vein pressure) = veno-occlusive diseases
        Aet.: for example liver damages due to immunosuppressives
    Pathogenically it is mostly impossible to draw a clear distinction, since many liver damages affect all vessel segments simultaneously.

3. Posthepatic block
    - Budd-Chiari syndrome [I82.0]= occlusion of the hepatic veins by thromboses, tumour compression or congenital membranous occlusions (Asia)
    - Cardiac ascites: Constrictive pericarditis, right-heart failure, severe cardiac insufficiency

PPh: assessment of portal pressure:
    - Duplex-Sonography
    - Invasive pressure-measuring (wedged hepatic vein pressure –measuring, e.g. by transjugular access; direct intraoperative pressure measuring)

The pressure in the portal vein of a healthy person totals 3 - 6 mm Hg. If the portal vein pressure rises to values >12 mg Hg, complications of portal hypertension may occur. The total blood circulation of the liver is approx. 1,500 ml of blood/min . 2/3 of this blood come from the portal vein, 1/3 from the hepatic artery. Normally, half of the O_2-supply of the liver comes from arterial blood and half from portal blood. Increased resistance in the portal blood flow ("backward flow") + increased arterial blood flow in the splanchnic vessels area ("forward flow") lead to portal vein hypertension. The porto-systemic (transhepatic) pressure gradient is defined as the difference between pressure of the portal and pressure of the V. cava inf. (standard value up to 5 mm Hg). With values > 10 mm Hg the formation of esophageal varices is probable, with values > 12 mm Hg there is an elevated rupture danger. As a result of portal hypertension collateral circulation develops between the portal vein and the caval vein system:
- porto-gastro-oesophageal collateral → oesophagus-/fundus varices
- umbilical collateral : venous connection between umbilical veins and epigastric veins (Cruveilhier-von-Baumgarten syndrome) → Di.: Colour duplex; clinical: "Caput medusae"
- mesenterico-haemorrhoidal collaterals
- gastro-phreno-(supra)renal collaterals

By these extrahepatic shunts the first-pass-metabolism of the liver is omitted: reduced decontamination of potentially toxic substances
    Cl.:   ▶ collateral circulation:
        - **Oesophagus- and corpus-/fundus varices, possibly with haemorrhage** [I85.0/I86.4] (also see chap. gastrointestinal bleeding)
        1/3 of the patients with liver cirrhosis have variceal bleeding - risk factors for the occurrence of a variceal bleeding are: Previous variceal bleeding, endoscopy finding (degree 3 -4 varices, "red colour sign"), persistent consumption of alcohol. The mortality of the first haemorrhage correlates with the Child-stage: Child A: < 10 % - B Child: approx. 25 % - Child C: approx. 50 %. Without recurrenceprophylaxis, 70 % of the patients have relapsing haemorrhage within a year (most frequently within the first 6 weeks after the first haemorrhage).

        Consider: Not every person suffering from cirrhosis has esophageal or variceal bleedings ; in 25 % of the cases ulcer is the cause, in 25 % it is erosive gastritis.

        - Visible collateral veins on the abdominal skin, rarely periomphalic as a so-called "Caput medusae externum" (only in open V. umbilicalis = 1 % of cases). More frequently: "Caput medusae internum" at the inner side of the abdominal wall (visible in the colour duplex).
    ▶ Congestive splenomegaly, possibly with hypersplenism
        Pot. thrombo-/leukocytopenia, anaemia (see chap. hypersplenism)
    ▶ **Ascites [R18]]:** Accumulation of serous fluid within the peritoneal cavity

Distinctive features	Transudate	Exudate
– Specific gravity	< 1.016 g/l	> 1.016 g/l
– Protein content	< 2.5 g/dl	> 2.5 g/dl
– Serum-ascites albumin quotient	> 1.1 g/dl	< 1.1 g/dl
DD:1. Portal ascites (80% of cases) 2. Cardiac ascites	Liver cirrhosis, right-heart insufficiency, Budd-chiari syndrome, constrictive pericarditis	

3. Malignant ascites (up to 10% of cases)		Ascites often haemorrhagic *) , Tumour marker: CEA, CA 125, CA19-9, <u>malignant</u> <u>cytology</u>
4. Inflammatory ascites		Bacterial peritonitis,leukocyte↑ positive culture (Bacteria, tuberculosis)
5. Pancreatogenic ascites		Pancreatitis: Amylase, lipase↑
6. Ascites hypo-albuminaemic	Nephrotic syndrome Exudative enteropathy	

*) <u>Remark</u>   Haemorrhagic ascites with liver cirrhosis is in 25 % caused by a primary liver cell carcinoma.
   Further: Chylous ascites (lab: triglycerides) and uroperitoneum (lab.: creatinine) , e.g. after abdominal /urologic
           surgical procedure
<u>Pathogenesis of ascites in liver cirrhosis:</u>
1. Portal sinusoidal hypertension with hypervolemia of the splanchnic vessels
2. Increased lymph production
3. Hypoalbuminemia with reduction of colloidosmotic pressure
4. Increased sodium reabsorption in the proximal tubule→ renal sodium- and water retention   amplified by
secondary hyperaldosteronism (increased synthesis of aldosteron + reduced hepatic inactivation of aldosteron

<u>CL</u>: - Abdominal girth increase, weight gain
    - Curved abdomen
    - Flanks projecting in lying position
    - Elapsed navel, possibly even umbilical hernia
    - Disproportion between emaciation of the limbs and abdomen with ascites
    - Possible dyspnoea due to diaphragmatic eventration

<u>Di.</u>: Detection of ascites:
• <u>Clinic</u> (lower detection limit approx. 1.000 - 1.500 ml)
    - Ballottement (Fluctuation wave)
    - Flank attenuation and attenuation change with changed position
    - Percussion in genucubital position
• <u>Sonographic</u> (lower detection limit approx. 50 ml): Preferred spots: Paravesical, perisplenic or -hepatic
• As a secondary finding in <u>CT/MRI</u>

<u>Examination</u> of ascites (after diagnostic puncture under sonographic control):
- laboratory-chemical (protein content, LDH)
- <u>Serum-ascites-albumin-gradient (SAAG) in g/dl</u> (albumin in the serum minus albumin in the ascites)
- bacteriologic
- cytological (leukocytes, erythrocytes, tumour cells)

Above all, the diagnose of spontaneous bacterial peritonitis (SBP) is of clinical relevance within a relapsing,
possibly therapy-refractory ascites. Here, the number of granulocytes (> 250 cells / µl) is of primary diagnostic
importance.
Pathogen detection is only possible in part of the cases.

<u>Remark:</u> The assignment of individual causes of ascites to transudate - exudate is not as clear in all cases as
shown in the chart, since the protein content can be individually different.

<u>Complications of ascites:</u>
▶ Reflux esophagitis, dyspnoe, orthopnoea
ea, intestinal wall hernias, hydrothorax
   ▶ **Spontaneous bacterial peritonitis = SBP** (about 15% of all patients with portal ascites)
     Pathogens: E. coli (50%), gram-positive cocci (30%), klebsiellae (10%)
     <u>Di.:</u> fever + abdominal pain are exceptional, clinically usually non infective; Ascites examination: > 250
     granulocytes/µl, bacteria detection in the ascites (test with aerobic + anaerobic blood culture bottles) is often
     negative
   ▶ Increased risk of variceal bleeding
Hepatorenal syndrome (HRS): [K76.7]
<u>Def.:</u> Progradient and irreversible decrease of gromelural filtration rate in patients with liver cirrhosis or fulminant
progressing hepatitis which originates in a severe vasoconstriction of the renal circulation. HRS is observed in
about 10% of patients with advanced liver cirrhosis and ascites and is a diagnosis of exclusion, since other
causes of a reduced glomerular filtration rate have to be excluded.

Two clinical presentations:

Type I: Rapidly worsening renal function with a doubling of the initial serum-creatinine to >2,5 mg/dl within 2 weeks

Type II: Slow gradual worsening worsening renal function

Pg.: According to the Underfill-theory, the combination of portal hypertension with arterial vasodilation in the splanchnic vessel area changes the intestinal capillary pressure by an increase in permeability, which thus leads to fluid accumulation in the abdominal cavity. With progressing disease, a distinct reduction of renal passing of free water and renal vasoconstriction are found. These mechanisms lead to dilutional hyponatremia and to HRS.

CL: Clinical symptoms of decompensated liver cirrhosis with ascites, oedema, icterus and hepatic encephalopathy.

Triggering factors of HRS:
- Gastrointestinal bleeding
- Paracentesis without expansion of plasma volume
- Intensive therapy with diuretics
- Spontaneous bacterial peritonitis (SBP)
- Overdose of lactulose (diarrhoea, hypovolemia)
- Nephrotoxic drugs (e.g. NSAID)

Di.: Decompensated liver cirrhosis + worsening of the renal function

Diagnostic criteria for HRS:
• Serum creatinine > 1.5 mg/dl or creatinine-clearance < 40 ml/min
• Absent symptoms of shock, bacterial infections , of fluid loss; no previous therapy with nephrotoxic drugs
• No lasting improvement of the renal function after stopping the intake of diuretics and after the expansion of plasma volume
• Proteinuria < 500 mg/d
• Exclusion of an obstructive uropathy by sonography

Th.: 1. Elimination of triggering factors
     2. Liver transplantation (best therapy option)
     3. If not possible
        • Placement of a transjugular intrahepatic portosystemic shunt (TIPS)
        In patients with sufficient residual renal function (Child-Pugh-Score < 12 points, serum bilirubin not over 10 mg/ dl, TIPS is worth considering
        • In patients with bad general condition or with contraindications against TIPS, pharmacologic treatment with vasopressin-analogues (octreotid or terlipressin) or alpha- adrenergic substances (noradrenaline or midodrin) combined with albumin is recommended.
        Duration of therapy: 5-15 days
        Target criterion : Reduction of creatinine to < 1.5 mg/dl.

Prg.: The prognosis of hepatorenal syndrome is poor, especially for patients with type 1 hepatorenal syndrome, in which the survival period without therapy is under 1 month. In type 2, the chance of surviving 2 years is 20%. In every other patient with type 1 hepatorenal syndrome treated with terlipressin and albumin, a normalisation of the renal function with a prolonged survival period of 3 months can be achieved.

▶ **Hepato-pulmonary syndrome (HPS):** Pulmonary function disorders with hypoxemia in liver cirrhosis.

Di.:   A) of a portal hypertension:
        • Evidence of esophagal-/fundus varices (endoscopy)
        • Evidence of splenomegaly, ascites (sonography)
        • Evidence of portocaval collaterals, slowing down and possibly retrograde flow in the portal vein (colour duplex)
        • assessment of portal pressure (see above)
        • Possible angiography with imaging of the collateral vessels and the phases of perfusion

        B) of the underlying disease: for example.
        • Liver cirrhosis (clinical picture + laboratory,. laparoscopy)
        • Thromboses in spleen, portal or hepatic veins (colour duplex)

Th.:   A) therapy of the causal disease

        B) therapy and prophylaxis of bleedings from esophageal-/fundus varices:

        • Circulation stabilisation: substitution of volume, fresh frozen-plasma (FFP), balance of blood losses up to a Hb of about 9 g/dl (see chap. gastrointestinal bleeding and chap. shock)

        • Haemostasis:
        1. Endoscopic haemostasis
          - Ligature treatment (with multi-band-ligature system) is the method of choice, as it shows no serious complications
          - Sclerotherapy (= sclerosing with polydocanol or highly concentrated alcohol) has a complication rate of
          > 10% (perforation, strictures, pleural effusions, pericardial effusions, fever + bacteriaemia) and has therefore been replaced by ligature as the standard therapy
            - treatment of haemorrhages with tissue adhesives is a therapy of reserve

2. Pharmaceutical lowering of the portal pressure: Somatostasin or -analogues (Octreotid) or Terlipressin (Vasopressin should not be used any more because of considerable SE)

Effects: Decrease of portal pressure by vasoconstriction, effective in acute bleeding

Dos: - Somatostasin 250- 500µg i.v., afterwards permanent infusion of 25 - 50 µg/h
- Octreotid (Sandostatin ®) 50 µg i.v., after that permanent infusion of 25 - 50 µg/h
   Beware: Different dosage of Sandostatin ® and somatostasin
- Terlipressin (Glycylpressin ®) 1 - 2 mg i.v. every 4 - 6 h
Combination with nitrates because of cardiovascular SE !

CI: CHD, arterial hypertension

3. Balloon tamponade of varices (method of reserve):

Ind: Massive variceal bleeding, failing of endoscopic and medicamentous haemostasis.

- Sengstaken-Blakemore probe or Minnesota-probe:

With varices of the terminal oesophagus and cardia region

- Linton-Nachlas-probe:

With varices in the gastric fundus

Remark.: The oesophagus balloon of the Sengstaken-Blakemore probe is filled with a pressure of approx. 40 mm Hg and should be deblocked every 5 – 6 h for 5 minutes (danger of pressure necrosis).

Complications. (frequently: 10 - 20 %): Pressure necrosis, airway obstruction in case of the oesophagus balloon sliding up (for this reason an active pull by means of weight is not advised), aspiration of blood and secretion (regular aspiration), aspiration pneumonia, cardia rupture.

4. Placement of a self-expanding metal stent into the distal oesophagus (Ella stent) for 1-2 weeks (method of reserve)

Further possibility of therapy as ultima ratio with unstoppable haemorrhage (failure of conservative therapy): TIPS(S): trans-jugular intrahepatic porto-systemic stent-shunt

- Prophylaxis of a hepatic coma after variceal bleeding: Aspiration of the bloody stomach contents, intestine cleaning, protein restriction, dose of lactulose orally and as enema (see therapy of hepatic encephalopathy).

- Prophylaxis of a variceal bleeding:

a) Primary prophylaxis (Prevention of the 1st haemorrhage): Since the risk of the 1st haemorrhage is only 30 %, a primary prophylaxis is indicated in elevated haemorrhage risk (big varices) Drugs of choice: non-selective beta-blockers (e.g. propanolol; target dosage: lowering the heart rate by 20%) lower the haemorrhage risk by about 50 %; they do not, however, stop the progress of the varices. In high risk of haemorrhage (endoscopic "red colour signs" , advanced Child-stage), an endoscopic ligature may be indicated even without previous bleeding.

TIPS(S) and shunt surgery are not indicated .

b) Secondary prophylaxis (Prevention of a haemorrhage relapse after the 1st haemorrhage). Since there is a great risk of a relapse haemorrhage (within approx. 35 % within10 days, up to 70 % within a year), a secondary prophylaxis is indispensable .

1. Banding in several sessions, supplementary dose of a non-selective beta-blocker which lowers portal pressure (see above).

2. Porto-Systemic Shunt Methods:

- Aim: Lowering of the portal pressure
- Precondition: Adequate liver function: Child A and B (Patients with Child C are candidates for a liver transplantation)
- Ind: Relapsing bleedings after administration of beta-blockers and sclerosing therapy or banding
- Methods:
  ▶ trans-jugular intrahepatic porto-systemic stent-shunt = TIPS(S)
  ▶ Method of reserve : Shunt surgery for example.
   - Selective porto-systemic shunts, for example distal splenorenal shunt (Warren-shunt)
   - Complete porto-systemic shunts, for example portocaval end-to-side anastomosis (PCA)

	TIPS	PCA	Warren	Ligation
Clinic lethality	approx. 5 %	approx. 10 %	approx. 10 %	< 1 %
Porto-systemic Encephalopathy	approx. 35 %	approx. 35 %	approx. 15 %	approx. 15 %
Thrombosing of the shunt	Non-coated stents approx. 50 %	5 %	approx. 20 %	—
Haemorrhage relapses	Up to 20 %	approx. 5 %	approx. 10 %	Up to 50 %

The 5 year survival rates for the various shunt methods do not differ considerably from eachother (even though the information from miscellaneous authors show considerable varieties). The long-term prognosis particularly depends on: Child-stage, aetiology of the portal vein hypertensionand complications.

C) Treatment of ascites :

- Mild cases:
- Salt restriction : Limit NaCl- supply to 2 g (88 mmol/d), potassium substitution
- Do not level out a hyponatriemia (usually, hypotonic hyperhydration) by administration of sodium, since otherwise, a wash out of ascites is made much more difficult. Hyponatriemia mostly improves spontaneously with general clinical improvement.
- Dose of aldosterone antagonists: Spironolactone - Initial dose: 100 mg/d; maximum dose: 400 mg/d (SE: See chap. diuretics)

The therapy effect of spironolactone is seen first after 1 week and perceptible by increased sodium uresis and weight loss.
- strict weight and electrolyte monitoring (in the serum and urine)
- make a fluid balance chart, so that the patient does not drink more fluid than needed for a balance

- moderate cases:
- In addition dose of loop diuretics: furosemide initially 20 –40 mg/d or torasemid (5-20 mg/d). (see chap. diuretics)

Caution: go for slow ascites dismission (daily weight loss not > 500 g!).

SE and CI of a diuretics therapy
1. worsening of the kidney function due to hypovolemia (hepatorenal syndrome = HRS2. worsening of hepatic encephalopathy (up to hepatic coma)
3. Electrolyte disturbances: Hyponatremia < 125 mmol/l, hypokalemia

In a fractional sodium excretion (FE_{Na}) < 0.2 , a therapeutic success with diuretics is not probable any more.

$$FE_{Na} = \frac{U_{Na} / S_{Na}}{U_{crea} / S_{Crea}}$$

- Therapy refractory ascites:
Def.: Missing response to sodium chloride reduction and maximum diuretic therapy. Prognosis unfavourable (mortality 50 %/6 mon.) → Therapy options:
- Paracentesis: Therapeutic ascites puncture + infusion of salt-poor albumin solution

(6 – 8 g albumin per liter of ascites); afterwards ascites prophylaxis with diuretics (see above)
- TIPS(S) can reduce or eliminate ascites in about 70%
CI: Child-stage C, hepatic encephalopathy, serum-bilirubin > 5 mg/dl
- Liver transplantation

Remark.: the peritoneo-venous shunt by Le Veen for ascites reinfusion is hardly used any more because of frequent complications, infections (shunt occlusion infections).

Caution Prior to diagnosis of a therapy-refrectory ascites, always exclude spontaneous bacterial peritonitis (cell number in ascites).

- Therapy of spontaneous bacterial peritonitis:
3rd generation Cephalosporins (for example cefotaxime, ceftriaxone) or gyrase inhibitor from the group 2/3. Since the mortality without therapy is > 50 % , therapy is started immediately after ascites puncture and injection of cultures. Since the relapse rate of 80% is high, a permanent prophylaxis with gyrase inhibitors is recommended

- Hepatorenal syndrome:
If the liver function adjusts itself, the hepatorenal syndrome also betters. Put down diuretics . Stabilisation of the intravascular volume. Ultima ratio: Liver transplantation.

HEPATIC ENCEPHALOPATHY (HE)

Syn: Portosystemic encephalopathy (PSE)

Def.: Rel. frequent, potentially reversible complication of liver cirrhosis due to retention of neurotoxic substances in the blood.

Aet.: Liver cirrhosis of different genesis

<u>Pg.:</u> Inadequate decontamination of CNS-toxic substances by the liver (ammonia, mercaptan, aromatic alcohols, fatty acids, γ-aminobutyric acid = GABA, endogenous benzodiazepines) due to
- Hepatic insufficiency within the setting of liver cirrhosis
- Partial liver-bypassing of portal blood (via collaterals and therapeutically placed shunt)→ hence the term <u>portosystemic encephalopathy</u> with reduced first-pass-clearence of the liver

<u>Triggering factors for a worsening of the PSE with danger of hepatic coma:</u>
• <u>Increased ammonia formation in the intestine:</u>
  - after gastrointestinal (for example varices-) <u>bleeding</u> (1,000 ml blood = 200 g protein)
  - After <u>protein-rich</u> meal and /or constipation
• <u>Increased diffusion of free ammonia into the brain with alkalosis</u>
• Intensified protein catabolism in febrile <u>infections</u>
• <u>Iatrogenic:</u> Therapy with benzodiazepines and other sedatives, analgesics, too intensive a diuretic therapy with hypovolemia and electrolyte disorders

<u>Cl.:</u> All biochemical tests effectuated for the diagnosis of a beginning PSE do not achieve as much as a careful clinical observation (with handwriting specimen, calculating test, numerical test).

Flickering test: The human eye can only perceive a flickering beneath a frequence of about 39 Hz.
In patients with PSE, the critical flickering frequence (CFF) is reduced prematurely (Hepatonorm analyzer®)

<u>Stages of PSE:</u>
St. 0:   Asymptomatic HE, only detectable by means of pathologic psychometric tests
St. I: <u>Beginning fatigue,</u> confusion, concentration difficulties, slowing down of the movements, mood disturbances, incoherent speech, flapping tremor (Asterixis) = nonrhythmic asymmetric trembling of the hands
St. II: <u>Increased somnolence</u> and apathy, changes in handwriting tests and EEG, flapping tremor
St. III: <u>Patient sleeps</u> almost all the time, is however arousable, preserved corneal and tendo reflexes, foetor hepaticus (odor of raw liver), flapping tremor still present, EEG changes
St. IV: <u>Hepatic coma:</u> Deep sleep, patient does not react to pain stimuli anymore , lost corneal reflexes, pronounced foetor hepaticus, flapping tremor is mostly missing, EEG changes

<u>Lab.:</u> Ammonia in the blood > 100 µg/dl, possibly respiratory alkalosis due to hyperventilation

<u>Th.:</u> A) <u>Causal treatment of liver cirrhosis</u>

B) <u>Symptomatic treatment:</u>
  • <u>Elimination of precipitating factors:</u> for example in gastrointestinal bleeding haemostasia intestine cleaning and prophylactic administration of antibiotics; important: therapy of infections.
  • <u>discontinuation of diuretics and tranquilizers intake,</u> dose of benzodiazepine antagonists (Flumazenil) in case of after-effects of benzodiazepines
  • <u>Reduction of CNS-toxic protein metabolites of the intestine</u> (ammonia, GABA, mercaptans):
    - <u>Reduction of protein catabolism</u> by adequate calorie supply (approx. 2,000 kcal/d) in the form of carbohydrates, parenterally in the form of glucose.
    - <u>Protein reduction:</u> In mild hepatic encephalopathy to 1 g/kg BW/d. After variceal bleedingand with imminent hepatic coma short term total protein abstinence. In case of clinical improvement gradual increase of protein supply according to the individual protein immune tolerance, with preference of vegetable protein and lactoprotein. I.v.- administration of L-ornithin-aspartate lowers the ammonia level and is thus believed to have a favourable influence on HE.
    - <u>Cleaning the intestine</u> from ammonia-producing substances: saline laxatives (magnesium sulphate 10 - 20 g orally) + high intestine enemas (by adding lactulose).
    - <u>Suppression of the ammonia-producing intestinal flora by:</u>
      a) <u>Disaccharides:</u>
        - <u>Lactulose :</u> Nonabsorbable disaccharide of galactose and fructose→ is split by intestinal bacteria in the colon under <u>formation of lactic acid .</u> Inhibition of the bacterial urease in intestine and hence of the ammonia formation. With dropping pH of the intestine milieu the absorbable NH3 more and more changes to the hardly absorbable NH_4^+ ion. Lactulose has a mild laxative effect and is suitable for long-term therapy of hepatic encephalopathy.
          <u>SE:</u>   Bloating, diarrhoea, nausea
          <u>Dos:</u>   3 x 10 - 40 ml/d orally; aim: 2-3 soft stools/d
               In coma: 100 ml by stomach probe + 20% solution by intestinal enema
        - <u>Lactitol:</u> Effects + SE like lactulose
      b) <u>Poorly absorbable antibiotics</u> are controversial, for example.
        <u>Neomycin</u> (for example ByComycin®) 2 - 4 g/d
        SE_ ototoxicity, nephrotoxicity, diarrhoea ⁺ apply only as long as absolutely necessary . Control of the kidney function.
  • Monitoring and correction of the water and electrolyte level, intensive-care measures

68

- **Liver transplantation (LTX)** in terminal liver disease
  - organ donation from deceased: Transmission of the liver of donators with dissociated brain death. By splitting, the liver can be distributed to 2 patients.
  In children there also is the possibility of a partial liver- live donation (for example by the parents) under observation of transplantation laws. Subsequent immunosuppression: for example application of Cyclosporin A, MMF, Tacrolimus
  Hospital mortality (30 days) approx. 15 %
  5 year survival rate approx. 80 % (in children > 90%)
  Hospital mortality of donors ≤ 0.5%

  Co.:1. Primary transplant failure (preservation-damage)

  2. Vessel or biliary duct complications:
  Secondary bleeding, occlusion of the transplant vessels, leakage or obstruction of the biliary duct

  3. Transplant rejection response (60 %):
  a) Acute: Periportal hepatitis, non-purulent cholangitis, venous endotheliitis
  Di.: Increase of the cholestasis parameters greater than increase of the transaminases, liver biopsy
  Th.: Corticosteroids, monoclonal antibodies against T-cell lymphocytes. IL-2 receptor antagonists (for example Basiliximab)
  b) Chronic: Non-purulent progressive destruction of the small biliary ducts = vanishing bile duct syndrome (VBDS)
  Di.:Increase of the cholestasis parameters, liver biopsy
  Th.:see above, possibly further liver transplantation

  4. SE by immunosuppressive therapy:
  - Infections (most frequent viral infections: CMV, HSV, EBV, VZV)
  - Drug-SE: Cortisol (Osteoporosis, infection tendency), Ciclosporin A (nephro-/neurotoxicity, hypertension), Tacrolimus (nephro /neurotoxicity, hypertension)
  - Late presentation of malignant tumours (post-transplant lymphoproliferative diseases = PTLD (see there) and B-cell lymphomas, skin tumours)

  5. Relapse of the underlying disease: While in Wilson's disease the genetic defect is being cured by transplantation, in all chronic types of viral hepatitis (B, C, D) there is the problem of a relapse (HC up to 100 %) In chronic hepatitis B the postoperative dose of anti-HBS immunoglobulins and a therapy with nucleoside-analogues can prevent a reinfection.

- Extracorporal detoxication: Prometheus or MARS (molecular absorbant recirculating system) for bridging the wait prior to transplantation. Disputed efficacy .

ACUTE LIVER FAILURE [K72.0]

Syn: ALF, Acute (fulminant) hepatic failure

Def.: Failure of the liver function in patients without any preceding chronic liver disease; clinical triadicterus, coagulation disorder, disturbance of consciousness.
According to the time period between failure of the liver function and beginning of the encephalopathy
3 courses of disease: fulminant (< 7 days) - acute (8 - 28 days) - subacute or prolonged (> 4 weeks)

Ep.: Relatively rare disease (Germany: 100 –150 cases/year)

Aet.: 1. Viral hepatitis (65 %):
- Frequency: HB (1%) > HA (0.2%) > HC (very rare)
- HD: > 2 %
- HE: Up to 3%. In pregnant women up to 20%
Also rarely: Double infection with 2 different hepatitis viruses
Rarely: Herpes viruses
Kryptogenous hepatitis (unknown origin)

2. Hepatotoxins (30 %):
- Drugs: for example paracetamol-intoxication (most frequent in the USA and GB);
-Halothan : Dose-independent idiosyncrasy in case of sensibilisation by previous halothan- anaesthetic (ab-detection possible)
- Drugs: ecstasy and other.
- Death cup (Amanita phalloides): Gastrointestinal discomfort 6 - 12 h after ingestion; after that discomfort-free period of 1 - 3 days, then acute liver failure due to amanita-toxines (detection in urine)
- Chemicals (for example carbon tetrachloride)

3. Other causes (5 %): Acute pregnancy fatty liver, HELLP-syndrome (haemolysis, elevated liver enzymes, low platelets), shock liver, M. Wilson, Budd-Chiari syndrome.

Cl.:　• Hepatic encephalopathy with disturbances of consciousness from somnolence to coma (4 stages, see Chap. hepatic encephalopathy)
- Jaundice, foetor hepaticus (odor of raw liver), flapping tremor
- Decreasing liver size (due to liver decay)
- Haemorrhagic diathesis by lack of blood clotting factors and disseminated intravascular coagulation (DIC)
- Arterial hypotension due to vasodilatation (systolic blood pressure increase in stage 4 encephalopathy indicates intracranial pressure enhancement)
- Hyperventilation (ammonia effect)

Co.:　- Brain oedema (up to 80 % of the patients with stage 4 encephalopathy)
- Gastrointestinal haemorrhages (> 50 %)
- Hypoglycemia due to reduced gluconeogenesis
- Acute renal failure
- Respiratory infections, urinary tract infections, sepsis

Lab.:　- Transaminases n/↑, bilirubin↑
- Ammonia↑
- Quick value < 20 %, blood clotting factors, thrombocytopenia
- Often hypokalemia, hypoglycemia
- Alkalosis: initially metabolic alkalosis (cause: reduced synthesis of urea and bicarbonate consumption), later mixed alkalosis due to hyperventilation

DD:　Rapidly progressing terminal liver failure in liver cirrhosis ("acute on chronic" liver failure)

Di.:　History taking, examination + investigation
Special diagnostics: EEG, intracranial pressure

Th.:　• Early transfer of patients to a transplantation centre
- Causal measures, for example.
  - pregnancy-associated acute liver failure: end pregnancy.
  - Decontamination measures after intake of hepatotoxine (gastrolavage, colonic irrigation, forced diuresis, coal perfusion, plasmapheresis).
  - antidote administration, for example:
    ° In case of paracetamol-intoxication : dose of acetylcysteine (Fluimucil®)
    ° In case of death cup-intoxication: dose of penicillin and Silibinin (Legalon SIL® - prevention of the toxin uptake in the liver).
  - Fulminant progression of the HB: attempt of antiviral therapy (lamivudin)
- Symptomatic (supportive) therapy:
  - Monitoring + substitution of electrolytes, glucose, blood clotting factors (FFP = fresh frozen plasma), substitution of AT to > 50 % of the desired value, maintain Quick-value > 20%; i.v.-nutrition
  - Prophylaxis of a hepatic coma: protein abstinence, high intestine enemas, administration of lactulose and neomycin orally (reduction of the ammonia-producing intestinal flora)
  - Ulcer prophylaxis with ranitidine (H2-blocker which also lowers the intracranial pressure)
  - Haemodialysis in acute renal failure
  - In brain oedema and normal renal function: Dose of hyperosmolar mannitol-solution; lift upper part of the body to 45 °-position; keep RR up (MAP must be 50 mm Hg above the intracranial pressure→ intracranial pressure probe); hyperoxygenation; hyperventilation to $paCO_2$ values of 30 - 35 mm Hg is only promising in the early stage; in late stage: dose of Thiopental (Reduction of the O_2-demand).
  -: Fructose, amino acids, corticosteroids, benzodiazepines are forbidden
- Liver transplantation: contact a transplantation centre early enough . More specific indication with aid of prognosis scores
- Temporary substitute of liver function in potentially reversible ALF until the neogenesis of the own liver has occurred: Auxiliary partial orthotopic liver transplantation (APOLT): Replacement of the patient's left hepatic lobe by a donor's transplant.
- Hepatocyte transplantation as bridging measure until liver transplantation (in clinical testing)
- Extracorporal detoxication: Prometheus or MARS (molecular adsorbent recirculating system) and other procedures

Prg.:　Dependent on: Aetiology of the acute liver failure, age and possibly pre-existing diseases and pace of ALF-progression (fulminant more favourable than protracted) . Low degrees of encephalopathy (1+2) have favourable prognoses.
A dropping hepatocyte-growth-factor (HGF) and a rising alpha- fetoprotein are prognostically favorable
Brain oedema is the most frequent cause of death (70 %). 50 % of the patients need a liver transplantation.
Patients who survive ALF mostly recuperate completely.

TUMOURS OF THE LIVER [D37.6]

A) Benign tumours

Classification:

1. Liver hemangioma: [D18.0]
   - In.: Most frequent benign liver tumour, 80% of hemangiomas are < 3cm in $\varnothing$
   - Cl.: Mostly asymptomatic incidental finding in sonography: echo-rich ("white") roundish-oval or lobule tumour; in the Echo-contrast centripetal filling; iris phenomenon in contrast media-sonography or angio-CT. Colour-flow Doppler sonography shows one or more vessels leading towards/away in the marginal area. In (smaller) hemangiomas there is an arterioportal shunt in up to 15% (shunt-hemangioma). Larger hemangiomas often do not show the typical CM- habits of small foci any more.
   - Co.: Seldom spontaneous rupture + haemorrhage into the abdominal cavity in big superficial hemangiomas

2. Liver cell adenoma = Hepatocellular adenoma (HCA) is relatively infrequent, mostly women of childbearing age, size of adenoma up to > 10 cm $\varnothing$
   - Aet.: intake of oestrogen-containing contraceptives
   - Sono: The small HCA < 5 cm $\varnothing$ is isoechogenic towards the liver tissue. The large superficially located HCA can cause complications (see below) and indicates the need of surgery. Venous signs in colour-flow Doppler- and contrast media- ultrasound (CMUS)
   - Hi.: Absence of central veins and biliary ducts, often necroses and haemorrhages
   - Co.: Infarction with acute abdominal pain, rupture of the tumour with life-threatening haemorrhage (10 % of cases)

3. Focal nodular hyperplasia (FNH) [K76.8]
   - In.: Predominatly in women
   - Aet.: Unknown; most of the FNH-patients have taken oestrogen-containing contraceptives; termination or continuation of this medication does, however, not influence the growth of FNH in most of the patients.
   - PPh.: Hamartom containing all cells of the normal liver tissue (in adenoma only hepatocytes). FNH usually shows a central scar with stellate septa.
   - Sono.: Mostly the same echogenity as the liver tissue; in colour- flow Doppler often radial vessels, in power-Doppler identfication of the supplying artery in 80%; contrast media-sonography: Central artery and wheel spoke pattern become visible in early artery

4. Biliary duct adenoma (infrequent) [D13.5]

5. Intrahepatic biliary duct cyst adenoma: [D13.4]
   Infrequent, women in the 5th decade of life, size up to 30 cm$\varnothing$

6. Intrahepatic biliary duct papillomatosis: [D37.6] rare precancerous lesion

Cl.: Benign liver tumours are mostly asymptomatic (sonographic incidental findings), hepatic cell adenomas can lead to complications (see above)

DD: Focal fat distribution disorders:
   1. focal multiple fat deposition (sonography: echo-richer, "lighter" area, sharply limited; CT: measurement of fat density)
   2. Focal minor fat deposition: (sonography: echo-poorer, "darker" area, sharply limited)

Di.: imaging step diagnostic :
   - colour-duplex sonography and power-Doppler-sonography, sonography with ultrasound contrast media = CMUS (most sensitive identification)
   - Spiral-CT after i.v.-contrast medium dose
   - MRI with liver-specific contrast media
   An adjustment of density to the surrounding liver tissue in the CT after contrast medium dose is typical of the first 3 tumours (this does not apply for liver metastases)
   - PET/CT (combination of CM-CT with PET): Very sensitive procedure for the diagnose of malignant tumours

Th.: Oestrogenic and anabolic steroids are contraindicated in liver cell adenoma and in FNH. Larger hepatic cell adenomas should always be removed surgically because of the danger of haemorrhage .

B) Malignant liver tumours

1. Hepatocellular carcinoma (HCC)[C22.0]:

- **Syn:** Primary liver cell carcinoma (HCC)
- **Ep.:** In Europe and USA increasing ; incidence 5/100,000 inhabitants p.a., m : f = 3 : 1; on the other hand, in tropical zones (Africa, Asia, China) partly most frequent malignant tumour in men. Incidence up to 150/100.000/year; frequency peak in Africa + Asia: 3. - 4th life decade; in Europe and USA: 5. - 6th life decade.
- **Aet.:** • Liver cirrhosis of any genesis (> 80 %): Per annum up to 4 % of the cirrhosis patients
   Cirrhosis patients have the largest HCC-risk on the basis of a chronic hepatitis B (about all HCC) or C (about 25% of all HCC) or haemochromatosis. Patients with neonatal HBV-infection are at high risk, too. HB- and HC- are carcinogens for the liver.
   • Aflatoxin B1 of the aspergillus flavus fungus that grows on cereals, nuts and other foods in damp climate

PPh.: • Growth: Solitary, multicentric, diffuse infiltration
  • Histology: Various differentiation, early metastatic spread
Cl.: • Pain on pressure in right epigastrium, emaciation, possibly palpable tumour, possibly flow murmur over the liver, ascites
  • Possible decompensation of a pre-existing liver cirrhosis
  • Possible paraneoplastic symptoms (fever, rise in inflammatory proteins)
  At the time of diagnose multilocular growth in 50 %, portal vein thrombosis in 25% and infiltration of liver veins and v. cava inferior in 10%
Lab.: Alpha-Fetoprotein (AFP)):
  Embryonal T antigen whose formation is strongly suppressed by gene repression of the assigned factor after birth→ normal serum concentration in adults < 15 µg/l. Physiologically elevated values are seen in pregnant women. In HCC, AFP is increased in 50 % of the cases, depending on value of rising specificity up to 95%
  Other causes of a pathologic AFP-increase: non-seminomatous testicular tumour (high sensitivity), sometimes gastrointestinal tumours (in 20 % of the cases); bronchial carcinoma. Undulating increased AFP levels are sometimes also found in chronic hepatitis . Suspected PHCC with increasing values > 20 µg/l
Di.: History (Liver cirrhosis, chronic HBV-/HCV-infection) - clinical picture - AFP increase - imaging diagnostic: colour duplex-sonography, MRI- and CT-procedure, intraoperative ultrasound.
  No fine-needle-puncture with potentially curative tumour findings (in 2 % implantation metastases may develop!).
Th.: ▶ Partial liver resection (in few resectable spots) and hepatectomy with liver transplantation (in individual cases) are the only curative therapy forms.
    ▶ Local ablative therapy procedures
    Ind: 1. In smaller foci potentially curative therapy
      2. Bridging measures until transplantation
      3. Palliative therapy
    · Radio frequency ablation = RFA
    · Percutaneous ethanol injection = PEI
    · Laser-induced thermotherapy = LITT
    · Magnetic drug therapy = MDT
    · Selective internal radiation therapy = SIRT = intravasal injection of radioactive microspheres into the tumour spot
    · MRI-monitored cryotherapy
    · Transarterial chemoembolisation = TACE
    ▶ Systemic chemotherapy has no life-prolonging effect
Prg.: If there are no curative therapy possibilities, the prognosis is poor. Average survival rate after diagnosis approx. 6 months
Early recognition: Screen patients with liver cirrhosis (esp. with HB- and HC-infection) every 6 months (AFP, sonography)
Pro.: Inoculation against B hepatitis lowers the incidence of HCC in endemic regions of B hepatitis (Taiwan-studies). Timely therapy of HB, HC, haemochromatosis and other causal diseases

2. Embryonal Hepatoblastom [C22.2] (rare tumour in children)

3. Angiosarcoma [C49.9]:
  Aet.: vinyl chloride (= monomer of the PVC), arsenic, thorotrast (formerly applied X-ray contrast medium made of thorium dioxide)

4. Liver metastases [C78.7] within the setting of extrahepatic tumour diseases: Most frequent form of malignant liver tumours, often with multiple occurrence
Di.: Sonography: Sonographically varying manifestation: Echo-rich or echo-poor, central echo amplification ("bull's eye"), echo-poor halo, rarely calcifications, compression/shifting of vessels; contrast-media-sonography: Irregular tumour vessels in the arterial phase CT, MRI,. PET
Th.: Individual liver metastases can be resected with curative aim if further metastases are absent (CT-arterioportography and intraoperative ultrasound prior to resection)
  In case of multiple liver metastases there are only palliative possibilities: systemic chemotherapy or locally ablative therapy methods (see HCC).

Alveolar Echinococcosis [B67.5]

Path.: E. multilocularis exists only in the northern hemisphere; it is transmitted mainly by foxes.
Inc.: 10 – 20 years
CL: Alveolar echinococcosis spreads infiltratively in the liver like a malignant tumour (with calcifications)
Di.: Evidence of infection of the liver: Sonography/CT + positive serology: Ab-detection
Th.: Curative resection only possible in ¼ of the cases + subsequent long term therapy with albendazole

DD: Cystic modification of the liver

1. Multiple dys-ontogenetic cysts (often also in kidney/pancreas)
2. 3 - 5 % of the individuals > 50 years show liver cyst solitaries [K76.8] : Mostly asymptomatic sonographic findings: Sonographic criteria: Round, anechoic, neatly limited, no detectable wall, marginal shadowing, distal sound amplification, distinct entering and leaving echo Cysts > 5 cm ⌀ may rarely cause complications (haemorrhages, infections)
3. Cystic echinococcosis: [B67.9] infection by Echinococcus granulosus
   Sonography/CT: Smoothly bordered space-occupying of varying echogenity, wall calcification, possibly proof of daughter cysts with double contours of the wall (cyst within a cyst, honey comb- or spoke-structure due to septa; antibody-detection
4. Liver abscess: [K75.0]
   - Bacterial = pyogenic abscess [K75.0] due to bacteriaemia in the portal vein, for example due to appendicitis, diverticulitis; postinterventional cholangitis. Most frequent pathogens are E. coli and klebsiella (70 %); pain in the right upper abdomen, ESR, intermittent fever, leukocytes, AP↑
     Sonography: Mostly echo-poor, gas-related mirror imaging, debris-echo, secondary inflammatory surroundings reaction, contrast media-ultrasound (CMUS)
   - Amoebic abscess [A06.4] due to infection with Entamoeba histolytica
     Di.: Pot. fever, ESR↑ , amoeba serology, sonography: (initially little, later higher) echo-poor space-occupying with internal echos, possibly gas inclusions, mostly round form with abscess wall
5. Liver haematoma [K76.8]
   Sonography: Modification of the morphology in the course of time : initially echo-poor, later increasing echogenity: trauma history. Examine complete spleen and liver; is there free fluid in the abdomen?
6. Peliosis hepatis: Late aftereffect of a Bartonella infection with blood cysts in the liver, also associated with hormone-therapy (oral contraceptives, anabolics)

Di.: Sonography, CT, MRI, pot. angiography-procedures

Th.:
   - solitary liver cysts are treated only in case of larger cysts (> 5 cm ⌀) which cause discomfort: Sonoguided puncture with placement of a catheter + sclerosing with 96% alcohol
   - Echinococcus cyst: PAIR (percutaneous alcohol injection and reaspiration)
     Cl.: Biliary infiltration (exclude beforehand by ERCP)
     Success rate of PAIR: 95%; obligate accompanying chemotherapy with albendazole. If PAIR is contraindicated operative cystectomy + accompanying chemotherapy
     In case of rupture, surgical removal or puncture of the echinococcus cyst, there is the danger of the peritoneal dissemination of protoscolizes and of an anaphylactic reaction. Therefore, in a possibly ultrasound-guided fine needle puncture at least 2 cm of liver tissue have to be between cyst and liver surface. A concomitant chemotherapy is indispensable in case of operation or puncture.
   - Pyogenic abscess:
     1. Antibiotics for at least 3 weeks (for example mezlocillin or cefotaxime + metronidazole)
     2. Sonography- or CT-guided abscess punction (with microbiology + cytology), irrigation with 0,9 % NaCl solution; in larger abscesses (> 4 cm ⌀) drainage + daily irrigation; in case of worsening: operation
   - Amoebic abscess: Metronidazole over 10 days is mostly successful; final treatment with Diloxanid (see chap. amoebiasis)

DISEASES OF THE GALL BLADDER AND BILE PATHWAYS (CHOLEPATHIAS)

CONGENITAL DISEASES

Bile duct atresia [Q44.2]

Classification according to Kasai:
I. Extrahepatic biliary atresia:
   Type I : D. choledochus - Type II: D. hepaticus communis - type III: D. hepatici
II. Intrahepatic biliary atresia
III. Biliary hypoplasias

Ep.: 1 : 12,000 childbirths

Cl.: Progressive icterus immediately after birth or little later

Di.: Sonography, MRC (= Magnetic resonance-Cholangiography), intraoperative cholangiography

Th.: Biliary-enteric anastomosis (for example by y-shaped jejunum loop, portojejunostomy); liver transplantation (transplantation of a liver part from living relatives)

Prg.: Without liver transplantation unfavourable.

Choledochal cysts [Q44.4]

4 types:
I: Common-channel-syndrome (most frequent): dilatation of D. choledochus and D. hepaticus
II. Isolated diverticulum of the D. choledochus
III. Stenosis of the papilla of Vater with choledochocele
IV. Carolis syndrome: Cystic dilatation of the intrahepatic bile ducts; recessive-autosomal; Type I with bileduct calculi and cholangitides; Type II with liver cirrhosis and unfavourable prognosis.

Cl.: Presentation in the 1st year (25 %), up to the 10th year (35 %) or adulthood: Relapsing icterus, colicky pain under the right costal arch, pruritus, palpable tender-elastic tumour in the right epigastrium

Di.: Sonography, MRC, ERC, intraoperative cholangiography

Th.: - Biliary-enteric anastomosis with cholestasis
   - Lysis/ESWL of gall stones, endoscopic papillotomy
   - Antibiotics in bacterial cholangitis
   - Liver transplantation

ACQUIRED DISEASES

GALL STONES (CHOLELITHIASIS) [K80.2]

In.: Prevalence in women about 15, in males about 7.5% (f : m = 2 : 1); prevalence in liver cirrhosis and in Crohn's disease
25 – 30%. Increase with age.

Stone types:
1. Cholesterol stones and mixed stones (which contain > 70 % cholesterol): 80 %
2. Bilirubinic (pigment-)stones (20 %)
   Causes: Chronic haemolyses, liver cirrhosis, partly unknown.
   Bilirubin stones and cholesterol stones do not leave any shadow in the radiogram Pigment stones sediment on the ground of the gallbladder, while cholesterol stones float in the gallbladder. In the CT the two stone types can be distinguished by measuring their density.

20 % of the patients have calcified stones following inflammatory processes. 10 - 15 % of the patients with gallbladder stones at the same time have stones in the choledochal duct .

Bile duct stones either develop primarily in the bile duct (mostly brown pigment stones) or have migrated from the gall bladder to the bile duct (mostly pigment stones with cholesterol nucleus). They lead to complications in 50% (obstructive jaundice, cholangitis, pancreatitis).

Aet.: Risk factors for the formation of cholesterol stones:
- Hereditary factors: for example frequent occurrence of cholesterol stones in so-called "biliary calculus families" or in Pima-Indians, absence of biliary stones in central African Massai
  Gene-mutations leading to cholesterol stones are : ABCG8 and ABCB4.
- Gender (f : m = 2 : 1 up to 3 : 1), pregnancy, oestrogen intake
- Age (Increase of biliary stones with age)
- Nutrition (cholesterol-rich, fibre-poor diet, parenteral nutrition, fasting)
- Obesity (20% of overweight doubles the risk for biliary stones)
- Intake of clofibrate-containing drugs
- Bile acid malabsorption (see there)

Note: 6 x F rule: "female, fair, fat, forty, fertile, family" !

Pg.: Bile contains approx. 80 % of water. Bile acids and glycerol phosphatides usually keep the insoluble cholesterol in solution in form of micelles. The normal (non-lithogenic) gall contains cholesterol, phospholipids and bile acids in a relation of about 5 : 25 : 70.
The high percentage of cholesterol and/or the reduced percentage of bile acids is typical of the lithogenic (calculus-forming bile, so that the gall is oversaturated with cholesterol, With it, the lithogenic index (LI) or cholesterol-saturation index (CSI) = Ratio of dissolved cholesterol to the maximum soluble cholesterol is > 1. The first step towards the formation of cholesterol stones is the production of cholesterol mono-hydrate crystals.
Hypomotility of the gallbladder with prolonged length of stay of the bile in the gallbladder or incomplete emptying of the gallbladder favours cholesterol calculus formation.

Cl.: A) Biliary calculus carrier without symptoms (= silent biliary stones): 75 %

B) Person symptomatic with biliary calculus with symptoms (= symptomatic biliary stones: 25 %

1. Biliary colics are typical complaints of a cholelithiasis and are triggered mostly by stone jamming/stone passage of the cystic duct. Small concrements can pass through the choledochus into the duodenum; the papilla of Vater – which additionally can be constricted by cramps or morphologic changes - (80 % of all individuals have a common exit of choledochus duct and D. pancreaticus) is a critical point. The colic pains last between15 minutes and 5 hours (rarely longer) and are located in the right and mesial epigastrium (always examine the stomach and duodenum), they often irradiateto the back (exclude pancreatitis!) and the right shoulder (DD: Myocardial infarction and pulmonary embolism). The colic can be accompanied by nausea, belchand brief icterus.

2. Unspecific epigastrium discomfort :
Pressure /fullness in the (right) epigastrium, meteorism, incompatibility of some foods and beverages (for example fatty, fried, bloating foods, coffee, cold drinks). Unlike the biliary colic these unspecific discomforts are found not only in cholelithiasis but also in other abdominal diseases (liver, stomach, intestine). In the majority of cases it is a question of purely functional discomforts in irritable bowel and/or irritable stomach syndrome.

3. Palpation:
Murphy 's sign: Sudden pain-related inspiratory arrest after the examiner has pressed the palpating hand into the gallbladder area during expiration (which had not been painful yet).

Co.: The majority of symptomatic biliary stone patients has to expect relapsing discomfort or complications in further life
.

1. Acute cholecystitis [K81.9], cholangitisand complications:
- Bacterial infection of the gall bladder and pathways; most frequent germs: E. coli, Streptococcus faecalis (Enterococci), klebsiellae, enterobacter, clostridium perfringens
  Charcot-triad in cholangitis: Pain in the right epigastrium, icterus and fever; often relapsing; lab.: leukocytes, γ-GT, AP, bilirubin (GPT) †
  Co.: Gallbladder empyema, gangrenous cholecystitis, liver abscess, sepsis

Note: in 90 % of cases a cholecystitis develops through a temporary obstruction of the D. cysticus or gallbladder infundibulum due to biliary stones; inflammations without stones are rare (for example in intensive care patients with parenteral nutrition or with salmonellae infection).

- Stone perforation (rarely):
  - Into the intestine tract: with obstruction of the duodenum (Bouveret-syndrome) or obstruction of the terminal ileum and gallstone ileus→ clinical triad: aerobilia (X-ray showing air in the gallintestine ileus + possibly stone shadow . The stone rarely perforates into the hepatic colic flexure
  - Covered perforation with possible sub-hepatic abscess (sonography)
  - Free perforation into the abdominal cavity with biliary peritonitis

- Mirizzi's syndrome: Very rare form of obstructive jaundice in which a stone in the neck of the gallbladder leads to a compression or scarred stenoses in the adjacent ductus hepaticus.

2. Chronic-recrudescent cholecystitis: shrinking gallbladder, "porcelain gallbladder", delayed complication: Gallbladder carcinoma

3. Stone migration and complications:
- Cystic duct obstruction [K82.0] (X-ray.: Negative cholecystogram)

Cl.: Biliary colic (acute occlusion)

Co.: Gall-bladder hydrops, bacterial cholecystitis, gallbladder empyema (palpable gallbladder), gallbladder gangrene, perforation

- Choledocholithiasis [K80.5]

  Co.: Cholestatic icterus, bacterial cholangitis, liver abscesses, secondary biliary cirrhosis, pancreatitis

DD:
- Pancreatitis (DD + Co. ! → Lipase; Amylase ↑)
- Ulcer (history, gastroscopy)
- Appendicitis with appendix turned upwards
- Nephrolithiasis with ureter colic, pyelonephritis (pathologic urine finding)
- posterior myocardial infarction (CKMB, troponin T/I, ECG, echocardiography,)
- Pulmonary embolism (history, clinical picture, pO_2, colour duplex of the legs))
- In icterus other causes of an icterus (see there)
- Fibrinous perihepatitis (Fitz-Hugh - Curtis syndrome)):
  Complication of a sexual infection with gonococci or chlamydia (younger women, pains right epigastrium, gynaecologic proof of the infection)
- Primary biliary cirrhosis: Cholestasis sign (AP, γ-GT, bilirubin), IgM increase, antimitochondrial antibodies = AMA, typical histology
- primary sclerosing cholangitis : Rare pipe-shaped wall sclerosis of the medium and larger biliary ducts
  Di.: MRC, ERC
- Tumours (liver, gallbladder, biliary ducts, papilla of Vater, pancreas, colon)
- Liver /biliary duct parasite (Askarides, Echinococcus, bilharziosis, fasciola hepatica, Clonorchis sinensis)
- Sphincter-Oddi- dyskinesia: biliary pain without detection of concrements
Invasive diagnostic hazardous

Di.: 1. History + clinical examination

2. Laboratory:
   - With cholecystitis: CRP + ESR ↑, leukocytosis
   - In obstruction of the choledochal duct: Increase of the cholestasis indicating enzymes (γGT, LAP, AP) and of conjugated bilirubin
   - with ascending cholangitis: possibly slight transaminase increase

3. Ultrasound examination:
   Most sensitive and most rapid assay technique of biliary stones
   Proof of an enlarged gallbladder, of form and wall modifications, a compressed choledochal duct (spread), proof of an enlarged/ congested choledochal duct: In patients with a gallbladder > 7 mm ∅; after cholecystectomy > 11 mm ∅. Proof of the gallbladder's contraction ability after meals.
   By means of sonografic high frequency (HF)-Signal-analysis we can find out whether we deal with cholesterol stones and whether these are calcified. However, the clinical relevance of gallstone composition is minor.
   In acute cholecystitis three-layered wall of the gallbladder perceptible

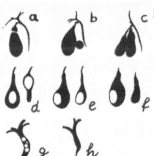

Findings at gall bladder and path:

a   divided by gallbladder septa

b   diverticulum of the gallbladder

c   doubled gallbladder

d   position change of the stone

e   constant spatial relation between tumour and wall

f   disappearance of an extracystic tumour

g   Choledocholithiasis with stone close to papilla, neg. Cholecystography

h   Stenosis of the papilla of Vater (State after cholecystectomy)

After sonografic diagnosis of gallbladder stones 3 questions arouse :
- Are there silent or symptomatic biliary stones?
- Is the cystic duct freely passable and the gallbladder contractible? → function-sonography: sonography on an empty stomach and 45 min after having a meal: If the gallbladder gets smaller after stimulus meal, the cystic duct is usually free.

4. MRC: Nuclear spin tomography imaging of the biliary duct

5. CT of the gallbladder:
   Most sensitive proof of a calcification of biliary stones

   Direct Cholangiography:
   - ERCP = endoscopic-retrograde cholangiopancreatography: Method of choice in suspected biliary duct stones! Therapeutic procedure with papillotomy and, if necessary, stone extraction, can take place directly. Complication rate of diagnostic ERCP < 2 %, of therapeutic ERCP up to 6%, of papillotomy up to 10% Risk of pancreatitis 5%, of sepsis 1%, lethality 0.4% (Carbonic anhydrase inhibitor, pancreatitis)
   - PTC = percutaneous, transhepatic cholangiography
   Because of higher complication rate reserve method, if ERC is not possible. Co.: Biliary peritonitis and haemobilia

6. Diagnostic of choledocholithiasis
   - Ultrasound, endosonography: D. choledochus widened?, direct detection of stones
   - MRCP and ERCP
   - Cholangioscopy by means of "Mother-Baby scope system"

Th.: ▶ Asymptomatic gallstones: Since only 25 % of the affected develop symptoms or complications in the course of 25 years, there is no necessity for treatment.
   Exception: Silent porcelain gallbladder→ surgery because of elevated carcinoma risk

   ▶ Symptomatic gallstones: Since the majority of patients shows relapsing discomfort or complications after one-time complaints, a therapy (operation) is indicated .

   A) Symptomatic therapy of a biliary colic:
   Slight colic:
   Butylscopolamin: e.g. 1 Buscopan®-supp. and/or Nitroglycerin: for example Nitrolingual 0.8 – 1.6 g Nitroglycerin: for example Nitrolingual 0.8 – 1.6 g as a bite-through capsule. Consider CI for Butylscopolamin (glaucoma, voiding dysfunction,)
   Severe colic:
   Strong analgesic, for example 50 mg meperidine (Dolantin®) + 20 mg Butylscopolamin (Buscopan®) i.v. + Novaminsulfon (Novalgin®) i.v.

   Beware: Morphine derivate (except for pethidine) that can set off a cramp of the Oddis sphincter

   - Fasting for at least 24 h, followed by special diet:: No fatty, no fried foods – the foods tolerated by the patient are allowed.
   - Antibiotics in case of suspected bacterial infection of the biliary ducts (cholecystitis, cholangitis):
   Most frequent germs: E. coli and enterococci (see above)
   Drug of choice: Fluoroquinolones of the group 2/3 (e.g. ciprofloxacine) or amino penicillin + β-Lactamase-inhibitor (for example Augmentan®)
   - In case of suspected anaerobe infection supplementary dose of metronidazole
   - Early elective cholecystectomy (see below)
   Therapy of obstructive jaundice caused by gallstones:
   In B) cholestasis caused by stones in the D. choledochus endoscopic extraction of the stones is indicated. In cholangitis within the setting of an obstructive jaundice (symptoms of inflammation like, for example, fever, leukocytosis, among other things), it is an emergency measure, because otherwise there would be the danger of a possibly lethal cholangiosepsis for which there is no conservative therapy.

   ▶ Therapy of choledochal stones:
   • Endoscopic papillotomy (EPT) and pot. removal with a Dormia sling or with a balloon catheter
   Co. after EPT: Pancreatitis (1%), bleeding (2%), cholangitis, with sepsis (<1%), perforation (0.1%); mortality about 0.1%
   If the concrement is too large (> 15 mm ⌀) and cannot pass the papilla, the stone is reduced to small pieces. Methods:
   - endoscopic mechanic lithotripsy
   - ISWL: endoscopically applied intracorporal shock wave lithotripsy (electro-hydraulic or laser-induced)
   - ESWL = extracorporeal shock wave lithotripsy
   • Method of reserve: Revision of the common bile duct within the setting of a surgical cholecystectomy
   ▶ Therapy of stones in the bile ducts of the liver (infrequent finding)
   ISWL within the setting of a cholangioscopy (perorally by means of " mother-baby-endoscope" or percutaneous transhepatic)

   C) Removal of biliary stones:

   I. Surgical:
   Cholecystectomy
   Advantage: Definitive cure, as a rule no relapse stones.
   Ind: Symptomatic gallstones are always a relative indication for cholecystectomy, with upcoming complications there is an absolute indication for operation. Early operation in the complication-free stage has a small mortality

(with elective intervention approx. 0.1 %). At the age > 70 y. as well as in the complication stage the mortality is considerably higher (10 % and more).

    a) laparoscopic cholecystectomy: Method of 1st choice - advantages:
        1. Avoidance of a bigger abdominal section→ no complications as far as the abdominal wound and later scars are concerned; cosmetic advantages
        2. No postoperative intestinal atonia3. Fast mobilisation, small risk of thromboembolism, shortened hospital stay

    b) Surgical cholecystectomy + possible restoration of the bile pathways (if a) should not be possible)

II. Non-surgical methods for stone removal are scarcely of any importance:

Disadvantage of UDC-lysis:
- Duration of therapy up to 2 years and success rate just about 70%
- High percentage of recurrent stones: 30-50% /5 y.
    1. Oral bile acid therapy = Systemic litholysis with ursodeoxycholic acid (UDC))
    2. Extracorporal shock wave lithotripsy = ESWL with subsequent litholysis

Prevention of recurrency: Avoidance of overweight, cholesterol-poor, fibre-rich diet, drink a glass of milk at night (leads to gall emptying), avoidance of therapy with clofibrinacid-derivates or oestrogens

▶ Therapy strategy with cholecystolithiasis:

I. Asymptomatic stones	No therapy
II. Symptomatic stones without complication	Cholecystectomy; if operation cannot be considered: (Lysis/ESWL)
III. Complication stadium	Cholecystectomy (absolute indication)

- Local litholysis (Bathing-therapy by nasobiliar probe)

So-called post-cholecystectomy syndrome [K91.5]

The term is misleading and false from surgical point of view because the discomfort of a cholelithiasis in case of accurate diagnosis and operation are not present any longer. If patients complain of discomforts in the upper abdomen after cholecystectomy, this may be for the following reasons:

1. Missed stenosis of the papilla of Vater, coledochal concrements, biliary duct stricture
    (therefore clarify the D. choledochus pre-/ and intraoperativly).
2. Other abdominal diseasesas as cause of the continuing discomfort (misdiagnoses with wrong indication for cholecystectomy).
3. Postoperatively newly presenting abdominal diseases

Haemobilia [K83.8]

Def.:   Haemorrhage from the bile duct system

Aet.:   Iatrogenic injury, traumas, liver puncture, stones, tumours, haemorrhage from a pancreatic pseudocyst, aneurysms.
Di.:   Evidence of occult blood in the stools, duodenoscopy, sonography, MRC, ERC, angiography

TUMOURS OF THE GALL BLADDER AND OF THE BILE DUCTS

A) BENIGN TUMOURS

Gall bladder polyps

Mostly sonografic incidental findings; prevalence 5%
- Cholesterol polyps are no pure epithelial tumours but cholesterol deposits in the mucosa
- adenomas and cystic adenomas which produce mucus; adenomatous hyperplasias
Gall bladder polyps which grow or which are ≥ 1cm ∅ already when first diagnosed should be removed by cholecystectomy, because they may turn carcinomatous
Bigger gallbladder polyps should be ablated by cholecystectomy (degeneracy risk).

B) MALIGNANT TUMOURS

In.: About 5/100,000 /year; gall bladder carcinomas: about 65% (f > m); hilar bile duct carcinomas (= Klatskin-tumours): about 25%; the residual ones are extrahepatic and intrahepatic carcinomas of the bile duct; highest frequency after age 60.

Gallbladder carcinoma [C23]

In.: f > m, frequency peak beyond the 70th year.

Aet.: Cholelithiasis and chronic cholecystitis are risk factors: In 80% of cases biliary stones are found simultaneously. An increased risk also exists for salmonella carriers; gall bladder polyps > 1 cm ⌀ may degenerate

PPh.: Mostly adenocarcinomas; the formation follows the dysplasia-carcinoma-sequence after the accumulation of genetic mutations (K-ras, p 16, p53 et al.)

Cl.: - No early symptoms, possibly incidental finding after cholecystectomy;
- The appearance of symptoms is a late finding
○ Palpable tumour in the gall bladder precinct
○ Obstructive jaundice (late symptom)

Lab.: Cholestasis parameters↑ (γGT, AP), pot. CA 19-9

DD: Cholelithiasis, cholecystitis

Di.: • Sonography, endosonography, intraductal ultrasound (IDUS)
• "One-stop-shop"-MRI including MRCP and MR-angiography
• Spiral-CT
• ERCP or percutaneous transhepatic cholangiography (PTC)
• Positron emission tomography (PET), as a PET-CT

Th.: Cholecystectomy is only sufficient in the incidentally detected carcinoma in situ (Tis) and T1N0M0 carcinoma. In more advanced stages it has to be checked whether a surgical therapy with curative target is possible.
In loco-regional disease (preoperative) neoadjuvant radiochemotherapy can make a subsequent resection possible.
For inoperable tumours palliative measures (for example stents), in order to restore the biliary drainage.

Prg.: If no R0-resection is possible (majority of the cases): Unfavourable.

Carcinoma of the bile duct and Klatskin tumour

Syn: Cholangio-cellular carcinoma (CCC), Cholangio-carcinoma[C22.1]

In.: about 3/100,000/year

PPh.: Mostly adenocarcinomas, in childhood embryonal rhabdomyoblastoma

Carcinomas of the hepaticus bifurcation = Klatskin-tumours:
Classification according to Bismuth into 4 types:
 Type I : Carcinoma concerns common hepatic duct without hepaticus bifurcation
 Type II: Hepaticus bifurcation also involved in the carcinoma
 Type III: Carcinoma reaches the segmental branches
 Type IV: Carcinoma expands to secondary segmental branches on both sides

Aet.: Risk diseases are choledochal cysts, cholelithiasis , primary sclerosing cholangitis and parasitic diseases of the bile ducts: trematodes, liver flukes (Opisthorchis = Distoma felineum; clonorchis = Chinese liver fluke)

Cl.: No early symptoms; Courvoisier's sign = painless icterus + palpable enlarged gallbladder for distal CCC/ papillary carcinoma or pancreatic tumour

DD: Carcinoma of the pancreatic head

Di.: See gall bladder carcinoma

Th. and Prg.: In about 50% of cases, r0-resections with improved chances of recovery are considered to be possible by means of expanded bile duct resection with hemihepatectomy and lymph node resection (in centres). Palliative measure in hilar CCC: stent therapy + photodynamic therapy (PDT) by means of photosensitizers + laser therapy.

II. SALT AND WATER HOMEOSTASIS

Ph.: Water makes up 60 % of the body weight in adult men, 50 % in women (more fat) and 75 % in infants.
Body water contents are divided into 2/3 intracellular, and 1/3 extracellular. Extracellular fluid consists of interstitial and intravascular fluid.

Intracellular fluid (ICF)	: 40 %
Extracellular fluid (ECF)	: 20 %
Interstitial fluid (ISF)	: 15%
Intravascular fluid (IVF)	: 5 %
(= plasma volume)	

transcellular fluid ("third space):
secretion into various cavities (e.g. cerebrospinal fluid, pleura-, peritoneal cavity, gastrointestinal tract).

Electrolytes:

Extracellular fluid: Cations are mainly sodium, anions mainly chloride and bicarbonate. Intracellular fluid: Potassium and phosphate-esters. There are small ion-shifts between interstitial and intravascular fluid due to different protein levels: Interstitial fluid is low in protein and hence slightly higher in chloride than plasma (which is high in protein) :Gibbs-Donnan-mechanism.

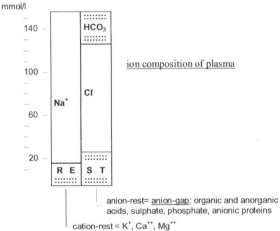

ion composition of plasma

anion-rest= anion-gap: organic and anorganic acids, sulphate, phosphate, anionic proteins

cation-rest = K^+, Ca^{++}, Mg^{++}

simple formula for calculations: anion-gap $\approx Na^+ - (Cl^- + HCO_3^-)$
normal range: 12 ± 4 mmol/l

Osmotic pressure and osmolality/osmolarity:

Plasma osmotic pressure is direct proportional to the amount of any parts(molecules/ions) in solution.

Osmolality describes the concentration of all parts in solution per kg water (osmolarity per litre)
Normal range: 280 - 296 mosmol/kg H2O

calculating osmolality of plasma or serum:
- mosmol/kg H2O = 1,86 x sodium + glucose + urea (in mmol/l) or
- mosmol/kg H2O = 1,86 x sodium + 0,056 x glucose + 0,17 x urea + 9
  (sodium in mmol/l, glucose and urea in mg/dl)

A constant range of osmolality at a physiological level is called iso-osmolality or isotonic. Within the extracellular fluid it is usually sodium determining the isotonic state. A change in anions doesn't have any significant effect on homeostasis, because the two main anions, HCO3⁻ and Cl⁻ can one replace another due to electric neutrality. A change in concentration of K^+, Ca^{++} and Mg^{++} has no effect on homeostasis. These electrolytes have a very specific action (e.g. cardiac electric path), and therefore small changes in concentration will cause death even before they can cause any change on osmolality. – Non-electrolytes like glucose and urea can cause a significant rise in osmolality (e.g. diabetic coma, renal failure).

Oncotic pressure:

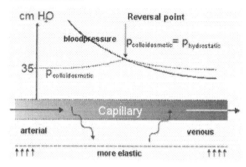

A special case of osmotic pressure is the colloidosmotic (or oncotic) pressure. It can be found on membranes, which are impermeable for colloids (e.g. proteins), but permeable for small molecules (e.g. electrolytes). Capillaries are examples for such membranes. The different protein levels between plasma and interstitial fluid cause a plasma oncotic pressure of ca. 35 cm water (mainly albumin). Interaction between hydrostatic and oncotic pressure in the capillaries is very important for any fluid exchange between plasma and in-terstitium.

Within the arterial branch of the capillary the hydrostatic pressure is higher than the oncotic pressure. Within the venous branch the oncotic pressure is higher. Impaired balance of this interaction leads to enhanced interstitial fluid = oedema.

Fluid balance:

Estimated fluid turnover of a healthy adult in 24 hours:

input (ml)		output (ml)	
fluid	1.000 - 1.500	kidney	1.000 - 1.500
solid food	700	skin + lungs =perspiration	900
oxidation water	300	bowel	100
	2.000 - 2.500		2.000 - 2.500

Note: Infants have a lower amount of extracellular fluid (ECF) (in comparison to the water turnover) than adults. They dehydrate much faster!

turnover of water in l/day:

infant : 0,7 → [1,4] → 0,7

adult : 2,0 → [13,0] → 2,0

ECF

In pathological circumstances loss of fluid can be considerably enhanced. The loss via skin and lungs (= perspiration) is almost 1 litre/24 hours at normal body and room temperature, but much higher in a pyrexial patient: Per 1° > 37°C there is an additional loss of 0,5 - 1,0 l. Perspiration via lungs causes electrolyte-free loss of water, whilst profused sweating causes an additional loss in electrolytes (NaCl). Sweat must not be replaced with water alone.

Gastrointestinal fluid loss (diarrhoea, vomiting, fistulae, nasogastric tubes) carries a particular hazard of electrolyte depletion: gastric secretion - Cl^- and H^+ (→ metabolic alkalosis!), bile and pancreas secretion - $HCO3^-$ (→ metabolic acidosis!). They also will lead to potassium depletion.

REGULATION OF WATER AND SODIUM HOMEOSTASIS

Regulation of sodium and water homeostasis targets a steady level of isotonia and isovolaemia of the intravascular space. ADH regulates water balance: increased plasma osmolality and/or reduced volume will stimulate ADH-secretion in the neurosecretory cells of the pituitary gland, which will lead to water retention and thirst (↑oral water input). Further hormonal triggers (Renin-Angiotensin-Aldosterone-system, natriuretic peptide: ANP, BNP and CNP) modify the renal sodium and water output.

Measured object	Circulating blood volume ↓	Circulating blood volume ↑	Osmolality ↑
receptors	Baro-receptors (renal juxtaglo-merular cells) ↓	Volume receptors (heart) ↓	Osmolality receptors (Hypothalamus) Baro-receptors (volume depletion) ↓
hormones	**Renin** Angiotensinogen → Angiotensin I A. Converting enzyme → ↓ Angiotensin II ↓	**ANP, BNP** and **CNP*** ↓	**ADH** (antidiuretic hormone) ↓

effect	1. vasoconstriction
	2. Aldosterone secretion
	   (adrenal cortex)↓
	renal sodium and water retention
	↓
		1. vasodilatation
		2. inhibition of Renin-Angiotensin-
		   Aldosterone-system
		3. ↓
		renal sodium and water excretion
		↓
			antidiuresis
			(water retention)
			↓

feedback	Circulating volume ↑	Circulating volume ↓	Osmolality ↓

* ANP = atrial natriuretic peptide (type A); BNP = brain natriuretic peptide (type B);
CNP = type C natriuretic peptide

DISORDER OF SODIUM AND WATER HOMEOSTASIS

Imbalance in isovolaemia and isotonia are intertwined with each other. Impaired isotonia is mostly caused by changes in sodium concentration (serum osmolality mainly depends on Na^+-concentration); severe hyperglycaemia and raised urea can also increase osmolality.
Volume regulation is faster than osmotic regulation.

A) CHANGE IN INTRAVASCULAR VOLUME:

1. Hypovolaemia [E86]

Refer to chapter 'hypovolaemic shock' !

2. Hypervolaemia [E87.7]

Aet.: chronic renal failure + excessive water intake

> **Note:** Acute hypervolaemia usually only develops due to combined impaired renal function and excessive water intake. Normal working kidneys don't have a problem to excrete excessive fluid rapidly.

Sym.: • cough, dyspnoea → Fluid lung and pulmonary oedema (widespread crackles)
    • CVP ↑, dilated cervical veins, pulse and blood pressure ↑
    • headache, convulsions
    • rapid weight gain
    • haemoglobin ↓

Th.: A) treat the cause
    B) symptomatic:
      - sit the patient in an upright position, legs down (this will reduce the hydrostatic pressure in the pulmonary blood vessels)
      - rapid acting loop diuretic: Furosemide 20 - 40 mg i.v. (may have to be repeated)
      - pulmonary oedema: reduce preload (glycerol nitrate) and hyperbaric ventilation with positive end-expiratory pressure (PEEP) and 100% O_2
      - hypertonic crisis: glycerol trinitrate and other anti-hypertensives
      - renal failure + excess water: dialysis, observe fluid balance, daily weight!

B. EXTRACELLULAR VOLUME CHANGES

Will eventually affect the intracellular space. There are 6 different types of disturbance:

	Serum-sodium Serum-osmolality mean cell haemoglobin concentration (MCHC)	mean cell volume (MCV)	haematocrit haemoglobin serum protein
Dehydration			
Isotonic	normal	normal	⇧
Hypotonic	↓	↑	
Hypertonic	↑	↓	
Hyperhydration			
Isotonic	normal	normal	⇩
Hypotonic	↓	↑	
Hypertonic	↑	↓	

DEHYDRATION [E86]

It depends on serum osmolality – i.e. usually on serum-potassium-concentration; we distinguish 3 forms of dehydration:

ECF	ICF	
		Normal
		Isotonic dehydration
		Hypotonic dehydration
		Hypertonic dehydration

1. Isotonic dehydration

Def.: Equal extracellular sodium- and water depletion

Aet.: 1. Renal depletion
- primary-renal depletion: polyuric phase of acute and chronic renal failure, salt-losing-nephritis
- secondary-renal depletion: therapy with diuretics, Addison's disease

2. Extrarenal depletion
- enteric depletion: vomiting, diarrhoea, fistulae
- shift into third space: pancreatitis, peritonitis, ileus
- percutaneous depletion: burns

Sym.: hypovolaemic symptoms:
thirst, tachycardia, collapse, oliguria

Lab.:
- haematocrit, haemoglobin, serum protein ↑
- serum sodium and -osmolality normal
- increased urine density (but normal renal function)

2. Hypotonic dehydration

Def.: salt depletion > water depletion → extracellular dehydration, intracellular oedema

Aet.: same as isotonic dehydration (as above); too often there is substitution of too much salt-free water!

Pg.: reduced extracellular volume stimulates ADH-secretion leading to renal water retention. Hyponatraemia causes intracellular increase in volume with subsequent cerebral symptoms.

Sym.:
- Hypovolaemia associated symptoms (like in isotonic dehydration) – severe collapses
- cerebral symptoms: drowsiness, confusion, convulsions

Lab.:
- haemoglobin, haematocrit, serum protein ↑
- serum sodium and -osmolality ↓
- urine-Na^+ < 20 mmol/l (extrarenal depletion)
  urine-Na^+ > 20 mmol/l (renal depletion)

83

3. Hypertonic dehydration

Def.: water depletion; reduced extra- and intracellular volume

Aet.: • insufficient water intake
  • water depletion via skin (perspiration), lungs (hyperventilation), kidneys (diabetic coma, diabetes insipidus), gastrointestinal tract
  • iatrogenic (excessive intake of osmotic active fluids)

Pg.: due to the osmotic gradient water depletion is mainly <u>intracellular,</u> only minor symptoms of hypovolaemia. Hypertonic dehydration causes mainly intracellular water depletion and hence shrinking of erythrocytes. Even a massive dehydration will only cause a minor rise of haematocrit.

Sym.:• thirst
  • dry skin and mucosa,
  • <u>fever</u>
  • drowsiness, confusion
  • oliguria,
  Note: circulation can remain stable for a long time!

Lab.: • haematocrit, haemoglobin, serum protein ↑
  • serum sodium and -osmolality ↑
  • urine osmolality ↑ in patients with normal renal function
  • urine osmolality ↓ (< serum osmolality) in diabetes insipidus (DI). ADH-application will raise urine osmolality in cases of central DI, but not in cases of renal DI

Th.:   **of dehydration**
   a) <u>treat the cause</u>
   b) <u>symptomatic:</u>
     1. <u>input/output chart, weight, regular monitoring of electrolyte homeostasis</u>

     2. <u>water replacement:</u>
        <u>estimation of water depletion (adult, 70kg):</u>
        - <u>thirst alone:</u> up to 2 litres
        - <u>additionally dry skin/mucosa:</u> 2 - 4 litres
        - <u>additional symptoms of impaired circulation</u> (pulse ↑, blood pressure ↓, CVP ↓): > 4 litres
          Symptoms of impaired circulation show earlier in hypotonic dehydration (additional water shift from extra- to intracellular!)
        Note: Don't give any plasma expanders for dehydration, since they will increase the extravasal fluid - deficit. Patients suffering from cardiac or renal failure should receive a slow fluid replacement → regular checks of CVP + weight! (hazard of pulmonary oedema).

     3. <u>correct the sodium homeostasis:</u>
        Minor imbalance of serum sodium (125 - 150 mmol/l) usually don't cause any symptoms. The most important thing to do, is the elimination of the cause (e.g. stop diuretics).
        Note: Long standing sodium imbalance cause changes in the cerebrospinal fluid. They must be corrected <u>SLOWLY over several days</u>. Rapid correction would cause a life threatening osmotic gradient between cerebrospinal fluid and extracellular fluid! This applies to longstanding hypo- and hypernatraemia.

        During treatment of severe symptomatic hyponatraemia the speed of rising the sodium level must not exceed 10 mmol/l/24h. The target level must not exceed 125 - 130 mmol/l.

        • <u>isotonic dehydration:</u>
          give isotonic, isoionic fluid
        • <u>hypotonic dehydration:</u>
          very slow and careful substitution of sodium → note: too rapid rise in sodium will cause a fast drop of pressure of cerebrospinal fluid! → risk of cerebral damage or haemorrhage.
        • <u>hypertonic dehydration:</u>
          Give fluid free of osmotic activity (5% glucose); but 1/3 of the fluid deficit must be replaced with isotonic, isoionic electrolyte solution.
          Again: the balancing must take place slowly over several days; a too rapid correction can cause raised pressure of cerebrospinal fluid, brain oedema and central pontine myelinolysis.

HYPERHYDRATION [87.7]

We distinguish between 3 forms – depending on serum osmolality:

ECF	ICF	
		normal
		isotonic hyperhydration
		hypotonic hyperhydration
		hypertonic hyperhydration

Aet.: Relative surplus of fluid and/or salt found in the following impairments:
1. chronic renal failure
2. cardiac failure
3. hypoproteinaemia:
   - protein depletion: nephrotic syndrome, exsudative enteropathy
   - reduced intake: starvation oedema
   - reduced albumin synthesis: liver cirrhosis
4. impaired regulation of homeostasis
   - secondary hyperaldosteronism
   - therapy using gluco- or mineral corticosteroids
   - Syndrome of inappropriate ADH-secretion = SIADH = Schwartz-Bartter-Syndrome:
     causes: • paraneoplastic (mostly bronchial carcinoma)
     • cerebral diseases
     • pulmonary diseases
     • hypothyroidism
     • drug induced (e.g. cytotoxic drugs)

   Note: SIADH does not cause oedema
5. other causes: e.g. TUR(P)-syndrome: rinsing fluid free of any electrolytes used during transurethral prostate resection [TUR(P)] can be absorbed by cells, leading to hypotonic hyperhydration.

Pg.: Depending on osmolality or serum sodium concentration, we divide into isotonic, hypotonic and (rarely) hypertonic hyperhydration. It depends on the relation of water/salt excess. Deviation from normal osmolality (serum-sodium-concentration) can lead to dangerous changes of intracerebral fluid:
hypoosmolality → increased cerebral fluid, cerebral oedema
hyperosmolality → cerebral fluid depletion

Sym.: • weight gain

• symptoms of hypervolaemia:
  - peripheral circulation: oedema
  - pulmonal circulation: dyspnoea, fluid lung, pulmonary oedema
• pleural effusion, ascites
• deviation from the normal osmolality or serum-sodium-concentration can cause additional cerebral symptoms: headache, convulsions, coma.
• blood pressure is often raised in hyperosmolality (hypernatraemia), but often too low in hypoosmolality (hyponatraemia).

Lab.:

		Specific urine density	serum-sodium serum-osmolality
hypertonic	hyperhydration	↑	↑
isotonic	hyperhydration	↓	normal
hypotonic	hyperhydration	↓	↓

• haematocrit, haemoglobin, serum protein ↓

Th.: A) cause: e.g. treat any underlying cardiac or renal failure

B) symptomatically:
1. balance input/output, check weight, check electrolytes

   Note: Hyponatraemia in hyperhydration (dilution associated hyponatraemia) must not be misdiagnosed as sodium deficiency, and hence must not be substituted with sodium! Cardiac failure associated with oedemas, liver cirrhosis associated with ascites, nephrotic syndrome and renal failure usually request water and salt reduction (+ diuretics).

2. Diuretics:
    - If hyperhydration is not life-threatening, and if there are no signs of pulmonary hypervolaemia: Start slow dehydration, observing potassium levels, e.g. using a combination of Thiazide-diuretic and a potassium sparing diuretic (to prevent hypokalaemia).
    - Life-threatening hyperhydration with signs of pulmonary hypervolaemia: Start a potent loop diuretic, e.g. Furosemide: 20 - 40mg i.v., repeat dose if necessary (see 'hypervolaemia' for further details)
3. Hyperhydration caused by chronic renal failure requires dialysis

OEDEMA [R60.9]

Def.: Pathological buildup of interstitial fluid – there are also physiological minor pre-tibial oedema after a long period of standing/sitting and premenstrual oedema.
Generalised oedema settles first in body parts most exposed to gravity: sacral area in the supine patient, pre-tibial and around the ankles in the mobile patient.

Aet.: 1. Increased hydrostatic pressure in the capillaries
    - Generalised: renal failure, right cardiac failure (see hyperhydration)
    - Localised: Venous obstruction (venous oedema): DVT, post thrombotic syndrome and chronic-venous deficiency (see there)
2. reduced plasmatic oncotic pressure due to hypoalbuminaemia (< 2,5 g/dl):
    - loss of protein: nephrotic syndrome, exsudative enteropathy
    - reduced intake: starving oedema
    - reduced albumin synthesis: liver cirrhosis
3. increased permeability of the capillaries:
    - generalised: acute postinfectious glomerulonephritis, angio-oedema (see below)
    - localised: allergic and inflammatory oedema, posttraumatic oedema, Sudeck' dystrophy
4. reduced lymphatic drainage: lymphatic oedema (refer to chapter)
5. drug induced oedema: Calcium antagonists, Minoxidil, NSAID, Glucocorticosteroids, Oestrogens, Antidepressants (ADH-effects), Glitazone etc.
6. cyclic oedema (usually premenstrual, occ. periovulatorial)
7. idiopathic oedema (mainly perimenopausal women)
8. artificial oedema: psychopathological auto-strangulation of an extremity (beware strangulation marks!)

Sym: • swelling of the dorsal feet, lower legs, shoes become too small, rings become too tight
• weight gain (peripheral oedema won't become visible until there is an interstitial fluid collection of several litres/ kilograms)
• pulmonary oedema: life threatening dyspnoea (angiooedema associated with inspiratory stridor)
• generalised oedema: occ. periorbital oedema

DD: • myxoedema due to hypothyroidism: dough-like consistency of the skin, leaving no dimple after pressure with the fingers (unlike real oedema which is caused by a collection of water).
• lipo-oedema: fatty tissue causing swelling of the legs with subsequent lymphatic oedema, feet are not affected; almost exclusively found in post pubertal women.

Di.: • history + symptoms
• laboratory: urine chemistry, creatinine, electrolytes, protein, albumin, electrophoresis, D-Dimer if DVT is suspected; brain natriuretic peptide if cardiac oedema is suspected (cardiac failure)
• scans (echocardiography, Duplex-sonography if venous oedema is suspected)

Th.: 1. causal therapy
2. symptomatic therapy
    - generalised oedema: diuretics, sometimes sodium- and fluid restriction
    - chronic venous deficiency: compression therapy
    - lymphatic oedema: see chapter
    - angio-oedema: refer to according chapter

ANGIO-OEDEMA [T78.3]

Syn: Quincke-oedema, angioneurotic oedema

Def.: acute oedema of the deeper connective tissue; usually lips, periorbital, tongue and throat.

    Beware: epiglottis oedema carries the risk of suffocation; tends to relapse!

Aet.:   1. histamine-induced angio-oedema and urticaria related angio-oedema (common!)
- idiopathic angio-oedema
- intolerance-angio-oedema; main trigger: acetylsalicylic acid
- angio-oedema caused by ACE-inhibitors, rarely by Angiotensin II-receptor antagonists
- IgE-induced angio-oedema = allergic angio-oedema
- mechanic angio-oedema (pressure, vibration, cold, light)

2. angio-oedema due to C1-esterase-inhibitor (C1-INH)-deficiency: rare
   a) hereditary angio-oedema:
   autosomal-dominant inherited fault of the complement system
   - usually type I: reduced synthesis of the C1-inhibitor
   - rarely type II: synthesis of a dysfunctional C1-inhibitor
   b) acquired angio-oedema
      type I: associated with malignant lymphoma
      type II: caused by auto-antibodies against C1-inhibitor
   chronic infections (Helicobacter pylori, Yersiniosis) can be the cause for angio-oedema.

Special form: capillary leak - syndrome associated with generalised oedema, ascites and cardiovascular shock (e.g. a rare complication of Interleukin 2-therapy).

DD:

	histamine-induced angio-oedema	hereditary angio-oedema due to C1-esterase-inhibitor deficiency
history	often begins in adult life often history of urticaria	positive family history usually begins before the age of 20 no history of urticaria
symptoms	angio-oedema usually periorbital and affecting the lips no gastrointestinal symptoms	facial angio-oedema, extremities and trunk affected abdominal pain no urticaria
lab results	no specific lab results	type I: C1-INH reduced; type II: pathological C1-INH function
symptoma-tic therapy	corticosteroids and antihistamines i.v.	acute: C_1-inhibitor, e.g. Berinert®HS, (2nd choice: fresh frozen plasma = FFP) prophylaxis: Danazole current trials: Icatibant
	respiratory support if indicated	

Note: in C1-INH-deficiency complement factor 4 is reduced (screening-marker).

Th.:   A) causal: e.g. omit any triggers, e.g. acetylsalicylic acid or ACE-inhibitors; avoid any allergens. Hereditary angio-oedema can partially improve after H. pylori-eradication or successful Yersiniosis treatment.
B) symptomatic: see DD

DD: **HYPONATRAEMIA** [E87.1]

Def.:   serum-sodium < 135 mmol/l in adults (< 130 mmol/l in children)

Pre: Common electrolyte imbalance. Usually this is due to excess water or water intoxication.

PPh: Physiological normonatraemia is regulated by water intake (thirst) and water output (renal function and antidiuretic hormone ADH). If renal function is normal, hyponatraemia is usually caused by ADH-induced reduction of renal water output → 2 causes:
1. hypovolaemia caused, baroreceptor mediated ADH-stimulation: e.g. decompensated liver cirrhosis, decompensated heart failure, hypovolaemia of different origin
2. syndrome of inadequate ADH-secretion (SIADH). This is often caused by medication, e.g. Neuroleptics.
In the elderly 10 – 20 % have hyponatraemia, which is often asymptomatic. There are various causes:

1. hypoosmolar hyponatraemia:
   (< 280 mosmol/kg H2O)
   a) excess sodium and water (oedema):
      syn: dilution hyponatraemia, hypotonic hyperhydration
      - extrarenal (urinary sodium < 20 mmol/l): heart failure, liver cirrhosis, nephrotic syndrome
      - renal (urinary sodium > 20 mmol/l): renal failure
   b) sodium and water deficiency (volume deficiency):
      syn: hypotonic dehydration

- extrarenal (urinary sodium < 20 mmol/l).
  diarrhoea, vomiting, burns, trauma, peritonitis, pancreatitis
- renal (urinary sodium > 20 mmol/l):
  diuretics, salt-losing nephropathy, mineral corticoid deficiency

c) normal sodium and fluid volume:
   (no oedema and no volume deficiency symptoms)
   glucocorticoid deficiency, hypothyroidism, psychogenic polydipsia, medication causing water retention, SIADH (= syndrome of inadequate ADH-secretion)

2. isoosmolar hyponatraemia = pseudohyponatraemia (280 - 296 mosmol/kg H2O)
   Marked hyperlipidaemia or increased plasma proteins contribute significantly to plasma volume, thus reducing the water content per unit volume. Whole plasma sodium is reduced, but it is normal in the plasma water. Photometric tests show reduced sodium levels, ion specific electrodes however reveal normal levels.

3. hyperosmolar hyponatraemia:
   (> 296 mosmol/kg H2O)
   hypertonic infusions (glucose, Mannit), hyperglycaemia

Sym: moderate hyponatraemia is usually asymptomatic; marked hyponatraemia: lethargy, confusion, reduced level of consciousness, anorexia, convulsions

Di.: 4 important questions:
- does the patient take any medication which might cause fluid retention?
- is there any underlying heart failure, nephropathy or hypothyroidism?
- does the patient take diuretics?
- persistent hyponatraemia: might SIADH be the cause?

Th.: 1. causal
2. symptomatic:
- hypovolaemic hyponatraemia: volume substitution with isotonic NaCl-infusion
- isovolaemic hyponatraemia: very slow and only partial sodium substitution over several days, indicated only in the presence of clinical symptoms or severe hyponatraemia (serum sodium < 120 mmol/l) (risk of central pontine myelinolysis if the substitution is too fast!)
- hypervolaemic hyponatraemia: reduce fluid intake!

DD: | HYPERNATRAEMIA | [E87.0]

Def.: serum-sodium > 145 mmol/l
1. symptoms of water deficiency:
   = hypovolaemic hypernatraemia
   - urine-osmolality > 800 mosm/kg:
     extrarenal water loss and/or inadequate water intake
   - urine-osmolality < 800 mosm/kg:
     renal loss of water:
     a) raised urine-osmolality post ADH-application = central diabetes insipidus
     b) no rise of urine-osmolality post ADH-application: nephrogenic diabetes insipidus or osmotic diuresis
2. symptoms of hypervolaemia (rare)
   = hypervolaemic hypernatraemia caused by excessive NaCl-infusion

Sym: - symptoms of the underlying problem which causes hypernatraemia

- muscle reflexes ↑, irritability, muscle fasciculation, convulsions

Th.: 1. causal
2. symptomatic:
- hypovolaemic hypernatraemia: volume substitution with 5 % glucose solution + 1/3 of the fluid deficiency with isotonic electrolyte solution
- hypervolaemic hypernatraemia: stop infusion of any hypertonic solution. If serum sodium > 160 mmol/l: 5 % glucose solution + Furosemide. In case of renal failure: dialysis.

| CHLORIDE |

Normal range: 97 - 108 mmol/l (serum)

Any changes in serum chloride concentration are usually parallel to those in sodium. Isolated changes can be found in a faulty acid-base-balance (as described below).

POTASSIUM

Ph.: normal range: children: 3,2 - 5,4 mmol/l
                            adults: 3,6 - 5,0 mmol/l

Daily potassium intake on a mixed balanced diet is ca. 50 - 150 mmol/day, excretion is 90 % renal and 10 % enteral. In the case of renal failure there will be a higher excretion via the large bowel.
Only 2 % of the total bodily potassium is located extracellular (Ke), 98% is intracellular (Ki). The Ki/Ke-concentration gradient determines the membrane potential. It is achieved by active transport (Na^+/K^+-ATPase).
The basic membrane potential is ca. - 85 mV, the threshold to cause an action potential is 50 mV.

PPh: • acute hypokalaemia increases the gradient Ki/Ke, and subsequently reduces neuromuscular excitability; in severe cases may cause muscular paralysis due to hyperpolarisation block.
  • acute hyperkalaemia initially leads to an increased neuromuscular excitability; in severe cases it causes muscular paralysis due to depolarisation block. Hyperkalaemia is negative inotrope (contraction ↓) and negative dromotrope (excitability ↓).
  • chronic potassium imbalance cause less severe neuromuscular disturbance, since extracellular changes in potassium concentration will cause a parallel change in intracellular potassium (with a subsequent partial normalisation of the Ki/Ke-gradient). Patients suffering from chronic hypo- or hyperkalaemia may not have any ECG changes.
  • Distribution of potassium between intra- and extracellular space depends on the following determinants:
    1. acid-base-balance: Acidosis of the extracellular fluid causes a shift of H^+ into the cells in exchange with potassium → acidosis causes hyperkalaemia. Alkalosis results in hypokalaemia.
    2. Insulin, aldosterone and adrenaline promote potassium shift into the cells. glucose-/insulin infusions are temporarily effective as a treatment for hyperkalaemia.
    3. Magnesium deficiency results in potassium loss from cardiac muscle and skeletal muscle cells (inhibition of Na^+/K^+-ATPase).

Note:
 ▸ 98 % of the total bodily potassium is found intracellular. Serum levels are not representative for the total body potassium.
 ▸ It is important to assess the function of the potassium dependant organs. ECG is useful in cases of acute potassium imbalance.
 ▸ Measuring the urinary potassium output can differentiate between renal or enteral loss of potassium.
 ▸ Potassium concentration inside erythrocytes is 25fold higher than serum potassium. Therefore phlebotomy has to be performed carefully to avoid haemolysis, and should be centrifuged within an hour, otherwise false increased levels may be found. For the same reason older erythrocyte concentrates show higher potassium levels.

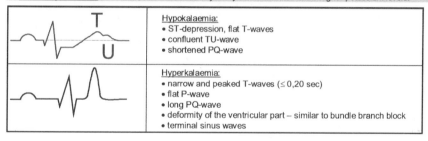

	Hypokalaemia: • ST-depression, flat T-waves • confluent TU-wave • shortened PQ-wave
	Hyperkalaemia: • narrow and peaked T-waves (≤ 0,20 sec) • flat P-wave • long PQ-wave • deformity of the ventricular part – similar to bundle branch block • terminal sinus waves

Hypokalaemia [E87.6]

Def.: Serum potassium in adults: < 3,6 (children < 3,2) mmol/l

Aet.: A) hypokalaemia due to loss of potassium (external)
    1. reduced intake
    2. intestinal loss:
        - diarrhoea, laxative abuse, fistulae, vomiting
        - villous adenoma (increased enteral mucus secretion)
        Note: Chronic laxative abuse is the most common cause for unexplained hypokalaemia. Young women presenting with unexplained symptoms (lethargy, constipation!) always must be asked for the use of laxatives! Hypokalaemia increases constipation!

3. Renal loss:
    a) primary renal loss of potassium caused by renal diseases:
       - chronic interstitial nephritis
       - polyuric phase of acute renal failure
       - renal tubular acidosis
       - Bartter-Syndrome

    b) secondary renal loss of potassium:
       - therapy with diuretics (most common cause!):
        Replace potassium or combine with potassium sparing diuretics.
       - primary or secondary hyperaldosteronism
       - pseudo-hyperaldosteronism due to liquorice abuse (Glycyrrhizin acid)
       - hypercortisolism
       - steroid therapy (gluco- or mineral corticosteroids)
       - Amphotericin B therapy

B) hypokalaemia due to impaired distribution (internal)
   Shifts of potassium from the extra- to intracellular spaces:
   - alkalosis
   - insulin treatment of diabetic coma
   - hypokalaemic paroxysmal paralysis (rare hereditary disease)

Sym.: Severity of symptoms is proportional to the acceleration of hypokalaemia. Chronic hypokalaemia usually is a result in asymptomatic patients found coincidentally in a routine blood test.
1. floppiness and paralysis, gentle percussion over muscles will lead to muscle contraction
2. constipation and paralytic ileus (including paralysis of the bladder)
3. weak or absent reflexes
4. ECG: low T-waves, ST-depression, U-waves (U-waves higher than T), TU-confluence, QT normal or prolonged, extrasystoles (extrasystoles during therapy with Digoxin is highly suspicious of hypokalaemia or Digoxin overdose).

   Note: Hypokalaemia increases Digoxin toxicity! Digoxin toxicity can be reduced by raising potassium- (and magnesium-) to high levels (within normal range)!

5. hypokalaemic nephropathy:
   tubulopathy associated with ADH resistant polyuria + polydipsia (renal diabetes insipidus). Chronic hypokalaemia can lead to interstitial nephritis.
6. metabolic alkalosis

Di.: • history + symptoms
   • serum and urine potassium levels: urine > 20 mmol/l: renal loss
                                      urine < 20 mmol/l: enteral loss
   • acid-base- balance
   • hypokalaemic hypertension: consider Conn-syndrome and renal artery stenosis!

Th.: A) causal: eradicate cause, e.g. stop any laxatives, non-potassium-sparing diuretics etc.

   B) symptomatic: potassium substitution - observe pH!
      • nutrition rich in potassium (fruit juice, bananas etc.)
      • potassium chloride: associated metabolic alkalosis can be treated this way.
      If renal function is normal, there is no risk of overdose in oral treatment with potassium chloride; potassium chloride in form of tablets is obsolete, since it can cause small bowel ulcers; hence potassium should be taken with food with plenty of fluid (e.g. dissolvable tablet).
      Parenteral: calculate K^+-deficit using a normogram (consider pH!). Intravenous potassium application – regular follow-up checks of potassium levels, ECG-monitoring.

   Note: 1 mmol extracellular serum-potassium deficit equals a lack of 100 mmol potassium. Don't give more than 20 mmol/hour parenteral (maximal daily dose 3 mmol/kg). Dilute the potassium solution sufficiently, since it is venotoxic! Hypokalaemia and acidosis require regulation of the potassium deficit first, only then may correction of acidosis be considered (if performed the other way round, hypokalaemia can deteriorate!).

Hyperkalaemia [E87.5]

Def.: Serum potassium in adults > 5,0 (children > 5,4) mmol/l

Aet.: A) External balance abnormality:
    1. Excessive potassium intake

> *Note:* Renal secretion capacity is more than twice the usual intake of potassium (100 mmol). Externally caused hyperkalaemia in normal renal function is almost impossible. High potassium intake will trigger a rise in aldosterone, and in turn an increased secretion into the tubular lumen.
> In advanced renal failure, too much fruit or potassium based salt can cause life threatening hyperkalaemia. Never give more than 20 mmol K^+/hour parental, when there is no facility to check follow up potassium levels!

    2. Reduced renal potassium secretion:
       - acute renal failure: Anuria causes a rise in serum potassium by ca. 1 mmol/l a day, caused by cellular catabolism.
       - chronic renal failure: As long as creatinine clearance is > 20 ml/min, i.e. in the absence of oliguria, there is usually no hyperkalaemia. Potassium will be secreted in the tubuli and the large bowel. Uncontrolled potassium intake (fruits, potassium based salt) or taking certain hyperkalaemia inducing medication will result in life threatening hyperkalaemia.
       - hyporeninaemic hypoaldosteronism in diabetes mellitus
       - Addison's disease (mineral corticoid deficiency)

    3. iatrogenic hyperkalaemia:
       - medication, which inhibit the Renin-Angiotensin-system (RAAS): ACE-inhibitors, Angiotensin II-receptor-antagonists, Spironolactone, NSAID
       - medication, which inhibit potassium secretion in the distal tubulus: Cyclosporin A, Amiloride, Triamteren (= potassium sparing diuretics), Cotrimoxazole, Pentamidine

> *Note:* A combination of ACE-inhibitors and potassium sparing diuretics require regular checks of potassium levels even in normal renal function!

B) internal balance abnormality (disturbed distribution)
    shift from intracellular potassium into the extracellular space.
    1. acidosis, diabetic coma (insulin deficiency),
       severe Digoxin intoxication (block of Na^+/K^+-ATPase resulting in a passive K^+-shift into the extracellular space)
    2. release of potassium in cell necrosis
       - large soft tissue damage resulting in myolysis, rhabdomyolysis, burns
       - haemolytic crisis, transfusion of cold blood
       - cytostatic treatment of malignancies
       - hyperkalaemic periodic paralysis (Gamstorp-syndrome)
       - delayed release of complete arterial obstruction (tourniquet-syndrome)

C) pseudo-hyperkalaemia:
    - haemolysed blood specimen:
       Also: high levels may be caused by artificial haemolysis (prolonged use of tourniquet, patient pumping with hand whilst the blood is being taken, rapid blood aspiration via small needles, delayed centrifugation)!
    - excessive thrombocytosis or leucocytosis (CML) may cause potassium release within the specimen - Di.: serum potassium↑, but plasma potassium is normal!

Sym.: Often only very few symptoms!

> *Note:* There is no reliable symptom, indicating the risk of hyperkalaemia!

    • sometimes neuromuscular symptoms: perioral paraesthesias, muscle twitching, paralysis
    • ECG: - peaked T-wave
           - broad QRS (similar to bundle branch block)
           - short QT, low P-wave
           - ventricular flutter/ fibrillation or asystole

Di.: - serum-potassium level checks + ECG
    - exclude renal failure (creatinine)
    - acid-base-balance
    - exclude any haemolysis (haptoglobin etc.), myolysis (CK etc.)

Th.: A) causal: e.g. stop all potassium retaining medication in patients suffering from renal failure!
    B) symptomatic:
       1. stop potassium intake, no food rich in potassium (e.g. bananas and other fruits)
       2. promote potassium shift into the intracellular space:
          • glucose and insulin: e.g. 50 ml 40 % glucose solution + 10 IU normal insulin i.v. over 30 min., followed by further blood sugar tests; lasts: 4 – 6 hours
          • sodium bicarbonate: 50 - 100 ml of 1 mol (8,4 %) solution over 30 min. i.v.; period of action: 2 hours

- Calcium gluconate only works for a short period (30 min.) and it is contraindicated in patients on Digoxin and those with hypercalcaemia. Dose: 10 ml 10 % Calcium gluconate
- Salbutamol (Beta-2-Sympathomimetic) inhalation can temporarily reduce the potassium level (SE + CI: see chapter 'Asthma').
3. Removal of potassium
- cation exchange resins; they exchange sodium or calcium against potassium inside the bowel. Period of action: 4 – 6 hours
2 types: sodium type (Resonium A®), contraindicated hypernatraemia and hypertension; Calcium type (Calcium Resonium®), contraindicated in hypercalcaemia. Oral use or as an enema. Consider SE (rarely intestinal necrosis)
- forced diuresis with loop diuretics
- dialysis in acute renal failure and chronic renal failure

Note: Serum potassium levels > 6,5 mmol/l are acute threatening and require rapid reduction!

MAGNESIUM

Total body magnesium is ca. 12,4 mmol (0,3 g)/kg, daily requirement: 15 - 20 mmol/day (360 - 480 mg/day)
Body magnesium distribution:
- 1 % in plasma (30 % bound to albumin, ca. 70 % ionised)
  normal plasma level: 0,75 - 1,05 mmol/l
- ca. 60 % in bone
- ca. 40 % in skeletal muscle

Intracellular magnesium is mainly ATP bound (MgATP), and it is equal to free magnesium ions. Magnesium is involved in the activation of various enzymes, e.g. Na^+/K^+-ATPase, hence it has an influence in potassium distribution. Magnesium inhibits the intracellular calcium availability ("natural calcium blocker").

Hypomagnesaemia [E83.4]

Def.: Serum magnesium < 0,75 mmol/l

Aet.: 1. Congenital magnesium loss illness (rare):
- intestinal hypomagnesaemia with secondary hypocalcaemia and convulsions; autosomal-recessive
- renal primary hypomagnesaemia; autosomal-dominant; gene mutation FXYD2
- mutations of the CASR-gene (Ca^{2+}-Mg^{2+}-sensing-receptor). There are activating and inhibiting mutations.
- familiar hypomagnesaemia with hypercalciuria and nephrocalcinosis (FHHNC); mutation of the Paracellin 1-gene (PCLN1); incidence ca. 1 : 100.000
- Gitelman-syndrome
2. Secondary (acquired) hypomagnesaemia: most cases
- malnutrition (alcoholism, parenteral nutrition)
- malabsorption syndrome
- pregnancy (more potassium needed)
- enhanced renal secretion: polyuria, diuretic therapy, Cyclosporin A, Cisplatin, Aminoglycosides
- acute pancreatitis
- laxative abuse
- endocrine diseases: diabetes mellitus, hyperthyroidism etc.

Sym.: Clinical symptoms of hypomagnesaemia are non-specific, because it can be associated with hypokalaemia and/or hypocalcaemia
- CNS/Psyche: irritability, depression, tetanus, paraesthesia etc.
- cardiac: extrasystoles, increased Digoxin sensitivity, coronary arteries are more prone to spasms causing angina
  ECG: ST-depression, flat T-waves, prolonged QT-wave
- gastrointestinal: bowel spasms etc.

Di.: symptoms (non-specific), serum magnesium ↓, 24-hour urine magnesium
exclude hypokalaemia/hypocalcaemia

Th.: • causal
• symptomatic: magnesium supplementation
  Ind:1. oral substitution
       Dose: 10 - 30 mmol magnesium/day
     2. Therapy in normal magnesium levels: e.g.
       - ventricular arrhythmia caused by Digoxin
       - extrasystoles can be improved by a rise in magnesium- and potassium to high normal levels

- torsade-de-pointes ventricular tachycardia
- eclampsia associated with generalised convulsions
- premature labour
Dose: see according chapter

Hypermagnesaemia [E83.4]

Def.:   serum magnesium > 1,05 mmol/l

Aet.:   the most common cause is renal failure and antacids containing magnesium
         also: rhabdomyolysis; parenteral magnesium therapy etc.

Sym.:   usually asymptomatic finding in blood test. Concurrent hypocalcaemia and/or hyperkalaemia may cause symptoms:
         - muscle weakness, nausea, facial paraesthesia
         - hypoventilation
         - somnolence, unconsciousness
         - ECG: prolonged PQ-interval, broad QRS

Th.:   parenteral magnesium overdose: give i.v. calcium as an antidote.
         hypermagnesaemia /hyperkalaemia due to terminal renal failure: dialysis

CALCIUM

PPh: refer to chapter 'parathyroid gland'

Hypocalcaemia [E83.5]

Def.:   total serum calcium < 2,2 mmol/l; ionised calcium < 1,1 mmol/l

Aet.:   1. if ionised calcium is normal: hypoalbuminaemia of various origin
         2. if ionised calcium is low:
             a) and normal magnesium level
                 • PTH low, phosphate high:
                     - hypoparathyroidism
                     - temporary post parathyroidectomy for primary hyperparathyroidism
                 • PTH high, phosphate low:
                     - vitamin D-deficiency
                     - drug induced (anticonvulsants)
                     - pancreatitis
                 • PTH high, phosphate normal or increased:
                     - pseudo-hypoparathyroidism
                     - rhabdomyolysis
                     - hypernutrition
                     - renal tubular acidosis
                     - chronic renal failure
             b) and low magnesium level
                 • alcoholism
                 • malabsorption
                 • drug induced (loop diuretics, Gentamicin, Cisplatin)

Sym.:   - tetany: painful cramps with full consciousness, frequently paraesthesia, muscle spasm may lead to a characteristic
           position of the hand, the so-called 'main d'accoucheur', laryngeal spasm
         - Chvostek' sign: gentle percussion over the facial nerve at its point of exit from the parotid gland produces ipsilateral
           twitching contraction at the angle of the mouth.
         - Trousseau' sign: application of a blood pressure cuff inflated up to the arterial mid pressure for several minutes will
           cause 'main d'accoucheur'.
         - ECG: prolonged QT-interval

DD: hyperventilation tetany (total calcium normal, reduced ionised calcium due to respiratory alkalosis) - therapy: calm
         down the patient, breathe into a bag

Di.:   symptoms + serum calcium < 2,2 mmol/l

93

Th.: A) causal

  B) symptomatic:
   - tetany: calcium i.v. (e.g. slow i.v. infusion of 10 ml of 10% calcium gluconate); the response is short-lived
   - long term results can be achieved by oral calcium substitution and additionally vitamin D
   (for further details refer to the chapter ' parathyroid gland')

Hypercalcaemia [E83.5]

Def.: total serum calcium > 2,7 mmol/l; ionised calcium > 1,3 mmol/l

Ep.: Ca. 1 % of all in-patients reveal hypercalcaemia.

PPh: refer to 'parathyroid gland'

Aet.: 1. Tumour induced hypercalcaemia (ca. 60 %). Most frequent tumours: bronchial-, breast-, prostate carcinoma, multiple myeloma. Tumour induced hypercalcaemia leads to a suppression of PTH.
   - Osteolytic hypercalcaemia caused by bone metastasis (e.g. breast cancer) and by multiple myeloma. Tumour cells will lead to an indirect stimulation of the osteoclasts via the release of cytokines (TGFα, TNF, IL-1 etc.) → osteolysis → hypercalcaemia.
   - Paraneoplastic hypercalcaemia due to ectopic production of PTH-related peptides (PTHrP) (tumour induced-e.g. bronchial carcinoma). Ca. 90 % of all patients who suffer from tumour-induced hypercalcaemia, also have increased PTHrP levels, irrespective of the presence of any bone metastasis. PTH levels are low.

  2. Endocrine origin: primary hyperparathyroidism (20 %), tertiary hyperparathyreoidism, hyperthyroidism, adrenal cortex deficiency

  3. drug induced: vitamin D- or vitamin A-intoxication, Tamoxifen, Thiazid-diuretics, calcium based phosphate binding drugs, calcium based cation exchanging drugs, Lithium

  4. Immobilisation

  5. Sarcoidosis (epitheloid cells of the granulomas produce 1,25(OH)2D3)

  6. Familial hypocalciuric hypercalcaemia (FHH): Inactivating mutation of the calcium-sensing-receptor in the parathyroid gland. It makes up to ca. 1% of patients suffering from pHPT. The main finding is a relative hypocalciuria, but mild hypercalcaemia and slightly increased PTH. No therapy neccessary.

  7. Thrombocytosis and essential thrombocytaemia may be associated with hypercalcaemia.

Sym.: A) Symptoms of the underlying disease (e.g. tumour)

  B) Hypercalcaemia symptoms: Half of the patients has no specific symptoms of hypercalcaemia (coincidental lab result)
   - kidney: polyuria, polydipsia (= renal diabetes insipidus); may cause dehydration and anuria; nephrolithiasis, nephrocalcinosis
   - gastrointestinal: nausea, vomiting, constipation, rarely pancreatitis
   - cardiac/skeletal muscles: arrhythmias, short QT-interval, inertia, muscle weakness, pseudo paralysis
   - CNS/Psyche: psychosis, somnolence, coma

   Serum calcium > 3,5 mmol/l carries the risk of a hypercalcaemic crisis:
   - polyuria, polydipsia
   - vomiting, dehydration, pyrexia
   - psychotic symptoms, somnolence, coma

Di.: 1. of hypercalcaemia: serum calcium ↑
  2. of the origin of hypercalcaemia:
   - PTH intact ↑ in primary hyperparathyroidism, ↓ in tumour-induced hypercalcaemia
   - parathyroid hormone related peptide (PTHrP) ↑ in tumour-induced hypercalcaemia
   - 1,25-(OH)2-vitamin D3 ↑ in hypercalcaemia due to sarcoidosis
   - 25(OH)D3 ↑ in vitamin D-intoxication
   - tumour search (CXR, mammography, abdominal ultrasound, check serum/urine for monoclonal immunoglobulines and light chains)

Th.: of a hypercalcaemic crisis:

  A) causal

  B) symptomatic:
   - general means:
    1. most important means is forced diuresis (5 l/day or more) using normal saline and Furosemide under regular checks of water and electrolyte balance (substitution of potassium).
    2. stop calcium intake (e.g. mineral water)

     Beware cardiac glycosides and Thiazide diuretics !

3. Bisphosphonates: drug of choice in tumour-induced hypercalcaemia
   inhibits osteoclast activity
   dose: e.g. Pamidronic acid (Aredia®) 45 - 90 mg i.v. over 2 hours or 2 - 4 mg Zoledronate (Zometa®) i.v. over 5 min.; can be repeated after 3 - 4 weeks

   Note: Calcitonin action is too short-lived.

- additional means:
  - glucocorticosteroids are vitamin D antagonists→ use for vitamin D-induced hypercalcaemia (vitamin D-intoxication, Boeck's disease); they are also useful in multiple myeloma.
  - renal failure: haemodialysis using calcium free dialysate

Prg.:   death rate of the hypercalcaemic crisis is up to 50 %

ACID-BASE-BALANCE

Ph.:   physiological blood hydrogen ion concentration (isohydria) of pH 7,37 - 7,45 can be maintained despite constant buildup of acid metabolites. This is done by 3 regulatory mechanisms:
1. buffer
2. respiratory regulation (expiration of CO_2)
3. renal regulation (excretion of hydrogen ions)

1 + 2: buffer:
The organism has 2 extracellular and 2 intracellular buffer substances:
- extracellular: bicarbonate ($HCO3^-$): 75 %
                   plasma proteins: 24 % (1 % phosphate)
- intracellular: phosphate ($HPO4^{2-}$)
                   haemoglobin

The ability to exhale CO_2 via the lungs and secrete $HCO3^-$ via the kidneys, makes the carbon acid-bicarbonate-system the most important regulator. Its function is described by the Henderson-Hasselbalch' equation:

$$pH = 6,1 + \log \frac{(HCO3^-)}{(H2CO3)} = 6,1 + \log \frac{20}{1} = 6,1 + 1,3 = 7,4$$

3: renal regulation of hydrogen ion stability is more slow and idle than the respiratory system, and it consists of **3 mechanisms:**
- bicarbonate absorption:
  bicarbonate is eliminated by buffering and subsequent pulmonal exhalation of CO_2; hence it must be regenerated in the kidneys. Carboanhydrase plays an important role: For each regenerated $HCO3^-$, one H^+ will be secreted. To achieve electric neutrality, Na^+ will be reabsorbed.
- production of titrable acid
- creation of ammonium ions, to neutralise excessive hydrogen ions in the tubulus.

PPh: 3 groups of disturbance of the acid-base-regulation:
1. respiratory imbalance are caused by increased or reduced CO_2 expiration. This changes the denominator of the buffer equation.
2. metabolic imbalance is associated with a change in bicarbonate concentration. This changes the numerator of the equation.
3. mixed imbalance caused by a combination of metabolic + respiratory disturbance

Mechanisms of compensation:
Hydrogen ion concentration needs to be stabilized (isohydria). The compensation mechanism will try to regulate any changes in bicarbonate concentration by changing the CO_2-concentration into the same direction, and also the other way round.

Note: Respiratory disturbance will be metabolically compensated. Metabolic disturbance will be respiratory compensated.
A pH within the range of 7,37 - 7,45 indicates a compensated disturbance, otherwise it indicates a non-compensated disturbance. A normal pH is not to be confused with a normal acid-base-balance. A normal pH only indicates that the compensation mechanism are still functioning.

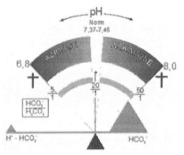

Di.: If 2 of the 3 variables of the equation are known, then the 3rd variable can be calculated. This way the place of the cause for the imbalance can be found.

Arterial normal levels:

pCO₂ (blood)	male 35 - 46 / female 32 - 43 mm Hg
standard bicarbonate	21 - 26 mmol/l
pH	7,37 - 7,45

Base excess (BE)

Deviation from the normal levels of buffer bases.
Normal range is ± 2,0 mmol/l.

Impaired acid-base-balance and their compensation mechanisms:

	Alkalosis decomp.	Compens.	Acidosis compens.	decomp.	
metabolic: pH HCO3⁻ pCO₂	↑ ↑ n (↑)	n ↑ ↑	n ↓ ↓	↓ ↓ n (↓)	metabolic: pH HCO3⁻ pCO₂
respiratory: pH HCO3⁻ pCO₂	↑ n (↓) ↓	n ↓ ↓	n ↑ ↑	↓ n (↑) ↑	respiratory: pH HCO3⁻ pCO₂

Note: changes in pCO₂ and HCO3⁻ are proportional in simple imbalance of the acid-base-balance due to compensation (this rule is not valid for mixed imbalance).
Marked deviation of HCO3⁻ (< 15 or > 40 mmol/l) indicate a metabolic cause (metabolic acidosis / alkalosis).

ACIDOSIS

General effects:
1. potassium shift from the intracellular into the extracellular space can often cause hyperkalaemia. Elimination of the acidosis will get the potassium levels back to normal (or even turn into potassium deficiency).
2. Acidosis reduces the sensitivity of vascular muscles to Catecholamines (e.g. cardiogenic shock) and has a negative inotrope effect.
3. marked acidosis causes renal malperfusion (shock + acidosis → anuria).
4. acid urine pH
5. the blood-brain-barrier is permeable to CO₂ (unlike metabolic acids and bicarbonate); respiratory imbalance causes a more rapid pH-change of the cerebrospinal fluid than metabolic disturbance

---

Metabolic Acidosis [E87.2]

Aet.: 1. Addition acidosis:
• Endogen acid production:
- Ketoacidosis: diabetic pre-coma/coma (accumulation of β-hydroxybutyrate and acetoacetate), starvation, alcoholism

- Lactate acidosis: lactate accumulation due to shock, hypoxia, rarely due to side effects caused by Biguanides (Metformin) or Propofol (Propofol-infusion syndrome), severe Thiamine deficiency etc.
- External acid intake:
  Intoxication with salicylates, methyl alcohol, glycol etc.

2. Retention acidosis:
   Reduced renal acid elimination
   - renal failure
   - distal tubular acidosis (type I) due to reduced H^+-ion secretion

3. Subtraction acidosis:
   - enteral bicarbonate loss, e.g. diarrhoea, uretero-enterostomy
   - renal bicarbonate loss:
     - proximal tubular acidosis (type II)
     - Carboanhydrase inhibitor therapy (Acetazolamide)

Sym.: deep Kussmaul' respiration (= compensation)

Di.: symptoms + blood gases
$HCO3^-$ ↓, compensatory pCO_2 ↓
pH normal (compensated) or ↓ (decompensated)

There are 2 constellation according to Cl^- and anion gap:
- hyperchloraemic acidosis and normal anion gap: e.g. subtraction acidosis
- normochloraemic acidosis and increased anion gap: e.g. addition- and retention acidosis

Note: Cl^- and $HCO3^-$ contribute 85 % to the serum anions, the rest (proteinate, sulphate, phosphate, organic anions) are called anion gap.

Simple equation: anion gap $\approx Na^+ - (Cl^- + HCO3^-)$

Normal range: 12 ± 4 mmol/l

Respiratory Acidosis [E87.2]

Aet.: Respiratory failure resulting in alveolar hypoventilation; it may occur as a result of many respiratory disorders (see chapter pulmonary diseases

Sym.: • Hypoventilation as part of the respiratory failure
• weakness, confusion, coma

Di.: symptoms + blood gases:
pCO_2 ↑, compensatory $HCO3^-$ ↑
pH normal (compensated) or ↓ (decompensated)
pO_2 ↓

ALKALOSIS

General effects:
1. potassium shift from the extracellular to the intracellular space and tubular secretion of potassium can cause potassium deficiency.
2. Alkalosis causes a reduction of ionised Ca^{++} (tetany).
3. urine pH is usually alkaline. An exception is the metabolic alkalosis due to extrarenal potassium loss: In this case the kidneys try to retain K^+, and hence they secret a slightly acid urine ("paradox aciduria").

Metabolic Alkalosis [E87.3]

Aet.: 1. loss of gastric acid (e.g. vomiting)
2. therapy with diuretics causing hypokalaemia: potassium deficiency will lead to increased renal secretion of H^+.
3. mineral corticoids surplus (Conn-syndrome, mineral corticoid therapy): mineral corticoids stimulate the secretion of K^+ and H^+ in the distal renal tubulus.
4. increased bicarbonate intake

Sym.: • reduced shallow respiration (= compensation mechanism)
• tetany
• cardiac symptoms: extrasystoles

<u>Di.:</u> symptoms + blood gases:
HCO3⁻ ↑, compensatory pCO_2 ↑
pH normal (compensated) or ↑ (decompensated)

According to the renal Cl⁻ secretion after saline application, we divide into 2 groups:
- <u>chloride sensitive form:</u>
  <u>Urine chloride secretion in 24 hours < 10 mmol/l:</u> loss of gastric acid, diuretic therapy → alkalosis can be corrected with infusion of 0,9 % saline.
- <u>chloride resistant form:</u>
  <u>Urine Cl⁻-secretion in 24 hours > 20 mmol/l:</u> excessive mineral corticoids

Respiratory Alkalosis [E87.3]

<u>Aet.:</u> <u>enhanced alveolar ventilation:</u>
- <u>psychogenic hyperventilation</u> (most common cause !) – refer to chapter 'hyperventilation syndrome
- <u>compensatory hyperventilation in hypoxia</u>
- hyperventilation due to CNS pathology
- other rare causes: septic shock, hepatic encephalopathy etc.

<u>Sym.:</u>
- <u>hyperventilation</u> as causal symptom
- <u>hyperventilation tetany</u>, paraesthesia, muscle tremor
- In severe cases there may be <u>reduced cerebral perfusion</u> causing irritability, lack of concentration, impaired consciousness

<u>Di.:</u> symptoms + blood gases:
pCO_2 ↓, compensatory HCO3⁻ ↓
pH normal (compensated) or ↑ (decompensated)

Note: hyperventilation is found in respiratory alkalosis(= cause) and metabolic acidosis (= compensatory); hyperventilation due to respiratory alkalosis is usually psychogenic in otherwise physically completely healthy individuals.

TREATMENT OF ACID-BASE IMBALANCE

A) <u>causal treatment:</u> <u>eliminate the cause!</u>
B) <u>symptomatic treatment</u> should be performed with care under close supervision (frequent blood tests).

- <u>respiratory acidosis:</u>
  <u>ventilated patients should receive increased ventilation in order to exhale</u> CO_2
- <u>metabolic acidosis</u> start being dangerous from pH < 7,15
  <u>give bicarbonate</u> → : HCO3⁻ + H⁺ → H2O + CO_2 → pulmonary expiration
  this is only advisable when respiration is not impaired in any way, in order to exhale the accumulating CO_2.

  $$\text{needed NaHCO}_3 \text{ in mmol} = \text{negative base excess} \times \frac{\text{kg body weight}}{3}$$

Note: One should wait giving bicarbonate in quickly reversible diseases (e.g. diabetic ketoacidosis). Infuse bicarbonate slowly in several steps → risk of hypokalaemia! Plasma bicarbonate should not get any higher than > 15 mmol/l.

- <u>respiratory alkalosis:</u>
  This common psychogenic hyperventilation mainly needs calming the patient, increase inhalation of CO_2 via breathing into a bag under medical supervision.
- <u>metabolic alkalosis</u> is dangerous from pH > 7,55
  - chloride sensitive version (e.g. loss of gastric acid): Infusion of 0,9 % saline. If there is a risk of sodium overload, arginin chloride is an alternative.
  - hypokalaemia: replace potassium

Enteral nutrition (EN)

Def.: EN compared to parenteral nutrition (PN) is more physiological and more uncomplicated.
Basically every nutritional therapy should obey the following priority: Normal nutrition is the most important target, and should be put before liquid nutrition, EN via nasogastric tube, combined EN + PN and total PN. PN should be replaced by some type of EN as soon as possible.
If EN will be need for longer than 2 - 4 weeks, then a percutaneous endoscopic gastrostomy (PEG) is indicated. Usually a standard high molecular formula diet with fibre is sufficient.

Ind: 1. General:
Existing or threatened malnutrition; inadequate oral intake (< 500 kcal/day) predicted for more than 7 days
2. Special:
Crohn's disease (acute), short bowel syndrome; severely underweight patients due to various diseases (e.g. anorexia nervosa), patients suffering from wasting diseases (e.g. tumour, AIDS etc.), cachectic geriatric patients – malnutrition causes delayed wound healing, frequent nosocomial infections, longer need for artificial respiration, longer hospital stay and is associated with a higher mortality.

CI.: EN against the patient's will.
polytrauma, shock, metabolic acidosis (pH < 7,25), hyperlactataemia, severely impaired coagulation, acute severe metabolic impairment (diabetic, hepatic or uraemic coma etc.); vomiting, risk of aspirations, ileus, haemorrhage, perforations, peritonitis, severe malabsorption syndrome etc.

SE + a specific risk of nutrition via nasogastric tube is aspiration pneumonia →
prophylaxis:
bedding the patient with a slightly raised upper body (30°), slow infusion, tube may have to be inserted with the tip ending in the jejunum
patients receiving EN have runny stools, which must not be confused with diarrhoea. A common problem in EN is diarrhoea, caused by a too rapidly enhancement of intake, bacterial contaminated feed, too much lactose, fat intolerance, too rapid rate, too large volume or too cold feed; but it can also be caused by a pathological change of the bowel flora. Vomiting can be another problem. In case of nausea or vomiting, a temporary reduction of feed, application of prokinetics or the insertion of a naso-jejunal tube can be helpful. In some patients there is a rise in transaminases and alkaline phosphatase caused by a high caloric nutrition; this usually resolves spontaneously. Electrolyte imbalance is possible (→ regular U+E checks); „tube-feeding-syndrome": severe dehydration and associated renal impairment, caused by hyperosmolar feed and subsequent diarrhoea + insufficient fluid intake.
EN requires regular monitoring of the patient's general condition and weight; staff must watch out carefully for reflux, bowel sounds, stool and flatulence.

Ingredients:
calories: standard feed: 1 kcal/ml; high caloric feed: 1,5 - 2,0 kcal/ml
Enteral feed should mirror the combination of a normal oral diet. That is: 15 - 18 % of calories should be proteins, 30 - 35 % long or medium chain triglycerides and ca. 50 - 55 % carbohydrates. In addition to this, there must be sufficient electrolytes, trace elements and vitamins. Industrially manufactured tube feed is standardised. Carbohydrates should be a balanced mixture of Poly-, Oligo- and Disaccharides. Lactose is not suitable because of the frequent incidence of Lactose intolerance. Fats should be composed of 1/3 from polyunsaturated fatty acids, monounsaturated fatty acids and saturated fatty acids each. Low molecular feed is used in severe malabsorption (e.g. short bowel syndrome).

Enteral diets:

Tube nutrition	Alternative names	Examples
High molecular tube nutrition Standard high molecular diet Modified high molecular diet	**Nutricient defined diet, Formula** simple, fully balanced or standard-formula Alternative Formula, Special diets	Biosorb® for tube, Ensure® Abbott, Fresubin® liquid, Osmolite Abbott, Salvimulsin® standard Rich in fibres: Biosorb® Plus for tube, Fresubin® Plus for tube, Osmolite® with fibres. High caloric: Biosorb Energy, Liquisorb cal., Nutrodrip Energy MCT-containing: Biosorbin® MCT, Fresubin® MCT
Low molecular tube nutrition Elementary diet Oligopeptide-diet	**Chemically defined diet** Astronaut diet, Synthetic diet Peptide-diet	Survimed® OPD, Salvipeptid®

In order to improve the immune system, a special immuno-nutrition has been developed, containing omega-3-fatty acids, amino acids, vitamins/trace elements.

Daily caloric requirements: adults: depending on the general nutritional state and underlying disease: 30 - 35 kcal/kg and at least 1 g protein/kg

Application technique:

1. Oral (drink):

   Ind: conscious patient without dysphagia and without oesophageal disease

2. Nasogastric tube:

   Ind: short term tube nutrition

   SE/Co.: sensation of a pharyngeal foreign body, reflux oesophagitis, pressure ulcer, dislocation, psychological stress etc. The position of the tube should be checked via CXR before feeding is commenced. Feeding can be applied as a bolus or continuously via infusion pump, when the tip of the tube is located in the stomach. Duodenal or jejunal position requests a continuous feed, in order to avoid a dumping-syndrome. Continuous application is also indicated in low molecular feed. Continuous infusion: start at a rate of 20 - 40 ml/h ,and increase to the target dose within 2 - 3 days. A high duodenal-gastric reflux, diarrhoea, flatulence, bloatedness and subileus will limit the feed volume. Duodenal-/jejuna tubes are used in patients suffering from diseases of having undergone surgery of the upper GI-tract, or those suffering severe gastro-oesophageal reflux during gastric feeding. Check position via CXR or endoscopy.

3. Enterostomy:

   3.1 PEG = Percutaneous-endoscopic gastrostomy (most common type)

   3.2 PEJ = Percutaneous-endoscopic jejunostomy

   Ind: long term nutrition > 3 - 4 weeks, oesophageal or hypopharyngeal tumour, neurological diseases causing ongoing dysphagia

   CI: clotting disorder/haemophilia, peritonitis, peritoneal carcinosis, ascites

   After tube insertion, patients at risk should receive prophylactic antibiotics (e.g. 1 g Ceftriaxon i.v. stat). 3 hours after insertion, feeding may be commenced. Insertion should take place in a sitting position, or the upper body raised by 45 °, if there is a risk of aspiration.

   Complication of PEG/PEJ: rarely unintended misplaced puncture, bleeding, peritonitis, peristomal wound infection (3 - 30 %) → diligent skin-/ stoma care, change of dressing, briefing of patients and relatives

Postoperatively the EN shouldn't be started before onset of motoric bowel activity, patient tolerance must be considered. Volume should be increased in daily steps of 250 - 500 ml, depending on the underlying problem, position of the tube and individual tolerance. Tube feed must not be too cold, it should have reached room temperature before feeding. If diarrhoea, bloatedness or vomiting occurs, then EN should be reduced or the patients is only allowed tea for 1 - 2 days. After that EN can be carefully restarted. As long as optimal energy- and fluid balance hasn't been achieved, any necessary fluid, electrolytes and nutritients must be applied parenteral.

EN should follow the normal circadian nutritional rhythm, including a nocturnal pause of 6 - 8 hours. Mobile pumps will facilitate home- and mobile feeding.

Tube feeding for special indications:

There are standard high molecular diets, but also commercially produced modified feeds. There are no good quality controlled, prospective studies, proving the benefits of special feed compared to standard high molecular isocaloric feeds.

There are special diets for diabetics, using alternatives to glucose. These substitutes however can cause gastrointestinal adverse reactions. Often a high-fibre standard-feed going with an adaption of the antidiabetic therapy is just as good, if not better, and is better tolerated. A special tube feed for diabetic patients in intensive care is not necessary, since glucose metabolism can be controlled by an Insulin sliding scale.

EN in chronic inflammatory bowel disease:

Children suffering from Crohn's disease and growth retardation do benefit from EN. Acute flare up of ulcerative colitis: Enteral feeding is not an indication for EN.

EN in tumour patients:

Malnutrition is an important risk factor for complications and for a poor prognosis. Therefore a prophylactic insertion of a PEG is indicated before elective radio-chemotherapy of ENT-tumours or oesophagus tumours, because of the predictable mucositis and dysphagia.

EN in liver cirrhosis:

Ensure sufficient protein intake: 1,2 - 1,5 g/kg/day (reduce in symptomatic encephalopathy)

PARENTERAL NUTRITION (PN)

PN is an unphysiological nutrition, which can lead to severe complications.

Ind: If at all possible- only short term, if EN is not possible or contraindicated. The change from PN to oral nutrition can be facilitated by using an intermittent balanced formula.

PN is available as an „all-in-one"-solution: All three components combined; the fat is in a separate compartment until to the components are mixed just before feeding (glucose and Amino acids combined, fat separate).

Daily requirement: mobile patients 30 – 35 kcal/kg/day; immobile patients 25 kcal/kg/day
1 kcal = 4,2 kJ

The following can be given parenteral (adult dose):
1. carbohydrates (hypertonic glucose), e.g.
   glucose 20 % = 0,8 kcal/ml
   glucose 40 % = 1,6 kcal/ml
   Dose: 100 - 400 g glucose/day as a steady infusion over 24 hours. Non-diabetic patients usually don't need insulin.
   Indication for insulin: persistent hyperglycaemia or glucosuria caused by glucose infusions. Sepsis: target normoglycaemia (improves the prognosis).
   Dose depends on blood sugar and glucose intake (rule of thumb: 1 U normal insulin covers ca. 5 g of glucose).
   Other sugars are not recommended, because they all will be metabolised via glucose, and there is less renal reabsorption, leading to a loss in calories!

2. amino acid mix:
   Daily requirement ca. 1,0 - 1,5 g/kg (ca. 100 g/day)

3. lipids:
   Products deriving from soya beans oil containing medium chain triglycerides (MCT). Lipid solution 10 % = 1,0 kcal/ml, lipid solution 20 % = 2,0 kcal/ml. Lipid emulsions should be started after > 3 days of TPN, in order to avoid any deficiency symptoms in essential fatty acids (e.g. hyperkeratotic dermatitis).
   Monitoring of triglycerides, which should be < 250 mg/dl.
   Dose: as in oral nutrition, ca. 30% of total caloric intake. Initially 1 g/kg/day; can be increased later up to 100 g/day.
   CI: immediately postoperative, shock, acidosis, pregnancy, hepatic failure.

4. Electrolyte substitution
   Any losses in electrolytes must be replaced:
   Total parenteral nutrition (TPN): 1 litre combined solution (amino acids + carbohydrates) of ca. 1.000 kcal (4.187 kJ) and the following medium amount electrolytes:
   Na^+ (+ Cl^-)   :   50 mmol
   K^+ :   30 mmol
   Ca^{++} :   3 mmol
   Mg^{++} :   3 mmol
   Phosphate   :   15 mmol
   stipulation: no imbalanced electrolyte homoeostasis
   Note: phosphate and calcium must not be mixed in the same bottle!

5. additional vitamins and trace elements
   In order to avoid any adherence of fibrin on the catheter, large veins have to be used (superior vena cava); low dose heparin (1 IU/1 ml) may be added.
   daily balance of input/output and tight observation of water-/ electrolyte homeostasis:
   - symptoms (skin tension, mucosa, oedema?, thirst?, fever?)
   - CVP
   - body weight
   - Hb, Hct, thrombocytes, serum protein, serum-electrolytes
   - Urine balance (volume, osmolality, glucose, electrolytes)

Co.:   • thrombosis, mainly originating from the tip of the catheter.
        • infections; infected catheter (mostly skin flora) associated with the hazard of sepsis
        Note: TPN reduces lymphatic tissue of the small bowel (GALT = gut associated lymphoid tissue) and also the secretorial IgA.
        • impaired water-/electrolyte homeostasis (Na^+, K^+, Ca^{++}, Mg^{++}, Phosphate)
          parenteral hyperalimentation (particularly in the cachectic) patient also requires regular checks of serum phosphate levels: forced input of carbohydrates enhances any phosphorilysation → hypophosphataemia causing polyneuropathy.
        • hypertriglyceridaemia, reversible fatty liver, "fat overloading syndrome"
          - transaminases, bilirubin ↑
          - thrombocytopenia + non-functioning thrombocytes and possible haemophilia
          - reduced O_2-diffusion capacity
        • hyperglycaemia
        • rarely lactate acidosis caused by a surplus of carbohydrates or vitamin B1 deficiency (Thiamine)
        • cholelithiasis
        • in long term (several months) TPN there may be osteopathy and bone pain (metabolic bone disease).
        • symptoms caused by the lack of trace elements (in long term TPN).

III. NEPHROLOGY

Internet-Info: *www.asn-online.org* – American Society of Nephrology
                   www.isn-online.org – International Society of Nephrology

Diagnostics:
A) history:
1. dysfunction of diuresis or micturition:
   Polyuria: > 2.000 ml urine/day
   Oliguria: < 500 ml urine/day
   Anuria: < 100 ml urine/day
   micturition frequency, a frequent symptom in cystitis
   painful micturition (cystitis and urethritis)
   Stranguria: very painful pelvic cramps during micturition (cystitis and urethritis)
   Dysuria: difficulty in passing urine, weak urine stream (e.g. prostate hyperplasia)

2. loin pain:
   - acute colic-like onset (typical radiation to the groin, accompanied by urgency and haematuria): e.g. ureteric
     calculus
   - ongoing dull loin pain and/or loin tenderness: e.g. pyelonephritis

3. oedema (glomerulonephritis, nephrotic syndrome, renal failure)

4. headache (causes: hypertension, pyelonephritis, renal failure)

5. fever (e.g. acute pyelonephritis)

6. history of urinary tract diseases

B) Clinical examination:
- pale ? (e.g. renal anaemia)
- Café-au-lait coloured skin ? (anaemia and deposits of urochromes in uraemia)
- uraemic foetor ?
- oedema ?
- arterial hypertension ?
- paraumbilical stenosis murmur? (e.g. in renal artery stenosis)
- pericardial rub? (e.g. as part of uraemia)
- soft heart murmurs + congested neck veins? (e.g. pericardial effusion in uraemia)
- tachypnoea and crackles? (e.g. a sign of alveolar pulmonary oedema in renal failure and hyperhydration)
- palpable renal tumour ? (e.g. Wilms-tumour, cystic kidneys etc.)

C) Laboratory investigations:

1. urine investigation:
▶ **first assessment:**
The urinary urochrome content, and hence the intensity of urine colour is reversely related to volume, and directly proportional to concentration:
- Thirst: dark amber-coloured + high specific gravity (maximal 1.035 g/l) or high osmolality (up to 1.200 mosm/kg)
- drinking large amounts of water: very light urine of low specific gravity (up to 1.001 g/l) or low osmolality (up to 50 mosm/kg)
  Exception: Diabetes mellitus: enhanced diuresis and light urine colour, but relatively high specific gravity due to glycosuria. Proteinuria can also increase specific gravity.

Urinary pH:
pH is 4,8 - 7,6 depending on nutrition.
- acid urine: high meat consumption, acidosis etc.
- Alkalic urine: vegetarian nutrition, secondary when the urine hasn't been processed fast enough for investigation, UTI caused by ammonium producing bacteria (Proteus), alkalosis

▶ **Proteinuria** [R80]:
The glomerular filtrate of a healthy kidney only contains low molecular proteins, 90% of which are re-absorbed in the proximal tubule. Proteinuria means excretion of > 150 mg protein/24 hours or a deviation of the physiological protein pattern. Microalbuminuria means an excretion of Albumin of 30 - 300 mg/24 hours or 20-200 mg/l (typical early symptom of diabetic or hypertensive nephropathy). In women there may be a weakly false positive urine analysis caused by vaginal discharge.
Orthostatic proteinuria: Minor proteinuria only at daytime, nocturnal urine is protein free (usually a harmless finding in young men)

Causes of proteinuria:

Proteinuria	Protein type	Causes
30 -300 mg/day 20 – 200 mg/l	Microalbuminuria	early diabetic and hypertensive nephropathy
up to 1,5 g/day	low molecular proteins: high molecular proteins:	tubulopathies minor glomerulopathies
1,5 - 3,0 g/day	low- and high molecular proteins:	chronic glomerulonephritis, transplanted kidney, nephrosclerosis
> 3,0 g/day	high molecular proteins:	nephrotic syndrome

a) Global protein test:
- dipsticks almost only test for Albumin; other proteins, like Bence-Jones-Protein (= L-chain in monoclonal gammopathy) will not be detected! Microalbuminuria can be detected via fast tests based on an immunological base. The usual protein dipsticks are only able to detect macroalbuminuria (> 200 mg/l).
- Biuret- and Trichloracetyl test will detect a larger variety of proteins.

b) Electrophoresis differentiation of proteinuria:
Division of urine proteins depending on molecular weight, using micro-sodium dodecyl sulphate-polyacrylamid-gel-electrophoresis (micro-SDS-PAGE). We can distinguish the following proteinuria patterns:

1) high molecular glomerular proteinuria (P.)
- Selective-glomerular P. (selecting proteins according to molecular size): mainly excretion of albumin (= marker protein of glomerular proteinuria) and transferrin
cause: minor glomerular damage, e.g. "minimal-change-nephritis"
- Unselective-glomerular P. (without size selection): excretion of IgG, Albumin
cause: severe glomerular damage

2) low molecular tubular proteinuria:
glomerular filtration and tubular re-absorption of low molecular β2-microglobulin. Tubular lesions will lead to increased urine levels.

3) Glomerular-tubular mixed proteinuria:
cause: glomerulopathies involving the tubules

4) Pre-renal proteinuria ("overflow"-proteinuria):
A surplus of light chains, myoglobin or haemoglobin can exceed the tubular re-absorption capacity.
- Bence-Jones-proteinuria = L-chain-excretion in monoclonal gammopathy
Quick test: salicylsulphonic acid test (cloudy urine when heated to 50 - 70 °C).
Differentiation: Immune electrophoresis
- Myoglobinuria (after muscle trauma)
Haemoglobinuria (haemolytic crisis)   } red/ brown urine

5) Postrenal proteinuria (detection of tubular secreted proteins (e.g. Tamm-Horsfall-protein)

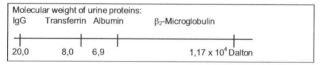

Molecular weight of urine proteins:
IgG   Transferrin Albumin β_2-Microglobulin

20,0   8,0   6,9   1,17 x 10^4 Dalton

▶ **Glycosuria [R81]:**
In diabetes mellitus the normal renal threshold for glucose (8,9 - 10,0 mmol/l) is exceeded by hyperglycaemia; renal glycosuria (in certain tubular renal diseases): the renal threshold for glucose is pathologically too low (glycosuria in the presence of normoglycaemia). Pregnancy can cause a physiological decrease in the renal glucose-threshold.

▶ **Sediment:**

a) Haematuria[R31]:
range: up to 5 erythrocytes/µl (= sensitivity limit of dipsticks)

Note: Dipsticks uses the peroxidative ability of haemoglobin and myoglobin for their detection. They can not differentiate between haematuria, haemoglobinuria and myoglobinuria! Positive dipsticks always require a microscopic examination.

- Microscopic haematuria: > 5 erythrocytes/µl, but no visible red discolouration
- Macroscopic haematuria: visible red discolouration
- Phase contrast microscopy:
Dysmorphic red blood cells = a sign of renal origin (glomerular disease): Akanthocytes: Ring shape with diverticulae („Mickey-Mouse-ears")

Iso- or eumorphic (morphologically normal) erythrocytes = a sign of postrenal origin.

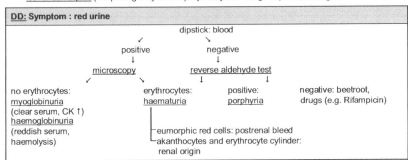

DD: Symptom : red urine

dipstick: blood

positive   negative

microscopy   reverse aldehyde test

no erythrocytes: erythrocytes: positive: negative: beetroot,
myoglobinuria   haematuria   porphyria   drugs (e.g. Rifampicin)
(clear serum, CK ↑)
haemoglobinuria
(reddish serum,   ─eumorphic red cells: postrenal bleed
haemolysis)   └akanthocytes and erythrocyte cylinder:
     renal origin

Causes of positive dipstick tests for blood:

▶ Haematuria:
1. women: contamination at time of period?
2. pre-renal: haemorrhagic diathesis, anticoagulants
3. renal: glomerulonephritis, pyelonephritis, hypernephroma, papillary necrosis, renal infarction, renal tuberculosis, trauma; familial benign haematuria (thin basement membrane disease; autosomal-dominant hereditary)
4. post renal: urolithiasis, tumours, cystitis, trauma

In approx. 15 % no cause for the microscopic haematuria can be found.

Visible red discolouration (macrohaematuria) occurs at 0,2 ml blood per 500 ml urine.

After a while the urine turns coffee-brown, only if pH is low: haemoglobin turning into acid haematin.

3-specimen-test: urine to be passed into 3 specimen bottles: beginning, middle and end of micturition.

Initial and terminal macrohaematuria (blood at the beginning and the end of micturition is likely to originate from the urethra; blood in all 3 specimen will rather originate from the bladder or even further proximal, clotting and colic-like pain indicate an origin proximal of the bladder – if in doubt, a bladder puncture will show if urine above the urethra contains any blood.

Note: Macrohaematuria should be investigated as long as bleeding is present! (US, CT, urography, cystoscopy → which kidney is bleeding?!).

Glomerular origin (e.g. glomerulonephritis) is likely if the following are noted:
- Akanthocytes under phase contrast microscopy
- simultaneous erythrocyte cylinders
- simultaneous high molecular proteinuria

▶ Haemoglobinuria[D82.3]:
intravascular haemolysis (incompatibility post blood transfusion, haemolytic crisis in haemolytic anaemia, termination of pregnancy using soap, march-haemoglobinuria after a long march etc.)
▶ Myoglobinuria: after muscle injury

b) Leucocyturia[R82.8]:
range: up to 10 leucocytes/µl → leucocyturia: > 10 leucocytes/µl
The detection limit of dipsticks is ca. 20 leucocytes/µl. Dipsticks give 40 % false positive rates
in women due to vaginal discharge (low specifity in women).
• a very high amount of leucocytes causes a cloudy urine: pyuria.
• leucocyturia is mainly found in UTI. Leucocyte cylinders are a sign of renal origin, usually in pyelonephritis.
• leucocyturia in sterile urine: pregnancy, UTI already receiving antibiotics, gonorrhoea, non-gonorrhoic and postgonorrhoic urethritis, urogenital-Tbc, Reiter-syndrome, analgesic nephropathy etc.

c) Epithelia:
Polygonal cells: usually of renal origin
Flat epithelium and flagellum epithelium: from the lower urinary tract (not significant)

d) Cylinder:
they are proof of renal tubule origin.
• Hyaline cylinders:
hyaline cylinders are a poor diagnostic tool with proteinuria; they can be found in healthy individuals (e.g. after vigorous exercise).
• Erythrocyte cylinders:
Pathognomonic for glomerulonephritis
• Leucocyte cylinders (positive peroxidase reaction):
can be found in chronic pyelonephritis in > 80 %

- Epithelium cylinders:
  These are coagulations of shed tubule epithelium, later converted into granulated cylinders and waxy cylinders. They are not-specific for any particular renal disease.
  cause: e.g. acute anuria, renal cirrhosis, nephrotic syndrome
  e) urine crystals (not significant)

▶ **Bacteriuria[N39.0]:**

urine sampling technique: mid stream urine (MSU): clean periurethral area with water, then catch the urine passed in the middle of micturition (cleaning/flushing effect !) into a vessel.
disadvantages: often contaminated.
advantages: screening test; sterile results are unequivocal.
2 MSU samples showing 10^5 pathogens/ ml ("figure of Kass" = "significant bacteriuria") are highly suspicious for UTI; it must be sent immediately to a microbiological laboratory for further investigation: use an appropriate transport medium, transport must be fast in refrigerated bottles!
Equivocal bacteriological results of MSU should be followed by a suprapubic bladder puncture.
Any bacteria found in suprapubic puncture urine is pathological (number of bacteria is not significant!).
Transurethral catheter urine should only be used, if suprapubic puncture is not possible (sterile technique!). The distal urethra is often not free from bacteria, hence catheter urine frequently is contaminated. Catheterization leads in 2 % to a UTI! This often can be avoided by using an invagination catheter.

2. substances excreted in urine

- Creatinine:
  Investigation techniques:
  1. Non-specific dye reaction (Jaffé-reaction)
  2. Specific enzymatic method. False high results in Jaffé-reaction can be seen in high blood sugar and ketoacidosis (diabetic coma) and haemolysis.
     Creatinine is derived from muscle from creatine phosphate metabolism, and healthy glomeruli manage to filter it almost completely. Apart from excessive consumption of meat, serum creatinine levels do not depend on nutrition, and they correlate with the glomerular filtration rate (see graphic).

  Note: serum creatinine levels do not rise above upper range limits (97 µmol/l), unless glomerular filtration rate (GFR) is reduced by more than 50%!
  Individuals with low muscle mass (children, women, older frail people) have lower levels. Hence, there may well be a slightly reduced GFR in older people, even if creatinine levels are still normal.
  Muscle trauma and acromegaly (increased muscle mass) may cause slightly increased creatinine levels with a normal GFR.
  Slightly increased serum creatinine is associated with 70 % higher risk of premature death (Cardiovascular Health Study).

- Urea:
  Tests: dye reaction or enzymatic (Urease method)
  Urea is the end product of protein metabolism. serum urea levels depends on various factors:
  - renal: Amount of glomerular filtrate and re-absorption of urea, which can vary between 40 % (diuresis) and 70 % (antidiuresis).
  - Extrarenal: increased protein intake and catabolism (fever, burns, cachexia) increase urea levels.
  Unless glomerular filtrate is below 25 %, the upper normal range limit of 8,3 mmol/l will not be exceeded.

- Cystatin C is produced by nuclear cells. Serum concentration of Cystatin C correlates with GFR. It is supposed to be more sensitive for early impaired renal function than creatinine clearance (but it is not a routine test).

3. Clearance methods:

Clearance: Plasma volume, which is cleared of a certain substance within a certain time via the production of urine.
To determine the GFR, certain substances are chosen, which are filtered in the glomeruli, but are not significant subjected to tubular secretion or re-absorption (Inulin, creatinine). Paraaminohippuric acid (PAH) is removed from plasma at a rate of 90% within a single renal passage, via glomerular filtration and tubular secretion. Hence it is a suitable marker for renal plasma flow. Clearance investigations do not contribute in any way to the differential diagnose of renal diseases. The importance of clearance measurements lies in their sensitivity and ability to diagnose functional problems. Minor renal impairment can be detected at an early stage, when all other investigations are still normal.

- measuring glomerular filtration:
  - Inulin clearance: this is a fairly cumbersome method (infusion of a test substance)
  - ^{51}Chrome-EDTA-clearance: apart from a minor deviation of 6% it corresponds with the Inulinclearance; range: 100 - 150 ml/min.
  - Creatinine clearance:
    advantage: simple (no infusion). Abnormal values even in early impaired renal function, when serum creatinine is still normal.

<u>Test performance:</u> collection of 24 hour-urine, measure creatinine concentration in serum and urine → calculation:

$$C \ (ml/min) = \frac{U \cdot UV}{S \cdot t}$$

U = urine creatinine concentration
S = serum creatinine concentration
UV = urine volume in 24 hours
t = collection time in minutes (24 x 60 = 1440)

glomerular filtration declines with increasing age.
range: = 110 (m) and = 95 (f) ml/min. These levels are valid up to 30 years of age; after that: deduct 10 ml/min for each decade.

serum creatinine (mg/dl)

Creatinine-
Clearance

1,3

50   100 (ml/Min.)

<u>Equation to estimate creatinine clearance (Cockroft and Gault):</u>

1. $ClCrea = \frac{(140 - age) \ x \ bodyweight(kg)}{Factor \ x \ serum \ creatinine}$

Factor = 72 if creatinine is being measured in mg/dl
Factor = 0,82 if creatinine is being measured in μmol/l
For women multiply the result by 0,85.

<u>Calculation of glomerular filtration rate (GFR) using the MDRD-equation:</u>
GFR (ml/min/1,73m^2) = 186 x (serum creatinine)$^{-1,154}$ x (age)$^{-0,203}$

(Internet-calculator: *www.nephron.com/mdrd/default.html*)

For women multiply with the factor 0,742, for black patients multiply with factor 1,21.
This equation produces accurate results, when GFR is below 60 ml/min/1,73 m² body surface. It is much less accurate, if the GFR is higher.

Note: serum creatinine alone is not a valid marker for glomerular filtration rate.

• <u>measuring the effective renal plasma flow:</u>

Paraaminohippuric acid-(PAH-clearance)   Normal: 500 - 800 ml/min

Filtration fraction = $\frac{Inulin \ clearance}{PAH- \ clearance}$ Normal: 0.16 – 0,21

Filtration fraction is the part of renal plasma flow, which is filtrated in the glomeruli. Reduced filtration fraction can be found in glomerular diseases (e.g. acute glomerulonephritis); increased values are a sign of vascular renal disease.

4. Immunological investigations, e.g.:
- ADB- and ASL-titre in suspected acute post streptococci-glomerulonephritis
- decreased C3 in various types of glomerulonephritis
- Anti-ds-DNA-antibodies in systemic lupus erythematosus
- Anti-basal membrane antibodies (Anti-GBM-antibodies) in Goodpasture-syndrome
- c-ANCA (= Anti-proteinase 3-antibodies) in Wegener's granulomatosis
- p-ANCA (= Anti-myeloperoxidase-antibodies) in microscopic panarteriitis
- Anti-C3-convertase-antibodies in membrano-proliferative glomerulonephritis/type II
- detection of monoclonal immunoglobulines in multiple myeloma

D) Radiology diagnostics:
▶ (Doppler duplex-)sonography:
- position and size of kidneys (normal length: 9 - 13 cm in adults)
- cysts, calculi, tumours
- obstructions of the collecting system
- for close monitoring (no radiation exposure)
- to assess arterial/venous perfusion
- for guided renal puncture

▶ X-ray:
• plain radiographs (also tomography): calcified stones?
• Intravenous urography has been replaced by CT (fewer side effects, better quality of pictures):
- size/shape of kidney?
- anatomical anomalies?
- malformation of the collecting system in pyelonephritis?
- papillary defects in analgesics nephropathy, shift of the collecting system in tumours
- obstructions, stones?
- bilateral equal excretion ? (early/late radiographs)

- micturition-cysto-urethrography (vesico-ureteric-renal reflux, residual urine?)

  Contraindications for use of iodine contrast medium:
  1. iodine allergy (therapy of an anaphylactic shock: see "shock")
  2. thyrotoxicosis
  3. IgM-paraproteinosis (risk of renal failure)
  4. hepatic- and renal failure (when serum creatinine levels are > 3 - 4 mg/dl, there is only minor renal contrast medium excretion!)

Note: In order to avoid any renal complications, patients should receive sufficient hydration before the application of a contrast medium! Additional administration of oral Acetylcystein in patients with impaired renal function is supposed to have a prophylactic effect.

- angio-MRI or -CT, intravenous digital subtraction angiography (DSA), arteriography: renal artery stenosis? vascularisation of a tumour? renal vein thrombosis?
- computer tomography: tumours? abscess?
- MRI and MR-urography if contrast medium is contraindicated

▶ **Radionuclide studies:**
Dynamic renal function scintigraphy using ^{99m}Technetium-MAG and ^{123}Iodine-Hippuran
Indication: Identification of ectopic renal tissue; allergy to contrast medium, separate functional diagnostic for either side; obstruction in mobile kidneys?(examine sitting and lying patient), reduced perfusion in anuric or transplanted kidneys?
Simultaneous administration of an ACE-inhibitor can detect renal artery stenosis requiring treatment. Bladder activity can be used for reflux diagnostic.

E) Renal biopsy:
Useful to differentiate between various glomerular diseases, and in suspected transplant rejection
CI: patient only has one kidney, clotting disorder, malignant hypertension etc.

GLOMERULONEPHRITIS (GN)

Def.: The term glomerulonephritis describes a variety of immunological diseases, causing intraglomerular inflammation and cellular proliferation.
- Primary GN: diseases of the glomerulus occurring in isolation, without any signs of a systemic disease.
- Secondary GN: renal involvement in various systemic diseases: e.g. connective tissue disorder, vasculitis, endocarditis lenta etc.

Pathology cannot distinguish between primary and secondary GN; this is only possible using histology, clinical symptoms and serological markers.
Apart from glomerulonephritis, there is also non-inflammatory glomerular disease – e.g. in amyloidosis, diabetes mellitus (diabetic glomerulosclerosis), eclampsia etc.

Ep.: Most patients suffering from glomerulonephritis will develop chronic nephropathy which is associated with the risk of premature cardiovascular disease and progressive renal failure. At a rate of 15% glomerulonephritis is the second most common cause of terminal renal failure, coming after diabetic nephropathy, which contributes to 35%. Many cases of glomerulonephritis follow a mild, asymptomatic course, often unnoticed by the patient, and hence remaining undiagnosed

Pg.: Humoral and cell mediated immune mechanisms play a role in the pathogenesis of glomerular inflammation.
1. Anti-GBM-antibody glomerulonephritis: linear deposits of IgG against Goodpasture-antigen. This auto-antigen is a normal part of the non-collagen containing domain of the Alpha3-chain of type IV collagen.
2. Immune complex glomerulonephritis: Immune complexes are distributed over the whole glomerular capillary wall, like in lupus nephritis or post infectious glomerulonephritis.
3. Anti-neutrophil cytoplasmatic antibody-(ANCA)-associated glomerulonephritis: ANCA induce glomerular damage via interaction with components of the neutrophil granules.
4. Activation of cell mediated immunological processes can induce glomerular damage. In humans T-cells have been found in proliferative and non-proliferative glomerulopathy.

After initial glomerular damage, a number of pro-inflammatory mediator systems are activated in infiltrating cells and in glomerular cells: Complement activation, influx of circulating leucocytes, cytokine synthesis, release of proteolytic enzymes, activation of coagulation cascade and production of pro-inflammatory lipid mediators. In proliferative glomerulopathies there is a rise in the number of glomerular cells, and proliferation of glomerular cells as a reaction to growth factors (e.g. epidermal growth factor (EGF) and platelet-derived growth factor (PDGF). Proliferating cells are mesangial cells and endothelial cells.

According to the clinical course, we can distinguish the following syndromes of GN:

I. Asymptomatic abnormalities of the urine sediment:

Def.: Macroscopic or microscopic asymptomatic haematuria with normal glomerular filtration rate, systemic disease affecting the kidneys has been excluded. Many, but not all patients with

asymptomatic haematuria also present with proteinuria, which is usually < 1,5 g/day. There is no arterial hypertension.

cause: 1. IgA-nephropathy (Berger disease)
2. thin basement membrane disease
3. Alport-syndrome
4. benign isolated proteinuria
5. idiopathic transient proteinuria
6. functional proteinuria caused by fever, cold, emotional stress, cardiac failure,
sleep apnoea-syndrome
7. positional proteinuria (usually < 2g/24 hours), which is only present in an upright position.

Proteinuria 4 – 7: Biopsy reveals a normal renal parenchyma or minor unspecific changes of the podocytes or the mesangium. They all have a good prognosis.

II. Acute nephritic syndrome and rapidly progressive glomerulonephritis:
1. Acute post infectious glomerulonephritis
2. Rapidly progressive glomerulonephritis (RPGN)

III. Chronic-progressive glomerulonephritis:
There is only minor malaise, but erythrocyturia, proteinuria and in most cases also hypertension and a slowly progressing renal failure. Renal biopsy is not indicated at this stage anymore, because results won't change therapeutic management.

Di.: Basic investigations (clinical symptoms, laboratory test) can be performed as an outpatient. renal biopsy is the gold standard to classify the presenting glomerulonephritis. Histology results will decide therapy and prognosis.

IgA-NEPHROPATHY – BERGER'S DISEASE   [N02.8]

Ep.: Most common form of idiopathic glomerulonephritis (15-40% of all cases of primary idiopathic glomerulopathy).

Aet.: • Idiopathic
• Secondary: hepatic diseases, chronic-inflammatory bowel diseases, psoriasis, sarcoidosis, SLE, RA etc.

PPh.: Diffuse or focal-segmental mesangio proliferative glomerulonephritis; immune histology: mesangial deposits of IgA1.

Sym.: 1-3 days after unspecific upper respiratory tract infections there may be intermittent macroscopic haematuria, which will resolve spontaneously. In most cases there is an asymptomatic microscopic haematuria with or without proteinuria. 40 % of all patients (and all patients suffering from renal failure) will present with arterial hypertension.

Lab.: urine results: Sediment: Erythrocyte cylinders and dysmorphic erythrocytes in phase contrast microscopy; unselective glomerular proteinuria (in most cases < 3 g/day).
Up to 10 % of all patients have nephrotic syndrome.
Raised serum IgA-levels (40 %)

DD: Acute post infectious GN (2 - 3 weeks post infection, e.g. Streptococci)

Di.: symptoms and biopsy

Th.: only symptomatic:
• Patients with proteinuria < 1 g/24 hours and normal serum creatinine don't need any special therapy. Prophylaxis for respiratory infections may reduce the frequency of macroscopic haematuria episodes. For hypertension give ACE-inhibitors or A2-blocker (target: < 130/80 mm Hg).
• Patients with proteinuria > 1 g/24 hours with or without hypertension: therapy with ACE-inhibitors and/or A2-blocker.
• In case of proteinuria > 1 g/24 hours and progressive renal failure: therapy with corticosteroids, may have to be high dose following the Pozzi-scheme. Azathioprin or Cyclophosphamid may be used.

Prg.: The grade of proteinuria is highly significant for prognosis:
Proteinuria > 3 g/24 hours: loss of renal function of 9 ml/min GFR per year
Proteinuria 1,0 - 3,0 g/24 hours: loss of GFR at 6 - 7 ml per year.
Within 25 years of diagnosis, ca. 25 % will develop terminal renal failure. Post kidney transplantation the relapse rate is ca. 40 %.

Thin Basement Membrane Disease (benign haematuria)

Ep.: as frequent as IgA-nephropathy in patients with asymptomatic haematuria.

Aet.: Family history (hereditary or sporadic). In hereditary cases, this is an autosomal dominant condition with a genetic deficiency of the alpha4-chain of type IV collagen.

Sym.: Persistent haematuria, occasionally intermittent haematuria with exacerbation triggered by URTI.

Di.: symptoms and biopsy
Light- and immune fluorescence microscopy show a normal kidney. Electronic microscopy reveals a conspicuously thin glomerular basement membrane, (in most cases < 300 nm in adults).

Th.: no specific therapy

Prg.: generally good prognosis, but a small percentage can develop hypertension and progressive renal failure. For these patients the use of ACE-inhibitors is advisable.

Alport-Syndrome (Hereditary nephritis)

Ep.: most common hereditary nephritis

Aet.: X-chromosome dominant hereditary. The genetic defect is found within the gene for the alpha5-chain of type IV-collagen of the glomerular basement membrane. Significant genetic heterogeneity, which can be seen in the phenotype variations of this disease.

Sym.: Male patients: in most cases microscopic haematuria, proteinuria and progressive renal failure.
Female carriers: in most cases just minor disease without renal failure.

There are also autosomal dominant and autosomal recessive forms, where the mutation is located on the gene for the alpha3-chain of type IV-collagen. Men and women are equally affected.

Extrarenal manifestations:
Neurosensorial deafness (60 %) and bilateral anterior lenticonus = lens anomaly (up to 30 %) and relapsing corneal lesions

Di.: symptoms and biopsy (electron microscopy: thickening, fragmentation and lamella formation of the lamina densa of the glomerular basement membrane).

Prg.: Men suffering from this disease usually develop terminal renal failure. There is no causal therapy, therefore treatment is just symptomatic. After kidney transplantation ca. 5% will develop anti-GBM-disease in the transplanted kidney.

ACUTE POSTINFECTIOUS GLOMERULONEPHRITIS [N00.9]

Aet.: Immune complex nephritis particularly after infection with ß-haemolytic streptococci group A (often type 12) = acute post streptococcal- GN, rarely other infections.

PPh.: Endocapillary diffuse proliferative GN: swelling and multiplying of mesangium cells and endothelial cells, detachment of the endothelium from the basement membrane; severe narrowing of the capillary lumen; there may be antigen-antibody-complexes or C3-complement seen as "humps" on the external side of the basement membrane; leukocyte-/monocyte infiltrations.

Pg.: Immune complex nephritis

Sym.: Convalescence after a Streptococci infection (pharyngitis, tonsillitis, skin infections) is suddenly interrupted: After a symptom-free interval of 1 - 2 weeks, the patient feels ill again. 50% run an asymptomatic course and go undiagnosed or are just coincidentally diagnosed.

Leading symptoms:
- always: microscopic haematuria + proteinuria (< 3 g/24 hours)
- sometimes: oedema, hypertension
  Volhard' Triad: haematuria, hypertension, oedema
- other possible symptoms:
  - macrohaematuria (red-brown urine)
  - facial oedema, headache, myalgia, fever
  - loin pain
  - epileptic fits, somnolence (intracranial oedema)
  - hypertonic crisis dyspnoea and pulmonary oedema

Lab.: - urine: erythrocyturia, erythrocyte cylinders, proteinuria (< 3 g/24 hours); unselective excretion of high molecular proteins

> **Note:** Infectious haematuria as it can be frequently seen in scarlet fever and other infectious diseases is relatively harmless, and resolves spontaneously. Post infectious haematuria after streptococci infections however are a sign of serious GN!
> Erythrocyturia can be equivocal, erythrocyte cylinders are likely to be a sign of GN.

- ASO-titre is raised in 50 %
- Anti-DNAse-B = ADB-titre is raised in 90 % in streptococcal skin infections
- serum complement (C3) is reduced during the 1. week
- urea, creatinine can be slightly raised

Sonography: Relatively large swollen kidneys

DD: Rapidly progressive GN (urea, creatinine ↑), IgA-nephritis (macrohaematuria)

Di.: history + clinical symptoms + laboratory, maybe biopsy (ind: increased urea/creatinine → to exclude a rapid progressive GN)

Th.: acute GN:
1. bed rest, avoid physical exertion; low salt/low protein diet, close monitoring of weight/laboratory results

> **Note:** check creatinine at least twice a week: when creatinine rises above normal levels, this is an indication for biopsy, to exclude rapidly progressive glomerulonephritis!

2. treat any Streptococci infection with Penicillin: 3 Mega IU/day for 10 days. There is no clear evidence for the therapeutic benefit of tonsillectomy under Penicillin cover during a symptom-free interval.
3. treat any complications: e.g.
   If there are signs of fluid
   Any signs of fluid collection (weight gain, oedema, rise in CVP, pulmonary congestion, intracranial oedema, hypertension, oliguria): reduce sodium and water intake, and reduce output by giving loop diuretics, e.g. Furosemide); treat hypertension, e.g. with ACE-inhibitors.
4. Patient follow up for several years (to detect any chronic course).

Prg.: 1. cure: children: > 90%

   adults: 50%

2. ongoing urinary symptoms (e.g. microscopic haematuria, proteinuria): further follow up will show if there is any deterioration of renal function (progressing to the chronic stage) or if these are non-significant residual symptoms without impaired renal function.
3. rarely death due to acute complications (e.g. hypertonic crisis and subsequent left ventricular failure and pulmonary oedema)

RAPIDLY PROGRESSIVE GN   [N01.9]

Def.: Relatively rare GN with rapid progressive deterioration of renal function. Without treatment the glomerular filtration will drop by 50 % in 3 months, and terminal renal failure will develop within 6 months.

PPh.: Extracapillary proliferative GN with formation of diffuse crescents (> 50 % of glomeruli), sometimes necrotising vasculitis.

Ep.: < 1/100.000/year

Aet.: 1. Symptomatic RPGN: renal manifestation of a vasculitis (e.g. Wegener's granulomatosis)
2. Idiopathic RPGN

Types:
- Type 1 (ca. 10 %): Anti basement membrane-RPGN: Serological detection of antibodies against glomerular basement membrane (GBM-antibodies). Histological detection (using immune fluorescence) of IgG and C3-complement as linear deposits on the glomerular basement membrane (linear immune fluorescence).
  - without pulmonary involvement (rare)
  - with pulmonary involvement = Goodpasture-syndrome: Related antigens (alveolar and glomerular basement membrane (C-terminal domain NC1 of the α3-chain of type IV-collagen) are the cause for a combination of RPGN + pulmonary haemorrhage (haemoptysis, crackles, shadows on CXR), a very rare disease, mainly affecting men < 40 years.
- Type 2 (ca. 40 %): Immune complex-RPGN: Granular deposits of immune complexes on the glomerular basement membrane, sometimes in form of humps (granular immune fluorescence).
  Cause: often post infectious, also SLE (anti-DNS-antibodies; see "lupus nephritis"), Schönlein-Henoch-nephritis

- Type 3 (ca. 50 %): ANCA-associated vasculitis (without deposits of immunoglobulines or complement)
  - renal involvement of a microscopic polyangiitis (mPA)
    Syn: microscopic PAN (mPAN)
    Detection of antineutrophil cytoplasmatic antibodies with perinuclear fluorescence pattern (p-ANCA) often the target antigen is myeloperoxidase: anti-myeloperoxidase-antibody (MPO-ANCA)
  - renal involvement of Wegener's granulomatosis
    Detection of antineutrophil cytoplasmatic antibodies with cytoplasmatic fluorescence pattern (cANCA) = antiproteinase-3-antibodies (PR3-ANCA)

Sym.: - pale patient, hypertension, often significant proteinuria, occ. nephrotic syndrome, CRP + ESR ⇑
- rapidly progressive renal failure; US reveals kidneys of normal size
- additional pulmonary haemorrhage in Goodpasture-syndrome
- detection of circulating anti-GBM-antibodies (type 1), circulating immune complexes (type 2), cANCA or pANCA (type 3)

DD: • acute renal failure (history of trigger event)
• acute sterile interstitial nephritis (history of trigger drugs)

Di.: • symptoms of rapidly rising urea/creatinine, immunological diagnostics
• renal biopsy (absolute indication for biopsy)

Th.: A rapid decline of the GFR is to be considered a medical emergency, justifying an immediate biopsy. Immediate immune suppressive therapy will improve the outcome!

Type 1: Anti-GBM-RPGN:
Plasma exchange for 2-3 weeks daily or every other day until anti-GBM-antibody serum concentration is low enough and stable; give 1 g Methylprednisolone/day for 3 days, then reduce steroids gradually. Oral Cyclophosphamid 2 mg/kg; after 2 - 3 months change to less toxic medication like Azathioprin. Therapy lasts about 6 - 9, max. 12 months.

Type 2: Immune complex - RPGN:
Methylprednisolone 1 g per day for 3 days i.v. with subsequent oral steroid treatment (gradual dose reduction) plus Cyclophosphamid as bolus (500 mg/m^2 body surface (day 1 and repeat after 28 days for 6 months), followed by another biopsy in order to plan further treatment.

Type 3: ANCA-associated RPGN:

	serum-creatinine (mg/dl)	organs at risk	remission induction	remissions maintenance
Limited	< 1,4	no	CS and/or MTX	CS and/or MTX
early generalised	> 1,4	no	CS + MTX or CS + CYC	CS + AZA or CS + MTX
active generalised	< 6,0	yes	CS + CYC	CS + AZA
Severe	> 6,0	yes	CS + CYC + PP	CS + AZA

CS = Corticosteroids; MTX = Methotrexat; CYC = Cyclophosphamid; AZA = Azathioprin; PP = Plasmaphoresis

Prg.: Early therapy start (i.e. in the presence of some remaining renal function) will lead to improved renal function in > 60 % of all cases!
Type 1/anti-basement membrane-RPGN is self limiting and does not relapse; type 2 and 3 however are prone to relapse, and hence require longer treatment.

NEPHROTIC SYNDROME   [N04.9]

Def.: - significant proteinuria (> 3 - 3,5 g/day)
- hypoproteinaemia
- hypalbuminaemic oedema (serum albumin < 2,5 g/dl)
- hyperlipoproteinaemia (raised cholesterol and triglycerides)

Aet.: 1. glomerulonephritis associated with nephrotic syndrome:
• **Glomerular minimal lesions** = **minimal change-glomerulopathy (disease)** = **MCD**: Most common cause of nephrotic syndrome in children. In adults it is the cause for nephrotic syndrome in 20 %.
Aet.: a) idiopathic - b) secondary in malignant diseases, NSAID, food allergy, post vaccination
Path.: light microscopy: normal; electronic microscopy: Diffuse extinction of foot processes of visceral epithelial cells.

- **Focal segmental glomerulosclerosis (FSGS)**:
  Ep.: Cause of nephrotic syndrome in ca. 15 % (cause for nephrotic syndrome in ca. 50 % in black adults).
  Aet.: -idiopathic: in 30 % genetic. Congenital podocyte disease:

disease	location	hereditary mode	gene	protein
congenital nephrotic syndrome	19q13	recessive	NPHS1	Nephrin
Steroid resistant nephrotic syndrome	1q25-31	recessive	NPHS2	Podocin
familial FSGS	19q13	dominant	ACTN4	Alpha-Actinin-4

  - secondary:
    - Heroin abuse
    - HIV-associate nephropathy
    - morbid obesity
    - vesico-ureteric reflux
    - post cholesterol embolism
    - malignant diseases
    - in all chronic nephropathies associated with a > 70% loss in nephrons (glomerular hyperfiltration and intraglomerular hypertension)
  Histology: Segmental glomerular changes, preceded by a loss in podocytes of the capillary basement membrane. Adhesions between glomerular capillary loops and Bowman's space. Special form: Collapsing FSGS (idiopathic or due to HIV-infection): collapse and sclerosis of glomerular capillaries: poor prognosis

- **Membranous GN (25 %) [N05.2]**:
  Most common cause of nephrotic syndrome in adults.
  Aet.: a) idiopathic (75 %) - b) secondary (25 %): infectious diseases (Hepatitis B and C, HIV, Syphilis, Malaria), autoimmune diseases (e.g. SLE) , malignant diseases, medication (e.g. Gold or Penicillamin) etc.
  Pg.: immune complexes and complement-C5b-9-complexes (detectable in urine).
  Hist.: swelling of the glomerular basement membrane, due to subepithelial deposits of immune complexes and complement on the outside of the glomerular basement membrane; formation of protruding 'spikes' between immune complex-deposits, originating from the basement membrane; 4 stages, where in stage IV the immune complexes are completely encapsulated by basement membrane. Immune histological granular deposits of IgG4 and complement C3 and C5b-9 along the glomerular capillary loops.
  Sym: Nephrotic syndrome (80 %), microscopic haematuria (50 %), hypertension (25 %)

  3 risk groups:
  High risk (poor prognosis):
  Male; > 50 years; proteinuria > 8 g/day; arterial hypertension; histological interstitial fibrosis and pronounced glomerulosclerosis; reduced GFR.
  Medium risk:
  Proteinuria > 3,5 g/day but < 8 g/day for a period of >6 months. Normal or almost normal serum creatinine, normal or almost normal endogenous creatinine clearance.
  low risk (good prognosis):
  Young age < 16 years; female; normal blood pressure; proteinuria < 3,5 g/24 hours; normal GFR; no tubulointerstitial fibrosis.

- **Membranoproliferative GN = MPGN = Mesangio-capillary glomerulonephritis [N05.0]**:
  Ep.: rare, affects children and young adults
  Aet.: - idiopathic
    - secondary: malignant diseases, autoimmune diseases (e.g. SLE), infectious diseases (Hepatitis B or C etc.), complement deficiencies etc.
  Pa.: typical for MPGN: Interposition of a mesangial matrix and mesangial cells into the subendothelial space of the peripheral glomerular capillary loops: „Mesangio-capillary" GN. Double contour of the basement membrane and C3-deposits.
  Histology of idiopathic MPGN:
    - type I (80 %): Subendothelial and mesangial deposits of immune complexes;
    relapse post kidney transplant: 30 %
    - type II (> 15 %): dense intramembranous deposits ("dense deposit disease") and significant swelling of the basement membrane. serum: detection of anti-C3-convertase-antibodies. Relapse post kidney transplantation: 100 %.
    - type III: Variation of type I (rare)
  Pg.: type I and II: characteristic systemic consumption of complement; this is due to immune complex and IgG-autoantibodies against C3-convertase of the alternative complement path („nephritic factors").
  Sym.: Nephrotic syndrome, hypertension, haematuria and persistent hypocomplementaemia (type I: C3 + C4 ↓; type II: only C3 ↓)

112

2. <u>Diabetic nephropathy</u> (refer to according chapter)

3. <u>Rare causes:</u> Multiple myeloma, amyloidosis, renal vein thrombosis etc.

Pg.:

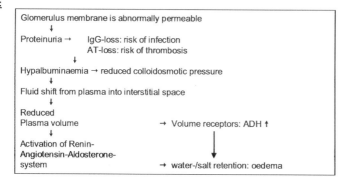

Glomerulus membrane is abnormally permeable
  ↓
Proteinuria → IgG-loss: risk of infection
      AT-loss: risk of thrombosis
   ↓
Hypalbuminaemia → reduced colloidosmotic pressure
  ↓
Fluid shift from plasma into interstitial space
  ↓
Reduced
Plasma volume   → Volume receptors: ADH ↑
  ↓
Activation of Renin-
Angiotensin-Aldosterone-
system   → water-/salt retention: oedema

Synthesis of albumin does play a role; the extent of hypalbuminaemia is determined by the balance between proteinuria and albumin synthesis.

Sym.:
- 4 leading symptoms of nephrotic syndrome (as above)
- symptoms of the underlying disease
- maybe acquired IgG-deficiency with its associated tendency to infections (severe protein loss)
- advanced stages: symptoms of renal failure, maybe hypertension
- thromboembolic complications (renal loss of antithrombin)

Lab.:
- <u>serum electrophoresis:</u> albumin and γ-globulin ↓,
  relative rise of α2- and ß-globulins
- renal failure: urea, creatinine ↑, creatinine-clearance ↓
- occ. IgG and antithrombin (AT) ↓
- cholesterol + triglycerides ↑
- <u>urine:</u>
  High specific weight due to high protein level; glomerular permeability for proteins of various molecular size can be determined via differential protein-clearance.

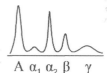

$A\ \alpha_1\ \alpha_2\ \beta\ \ \gamma$

  Relatively <u>"low molecular" proteinuria</u> is called <u>"selective proteinuria"</u> unlike <u>"non-selective"</u>, where <u>high molecular proteins</u> (up to beta-lipoproteins – molecular weight 25×10^5 Dalton) are excreted. Patients presenting with non-selective proteinuria and high excretion of beta-lipoproteins and alpha2-macroglobulins don't respond to steroids. Their glomeruli already show severe damage of the basement membrane in light microscopy.

Di.:
- symptoms + laboratory results
- renal sonography
- <u>renal biopsy:</u> histology is required for diagnostic, therapeutic and prognostic reasons! Exception: MCD in children (responds very well to corticosteroids).

Th.:
A) treat underlying disease, remove any toxic cause, antiviral treatment of hepatitis C

B) symptomatic treatment
- ▶ **General:**
  - rest
  - <u>diet:</u> low protein diet (0,8 g/kg/day) and low salt diet (ca. 3 g NaCl/day)
  - <u>diuretic therapy:</u> combination of potassium sparing diuretic + Thiazide. If diuretics have no effect anymore and there is oedema: combination of Thiazide + loop diuretic; <u>monitor electrolytes</u> (particularly K^+ and Na^+) and fluid intake/output.

    > **Note:**
    > Overdose of diuretics will cause hypovolaemia, hyponatraemia and secondary hyperaldosteronism, leading to reduced efficacy of diuretics. AT-deficiency increases the risk of thrombosis → take great care when treating oedema + start thrombosis prophylaxis using low dose Heparin and supportive stockings. Give Warfarin/Cumarin for thromboembolic complications.

  - severe life threatening oedema: rise the colloidosmotic pressure gradually with infusions of hyperosmolar low salt human albumin solution.
  - <u>bacterial infections:</u> antibiotics + immune globulin infusion.
    vaccination against pneumococci and influenza virus.

- treat hypercholesterolaemia (Statins)
- aggressive treatment of hypertension (treat already borderline hypertension), since this causes additional renal damage! target: < 130/80 mm Hg. Use preferably ACE-inhibitors or ATII-blocker.
- monitoring of proteinuria, renal function and blood pressure

▶ **Special:**
Immunosuppressive therapy is only indicated, as long as renal function is largely preserved (creatinine < 2 mg/dl).

- **Glomerular minimal lesions (MCD):**
  - corticosteroids (therapy success up to 90 %, particularly children). Once corticosteroids have been stopped, ca. 50 % will relapse within 6 - 12 months → second steroid therapy.
  - In case of frequent relapses or unacceptable side effects of steroids: give Cyclosporin A or Cyclophosphamide. Second choice: Mycophenolatmofetil or Rituximab
  Minimal changes-nephropathy can settle spontaneously and usually doesn't cause terminal renal failure.

- **Focal-segmental glomerulosclerosis (FSGS):**
  - Patients presenting with proteinuria < 2 g/24 hours:
    ACE-inhibitors or ATII-blocker
  - Patients presenting with nephrotic syndrome but fairly good renal function:
    Prednisolone + Cyclosporin as an initial therapy for patients who can't tolerate high corticosteroid doses; the remission rate is up to 70 %. Cyclosporin should not be given in cases of advanced renal failure and biopsy showing pronounced interstitial fibrosis.
    Alternative drug in cases of steroid resistance and Cyclosporin toxicity: Mycophenolatmofetil
  - Recurrent FSGS of the transplant kidney: Plasmapheresis or protein absorption
    Prg.: Patients presenting with massive proteinuria (> 14 g/day) will develop terminal renal failure after 2 - 3 years. Those without nephrotic syndrome still have functioning kidneys after 10 years in 85 %. Post kidney transplantation the relapse rate is ca. 30 %.

- **Membranous GN :**
  Causal therapy for secondary forms
  low risk: strict blood pressure control (target < 125/80 mm Hg): ACE-inhibitors, ATII-blocker etc.
  Intermediary risk: Protein restriction to 0,8 g protein/kg body weight/day
  Immunosuppressive therapy, if creatinine rises > 1,5 mg/dl.
  high risk:
  - serum albumin < 25 g/l: prophylactic anticoagulation-therapy (risk of thromboembolism is ca. 50 %)
  - Immunosuppressive therapy : Corticosteroids + Chlorambucil (Ponticelli-scheme) or Cyclophosphamide. Alternative: Prednisolone + Cyclosporin (not in case of significantly reduced renal function and/or interstitial fibrosis).
  Reserve drug for therapy resistant nephrotic syndrome: Rituximab
  Pg.: 30 % spontaneous remissions, 35 % partial remissions with a stable renal function for several years, 25 % progression into renal failure, 10 % die from extrarenal causes.

- **Membrano-proliferative GN = MPGN = Mesangio-capillary GN:**
  Idiopathic MPGN: Children- try corticosteroids, adults-try Acetylsalicylic acid + Dipyridamol.
  Hepatitis B: treat the underlying disease.
  Hepatitis C-associated GN: see further down
  Prg.: 1/3 good with normal renal function, 1/3 slow progression, 1/3 rapid progression into terminal renal failure

- **Hepatitis C-associated GN :**
  1. Patients with moderate proteinuria and non- rapidly progressive renal failure :
     Antiviral therapy of Hep C + symptomatic therapy:
     Blood pressure control with ACE-inhibitors and/or ATII-blocker
  2. Patients with nephrotic syndrome and/or progressive renal failure:
     - Antiviral therapy of Hep C + symptomatic therapy: Furosemide, blood pressure control (as above)
     - Immunosuppressive therapies:
       · Plasma exchange (3 l: 3 x/week for 2 - 3 weeks)
       · Pulsed corticosteroid therapy (1 g Prednisolone for 3 days)
       · Cyclophosphamide or Rituximab

CHRONIC PROGRESSIVE GN-SYNDROME [N03.9]

Def.: Chronic stage of various glomerulopathies. In most cases there is no history of acute GN.

Sym.: - insidious onset
- haematuria (non-haemolysed blood), proteinuria
- occ. nephrotic syndrome
- hypertension
- symptoms of a slowly progressive renal failure

Di.: history / symptoms
renal biopsy is not indicated because it has no therapeutic consequence.

Th.: There is no causal therapy, hence special medication is not indicated (steroids, NSAID, immunosuppressants). Treatment is symptomatic, good blood pressure control being the most important part (refer to chapter "chronic renal failure").

Prg.: No cure; progression into terminal renal failure

UTI AND INTERSTITIAL NEPHRITIS (IN)

summary:

- UTIs:
  1. Asymptomatic bacteriuria
  2. Acute cystitis
  3. Acute pyelonephritis
  4. Chronic pyelonephritis

- Urethritis
- Hantavirus-infection

- non-bacterial tubulo-interstitial nephritis:
  - acute abacterial IN
  - analgesics-nephropathy
  - Balkan-nephritis

UTI (URINARY TRACT INFECTION) [N39.0]

Def.: Presence of infectious pathogens in the urinary tract. An isolated infection of the urethra distal of the internal urethral sphincter has to be distinguished from any infections higher up the urinary tract (refer to chapter „Urethritis"). A true UTI must be distinguished diagnostically from bacterial urine contamination, which is caused by a faulty urine collection method.

Ep.: Ca. 5 % of all adult women have asymptomatic bacteriuria. 30% of all pregnant women with untreated asymptomatic bacteriuria will develop acute pyelonephritis during pregnancy.
1 in 2 women will suffer cystitis at least once in their lifetime (the most common cause for certified days off sick in women). UTIs contribute to the 3 most frequent nosocomial infections (after wound infections and pneumonia).
The high frequency of UTIs in women is due to the short urethra being located next to the contaminated anal area.
The first peak of UTIs is in infants and toddlers, often due to vesico-ureteric reflux (always consider pyelonephritis in a case of fever of unknown origin or unexplained anaemia in young children!).
In adults pregnant or postnatal women are of increased risk (also recently married women → "honeymoon-cystitis"!).
Older women are more prone to UTIs.

Men usually get more UTIs as they get older; in most cases the cause is an obstruction (e.g. prostate disease).

Aet.: A) Predisposing risk factors:
1. urinary tract obstructions
   - anatomical anomalies of the kidneys and the urinary tract
   - obstructions (stones, tumours, prostate adenoma, urethral strictures, urethral valves)
   - impaired bladder function (paraplegia and other neurological impairment)
   - vesico-ureteric-renal reflux (VUR) [N13.7]:
     Def.: Usually the submucosal vesical ureteric opening is kept closed by the intravesical pressure. In case of VUR the submucosal route of the ureter is too short, and the openings are laterally located. This results in an insufficient valve mechanism.

- primary (congenital) VUR: 40 % of children suffering from relapsing UTIs (w:m = 4:1)
- secondary (acquired) VUR caused by infravesical obstructions or innervation disorders of the bladder.
  2. analgesic abuse
  3. metabolic disorders (diabetes mellitus, gout, hypercalcaemia, hypokalaemia)
  4. invasive investigation of the urinary tract and urinary catheter associated UTI:
     the risk of a nosocomial (acquired in hospital) UTI in patient with a transurethral catheter is ca. 4 % per day.
  5. immune deficiency, immunosuppressive therapy
  6. pregnancy
  7. other trigger factors:
     - soaking, hypothermia (incl. cold feet)
     - sexual activity ("honeymoon-cystitis" in women)
     - poor fluid intake and loss of fluid leading to oliguria

B) pathogens of UTIs:
  acute pyelonephritis: in most cases monoinfection usually caused by E. Coli
  chronic pyelonephritis and nosocomial UTIs, postoperative and post investigation (urinary tract): mixed infections are more common

  Percentage frequency of bacteria
- acute uncomplicated UTI (no predisposing risk factors):

E. coli:	70 - 85 %
Proteus mirabilis:	10 - 15 %
Women: also Staphylococci:	5 %
other bacteria	rare

- complicated UTI (with predisposing risk factors):

E. coli:	up to 50 %
Proteus mirabilis:	10 %
Klebsiella and other Enterobacteria:	15 %
Enterococci:	10 % (often MSU is contaminated with this !)
Staphylococci:	10 %
Pseudomonas aeruginosa:	5 %

  Nosocomial UTI: frequently unusual bacteria like Enterococci, Pseudomonas, Proteus, Enterobacteria and Citrobacter. 50% of all patients with a long term catheter have a UTI after only one week, after 1 month almost all of them, often mixed infections. Asymptomatic urinary Candida can be found in ca. 20 % those long term catheterised.

Pg.: Infection path:
1. mostly ascending (canaliculary) (98 %): usually bowel flora
2. rarely haematogenous when the kidneys have been previously damaged

The anterior urethra usually shows a physiological bacterial contamination; but the bladder should be sterile in a healthy individual. The barrier function of the bladder sphincter can be disturbed by the above mentioned predisposing factors.

PPh: Pyelonephritis [N12]:
1. Acute bacterial purulent pyelonephritis: uni- or bilateral wedge-shaped abscess within the renal parenchyma, between papilla and cortex (lines of granulocyte accumulation). Complications: Abscess and pyonephrosis (pus in the renal pelvis).
2. Chronic focal destructive pyelonephritis:
   Wedge-shaped scars involving the renal surface, deformed renal calyx, occ. papillary necrosis. Histology: focal chronic-destructive inflammation of the tubular calyx.

Sym.: UTIs can present with a variety of symptoms:

A) **Asymptomatic bacteriuria:**
   Coincidental detection of bacteriuria, normal urine sediment in asymptomatic people: only pregnant women, children and those with an underlying obstruction require treatment.

B) **Symptomatic UTI:**
   1. Uncomplicated: no predisposing risk factors; usually E. coli
   2. Complicated: existing predisposing risk factors; often unusual and/or resistant pathogens

   3 severity:
   I.   no renal involvement
   II. renal involvement (morphological renal changes on X-ray/ US, occ. impaired renal function
   III. obstruction which can't be eliminated, patients with a long term transurethral catheter or suprapubic catheter. A permanent cure of the UTI is not possible.

- **Acute cystitis [N30.9]:**
  painful inflammation of the bladder
  Aet.: - bacterial UTI (in 50%) mainly seen in (sexually active) women (honeymoon-cystitis) or in individuals with
            predisposing risk factors
        - rarely any other infections (Trichomonas, fungal etc.)
  Sym.:- dysuria (difficulty in passing water), alguria (painful urinating)
        - urine frequency; nocturia
         (DD: cardiac failure, prostate adenoma)
        - suprapubic pain, tenesmus (painful, spastic urge to urinate) - no pain in the loins!
  Co.: - haemorrhagic cystitis with macroscopic haematuria
       - ascending infection, pyelonephritis
       - recurrent cystitis
  DD: • tuberculous cystitis
       • parasitic cystitis due to infection with Schistosoma haematobium
       • radiogenic cystitis (post radiation)
       • drug induced cystitis: e.g. after NSAID, Cyclophosphamide or Ifosfamide (often haemorrhagic cystitis)
       • other diseases of the bladder (tumour, stone, foreign body etc.)
       • adnexitis, prostatitis, bowel diseases etc.

- **Acute pyelonephritis:**
  bacterial infection of the proximal urinary tract causes acute interstitial nephritis:
  Sym.: - fever, general malaise
        - dysuria
        - tenderness/pain over loins

  atypical symptoms:
  - fever of unknown origin (particularly in children and older people)
  - older men or patients with an indwelling urinary catheter: always consider the possibility of a urinary obstruction
  - gastrointestinal symptoms (nausea and vomiting, abdominal pain, ileus)
  - headache

  False diagnoses:
  - lumbago/lumbar spine-syndrome
  - abdominal diseases

- **Chronic pyelonephritis (CPN):**
  Chronic interstitial nephritis caused by an obstruction of the urinary tract or by urine reflux and secondary
  bacterial UTI. Most cases start in early childhood, often due to vesico-ureteric reflux.

  Note: CPN can only develop in the presence of predisposing factors (partially) obstructing the urinary tract.

  Sym.: Differentiation between acute pyelonephritis and acute flare-up of a chronic pyelonephritis is clinically
        impossible- without knowing the patient's medical history.
  Symptoms are often uncharacteristic:
        - headache, weakness, nausea, weight loss, dull back pain

  atypical symptoms:
  - fever of unknown origin   - unexplained anaemia
  - unexplained ESR ↑   - unexplained hypertension
  Complications of pyelonephritis:
  1. purulent nephritis and renal carbuncle (multiple confluent cortex abscesses)
  2. urosepsis: In 65 % post invasive investigations/operations of the urinary tract. This is a life threatening
      complication!
  3. paranephritic abscess (may be caused by purulent nephritis): loin pain, fever; X-ray: shadow within the psoas
      muscle, spine concavely shifted to the diseased side; ultrasound, urography: kidney shift, not moving with
      respiration.
  4. obstruction of the urinary tract can cause hydronephrosis and pyonephrosis and pyelonephritic renal cirrhosis.
  5. renal failure may develop in chronic pyelonephritis, particularly when the predisposing factors are not
      eliminated (rise in serum urea, creatinine, drop of creatinine clearance).
  6. tubular partial function impairment:
      - impaired ability to concentrate urine (polyuria + polydipsia)
      - renal sodium loss
      - renal potassium loss, renal tubular acidosis
  7. chronic pyelonephritis causes hypertension in 30 - 50 % and sometimes also complications
  8. delayed development in young children

Di.: **I. Laboratory:**

1. urine investigation:

    - **leucocyturia**, occ. leucocytes cylinders as a sign of pyelonephritis, occ. erythrocyturia

- **bacteriuria:** correct method of catching the urine sample is crucial for correct interpretation of the bacteriological report (refer to chapter "kidneys" - introduction) and immediate process or rapid transport (cold chain must not be interrupted) in a special medium for urine cultures: misdiagnoses and unnecessary therapy will follow when these stipulations are not met! Always obtain the urine sample before starting therapy with antibiotics! Don't obtain any urine samples from catheter bags!
  - significant bacteriuria:
    definition by Kass: fresh MSU revealing a bacterial count of 10^5 and more per ml urine implies a real bacteriuria (if possible send 2 samples), whereas contaminated urine shows lower bacterial count. Clinical symptoms indicating a UTI, or patients already on antibiotics: lower bacterial counts ($< 10^5$/ml urine) have to be considered pathological.
  - any detection of bacteria in bladder puncture urine is a real bacteriuria, that means contamination can be excluded (if the technique was sterile).
  - 10^5 (and more) bacteria/ml in MSU or bacteria detection in puncture urine requires MC&S (microscopy, culture & sensitivity) investigation.
  - any positive pathogen detection in association with clinical symptoms is an indication for a UTI requiring therapy. Recurrent UTIs: MSU showing growth of Enterococci or a mixed infection require confirmation (bladder puncture), because samples are frequently contaminated.
  - asymptomatic temporary bacteriuria does exist, particularly in women; it vanishes repeatedly without any therapy and obviously doesn't lead to any illness. Asymptomatic bacteriuria in pregnancy and in childhood always require treatment!

  ▶ causes of leucocyturia when urine is sterile ("sterile" leucocyturia):
  - genital contamination from discharge
  - antibiotic treatment for UTI has already been started (detection with dipsticks)
  - gonorrhoea (→ culture from a fresh urethral swab or use of a special transport medium)
  - non-gonorrhoic and postgonorrhoic urethritis
  - urogenital tuberculosis
  - Reiter-syndrome (triad: urethritis, conjunctivitis, arthritis)
  - analgesic nephropathy

2. blood investigation:
  - ESR/CRP ↑
  - test urea, creatinine and creatinine clearance
  - blood count: leucocytosis in purulent renal complications, anaemia in chronic pyelonephritis and renal failure
  - blood culture when urosepsis is suspected

II. radiography diagnostics:

1. ultrasound:
- position, form and size of kidneys
- detection of a congested renal pelvis
- detection of any calculi
- detection of shrinking parenchyma in the case of pyelonephritic renal cirrhosis

2. contrast CT: better than i.v.-urogram

3. i.v.-urogram:
- detection of any anatomical anomalies
- detection of any obstructions and stones (occ. calcification on plain X-rays)
- X-ray findings in chronic pyelonephritis: Deformities and plump renal calyx, narrowed renal parenchyma etc.

4. MRI: is an alternative in case of CI to iodine containing contrast medium

III. Micturition urosonography or micturition cysto-urethrography: detection of a VUR

■ **Diagnose of a vesico-ureteric-renal reflux (VUR):**
  - urography - 5 stages according to radiography results
  - micturition cysto-urethrography
  - cystoscopy to assess the ureter ending

■ **Diagnose of acute UTI:**
  - history, symptoms
  - bacteriuria, leucocyturia

■ **Diagnose of CPN:**
  - history, symptoms
  - bacteriuria, leucocyturia
  - impaired renal function
  - morphological renal changes (radiography)
  - detection of any predisposing factors

Th.: **Therapy of acute UTI:**
  a) causal therapy:
  - • elimination of any obstruction
  - • elimination or treatment of any predisposing factors.

  recommendations for primary VUR:
  - grade I + II: give long term antibiotic prophylaxis and wait (spontaneous healing in ca. 60 % /5 years)
  - grade III + IV: anti-reflux-surgery

  b) Symptomatic treatment:
  - • general:
    1. sometimes bed rest in acute pyelonephritis
    2. drink plenty of fluid, empty bladder frequently
    3. ensure regular bowel movements
    4. spasmolytics are necessary in some cases
    5. omit any nephrotoxic analgesics

  - • Antibiotics:
    Always expect antibiotic resistance (particularly in nosocomial UTI and in complicated UTI); hence an antibiogram is essential! There are exemptions to this rule, e.g. in single uncomplicated UTI in the community, which usually respond well to short therapy. Recurrent UTIs require urological investigations!
    Take a urine sample for bacteriological investigations: "Blind" treatment of the UTI with a broad spectrum-antibiotic; once the microbiology report has been received, the treatment has to be adapted according to the result.

    Choice of antibiotics:
    - Gyrase inhibitors (=Quinolones):
      group 1: e.g. Norfloxacin; group 2: e.g. Ofloxacin, Ciprofloxacin etc.
      In recurrent UTIs Quinolones are not effective in 15 % (in nosocomial UTIs even in 25 %). Hence the request for culture and sensitivity in recurrent UTIs.
      SE: Gastrointestinal symptoms, allergic reactions, impaired function of the central and peripheral nervous system, depressions, confusion, hallucinations, changes in blood count, raised liver enzymes, prolonged QT-interval and occ. cardiac arrhythmia, tendonitis and ruptured tendons (particularly when the patient is on corticosteroid treatment etc.)
      CI: children and adolescents who haven't reached their final height (cartilage damage), pregnancy, breast feeding; beware epileptic and older patients
      Interactions: raised Theophyllin- and Cumarin/Warfarin levels
    - Aminopenicillins or Ceftriaxon may be given in pregnancy.
    - Trimethoprim (with or without Sulfonamide) can be used for community acquired UTIs, but it is not advisable for nosocomial UTI (resistance of E. coli up to 25 %).

    Note:
    - always ask for any allergies
    - SE and CI
    - follow the manufacturer's dose guidelines for children
    - reduce dose in renal failure

    Criteria for a successful therapy:
    After 24 hours there should be a clinical improvement, and any fever should be settling; after 3 days urine should be sterile and any other findings should be normal. If the fever doesn't settle after 3 days, consider any complications (e.g. paranephritic abscess → CT !), arrange MC&S.

    Length of treatment: acute pyelonephritis - give antibiotic for 7-10 days. Uncomplicated cystitis in young women: a short course of 1 day is sufficient. Advantages: less SE, lower costs.

    5 days after completion of the antibiotics therapy: send another urine sample for bacteriological investigation.
    Recurrent UTI: intermittent treatment with antibiotics confirmed by MC&S, additional acidification of alkaline urine, e.g. Methionin.

- ▪ **therapy of asymptomatic bacteriuria:**
    Only for pregnant women, transplanted or immunosuppressed patients and obstructions of the urinary tract; also preoperative before transurethral prostate resection

- ▪ **therapy of chronic pyelonephritis:**
    - wait for the bacteriological investigation, if at all possible
    - then give antibiotics for one week
    - after several unsuccessful courses of oral antibiotics, admit the patient to hospital for intravenous treatment.
    - in case of a remaining asymptomatic bacteriuria, don't prescribe any further treatment; only treat any acute exacerbations associated with clinical symptoms with another course of antibiotics (after MC&S).
    - treatment of complications: treat any renal hypertension and renal failure etc.

Prg.: ▪ prognosis of acute UTI: good, cured after a course of antibiotics
- ▪ prognosis of recurrent UTI:
  the risk of transition into CPN is fairly low, as long as there are no predisposing factors.
- ▪ prognosis of CPN: no permanent cure

Pro.: re-consider indication for catheterization, correct handling of equipment and strict adherence to the hygiene rules for urinary catheters (which are the most frequent cause for nosocomial UTIs).

URETHRITIS [N34.2]

Def.: Isolated UTI of the anterior urethra, distal of the internal urethral sphincter; distinguished from any UTI of the higher urinary tract for etiological, clinical and prognostic reasons.

Ep.: common; high unreported number; asymptomatic Chlamydia-carriers: Up to 7% of all men and up to 20 % of women. Chlamydia infections: also consider testing for syphilis and HIV!

Aet.: 1. most frequently non-gonorrhoic urethritis (NGU) and postgonorrhoic urethritis (PGU) caused by:
- Chlamydia trachomatis, serotypes D - K (40 - 80 %) ; incubation period : 10 - 24 days
- Ureaplasma urealyticum (20 %)
- Mycoplasma genitalium
- Trichomonas vaginalis (4 %)
- Herpes virus type II (rarely type I)
2. E. coli and other bacteria, which can be a cause for UTI.
3. gonorrhoea (for details refer to chapter "gonorrhoea")

Sym.: - occ. urethral discharge, sometimes only one drop early in the morning "Bonjour-droplet"
- occ. urethral itching or burning when urinating

Co.: Men: prostatitis and epydidimitis. Women PID = pelvic inflammatory disease: infection of uterus, fallopian tubes, ovaries; perihepatitis after Gonococci- or Chlamydia infection = Fitz-Hugh-Curtis-syndrome; occ. ectopic pregnancy after Chlamydia infection. Complications in both genders: infertility (20 % after Chlamydia infection), reactive arthritis and Reiter-syndrome (triad of: arthritis, conjunctivitis, urethritis)

DD: Chronic interstitial cystitis: mainly women, chronic urine frequency, painful urination –

Di.: cystoscopy (mucosal haemorrhage after hydro-distension), histology (mast cell infiltration); Aet.: unknown; no causal therapy

Di.: • sometimes leucocyturia in early morning urine
- • pathogen- or antigen detection: culture from fresh urethral- or cervical swab, for some tests special transport mediums have to be used; Chlamydia and Gonococci: the most sensitive test is nucleid acid reamplification test (NAT): first catch urine or urethral swab for men; cervical swab for women.

Th.: ▪ general measures:
  drink plenty + urinate frequently (rinsing effect) – treat the partner - temporary sexual abstinence – sexual hygiene; urinate after intercourse - avoid use of intimate sprays and soap for the genital area; no intravaginal spermicidal substance; remove a wet swimming suit, make sure the feet are warm; drink acidifying fruit juice (currants, cranberries)

- ▪ antibiotics:
  - Chlamydia trachomatis, Ureaplasma urealyticum and Mycoplasma: Macrolides (e.g. Azithromycin), therapy for 2 weeks. Chlamydia can persist in intracellular form of inactive elementary bodies; for chronic infections longer therapy is recommended (3 weeks to 3 months).
  - Trichomonas: e.g. Metronidazole
  - therapy of gonorrhoea: refer to according chapter

Pro.: detect + treat infected sexual partners, avoid promiscuity, use condoms; screen all pregnant women for Chlamydia infection etc.

HANTAVIRUS - INFECTIONS   notifyable disease [A98.5]

Syn: war- or field nephritis

Ep.: Worldwide, part. Southeast Asia; first noticed 1951 in the Korea-war; in Europe mainly in Scandinavia, the Ardennes, Bosnia etc.

Aet.: Hantaviruses (RNA-viruses of the Bunyaviridae family) with various serotypes:

serotype	Courses	main reservoir	Areas
Hantaan	Haemorrhagic fever with renal syndrome (HFRS)	mice	Southeast Asia Southeast Europe
Seoul		rats	Worldwide
Dobrava: middle- and southeast European variation		mice	Southeast Europe, Balkan
Puumula	usually epidemic nephropathy with good prognosis	mice	Middle- and North Europe Germany: > 90 %
Sin Nombre, Bayou, Black Creek-Canal, New York Andes	Hantavirus-cardiopulmonary syndrome (HCPS) = Hantavirus pulmonary syndrome (HPS)	mice	USA, Canada
		rats	South America

Inf.: Reservoir: mice and rats, infection by inhalation of the virus containing excrements of these animals, forest workers, soldiers, hunters, farmers and refugees are at particular risk. There is no transmission from human to human. Only for the Andesvirus is there suspected human to human transmission.

Inc: 2 - 4 weeks (occ. 5 - 60 days)

Sym.: I. symptoms of HFRS - 3 phases:
1. sudden onset with high fever, shivers, headache, myalgia, occ. conjunctivitis, facial flush
2. lumbago, abdominal pain, nausea, vomiting, diarrhoea
3. interstitial nephritis with pronounced proteinuria, oliguria, raised urea and creatinine

II. symptoms of HPS:
fever, myalgia, nausea, abdominal symptoms, dry cough, interstitial pulmonary oedema, haemorrhagic pneumonia
Laboratory-triad: leucocytosis with a shift to the left + atypical lymphocytes + thrombocytopenia

Co.: The course of HFRS is more severe than the epidemic nephropathy, frequently with complications: thrombocytopenia, petechias, occ. haemorrhage, shock, pulmonary oedema, acute renal failure (up to 10 %). ARDS in cases of HPS

DD: Respiratory infections, leptospirosis, renal diseases of different origin

Di.: • (occupational-) history + symptoms (fever, lumbago, raised creatinine, thrombocytopenia)
• serology diagnostics (IgM-antibodies ↑), pathogen detection (PCR)

Th.: Therapy trial with Ribavirin; supportive therapy of any complications: e.g. haemodialysis for acute renal failure; therapy of ARDS (refer to chapter „ARDS")

Prg.: mild course of illness in epidemic nephropathy with a good prognosis.

Mortality of HFRS: ca. 5 - 10 %, of HPS: ca. 50 %.

Pro.: Prophylaxis; no vaccination available.

TUBULO-INTERSTITIAL NEPHROPATHIES

Def.: Group of various diseases, causing interstitial inflammation and renal-tubular cell damage.

1. Acute tubulo-interstitial nephropathies:

Aet.: • directly infectious: e.g. Hantaan-virus
• parainfectious: e.g. Streptococci (DD: post-infectious acute glomerulonephritis)
• immunological: systemic lupus erythematosus, Sjögren-syndrome, sarcoidosis etc.
• drugs: NSAID, Omeprazole, Allopurinol, Methicilline etc. – many other medications in individual cases

Sym.: • non-glomerular haematuria, proteinuria < 1 g/day, rarely > 1 g/day (mainly due to NSAID)
• fever, rash, eosinophilia, eosinophiluria (rarely all 3, often absent in NSAID induced disease)
• TINU-syndrome (tubulo-interstitial nephritis + uveitis): rare complication of EBV-infection n children/adolescents

Co.: acute renal failure (ARF)

DD: drug induced renal damage:
1. acute toxic (depends on dose), e.g. Aminoglycosides, Cephalosporins, Gyrase inhibitors
2. chronic toxic (depends on dose), e.g. Phenacetin or Paracetamol
3. hypersensitivity reaction (not depending on dose), e.g. Methicilline
4. immunological (not depending on dose): e.g. gold (can cause glomerulonephritis)

Di.:   drug history + symptoms + renal biopsy:
lymphoplasma cellular interstitial infiltrates of the renal cortex

Th.:   omit any potentially toxic drugs, give corticosteroids for TINU-syndrome; dialysis for ARF etc.

2. Chronic tubulo-interstitial nephropathies:

Aet.:   • drugs: in most cases analgesics (→ analgesics nephropathy)
- chemicals: Cadmium, lead
- metabolic disorders: gout (hyperuricaemia), hypercalcaemia, hypokalaemia, oxalate nephropathy, cystinosis
- haematological/immunological diseases: e.g. multiple myeloma, amyloidosis etc.
- other causes: congenital nephropathies, Balkan nephritis etc.

ANALGESICS-NEPHROPATHY   [N14.0]

Def.:   Chronic tubulo-interstitial nephritis caused by mixed anti-inflammatories abuse of (APC= ASS + Paracetamol + Caffeine) and Phenacetin containing analgesics, and their metabolites Paracetamol or non-steroidal anti-inflammatories. Drug history is essential: After a cumulative intake of ≥ 1.000 g Phenacetin or Paracetamol over several years it is a likely diagnosis.

Ep.:   Ca. 1 % of all patients suffering from chronic renal failure have analgesics-nephropathy.

Aet.:   Phenacetin, its metabolite Paracetamol and non-steroidal antiphlogistics block the synthesis of the vasodilating prostaglandins E2, leading to impaired circulation and papillary necrosis.

Sym.:   often no symptoms in the early stages; occ. headache, tiredness, dirty grey-brown skin and anaemia (cause: gastrointestinal blood loss, haemolysis, production of Methaemoglobin and Sulfhaemoglobin; later renal anaemia)

Co.:   • papillary necrosis: loin pain, haematuria, often fever, occ. detection of papillary tissue in the urine and papillary defects in the urogram.
- tubular damage and reduced urine concentration; tubular acidosis
- lipofuscin-like pigments are deposited in the medullary wedges and in the liver.
- occ. bacterial UTI
- renal failure
- late complications: increased risk for urotheliomas (→ 2 x/year urine cytology) and breast cancer (up to 10 %)

Urine: leucocyturia without bacteriuria (in complicated UTI with bacteriuria), occ. erythrocyturia, occ. mild proteinuria (tubular proteinuria type).

US + CT: calcified papillae and papillary necrosis, scarred cortex next to the medullary wedges, renal cirrhosis with irregular contours

DD: • Chronic tubulo-interstitial nephritis of other origin
- papillary necrosis of other origin: 1. diabetic nephropathy, 2. obstructive nephropathy, 3. urogenital-Tbc, 4. sickle cell anaemia

Di.:   symptoms + drug history + radiographic diagnostics
If you suspect that the patient conceals Paracetamol-abuse from you, the metabolic product N-Acetyl-Paraaminophenol (NAPAP) can be detected in urine.
US + CT without contrast: small kidneys with retractions on the cortex + detection of papillary calcifications

Th.:   omit the suspected drug; therapy of any renal failure

Prg.:   If the analgesics abuse detected before the onset of advanced renal failure (serum creatinine < 3 mg/dl), and the drug is stopped, then the disease process will be halted.

BALKAN-NEPHRITIS   [N15.0]

Chronic interstitial nephritis of unknown origin
Endemic in the Balkan countries. Starts at a young age with asymptomatic proteinuria; slow course over 2 - 3 decades; ends in 5 - 10 % in renal failure; late complications: urotheliomas.

CHINESE-HERBS NEPHROPATHY

Cause: various chinese herbs and their products, containing nephrotoxic aristolochia acid. In 50 % the result is irreversible renal failure. Late complication: urotheliomas.

PREGNANCY INDUCED NEPHROPATHIES [O26.8]

1. EPH-Gestosis (or pre-eclampsia): hypertension + proteinuria with or without oedema
This is a disease in late pregnancy. 30 % of pregnant women suffering from a pre-existing renal disease, will get pre-eclampsia; in case of a pre-existing impaired renal function or hypertension the rate is 60%.
Th.: watch fluid- and electrolyte balance, give diuretics, antihypertensives, magnesium. If blood pressure can't be controlled or when convulsions start: Caesarean section!
2. Acute pyelonephritis
1 in 3 women with asymptomatic bacteriuria will develop acute pyelonephritis during their pregnancy. Urine screening must be part of routine pregnancy care! Asymptomatic bacteriuria in pregnancy must be treated with antibiotics.
3. Pregnancy and pre-existing renal diseases

> **Note:** All renal diseases (apart from SLE) will cause a deterioration of the renal function in pregnancy. Women suffering from renal disease should avoid getting pregnant unless renal function and blood pressure are normal.

PARAPROTEINAEMIC RENAL DISEASES

(refer to chapter "Multiple Myeloma" and "Amyloidosis")

Def.: A group of renal diseases associated with deposition of intact immunoglobulins or immunoglobulin fragments (heavy chains and light chains). Myeloma-kidneys (cast-nephropathy), AL-amyloidosis (amyloid, made of light chains), AH-amyloidosis (amyloid made of heavy chains), light chain disease, fibrillary immunotactoide glomerulopathy and glomerulonephritis, associated with type-1-cryoglobulin deposit

▶ Myeloma-kidney (Cast-Nephropathy)

Ep.: Ca. 30% of all patients suffering from multiple myeloma will develop a myeloma-kidney.

Pg.: In this disease ca. 85g monoclonal light chains are synthesized per day, compared to 0,9g/day polyclonal light chains in a healthy person. The light chains accumulate in the lysosomes and they cannot be metabolised by the proteases. The resulting atrophy of the proximal tubular cells is one of the most important factors leading to renal failure. The non-resorbed light chains travel into the distal tubular and into the collecting duct; here they precipitate together with Tamm-Horsfall-protein and create urine cylinders; these finally lead to tubular obstruction.

Sym.: renal failure but normal sized kidneys and bland urine sediment. Proteinuria < 3 g/24 hours, detection of mainly Bence-Jones proteins and low amounts of albumin.

Di.: 1. Diagnosis of MM
2. renal biopsy

Th.: Refer to chapter MM

▶ AL-Amyloidosis

Def.: Fibrillary deposits of amyloid, made of light chains

Ep.: Whilst myeloma-kidneys can only be seen in multiple myeloma, AL-amyloidosis also is a symptom of monoclonal gammopathy of undefined significance (MGUS). Ca. 30 % of patients with AL-amyloidosis suffer from multiple myeloma, and 15 % of all patients with multiple myeloma will develop AL-amyloidosis.

Sym.: In 50 % of all patients the kidneys are affected; in 30 % a restrictive cardiomyopathy will be diagnosed.

renal manifestation:
severe nephrotic syndrome
- 50 % of all patients have a reduced glomerular filtration rate at the time of diagnose.
- hypertension in 25 %
- kidney size is normal or slightly increased.

Di.: Rectal biopsy and subcutaneous paraumbilical fat biopsy show in ca. 70 % amyloid deposits; renal biopsy may not be necessary in that case. renal biopsy is highly accurate, when clinical detection of renal involvement is present.

Th./Prg.: Prognosis for patients with AL-amyloidosis is poor; median survival of 12 ± 6 months. Cardiac involvement is the cause for 40 % of deaths.
Oral combination therapy with Melphalan and Prednisolone improves the survival rates for a few months. High dose chemotherapy with Melphalan and autologous stem cell transplantation seems to be more effective.

► Light chain disease

Ep.: Light chain disease causes non-fibrillary deposits of monoclonal light chains or their fragments in various organs. renal manifestation frequently dominates in this disease. 2/3 of the patients have multiple myeloma, but a third do not fulfil the criteria of multiple myeloma, and in 6 % no monoclonal protein can be detected in urine or serum using standard laboratory methods.

Sym.: - at the time of diagnosis most patients suffer from advanced renal failure.
- nephrotic syndrome (40 %), microscopic haematuria (30 %)
- heart and Liver affected in ca. 30 %

Th.: Alkylating substances, particularly Melphalan/Prednisone are used, but the success is limited. In younger patients a high-dose-chemotherapy with autologous stem cell transplantation looks promising.

Prg.: large distribution of the median survival rate ranging from as little as 1 year to 10 years

► Fibrillary immunotactoid glomerulopathy

Ep.: this disease represents 1% of all diagnoses of renal biopsies.

Aet.: unknown

Pa.: Electronic microscopy reveals, that PAS-positive material consists of randomly arranged (fibrillary glomerulopathy) or organised bundles (immunotactoide glomerulopathy) of micro fibrils and microtubules.

Sym.: - proteinuria is seen in almost all patients, more than 50% have nephrotic syndrome.
- haematuria, hypertension and renal failure in most cases.

Th./ Prg.: there is no effective therapy for fibrillary immunotactoid glomerulopathy; and many patients will develop terminal renal failure within 1-10 years.
Renal transplantation seems to be a promising therapy.

RENAL TUBULAR PARTIAL DYSFUNCTION

1. primary: usually congenital
2. Secondary: consequence of renal disease, particularly interstitial nephritis

   A) reduced function of amino aid transport:
   Cystinuria: autosomal-recessive inherited impairment of the proximal-tubular resorption of amino acids. 2 mutations: SLC3A1 and SLC7A9 → result: cystic kidney stones in children; typical hexagonal crystals in urine
   Further diseases: blue nappy syndrome (impaired intestinal transport of Tryptophan → bacterial metabolism leads to Indigo blue); homocystinuria, cystathionuria, glycinuria etc.

   B) reduced glucose resorption : e.g.
   renal glycosuria: Mutation SLC5A2. Harmless congenital dysfunction of the proximal-tubular resorption of glucose → glycosuria, but normal blood glucose (DD: Diabetes mellitus); very rare

   C) Dysfunctional water- and electrolyte transport:
   - Phosphate diabetes: Congenital defect of the phosphate transport: hyperphosphaturia, hypophosphataemia, vitamin D-resistant rickets in children
   - reduced urine concentration leading to polyuria
   - nephrogenic (renal) diabetes insipidus (NDI):
     • congenital form, 2 genetic variations:
       ▪ X-chromosomal-recessive NDI: mutated gene (located on Xq28) for the vasopressin-type 2-receptor
       ▪ autosomal-recessive NDI: defect water transport channel "Aquaporin 2" of the renal collecting tube
     • acquired form:
       kidney disease with tubular damage; drug induced side effects (Lithium carbonate)
   - sodium loss kidney if advanced renal failure has caused sodium loss syndrome, then a diet low in salt can cause a deterioration of the renal function (which will improve after increased salt intake). Hence the level of sodium intake must be calculated according to the loss (urine balance!).
   - potassium loss kidneys, in most cases combined with secondary hyperaldosteronism
   - renal tubular acidosis (RTA) - 4 types:
     ◦ type I-RTA = distal RTA: severe hyperchloraemic metabolic acidosis; kidneys are not able to reduce the urine-pH below 6; complications: vitamin D-resistant osteomalacia, nephrocalcinosis and hypokalaemia.
     ◦ type II-RTA = proximal RTA: re-resorption defect of bicarbonate (bicarbonate loss acidosis). Clinically less severe than type I (no osteomalacia and nephrocalcinosis)
     ◦ type III and IV RTA are rare. Type IV is the same as hyporeninaemic hypoaldosteronism.

   D) Syndromatic impaired tubular function, e.g.
   Debré-Toni-Fanconi-syndrome, congenital or acquired tubulopathy with hyperaminoaciduria, glycosuria, hyperphosphaturia, frequently with chronic acidosis and hypokalaemia; secondary form for example in cystinosis, multiple myeloma, interstitial nephritis etc.

BARTTER-SYNDROME [E26.8]

Def.: Autosomal-recessive inherited group of disorders renal tubular function with hypokalaemic alkalosis, salt loss and hypotension; associated with hypercalciuria and (in type I and II) with normomagnesaemia.

Bartter-Syndrome Type I

Manifestation in infants. All patients were premature babies of women with polyhydramnion and develop severe dehydration within the first few months of life.

Cause: Mutation of gene SLC12A1/15q15-q21, which codes sodium-, potassium-, 2-chloride-co-transporter of the large ascending branch of the Henle loop. This leads to: reduced re-absorption of sodium and chloride in the ascending part of the Henle loop with salt loss and hypovolaemia. Activation of the Renin-Angiotensin-Aldosterone system leads to hypokalaemic alkalosis. Calcium reabsorption depends on the activity of the sodium-, potassium-, 2-chloride-cotransporter in the ascending part of the Henle loop, and it explains the hypercalciuria in Bartter-syndrome.

Bartter-syndrome type II

Phenotypically there is no difference to type I patients.

Cause: Mutations of gene KCNJ1/11q24-q25, which codes the apical ATP-dependent potassium channel (ROMK). This channel is important for the reclaim of potassium into the cells and for the function of the sodium-, potassium-, 2-chloride-cotransporter.

Bartter-syndrome type III

Patients with Barter-syndrome type III differ phenotypically from the two previous types, as these patients do not develop nephrocalcinosis. 30 % have hypomagnesaemia.

Cause: Mutations of the gene CLCNKB/1p36 of chloride channels, which are responsible for the chloride reabsorption along the basolateral membrane of the tubular cells in the ascending part of the loop of Henle.

Bartter-syndrome type IV
Mutation of the gene BSND/1p31-32
Triad: Bartter-syndrome, renal failure, impaired hearing

Bartter-syndrome type V
Mutation of the gene CASR/3q13-21. causes hypocalcaemia (low PTH).

Gitelman-syndrome

Def.: Autosomal-recessive inherited disease, onset in young adults with hypokalaemic alkalosis, salt loss, hypotension, hypomagnesaemia and hypocalciuria.

Cause: mutation of the gene, which codes the sodium-chloride-cotransporter in the distal tubular of the nephron (SLC12A3-Gen/16q13)

Th.: potassium substitution for all 4 forms (prostaglandin synthesis inhibitors only have a short effect); give additionally Spironolactone or Triamterene (without Thiazide).

PSEUDO BARTTER SYNDROME [E26.8]

Often young women in medical professions, occ. anorexia nervosa

Aet.: - laxative abuse
- diuretic abuse

Sym.: as in Bartter-syndrome

Di.: history + symptoms, detection of traces of diuretics in urine

Th.: psychosomatic support, omit any triggering medication

ACUTE RENAL FAILURE (ARF) [N17.9]

Ep.: 1 - 5 % of all inpatients and > 10 % of those in intensive care

Def.: Acute onset, rapid decline of renal function, lasting several days, and can be reversible.

- leading to:- retention of urea/creatinine
- impaired fluid-, electrolyte- and acid-base-balance

Leading symptom: urine secretion ceases - oligo-/anuria and rising serum creatinine > 50 % of the original level.
oliguria: < 500 ml urine/day
anuria: < 100 ml urine/day
up to 30 % of the ARF follow a normo- or polyuric course; is these cases the only leading symptom is the rise of serum creatinine.

Et/Pg:

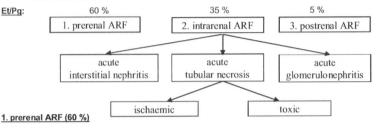

1. prerenal ARF (60 %)

PPh: In prerenal renal failure the renal tubular and glomerular structure are completely normal to start with. Reduced perfusion is the cause for the loss of renal function. Reduction of the effective blood volume causes a reactive activation of the Renin-Angiotensin-Aldosterone-system, and excretion of catecholamines and ADH. These hormonal counter-regulations lead to reduced natriuresis and a rise of urine osmolality. For patients with cardiac failure, liver cirrhosis and some with nephrotic syndrome, it is not unusual to develop prerenal renal failure with contraction of the intravascular space, despite clinical symptoms of hyperhydration. If these patients receive diuretics, the effective blood volume will be reduced even further with the associated risk of developing intrarenal renal failure. After the elimination of the trigger factor, prerenal renal failure is immediately reversed.

Causes :

1. drop in circulating blood volume
2. drop in HZV and mean arterial pressure; cardiovascular shock of various origin
3. systemic vasodilatation (e.g. in sepsis)
4. cytokine-mediated renal vasoconstriction
5. renal vasoconstriction in hepato-renal syndrome

2. Intrarenal ARF (35 %)

PPh: Morphologic correlate of acute tubular necrosis; these lead to obstruction of the tubules by epithelia detaching from the basement membrane. The cause for this tubular necrosis is an O_2-deficiency situation caused by reduced renal perfusion and impaired blood vessel autoregulation of the kidneys. GF is controlled by a tubulo glomerular feedback-mechanism. The damaged tubular cells lead to insufficient sodium reabsorption; this induces signals in the macula densa, causing constriction of the vas afferens. This mechanism and other factors cause a reduced GFR. Nephrotoxic substances may enhance the ischaemic renal changes. It takes 3-4 weeks for the total regeneration of the tubular epithelium.

causes of intrarenal ARF:

1. acute tubular necrosis:
   - ischaemic
   - toxic
   - septic (part of multiple organ failure)
   - hepato-renal syndrome
2. macrovascular diseases:
   - vasculitis
   - atheroembolism
   - thromboembolism
3. microvascular diseases:
   - rapidly-progressive glomerulonephritis
   - IgA-nephritis
   - haemolytic-uraemic syndrome (HUS)
4. acute interstitial nephritis:
   - allergic (NSAID, Betalactam antibiotics, etc.)
   - Parainfectious (e.g. Hantaan-viruses, CMV, EBV, Leptospira)

ARF caused by drugs and toxins:
Drugs and toxins can cause ARF via two different mechanisms: They are toxic via acute tubular necrosis or via the acute interstitial nephritis (e.g. Sulfonamide and Penicillin). Typical symptoms of acute interstitial nephritis are allergic reactions like rash, eosinophilia or fever.

Toxic triggers for intrarenal ARF:
• drugs: non-steroidal anti-inflammatories = NSAID (Prostaglandin synthesis inhibitors); antibiotics: Aminoglycosides, Cephalosporins, Gyrase inhibitors, Vancomycin, Amphotericin B etc.; cytostatics: Cisplatin, Methotrexate, Cyclosporin etc.; gold salts; diuretics

Note: Often there are additional other factors triggering the ARF like dehydration, cardiac failure, renal failure.

126

- radiological contrast medium: „contrast medium-nephropathy" (particularly in patients with pre-existing kidney disease and/or therapy with potential nephrotoxic medications and/or multiple myeloma or Waldenström's disease)

  Note: Patients with reduced renal function need a carefully considered indication, if contrast medium has to be used. Try to use alternatives; prefer non-ionic, low osmolar contrast medium + prophylactic measurements (sufficient hydration, stop any potentially nephrotoxic drugs. Currently we don't know if additional application of Acetylcystein the day before and the day of investigation is of any prophylactic value.

- Pigment-nephropathy → cause:
  - haemolysis (transfusion incident)
  - rhabdomyolysis: trauma (crush-syndrome), drug abuse, alcohol withdrawal delirium, excessive physical exercise, lipid lowering drugs (Statins, Fibrates) etc.
  - tubular obstructions: caused by light chains (in multiple myeloma), urate (in hyperuricaemia), oxalates (e.g. in Glycol intoxication)

3. Postrenal ARF (urinary tract obstruction) → causes:
1. congenital malformation of kidneys, ureter, bladder and urethra
2. acquired obstruction to urine flow in the renal pelvis, ureter, bladder or urethra
3. malignant tumours
4. gynaecological diseases and operative complications
5. misplaced or obstructed urinary catheters
6. drug induced (anticholinergics, neuroleptics) in patients with an underlying obstruction (e.g. prostate adenoma)

Sym.: Symptoms of ARF are unspecific. Usually these are the symptoms of the underlying disease causing the ARF. In some cases the leading symptom - oliguria – can be absent; therefore it is important to monitor renal function in diseases, which may lead to ARF! (fluid balance, urea/creatinine, urine investigation).

three phases of ARF:

- **Initial phase**
  The initial phase is asymptomatic or patients show symptoms of the underlying disease.

- **Phase of manifest renal failure**
  This phase is characterised by ongoing deterioration of the GFR with progressive rise of urea/creatinine. We differentiate between oliguric and non oliguric course (the latter with better prognosis) according to the grade of diuresis.
  Main risks: - overhydration, left ventricular failure and pulmonary oedema, brain oedema
  - hyperkalaemia, metabolic acidosis, uraemia

- **diuretic or polyuric phase**
  main risks: loss of water, sodium and potassium

Co.: 1. lungs:
- pulmonary oedema (fluid lung), pleural effusion
- pneumonia (e.g. due to ventilation)
- (ARDS) as part of multiple organ failure
2. cardiovascular:
- pericarditis
- arrhythmia caused by electrolyte imbalance
- hypertension.
3. gastrointestinal:
- haemorrhagic gastritis, ulcers
- gastrointestinal haemorrhage
4. central nervous system:
- encephalopathy with flapping tremor
- convulsions, drowsiness, confusion, coma
5. haematological system:
- rapid development of anaemia
  tendency to uraemic haemorrhage
6. infections:
- nosocomial (hospital acquired) infections
- sepsis (wound infections, catheter sepsis, UTIs)

DD: of oligo-anuria:

1. Functional oliguria (e.g. due to restricted fluid intake)
   serum urea is raised in functional oliguria and in ARF (in functional oliguria creatinine is only minimal raised).

Urine investigations	Functional oliguria	A R F
specific weight (g/l)	> 1.025	< 1.015
osmolality (mosm/kg)	> 1.000	< 600

Note: untreated functional oliguria (i.e. no fluid substitution) can lead to ARF!

2. prerenal ARF:
   history (!): shock? anaesthesia(protocol)? insufficient fluid substitution in a febrile patient, severe sweating, diarrhoea etc.?

3. intrarenal ARF:
   underlying renal disease? medication history; history of radiographic contrast medium? previous infections? systemic diseases? transfusions? haemolysis(symptoms)? myolysis?
   Oliguria + haemoptysis: always consider Goodpasture-syndrome!
   Exclude a Hantavirus infection (antibodies - IgM ↑).

4. postrenal ARF:
   Mechanical obstruction of the urinary tract (kidneys or bladder) due to an urological, gynaecological disease, operations near the bladder. Don't forget palpation of the bladder!
   US: fluid filled renal pelvis, "dry" bladder due to ureter obstruction; full bladder when the obstruction is located distal of the bladder (urethra, prostate). Estimate the bladder size by percussion/palpation; perform bladder puncture, when US is not available.

5. Chronic renal failure:
   history of renal disease? renal anaemia? Hypertension for many years? (US: renal cirrhosis - ARF: large kidneys)

Di.: 1. history + symptoms + urine volume
2. laboratory:
   - urine: urine microscopy and urine sediment, akanthocytes
   - blood: creatinine, urea, sometimes. endogen creatinine-clearance, sodium, potassium, calcium, blood gases, blood count, CK, LDH, lipase, electrophoresis, blood culture
   In case of intrarenal renal failure and suspected microvascular disease, antibody diagnosis is required (refer to relevant chapter).
   Fractioned sodium excretion is a way to differentiate between prerenal and intrarenal ARF. Fractioned excretion of sodium equals the sodium-clearance with reference to creatinine-clearance:

$$\text{fractioned excretion Na} = \frac{(\text{Urin-Natrium x Serum-Kreatinin})}{(\text{Serum-Natrium x Urin-Kreatinin})} \times 100$$

	urine-sodium (mmol/l)	urine-/plasma-osmolality (mosm/kg)	urine-/plasma-creatinine	Fractioned Na-excretion (%)
prerenal ARF	< 10	> 1,1	> 15	< 1
intrarenal ARF	30 - 90	0,9 – 1,05	< 15	> 1

The idea of the equation is, that in prerenal azotaemia the GFR is reduced, the tubules however are intact. Prerenal azotaemia produces concentrated urine with low sodium content; the problem of acute tubular necrosis is insufficient reabsorption of water and sodium; hence the urine has a low concentration with high sodium.

3. radiographic diagnostics:
   - US:
     ARF: large kidneys
     chronic renal failure: small, dense kidneys
     postrenal ARF: dilated renal pelvis, assess size of bladder
   - coloured doppler: diagnosis of dysfunctional arterial and venous perfusion
   - angio-MRI when thrombosis of renal blood vessels is suspected
   - spiral-CT in case of urinary tract obstructions (postrenal ARF) - **Beware** contrast medium !
   - percutaneous renal biopsy: Ind: suspected rapid progressive glomerulonephritis

Th.: 1. Treat the underlying disease, which caused ARF, e.g.:
   - optimise management of shock
   - stop any suspicious drugs in acute interstitial nephritis
   - revascularisation in renovascular obstructive diseases
   - urological treatment in postrenal ARF (e.g. transurethral or suprapubic catheter). This leads to intense diuresis, and may cause hypokalaemia.
2. symptomatic therapy of renal/prerenal ARF:
   - fluids- and electrolyte balance, monitor weight (daily)
   - fluid intake has to be adapted according to the fluid loss:
     a) insensible loss (≈ 1 l/day, more in a febrile patient, per °C above 37 °C: 500 ml more)
     b) renal excretion
     c) extrarenal loss (vomiting, diarrhoea, sweat, loss via nasogastric tube, wound secretion etc.)

Fluid intake in anuria is calculated from the extrarenal loss + 600 ml (≈ 1 l insensible loss minus 400 ml oxidation water and endogenous produced water). A daily weight loss of 200 - 300 g (catabolism) is a steady balance.

- nutrition: sufficient caloric intake (ca. 40 kcal/kg)
- reduce the dose of any medication which is subject to renal excretion. Therapy monitoring: check blood concentration of drugs. Don't give any nephrotoxic drugs.
- maintain diuresis in oliguric ARF with the use of loop diuretics (diuretic therapy increases the diuresis, but not the glomerular filtrate; diuretics do not improve the renal function).
- in the early phase of myoglobinuric ARF, but intact diuresis, you may attempt forced alkali diuresis (urine pH > 7).

3. kidney replacement therapy in ARF:
The aim of the extracorporal treatment of ARF is the therapy of azotaemia, adjustment of electrolyte- and fluid balance and correction of the metabolic acidosis. Extracorporal therapeutic methods are haemodialysis and haemofiltration; both can be performed intermittently or continuously. Continuous veno-venous haemofiltration, continuous veno-venous haemodialysis or haemodiafiltration are standard therapies in intensive care units. Once ARF has been diagnosed, extracorporal therapy should be started as soon as possible, before the onset of acute uraemia. Critically ill patients with ARF should be started on this therapy no later than serum creatinine being 4 – 6 mg/dl and serum urea being 120 – 140 mg/dl. Intensity of the extracorporal therapy is important for prognosis: An Italian study has shown for continuous haemofiltration in intensive care patients with ARF:, a rise of the filtration rate from 20 ml/kg/hour (ca. 1,3 l/hour) to 35 ml/kg/hour (ca. 2,3 l/hour) improves the survival rate.

Prg.: Despite advances in intensive care, mortality of intensive care patients with ARF has remained high over the last few years(ca. 60 %). This is due to severe underlying disease, particularly multi organ failure. The ARF itself has a negative influence on complications and prognosis, irrespective of the causal disease. ARF does not only lead to immediate complications, like impaired volume- and electrolyte balance; it also has a negative influence on all biological processes and organ functions.

Pro.: All patients at risk of prerenal ARF need haemodynamic optimisation and correction of the fluid balance; avoid any malperfusion of the kidneys; strict indication for radiographic contrast medium and potentially nephrotoxic drugs.
Possible nephroprotective substances in ARF: Acetylcysteine in case of contrast medium nephropathy, Mannitol in rhabdomyolysis, Selene and Glutamine in septic ARF. Close blood sugar monitoring using an Insulin sliding scale therapy reduces the incidence of ARF in the critically ill.

CHRONIC RENAL FAILURE [N18.9] AND URAEMIA   [N19]

Def.: Chronic renal failure is due to an irreversible reduction of the glomerular, tubular and endocrine function of both kidneys.

Ep.: Incidence in West Europe 10/100.000/year (USA: 60/100.000/year)

Aet.: The most frequent renal diseases causing terminal renal failure:
1. diabetic nephropathy (ca. 35 %)
2. primary and secondary glomerulonephritis (ca. 15 %)
3. chronic tubulo-interstitial diseases
4. vascular (hypertensive) nephropathies
5. polycystic renal diseases

PPh: Irrespective of the aetiology of the underlying disease, some chronic renal diseases will lead to progressive terminal renal failure over several years. The remaining healthy glomeruli will maintain a basic function of the kidneys due to increased intraglomerular pressure with hyperfiltration; the latter will be enhanced even more in case of arterial hypertension. Angiotensin II is an important mediator of this glomerular hyperfiltration, and (via increased production of cytokines and growth factor)causes glomerular hypertrophy and hyperplasia. Angiotensin II also causes enhanced glomerular permeability with loss of glomerular filter function. Proteinuria is the result, which in itself (being a direct nephrotoxin) leads to progressive glomerulosclerosis and consequent development of renal cirrhosis.

Chronic renal failure causes:
1. failure of excretory renal function
2. disorder of water-, electrolyte- and acid-base-balance
3. reduced endocrine renal function:
    reduced secretion of Erythropoietin, Renin, active vitamin D and prostaglandins
4. toxic organ damage due to retained urea and creatinine

1: Failure of excretory renal function
serum urea/creatinine will only increase, when more than 60 % of the functioning renal tissue has ceased working (GFR < 50 ml/min.). In this case plasma concentration of own and foreign substances (e.g. drugs) rises; rise of

plasma concentration of these substances leads to a rise of their concentration in the primary filtrate. This way a new balance between metabolic and excreted substances develops. serum creatinine concentration/endogen creatinine-clearance represent the glomerulus filtrate. There is an early impaired maximal concentration capacity of the kidneys. The reduced number of nephrons causes a surplus of substance solution for the single nephron (e.g. urea), which results in osmotic diuresis with nocturia, polyuria and polydipsia. Healthy kidneys manage to excrete the accumulated osmotic substances of ca. 900 mosmol/day, with (in case of maximal concentration) fine urine volume of ca. 750 ml. In renal failure with isosthenuria (specific urine weight fixed at ca. 1010g/l) for the same filter function, a volume of ca. 3 l urine is needed. Diuresis > 3 l per day does not lead to any further significant excretion of urea.

2: Dysfunctional water-, electrolyte- and acid-base-balance
sodium balance:
Fractional excretion of sodium (sodium excretion per single nephron) rises exponentially with diminishing glomerular filtration rate. When the glomerular filtration rate is below 10 - 20 ml/min., adaptation capacity of the ill kidneys is exhausted; salt- and water retention with progressive rise of the extracellular volume will develop; this in turn triggers the development of hypertension in the uraemic patient. For these reasons, diuretics are (at least in advanced renal failure) an essential part of any combined antihypertensive therapy. In some cases (tubulointerstitial nephropathy) there is early tubular dysfunction (salt losing nephropathy). In these patients an over-strict salt reduction will cause increasing lack of salt, which will improve after correction of the NaCl-deficit. Hence a general strict diet low in salt is not always indicated.

Steady potassium balance can be seen even in advanced renal failure; this is due to increased distal tubular potassium secretion per single nephron, and due to increased intestinal potassium secretion. In case of sudden potassium intake, there will be a shift of potassium from the extracellular space into the cells. In terminal renal failure always consider hyperkalaemia in the following cases:

- due to excessive potassium intake and/or acidosis the secretion capacity has been exceeded

- in oliguria diuresis is below 500 ml per day

- lack of sodium in the distal tubules does not offer sufficient sodium for exchange against potassium

Use of potassium sparing diuretics (which are contraindicated in renal failure) and hyporeninaemic hypoaldosteronism (e.g. in diabetic nephropathy): Always consider hyperkalaemia!

acid-base-balance:
Glomerular filtration rate < 30 ml/min. will often lead to metabolic acidosis. The kidneys are not able anymore, to eliminate the daily amount of 60-100 mmol H-ions (from protein metabolism), so they lost the ability of tubular creation of Ammonium ions.

Consequences of persistent metabolic acidosis:
1. increasing osseous calcium release
2. increasing gastrointestinal symptoms (nausea, vomiting, anorexia)
3. tendency to develop hyperkalaemia
4. subjective perception of dyspnoea
5. increased protein catabolism.

3: reduced endocrine renal function
Progressive renal failure will lead to impaired endocrine function, which in turn can have an influence on the renal haemodynamic (Renin and Prostaglandins). It also causes renal osteopathy (active vitamin D) and renal anaemia (Erythropoietin).

4: toxic organ damage due to retained urea/creatinine
The uraemic syndrome describes the influence of azotaemia on all organs; particularly on the cardiovascular circulation system, the central and peripheral nervous system, blood and skin. - Mild renal failure already causes significant increased cardiovascular risk!

Sym.: ▶ history → ask for:
previous infections, diabetes mellitus, system diseases, analgesics abuse, hypertension, hereditary renal diseases (e.g. cystic kidneys)

▶ early symptoms:
- increased excretion (polyuria) of pale urine (concentration deficiency with isosthenuria)
- high blood pressure
- oedema of the lower limbs, lid oedema
- loin pain and dysuric symptoms with fever (in pyelonephritis)

▶ late symptoms:
- tiredness, weakness, pallor (renal anaemia)
- headache, impaired vision
- anorexia, nausea (uraemic gastroenteropathy)
- pruritus
- muscle twitching

▶ symptoms of the terminal stage:
  - vomiting, weight loss
  - dyspnoea
  - diminished urine volume
  - uraemic encephalopathy (drowsiness, lethargy, cramps, coma)
  - increased risk of haemorrhage (thrombocytopathy, thrombocytopenia)
▶ Clinical examination:
  - pale skin and mucosa (anaemia)
  - café-au-lait-colour of the skin (uraemia)
  - uraemic fetor
  - oedema
  - paraesthesiae (uraemic polyneuropathy)
  - muscle fibrillations (myopathy)
  - hypertension
  - pulmonary congestion (fluid lung caused by sodium- and water retention)
  - pericardial rub or -effusion (uraemic pericarditis), pleurisy
  - renal cirrhosis (chronic glomerulonephritis) or nephromegaly (in cystic kidneys)
  - renal osteopathy
▶ laboratory investigations:
  - increased urea/creatinine (creatinine, urea; plotting creatinine levels on a graphic over a certain period will give an idea of the course of renal failure)
  - diminished creatinine-clearance
  - renal anaemia (low haemoglobin, haematocrit, erythrocyte count)
  - hyperkalaemia, hyperphosphataemia, serum calcium: various results
  - hyponatraemia (in overhydration or diuretics therapy)
  - 1,25-(OH)2-vitamin D3 (Calcitriol) deficiency
  - raised parathyroid hormone
  - metabolic acidosis
  - hypoproteinaemia and hypalbuminaemia in nephrotic syndrome
  - albuminuria/proteinuria
  - erythrocyturia (dysmorphic erythrocytes, erythrocyte cylinders in glomerulonephritis)
  - leucocyturia, bacteriuria in UTI
  - specific urine gravity (finally around 1.010 (isosthenuria) − if not increased due to proteinuria or glycosuria - osmolality < 600 mosmol/kg, urea < 1 g/dl; occ. glycosuria (reduced tubular reabsorption).
▶ ultrasound / doppler-ultrasound:
  Chronic glomerulonephritis or pyelonephritis: shrunken kidneys with irregular surface, narrow parenchyma, occ. detection of cystic kidneys, congested renal pelvis in obstructions of the urinary tract etc.
▶ Other radiographic investigations: Angio-MRI, Spiral-CT (Beware contrast medium !)

Stages of chronic renal failure (NKF- National Kidney Foundation):

Stage	Description	GFR (ml/min/1,73m^2)	Therapy
0	increased risk of renal failure	≥ 90	diagnostics; prophylaxis of renal failure
1	renal damage, but normal renal function	≥ 90	diagnostics + therapy of any accompanying diseases; reduce any progression + cardiovascular risk
2	renal damage with mild renal failure	60-89	as in 1
3	moderate renal failure	30-59	additionally diagnose + therapy for any complications
4	severe renal failure	15-29	prepare renal renal substitution therapy
5	renal failure	< 15	renal substitution therapy

Di.: history + symptoms + laboratory+ radiography

If a high serum creatinine level is found for the first time, all diagnostic and therapeutic efforts must aim to find a reversible cause of renal failure, or to find a treatable underlying disease. First of all one has to differentiate between acute renal failure and chronic renal failure as a cause for the raised creatinine.

Th.: In case chronic renal failure with raised creatinine has been diagnosed, everything must be done to delay a progression of the renal failure. Refer patients early to a nephrologist!

Factors for progression of renal failure:
1. intraglomerular hypertension and hypertrophy
2. systemic hypertension
3. proteinuria
4. protein intake
5. phosphate intake
6. hyperlipidaemia
7. metabolic acidosis
8. tubulointerstitial diseases

A) conservative therapy: start early!
   1. treat the underlying renal disease

   2. avoid any nephrotoxic substances :
      e.g. aminoglycosides, analgesics, NSAID; restrictive (strict) indication for any use of radiographic contrast medium. Preventive measurements to reduce the toxicity of contrast mediums: adequate hydration of the patients before the investigation and application of Acetylcysteine.

   3. correct the arterial blood pressure and aim for low-normal readings!
      Treat hypertension, as this will cause additional damage to the kidneys, hence increasing the cardiovascular risk.
      Control of arterial hypertension with medication is of significant importance regarding the progression of the renal failure. The following should be aimed for:
      • renal failure and proteinuria < 1 g/24 hours: 130/80 mm Hg
      • renal failure and proteinuria > 1 g/24 hours: 125/75 mm Hg

      Large multicentre intervention studies have shown the following: ACE-inhibitors should be first line treatment in non diabetic and diabetic (type 1) renal failure. For type 2 diabetics Angiotensin-receptor blockers should be preferred. In most cases a combination therapy of several antihypertensive drugs is necessary. There is a direct correlation between the extent of proteinuria and progression of the renal failure. Reduction of protein secretion is of major significance.

   4. Dietetic protein restriction:
      A diet low in protein is supposed to reduce hyperfiltration of the remaining nephrons (hyperfiltration theory) and reduce proteinuria.
      To date, no study results make any recommendations regarding protein restriction for patients with renal failure. Protein restriction of 0,8 g per kg in advanced renal failure (serum creatinine > 2,5 mg/dl) appears to make sense.
      - the diet must provide sufficient caloric intake (> 2.000 kcal)
      - the diet is not supposed to be strictly low in salt; salt loss syndrome even requires additional salt intake (as mentioned above)!
      - a diet low in salt is required in hypertension or oedema due to sodium retention.

   5. increase fluid intake to 2,0 - 2,5 l/day – increase diuresis to 2,5 l/day when water balance is stable! This way serum urea levels can be lowered, but not serum creatinine. With increasing renal failure loop diuretics will be required (e.g. Furosemide).
      Over time, the effect of loop diuretics can be diminished, due to compensatory increased resorption in the distal tubular. This is called diuretic resistance. Hyponatraemia and NSAID also lead to diuretics resistance. Instead of high dose monotherapy with a loop diuretic, the loop diuretic can be combined with a Thiazide. This sequential nephron block will avoid the diuretics resistance. This may cause increased loss of potassium + magnesium → monitor electrolytes!
      - give bicarbonate to compensate any acidosis:
      When serum bicarbonate falls below < 22 mmol/l, correction of the renal acidosis with medication should be commenced. These recommendations are based on the fact, that chronic metabolic acidosis increases bone resorption. Optimised correction of acidosis will delay the progression of secondary hyperparathyroidism in patients with high-turnover-osteopathy. It stimulates bony-turnover in patients with low-turnover-osteopathy.
      - monitor water-, electrolyte- and acid-base-balance and correct any imbalance: regular electrolyte monitoring (serum and urine), urine volume and weight. Salt deficiency is the most common cause of diminishing diuresis! Daily NaCl-intake has to be adapted according to its loss in urine.
      - prophylaxis and treatment of hyperkalaemia:
      Eliminate any underlying causes (e.g. diet rich in potassium? catabolism?), treat acidosis, give ion exchange salts (Na^+- or Ca^{++}-based, depends on the individual case): potassium sparing diuretics are contraindicated! ACE-inhibitors and Cotrimoxazole can contribute to hyperkalaemia.

   6. Consider any change in pharmacokinetics: Renal failure requires a reduction of the maintenance dose (normal first dose) for renal eliminated drugs. Select drugs, that don't require an alteration of the dosage despite renal failure. If in doubt, monitor the medication drug levels. (Cardiac glycosides in renal failure: refer to chapter "cardiac failure"!). Avoid any nephrotoxic drugs! Avoid any radiographic contrast medium!

   7. prophylaxis and treatment of renal osteopathy: refer to the relevant chapter in "annex"

   8. therapy of renal anaemia: give Erythropoietin. Target-Hb: 11–12 g/dl (refer to chapter "renal anaemia")

9. treat any cardiovascular risk factors

10. treat uraemic pruritus with selective UV-phototherapy (SUP): use UV-rays between UVA and UVB.

B) renal replacement therapy

Targets: Elimination of water + creatinine, urea, uraemia toxins), correct any disorder of electrolyte- and acid-base-balance; avoid any complications of chronic renal failure

Procedures:

a) extracorporal haemodialysis (HD) and peritoneal dialysis (PD):

- HD: this is the most frequently used (85 %) type of dialysis: dissolved substances travel along a concentration gradient through a semipermeable membrane from blood into the isotonic/isoionic dialysate fluid. The concentration gradient between blood and dialysate is maintained by the machine. In the presence of an additional osmotic or physical pressure gradient between blood and dialysate, then there is also the possibility to withdraw water from blood (and body) (= ultrafiltration).
  In extracorporal haemodialysis synthetic semipermeable membranes are used. In order to obtain a repeatedly easy access to blood vessels in patients undergoing chronic-intermittent dialysis, an arterio-venous shunt has to be constructed (e.g. Cimino-shunt between A. radialis and V. cephalica).
  Chronic-intermittent haemodialysis (HD) is performed 3 x per week in a dialysis centre or at home (home dialysis) for 4 - 8 hours at a time (depending on remaining renal function and body size).
  Note: Daily haemodialysis (2 hours/day) improves the patients' wellbeing. Hence continuous HD is becoming more and more popular.

- PD: Peritoneal dialysis: the peritoneum acts as the semipermeable membrane (with an exchange surface of ca. 1 m^2). The abdominal cavity acts as the container for the dialysate, which is being instilled by a Tenckhoff-catheter.

    Possible rinsing fluids:

    -potassium free glucose solution (adapted to the serum electrolyte levels); this will damage the peritoneum after a few years, depriving it of its ultrafiltration capacity

    -the glucose polymer "Icodextrin"

    disadvantages: loss of protein and glucose, increased risk of peritonitis, catheter infections

    advantages: regular removal of urea/creatinine, longer maintenance of the remaining renal function, no blood loss etc.

    2 variations:

    CAPD = continuous community based PD: manual PD with 4-5 bag changes/day with 2 or 2,5 l dialysate each.

    CCPD = continuous cyclic PD = NIPD (nocturnal intermittent PD): Automatic nocturnal PD (with the support of a so called cycler): 6 cycles per night with a total turnover of 12-15 l dialysate. Advantages compared to CAPD: patients are more mobile at daytime and also cosmetically less deprived (ideal for working people).

b) haemofiltration:

This method mimics the glomerular filtration: Blood flowing from a vein runs across a membrane, which is permeable for molecules of a weight of 35.000 Dalton. The pressure gradient causes a fluid similar to the primary filtrate; currently this fluid cannot be processed any further (tubular function can't be reproduced). The haemofiltrate containing urea/creatinine is discarded, and substituted with the same volume of isotonic/isoionic fluid. The cleared blood is returned to the patient. Haemofiltration is equally as good as dialysis; it provides less strain for the cardiovascular circulation. 14 -18 l are exchanged 3 times/week.

2 variations:

• continuous arterio-venous haemofiltration (CAVH):
  the physiological pressure gradient between arteries and veins is used for this method.

• continuous veno-venous haemofiltration (CVVH):
  Uses a pump

c) haemodiafiltration:

This method combines the advantages of haemodialysis (good elimination of low molecular substances) with those of haemofiltration (good elimination of medium sized molecular substances). Chronic renal failure requires 3 sessions/week lasting 3,5 -5 hours each. Long term results (mortality) are better than in haemodialysis.

Ind: 1. permanent dialysis therapy for chronic renal failure:

    indications to start dialysis

    - uraemic symptoms (nausea, vomiting, reduced fitness, insomnia, tiredness, pruritus)
    - uraemic pericarditis, uraemic encephalopathy
    - therapy resistant hypertension
    - hyperhydration with fluid lung/oedema
    - hyperkalaemia (serum potassium > 6,5 mmol/l: emergency indication)
    - renal acidosis, pH < 7,2; base excess > -10 mmol/l

133

- renal anaemia, Hb < 8,5 g/ dl despite adequate substitution with iron / Erythropoietin
- serum creatinine > 8 - 10 mg/dl
- serum urea > 160-200 mg/dl
- glomerular filtration rate (GFR, arithmetic mean of creatinine- and urea clearance) < 10,5 ml/min/1,73 m^2, in accordance with creatinine clearance of 9 - 14 ml/min
- GFR < 15,5 ml/min/1,73 m^2 and diabetes mellitus
- GFR < 15 - 20 ml/min and malnutrition

2. acute renal failure
   - anuria > 12 hours post conservative therapy
   - rising serum creatinine > 1,0 mg/dl in 24 hours
   - hyperkalaemia, azotaemia, acidosis, hyperhydration, uraemia symptoms
   - hyperuricaemia > 12 mg/dl (e.g. tumour-lysis-syndrome)

3. intoxication with dialysable/ultra filterable toxins

4. cardiac induced hyperhydration

Co.: • Shunt: stenosis, thrombosis, haemorrhages, infections, sepsis, steal-syndrome (with pain located in the fingers), aneurysms, cardiac failure
- CAPD-associated peritonitis and occ. deep tunnel infection along the catheter (in most cases caused by skin flora: > 70 % Staphylococci)
  Di.: cloudy drain with a high concentration of leukocytes, detection of pathogens
- hypotension caused by a too high ultrafiltration rate (in intermittent haemodialysis)
- muscle cramps
- disequilibrium-syndrome caused by a too fast removal of urea → can lead to cerebral oedema: headache, nausea, vomiting, impaired vision; in severe cases impaired orientation and coma, convulsions. Prophylaxis: gentle dialysis, particularly initial in acute renal failure.
- rarely hypersensitivity reactions, e.g. against membrane material
- hyperhydration and hypertension in uncontrolled fluid intake (monitor the weight daily!)
- life threatening hyperkalaemia in uncontrolled potassium intake
- Hepatitis B (active vaccination !) and Hepatitis C
- HIT II by anticoagulation with Heparin (→ monitor the thrombocytes)
- cerebral aluminium deposits (dialysis dementia) and bone aluminium deposits (beware aluminium containing antacids)
- cachexia due to catabolism
- polyneuropathy
- rarely amyloidosis with carpal tunnel syndrome and amyloid arthropathy
  (Cause: raised ß2-microglobulin levels?)
- psychological problems

prognosis during dialysis: 10-years survival rate for home dialysis is ca. 55 %; the older the patient, the worse the prognosis

C) Kidney transplantation (NTX):

Ind: NTX is the preferred treatment for patients with terminal renal failure. Even an optimized dialysis is never as good as the function of a transplanted kidney. NTX is definitely better than dialysis!

The transplantation acts of each country of the EU are the basis for the guidelines.

Blood group compatibility of the A-B-O-system is the basic condition. The following rules apply:

donor blood group	recipient blood group
0	→ 0
A	→ A, AB
B	→ B, AB
AB	→ AB

Current trials: blood group incompatible live kidney donation after elimination of the blood group-antibodies from the recipient via immunoadsorption.

CI: absolute: Metastatic malignancy, active systemic infection, HIV-disease, life expectancy < 2 years; relative: Advanced arteriosclerosis, lacking compliance etc.

Patients with preformed antibodies against donor HLA-antigens (immunised e.g. by previous transplantations, blood products or pregnancies) are at an increased risk of rejection. All patients are tested every three months for any preformed antibodies during the waiting period (panel-reactivity-test).
Just before the transplantation any so far unknown antibodies must be excluded via the lymphocyte-cross-match-test. Positive cross-match-test means CI for the transplantation.

Cadaver donation: waiting period: ca. 5 - 6 years
All patients are reported to Eurotransplant Leiden by their responsible transplantation centre. Eurotransplant brings together the waiting lists of the participating countries (Belgium, Netherlands, Luxemburg, Austria,

Slovenia, Germany) and finds any donated organs. There is a score-system (ETKAS) to allocate the organs; criteria for scores are; urgency, HLA-compatibility (tissue match), waiting period, distance to the place of explantation etc.

First warm ischaemia period: the time between ceasing of organ circulation and reaching a temperature of 4°C (few minutes). Cold ischaemia period: the time between cooling and start of anastomosis (up to 36 hours). Second warm ischaemia period: the time between the onset of anastomosis and onset of circulation.

Op.: The new kidney is placed extraperitoneally into the iliac fossa.

Live donation: ca. 20% of all kidney transplantations in Germany. Suitable donors are: close relatives or close friends.

Prognosis regarding renal function is –even in absent HLA-compatibility - better than after a cadaver donation (short ischaemia period, sometimes better postoperative compliance).

immunisation before transplantation
standard vaccinations (polio, diphtheria, tetanus), hepatitis B, Pneumococci, influenza

Postoperative prophylaxis:
-against Pneumocystis jiroveci (previously P. carinii): Cotrimoxazole, 480 mg per day or 920 mg every 2.-3. day for at least 4 months. Cotrimoxazole-allergy: give Pentamidin-inhalation 300 mg 1 - 2 x / month
-against CMV (positive donor and negative recipient): e.g. Valganciclovir, 3-4 months. (adapt dosage according to renal function)

Lifelong immunosuppression post transplantation:
Start with a triple combination (e.g. Calcineurin inhibitor, Mycophenol acid-derivate, corticoid) (various protocols in various transplantation centres), after a while this may be reduced to a combination of two drugs with a slow reduction of the steroids.

Immunosuppressants:
Steroids, Calcineurin inhibitors (Cyclosporin A = CyA, Tacrolimus), purine synthesis inhibitors (Azathioprin, Mycophenolate-Mofetil, Mycophenolate-Na), proliferation inhibitors (Sirolimus, Everolimus)

complications after transplantation:
1. postoperative complications:
   haemorrhage or thrombosis of the renal blood vessels, lymphoceles, ureter leaks, acute renal failure of the transplanted kidney etc.
2. rejection (Banff-classification):
   • normal
   • antibody-mediated rejection
   • borderline-changes
   • acute/active cellular rejection
   • chronic/sclerosing allograft-nephropathy
   • other changes
   acute rejections usually can be treated successfully with pulsed corticosteroid therapy or intensive immunosuppression (give Tacrolimus instead of CyA). Steroid resistant rejections are treated with ATG or OKT3.
   IL-2-receptor-antibodies (Basiliximab, Daclizumab) are used perioperative to prevent acute rejections.
3. side effects of the immunosuppressive therapy:
   Increased tendency to infections: UTIs, sepsis, Pneumocystis jiroveci, CMV, HSV, EBV, VZV, Polyoma-BK-virus.
   Up to 80% of patients with Polyoma-BK-virus-nephropathy will lose their transplant. Di.: Decoy-cells in the urine sediment (= epithelium cells with enlarged nuclei and intranuclear virus inoculations), PCR (virus-DNA), occ. renal biopsy. Th.: reduce the immunosuppression, antiviral therapy (Zidovir), i.v.-immunoglobulines etc.
   malignancies: skin tumours; post-transplantation lymphoproliferative diseases (PTLD) = EBV-associated B-cell-lymphomas (prophylaxis with EBV-antibodies?)
   Gingival hyperplasia (CyA), leukopenia, thrombocytopenia, nephrotoxicity (CyA), impaired wound healing (Sirolimus), pneumonitis (Sirolimus), steroid induced SE
4. The original disease affecting the transplanted organ (e.g. glomerulonephritis)

Prg.: 5-years-functional rates after cadaver donation is ca.70 %, after life donation up to 90 %
The 3 most frequent causes of death after kidney transplantation are:
- cardiovascular complications (ca. 50%)
- infections (ca. 20%)
- malignancies (ca. 10%), particularly B-cell-lymphomas and skin tumours
Important for the prognosis of the renal function after transplantation: optimal blood pressure (< 135/85 mm Hg, in the presence of proteinuria < 125/75 mm Hg) and treatment of any hyperlipidaemia and proteinuria; weight reduction and stop smoking.

The German transplantation act of 1997 lays down organ donation as combined duty of the transplantation centres and of the regional hospitals. All hospitals in Germany are obliged, to report any possible organ donors to a central coordination authority.

Stipulations for an organ donation:

1. Proof of cerebral death

The diagnosed, complete and final loss of all brain functions describes a certain internal sign of death of the human. This diagnosis has to be performed by two doctors, who are very experienced in the intensive care treatment of patients with severe intracranial damage.

The examiners must not be related to each other, and they must not participate in the organ removal or -transfer.

The diagnose of cerebral death follows the "guidelines on diagnosis cerebral death" of the "Bundesärztekammer" = National Medical Council.

2. donor's consent

When the deceased patient does not have a donor pass, or when he/she did not express any wish regarding organ donation orally, relatives will be asked to make a decision. They should consider the likely preferred wishes of the deceased.

3. no medical contraindications

Contraindications for organ donation: HIV-infection, i.v.-drug abuse, active tuberculosis, sepsis with multiresistant pathogens, incurable malignancies (apart from some brain tumours).

There is no age limit. Current organ function is what matters.

The German Foundation Organ transplantation (DSO) is the nationwide coordination centre for organ donations. It can be contacted via its organisation centres at all times; it sends staff (coordinators) to hospitals, to support hospital staff, and to deal with any questions arising around the organ donation and with the coordination. Donor data are anonymised before being reported to the international agency Eurotransplant in Leiden, Netherlands.

The coordinators also advise hospital staff during the whole process of organ donation with all medical and logistic questions.

Once Eurotransplant has determined the recipients for the donated organs, the DSO organises the organ explantation by a specialised surgical team, and organises organ transport to the transplantation centres, where the transplantations will be performed.

Internet-Infos: *www.eurotransplant.nl*

Renal Osteopathy   [N25.0]

Def.:   Various osseous changes caused by chronic renal failure:

1. High turnover-Osteopathy:
   - Secondary hyperparathyroidism

   30% of patients with raised serum creatinine of just 2 mg/dl develop symptoms of secondary hyperparathyroidism. Creatinine increase of 5 mg/dl: 80 % of all patients have symptoms of secondary hyperparathyroidism.

2. low turnover-osteopathy (particularly seen in dialysis patients):
   - aluminium induced osteopathy
   - aplastic bone disease (= adynamic osteopathy)

3. mixed disorders

1. High turnover-osteopathy and secondary hyperparathyroidism:

The following factors are important for the development of secondary hyperparathyroidism:

a) hypocalcaemia

b) low 1,25-(OH)2-vitamin D3-levels

c) skeletal resistance to the calcaemic effect of parathyroid hormone

d) phosphate retention: In advanced stages of renal failure (creatinine-clearance > 30 ml/min) hyperphosphataemia can be seen. Phosphate retention carries the risk, that the increased calcium-phosphate product (Ca x P > 5,7 mmol/l) can cause extraosseous calcification, and that hyperphosphataemia (via reduction of ionised serum calcium) can cause additional stimulation of PTH-secretion. Hyperphosphataemia also inhibits metabolization of 25-OH-vitamin D3 into the active 1,25-(OH)2-vitamin D3 in the kidneys.

Stimulation of PTH-secretion has 3 causes:

- disorder of the gastrointestinal calcium absorption with lowered ionised calcium, and resulting stimulation of PTH-secretion.
- reduced suppressing effect of 1,25-(OH)2-vitamin D3 on the biosynthesis and secretion of PTH in the parathyroid glands
- direct stimulation of parathyroid hormone secretion by hyperphosphataemia.

2.: <u>low turnover-osteopathy:</u>
- <u>aluminium overload</u> leads to reduced osteoblast differentiation and increased synthesis of bone matrix (due to the disturbed osteoblast function).
- <u>aplastic bone disease:</u> there is relative hypoparathyroidism; plasma-PTH is - compared to the low bone turnover - inadequately low.

Sym.: Radiological changes can be found in 30 - 40% of all patients; histological proof of renal osteopathy in almost all patients with chronic renal failure. But bony symptoms are only seen in 5 - 10 %.
The following 3 guidance symptoms can be a hint of the presence of renal osteopathy:
- diffuse bone pain of the spine, pelvis, ribs, hips, knees and ankles.
- pathological fractures of ribs, vertebrae and hips.
- muscle weakness, particularly proximal leg muscles (waddling gait).

Di.: <u>renal osteopathy:</u>
   - <u>laboratory:</u>
   Intact parathyroid hormone **t** (> 45 pmol/l or 450 pg/ml), <u>increased alkaline phosphatase, increased serum phosphate</u> are typical for secondary hyperparathyroidism.
   Intact parathyroid hormone < 10 pmol/l (100 pg/ml) can be seen in aplastic bone disease.
   Intact parathyroid hormone of 10 - 45 pmol/l (100 - 450 pg/ml) plus increased aluminium levels (or pathological Desferal-test) indicate aluminium induced osteopathy.
   - bone biopsy and histology
   - <u>radiographic signs of renal osteopathy are always signs of advanced disease:</u>
   hands, spine: subperiostal resorption, reduced cortical density, horizontal stripes in the vertebral bodies etc.

Th.: • <u>treat the hyperphosphataemia and monitor serum calcium:</u>
Pro.:   - <u>dietetic phosphate restriction</u> (reduce intake of dairy, boiled sausage, inner organs, egg yolk, pulses etc.).
     - <u>Oral therapy with calcium containing phosphate binder</u> (calcium acetate, calcium carbonate)
     <u>SE:</u> risk of developing hypercalcaemia, particularly in simultaneous treatment with Vitamin D-compounds and underlying low-turnover-osteopathy. Increased calcium-phosphate-product and a positive calcium balance cause metastatic calcifications, particularly cardiac calcifications with valve calcification, and noticeable calcifications of the coronary arteries; these will increase the cardiovascular risk of patients suffering from terminal renal failure significantly.
     <u>Additional SE of calcium carbonate.:</u> belching, constipation or diarrhoea.
     <u>Aluminium containing phosphate binders should be avoided</u>, since they can cause <u>aluminium intoxication</u> (encephalopathy, anaemia and osteopathy).
     <u>Dos:</u> e.g. calcium carbonate 2 - 3 g/day with meals
     Calcium carbonate also counteracts any acidosis.
     monitor serum calcium and phosphate: The product "calcium x phosphate" must not exceed 5,3 mmol/l.
     Ca. 15 % of all patients are forced to reduce the dose due to hypercalcaemia.
     - <u>calcium- and aluminium free phosphate binders:</u>
     · Sevelamer (Renagel®)
     · Lanthancarbonat (Fosrenol®)

• <u>treat with active vitamin D for patients in stage III and IV of chronic renal failure</u>
Patients with chronic renal failure in stage III and IV require drug therapy with a active oral vitamin D-compound (Calcitriol, Alfacalcidol, Doxercalciferol, Paricalcitol); it is indicated, when serum levels of 25 Hydroxyvitamin D is > 30 ng/ml (75 nmol/l), and serum concentration of intact parathyroid hormone is > 70 pg/ml (7,7 pmol/l) for stage III, and > 110 pg/ml (12,1 pmol/l) for stage IV.
The oral initial dose is 0,25 µg Calcitriol or 0,25 µg Alfacalcidol/d.
Doxercalciferol: 2,5 µg 3 x/week
Serum calcium should be < 2,37 mmol/l, and serum phosphate < 1,5 mmol/l.

<u>Monitoring:</u> Serum calcium and phosphate should be monitored monthly for a period of 3 months after the start of therapy; after that every 3 months. Serum-PTH-levels are monitored every 3 months.
When serum levels of the intact parathyroid hormone drop below the target level for the respective stage of renal failure, active vitamin D-compound should be stopped until PTH serum levels have reached the target level. After that vitamin D-compound should be restarted at half the daily dose. An alternative would be to take the vitamin D compound every other day.

• <u>therapy with active vitamin D for patients in stage V of chronic renal failure</u>
Patients in stage V of chronic renal failure (haemodialysis and peritoneal dialysis-patients) should receive therapy with active vitamin D (e.g. Calcitriol, Alfacalcidol), when serum concentration of intact PTH 300 pg/ml exceeds (33,0 pmol/l).
The following table shows target levels.

Parameter	Target
intact serum parathyroid hormone	150 – 300 pg/ml (16,5 – 33 pmol/l)
serum calcium	8,4 – 9,5 mg/dl (2,10 – 2,37 mmol/l)
serum phosphate	3,5 – 5,5 mg/dl (1,13 – 1,78 mmol/l)
calcium x phosphate	< 55 mg^2/dl^2 (< 4,5 mmol^2/l^2)

A drop in serum-PTH below 150 pg/ml carries the very high risk of subnormal bone formation. If serum-PTH levels drop < 65 pg/ml, then all patients will develop adynamic bone disease.

Some studies have shown an association between hypovitaminosis D and secondary hyperparathyroidism. Hence substitution with physiological doses of native vitamin D2 or D3 is advised for dialysis patients.

- Calcimimetics: Cinacalcet (Mimpara®)

  Mode of action: Activation of calcium-sensing-receptors in the cells of the parathyroid gland → this increases the sensitivity to extracellular calcium → PTH-secretion and calcium-phosphate-product drop.

  Cinacalcet is licensed in the USA and Europe for the treatment of secondary renal hyperparathyroidism in patients with renal failure receiving dialysis. The exact importance of Calcimimetics in relation to the standard therapy of secondary hyperparathyroidism yet has to be established.

  therapy and prevention of low turnover osteopathy:

  1. avoid any aluminium containing phosphate binders and aluminium containing dialysate → use Sevelamer
  2. treatment of aluminium-induced osteopathy with Deferoxamin (Desferal)
  3. avoid any excessive PTH-suppression

RENAL TUMOURS

1. Mesenchymal tumours: rare
2. Epithelial tumours:
   - Benign: renal cortex adenoma
   - malignant: renal cell carcinoma (Hypernephroma)
3. mixed tumours:
   - rarely benign forms (Angiomyolipoma)
   - malignant: Nephroblastoma

Adenocarcinoma [C64]

Syn: Grawitz-tumour

Vo.: incidence 10/100.000; m : f = 2 : 1; peak > 50 years of age; usually sporadically

Von-Hippel-Lindau-disease (type I and IIb):in ca. 30 % of all cases there is a renal adenocarcinoma. This is a hereditary multisystem disease, associated with various mutations of the VHL-tumour suppressor genes. Autosomal dominant inheritance. Also increased incidence of CNS-haemangioblastoma, retinal angiomatosis tumours.

Aet.: Unknown; risk factors: smoking, analgesic-nephropathy, acquired renal cysts in dialysis patients, occupational toxins (e.g. Cadmium, Trichlorethen). Papillary adenocarcinoma: gene mutation (MET, 7q13).

PPh: originates from the tubular or collecting duct epithelium, in > 80 % small cell carcinoma (the older term „Hypernephroma" is not correct)

staging (TNM-classification):

T0	no primary tumour can be found
T1	tumour up to 7 cm, limited to the kidney (T1a: < 4 cm, T1b: 4 - 7 cm)
T2	tumour above 7 cm, limited to the kidney
T3	invasion into larger veins, adrenal gland or perirenal fat without exceeding the Gerota-fascia
T3a	Invasion into adrenal gland or perirenal fat pad
T3b	Invasion into renal veins or V. cava below the diaphragm
T3c	Invasion into V. cava above the diaphragm
T4	invasion into Gerota-fascia
N0	no lymph node metastases
N1	metastasis of a solitary regional lymph node
N2	metastases in more than one lymph node
M0	no distant metastases
M1	evidence of distant metastases

stage grouping (AJCC)

stage I pT1N0M0
stage II pT2N0M0
stage III   pT3N0 or T1-3N1M0
stage IV   every pT4 / every pN2 / every M1

Sym.: Up to 70 % of adenocarcinomas are coincidental asymptomatic findings in US scans.

There are no typical early symptoms. The following symptoms are very variable and may present late symptoms!

- early tendency to invade the renal pelvis → guiding symptom: haematuria (60 %)

> *Note:* Haematuria is the common sign of all tumours of the kidneys and the upper urinary tract! Haematuria is always a serious symptom. Macroscopic haematuria requires immediate cystoscopy (during haemorrhage) to find out which side is affected (in haemorrhages above the bladder).

- loin pain (40 %)
- early invasion into the renal vein, and the haematogenic metastases in lungs, bone, liver, brain (25 % have haematogenic distant metastases at the time of diagnosis).
- unexplained fever, ESR ↑, anaemia
- varicocele of the left testicle when a tumour has invaded the left renal vein
- a palpable tumour means it is inoperable
- paraneoplastic syndromes can be seen due to hormone production by the tumour: hypercalcaemia (parathyroid hormone-related protein = PTHrP), hypertension (Renin), polyglobuly (Erythropoietin), Stauffer-syndrome (impaired liver function with raised AP).

DD: - of haematuria: particularly nephrolithiasis; adenocarcinoma can present with a colic, when blood clots get trapped in the ureter
- of loin pain
- of a sonographic renal tumour (rarely benign tumour, e.g. Angiomyolipoma)

Di.: • renal diagnostics:
(doppler-)ultrasound and angio-CT are first line diagnostic methods, occ. additional arteriography: enhanced vascularisation of the tumour; evidence of tumour invasion into the renal vein and V. cava inferior
• search for metastases: CXR (round shadow ?), skeletal scintigraphy, ultrasound, CT of liver + brain

Th.: Stage I - III radical nephrectomy using the no-touch-technique, i.e. to tie all blood vessels leading to and from the kidney, before any manipulation of the kidney: En-bloc-removal of tumour and kidney with the perirenal fat pad, adrenal gland, ureter and spermatic- or ovarian blood vessels; removal of all para-caval/para-aortal lymph nodes + occ. removal of tumour cones from the V. cava, operative removal of solitary distant metastases (liver or lungs). Small tumours, located peripherally in the renal cortex (particularly when the patient only has one kidney): an organ saving tumour resection can be considered.

▶ multiple metastases: there is no standard therapy. Various remission rates have been reported for these therapies:
- Interferon alpha combined with chemotherapy
- Bevacizumab (VEGF-antibodies)
- Tyrosinkinase inhibitors: Sorafenib (Nexavar®), Sunitinib (Sutent®)
- Temsirolimus
▶ bone metastases: local radiation and Bisphosphonates

Prg.: 5-years survival rate:
St. I and II: up to 90 %
St. III: without lymph nodes being affected: up to 60 %
    with lymph nodes affected: up to 30 %
St. IV: solitary metastases: up to 40 %
    otherwise < 5 %

Nephroblastoma

Syn: Wilms-tumour (WT) [C64]

Inc.: 7,5 % of all neoplasias in children, peak 3.-4. years, sometimes autosomal-dominant inherited renal tumour, occ. additional congenital malformation; in 5 % bilateral tumours.

Genetics: 2 Wilms-tumour (WT)-genes are unknown:
WT 1 located on gene location 11p13 coded for the "zinkfinger"-protein, a transcription-regulator; occ. WAGR-syndrome: Wilms-tumour, Aniridism, (growth)retardation
WT2 on gene location 11p15,5

Aet.: Unknown

Sym.: palpable abdominal tumour, abdominal pain, anorexia, vomiting, cc. fever, haematuria

Di.: ultrasound, CT, NMR, angiography

Th.: interdisciplinary strategy tailored to the according stage: Operation (generous nephrectomy), chemotherapy, radiotherapy; resection of solitary metastases

Prg.: 5-years survival rate for all cases is ca. 90 %.

KIDNEY CYSTS AND CYSTIC KIDNEYS

Kidney Cysts [Q61.3]

Kidney cysts can be solitary, multiple unilateral or bilateral. In most cases they are asymptomatic and found coincidentally without any therapeutic consequence.

- **Co.:**
  - large cysts: occ. back or abdominal pain
  - occ. polyglobuly
  - occ. hypertension
  - rarely combination with haemangioblastoma of the cerebellum or retina (Hippel-Lindau disease)
  - rarely proceeds into malignancy

- **DD:**
  - cystic dilation of the pelvis calyx
  - haematoma
  - haemangioma
  - echinococcus cyst
  - abscess
  - tuberculous cavity
  - dermoid cyst

- **Di.:**
  - ultrasound
  - in case of suspected malignancy: cytology and endoscopy

- **Th.:**
  - asymptomatic cysts don't require any treatment.
  - large cysts with complications: maybe puncture (with cytology) + sclerotherapy or cyst resection

Cystic Kidney Diseases [Q61.9]

Syn: Cystic kidneys

1. Autosomal recessive polycystic nephropathy (PKHD1 or ARPN or ARPKD [Q61.1])
   Syn: Infantile polycystic kidney disease, cystic kidneys type I after Potter
   Gene location: 6p21.1-p12
   incidence: 1 : 20.000
   Sym.: always bilateral + combined with congenital liver fibrosis. 50 % of all children with ARPN die in the perinatal period due to respiratory complications (pulmonary hypoplasia, respiratory distress syndrome).

2. Autosomal dominant polycystic nephropathy (ADPN or ADPKD)
   Syn: cystic kidneys type III after Potter, adult polycystic kidney disease
   incidence: frequent! 1 : 1000
   2 gene locations: ADPN1 (85 %): 16p13,3: Polycystic breakpoint gene ADPN2 (15 %): 4q21-q23
   manifestations age: after 20 years of age, in most cases bilateral, cysts measure up to several cm ⌀. Concurrent liver cysts (almost 100 %); cerebral arterial aneurysms (in ca. 10 %) with the risk of aneurysm haemorrhage (particularly in the presence of hypertension), pancreas cysts in 10 %.
   Sym.:- occ. loin pain, occ. macrohaematuria
       - pathological urine sediment (90 %): proteinuria, erythrocyturia
   Co.: - UTI, occ. with infected cysts (→ renal abscess)
       - renal hypertension (50 %, in renal failure 100 %)
       - renal failure (in ADPKD2 onset at an advanced age)
   2 common causes of death: renal failure and aneurysm haemorrhage

3. cystic kidney dysplasia:
   Syn: cystic kidneys type II after Potter
   Uni- or bilateral. Unilateral manifestation: first symptoms can be UTIs in childhood; severe bilateral manifestation causes congenital renal failure.

4. cystic nephropathies a part of malformation-syndromes: e.g. the autosomal-recessive inherited Meckel-syndrome, often with polydactylia + cerebral anomalies

5. other cystic kidney diseases: e.g. medullary sponge kidneys:
   Cystic dilatation of the collecting duct in the papillary area, where nephrolithiasis can develop → urography: pearl-necklace-shaped calculi in den renal papillae + star-shaped stripes of contrast medium in the medulla
   Co.: nephrolithiasis (with colic) + pyelonephritis

- **Di.:**
  - ultrasound > 3 cysts per kidney and positive family history
  - urography (search for any additional malformations of the urinary tract)
  - MR-Angio (search for any cerebral aneurysms)

<u>**Th.:**</u>   only symptomatic, e.g.
- <u>laparoscopic cystostomy</u>, with the aim to prevent any pressure damage of the remaining renal tissue!
- <u>treatment of hypertension</u> (which worsens renal function)! target level: 125/75 mm Hg
- treat any UTI!
- treat any renal failure
- genetic counselling for ADPKD
- clinical trials with <u>Everolimus</u>, a TOR-inhibitor, which is supposed to stop cyst growth

NEPHRONOPHTHISIS – medullary cystic kidney disease

The most common variation is the autosomal recessive inherited infantile and juvenile nephronophthisis. Most cases are due to the deletion of the NPHP1-gene on chromosome 2q12-q13 (juvenile form).
Medullar cystic kidney disease is inherited autosomal dominant. There are 2 genes for this disease, MCDKD1 and MCDKD2; they are located on chromosome 1q21 and 16p12. The gene for familial juvenile hyperuricaemic nephropathy (FJHN) is located on chromosome 16p12.

<u>**Sym.:**</u>   Progressive renal failure with renal anaemia.
Recessive juvenile nephronophthisis starts in childhood and leads to terminal renal failure in the second decade of life. The most common extrarenal manifestation is the tapeto-retinal degeneration; growth arrest in most cases.
Medullar cystic kidney diseases becomes manifest at the age of 30 – 40 years, and cause renal failure at the age of 50.

<u>**Di.:**</u>   (family-)history, symptoms, laboratory, US (small and dense kidneys)

<u>**Th.:**</u>   symptomatic; dialysis in terminal renal failure

MEDULLARY SPONGE DISEASE

Non-inherited congenital deformity of the kidneys with medullary ectatic dilatation of the collecting tubes in the pyramides. Incidence is ca. 5:10.000-100.000. In 75% these changes are bilateral. Hypercalciuria, nephrolithiasis and/or nephrocalcinosis (50 %) are the main symptoms. Nephrocalcinosis can cause renal failure.

<u>**Di.:**</u>   history, symptoms, X-ray (calcium deposits in the papillary tips)

<u>**Th.:**</u>   Symptomatic

UROLITHIASIS   [N20.0]

<u>**Def.:**</u>   urinary calculi can be located in the kidney (nephrolithiasis), in the ureter (with ureteric colic), in the bladder and rarely in the urethra.
They consists of a matrix (Uromucoid) and urine crystallisations.

<u>**Types of urinary calculi:**</u>
1. <u>calcium oxalate</u> (75 %)
2. „infection calculi" (10 %): <u>Struvite</u> (= magnesium-ammonium-phosphate), carbonate apatite
3. <u>urate stones</u> (5 %)
4. <u>calcium phosphate</u> (5 %)
5. rare stones: cystine stones in cystinuria, xanthin stones, 2,8-Dihydroxyadenin (DHA)-stones in the presence of the rare autosomal-recessive inherited defect of Adenin-phosphoribosyl-transferase (APRT)

<u>**Ep.:**</u>   prevalence: 5 % (Germany) up to 8 % (USA), m : f = 2 : 1; peak between 30. - 60. years
Hot and <u>arid areas</u> are endemic for stones. An affluent lifestyle is directly related to a <u>diet rich in protein</u>, and hence increased urine excretion of urate-, oxalate- and calcium, and increased risk of this disease (which is rare in poor countries like India and Pakistan).

<u>**Aet.:**</u>   Multifactorial disease, a surfeit of lithogenic substances in the urine caused by various <u>metabolic factors:</u>

• <u>increased excretion of lithogenic substances in the urine:</u>
- <u>hypercalcaemia</u> of various origin (primary hyperparathyroidism, vitamin D-overdose, immobilisation etc.)
- idiopathic hypercalciuria = hypercalciuria with normocalcaemia
- hyperoxaluria (> 0,5 mmol/d)
- hyperphosphaturia (> 35 mmol/d)
- hyperuricosuria (> 3,0 mmol/d) in hyperuricaemia
- cystinuria (> 800 µmol/d)

- reduced excretion of antilithogenic substances (inhibitor deficiency) in urine:
  - hypomagnesiuria (< 3 mmol/l)
  - hypocitraturia (< 3 mmol/l)
- critical urine-pH (≤ 5,5 and > 7,0)
- urine concentration too high (specific weight ≥ 1015 g/l)

other factors:
- urinary tract obstruction (caused by anatomical or functional changes)
- UTIs
- immobilisation
- nutrition rich in protein, lack of fluid intake, weight reduction

Note: Nephrolithiasis and UTI favour each other (gram-negative bacteria – apart from E. coli - divide urea into NH3 and CO_2 using the enzyme urease; this renders the urine alkaline, and changes the solubility of ions).

Sym.: a mobilised small urinary calculus irritates the ureter, and triggers the most common symptom:
- ureteric colic: severe pain coming in waves, associated with severe restlessness. Depending on the position of the stone, the pain is located either in the back and/or lateral lower abdomen; low ureteric stones cause radiation of the pain into scrotum/labia. Frequently there is nausea or vomiting, constipation and inability to pass wind (refectory ileus). Only very small amounts of urine are passed during colic; at the same time there is bladder tenesmus.
- haematuria (microscopic haematuria is almost always present, macrohaematuria in 1/3 of all cases)

Co.: most frequent and most important complication: UTI, which can develop into urosepsis. In very rare cases the obstructed urinary tract can cause fornix rupture.

DD: 1. other kidney diseases:
  - tumours of the kidneys and the urinary tract (ureteric blood clots can trigger colic – arrange renal ultrasound)
  - renal infarction (most frequently caused by embolism due to atrial fibrillation): proteinuria, haematuria, very high LDH, only minor changes of GOT and AP, occ. raise in blood pressure after a few days; doppler sonography
  - papillary necrosis, e.g. seen in analgesics nephropathy (papillary damage can be seen in a urogram)
  - renal vein thrombosis (proteinuria, left sided thrombosis: venous swelling of the left testicle in men) → doppler US

2. extrarenal diseases:
  - appendicitis (more insidious onset; a colic is more sudden, abdominal tenderness over McBurney's point etc.)
  - twisted ovarian cyst, ectopic pregnancy, adnexitis (gynaecological advice, ultrasound, pregnancy test)
  - ileus (paralytic: history; mechanical: hernias?, use stethoscope, arrange US, X-ray)
  - pancreatitis (amylase, lipase)
  - biliary colic (pain radiation into the right shoulder, US)
  - diverticulitis (history, palpation etc.)
  - LS-syndrome (occ. pain radiation into legs, Lasègue' stretching pain)
  - lumbar Herpes Zoster
  - testicular torsion (doppler; diagnosis must be complete within 6 hours, otherwise the testicle is lost!)

diagnostics:

Lab.:
- use urine-dipstick and test for pH, specific weight, erythrocytes, leukocytes, bacteria, protein (e.g. Combur 9®). Dipstick testing Cystine (e.g. Urocystin®)
- urine balance for calcium, urate, oxalate, phosphate, cystine (in children also test for DHA):
  - hypercalciuria/-aemia, parathyroid hormone ↑: primary hyperparathyroidism
  - hyperuricaemia, hyperuricosuria: urate stones in gout
  - bacteriuria, leucocyturia → UTI
- serum: calcium, uric acid, Creatinine
- stone analysis of any expelled or removed stones (infrared-spectroscopy or radio-diffractometry)

radiographic diagnostics:
- ultrasound, spiral-CT: small stones often can't be detected in US-scan, but any associated congestion of the renal pelvis can!
- urography: calcium-containing oxalate- and phosphate stones can be easily seen on plain AXR in 80 %! Urate stones and rare cystine stones can only be detected because of the contrast medium leaving an empty space. This empty space can also be caused by a tumour (or blood clots); therefore Spiral-CT is more and more used as the first investigation.
- MR-urography when contrast medium is contraindicated

Th.: 1. conservative therapy of the acute ureteric colic:
  Analgesics: e.g. Pethidin (Dolantin®) 50 mg i.v.
  Note: for minor colic Diclofenac (100 mg/day) can be given.
  ASS is contraindicated before any planned lithotripsy (risk of renal haematomas caused by lithotripsy).

Metamizole is very effective; because of its risk of agranulocytosis (incidence 1 : 1.000) it is not used anymore in the USA, Canada, UK, Sweden, Australia.
The benefit of additional N-Butylscopolamin (e.g. Buscopan®) controversial.

2. Endo-urological manipulations:

2.1. renal pelvis stones:
- extracorporal ultrasound wave lithotripsy (lithotripsy): preferred method!
  The stone is being located with US or X-ray; focusing of the waves onto the stone destroys the stone.
  conditions: it must be possible to locate the stone; urinary tract obstruction must be absent. Insertion of a ureteric splint before lithotripsy will ensure pain free passage of the stone fragments + urine drainage.
  success rate: > 90 %

- percutaneous nephrolithotomy = percutaneous nephrolitholapaxy (PNL or PCNL):
  percutaneous US guided endoscopy of the renal pelvis
  Under direct vision the stone is disintegrated using US, laser-pneumatic or electro-hydraulic procedures. The fragments are flushed out with a suction/pump or extracted with forceps/catching device. This is followed by endoscopic and radiological follow up to check if all stones have been removed; then a nephrostomy-catheter is inserted for drainage for a few days. Any remaining stones that can't be reached, can be removed via lithotripsy.
  Ind: larger stones in the renal pelvis, stone diameter > 30 mm
  SE: rarely postoperative haemorrhage, kidney loss, sepsis, perforations of neighbouring organs; lethality < 1 %

2.2. ureteric stones:
- stones < 5 mm ∅, absence of any signs of infection: conservative stone removal can be attempted. Stones < 5 mm are expelled spontaneously in ca. 90 % (spasmolytics, drink plenty, apply heat, exercise etc.). Monitor urine and temperature (risk of UTI and urosepsis!). Fever and/or anuria requires immediate hospital admission!

- lithotripsy (extracorporal US-wave lithotripsy)
  Ind: pyelon-, calyx stones measuring 5 – ca. 30 mm ∅; ureteric stones
  Co.: cutaneous changes, subcapsular renal haematoma, colic and congestions caused by remaining fragments

- Endo-urological stone treatment with ureteroscopy:
  There are laser-, US-, pneumatic or electrohydraulic lithotripsy. Mechanic tools like forceps or baskets will facilitate the stone removal.

Prg.: > 80 % of all stones up to 2 mm ∅ will be expelled spontaneously. For stones > 5 – 6 mm ∅ this only applies in exceptional cases. 60 % of ureter stones are expelled after ca. 17 days spontaneously (ca. 20 % of proximal, 45 % of mid and 70 % of distal ureter stones are expelled spontaneously).

Pro.: stone prophylaxis: Urinary stones tend to relapse; this fact requires a stone prophylaxis. Relapse rate without prophylaxis: 50 %, with consequent prophylaxis < 5 % !
Drink plenty! The specific gravity of urine should not exceed 1010 g/l→ home-monitoring of specific weight with dipsticks (e.g. MD Spezial®); avoid apple- and grapefruit juice.
diet: only small amounts of proteins of animal origin (little meat and sausage) and a diet low in salt and high in potassium, lose weight.

Further procedure after stone analysis:
- calcium containing stones: exclude hyperparathyroidism
  If hypercalciuria is caused by increased enteric resorption, it is called absorptive hypercalciuria; if it is caused by increased mobilisation of calcium from bone, it is called resorptive calciuria.
  The urinary calcium content can be lowered with Thiazides.

  Note: Even patients with calcium containing stones should aim for the daily calcium intake of 1.000 mg/day as suggested by nutritional science; particularly those at risk of osteoporosis. Diet low in calcium can even increase the incidence of renal stones (Nurses Health Study, Curhan-Study)!

- Urate stones: Litholysis and prophylaxis: only for urate stones: urine neutralisation (vegetarian diet, K⁺/Na⁺-hydrogen citrate = Uralyt U®) to a urine-pH of 6,5 - 7 (dipsticks); drink plenty, diet low in purine, maybe Allopurinol (refer to "gout").
- Oxalate stones: diet low in oxalate (no spinach, rhubarb) is not very useful, since oxalate is produced in the intermediary metabolism. Urine stone inhibitors like magnesium and citrate combinations are advised.
- Phosphate stones: always exclude hyperparathyroidism!
- Infective stones: Often magnesium-ammonium-phosphate stones; stone disease and infection trigger each other! Infected stones require stone removal + tailored antibiotic therapy according to sensitivity; urine acidification: Methionin (e.g. Acimethin®), apple juice, cranberry juice.
- cystine stones: urine alkalisation, Tiopronin etc. drugs

IV. RHEUMATOLOGY

INTERNET- INFOS: *www.rheumanet.org*

With most of the systemic inflammatory rheumatic diseases presented here the aetiology is unknown.

Autoimmune reactions are characterized by the occurrence of autoreactive B cells and T cells. Autoreactive cell clones are also found in healthy people, but they are not activated by other cells and do not react very specifically (i.e. production of auto-antibodies with low receptor binding avidity). A prerequisite for an adequate immune tolerance to our own body is that pro-inflammatory and regulatory mechanisms are in balance.

Apart from genetic predisposition, environmental factors (e.g. smoking and infectious diseases) play a role in the development of autoimmune diseases, where multiple steps often seem necessary before the actual disease can become manifest. When this happens, the T cells react in a variety of ways (e.g. T_{H1} pattern: predominantly cellular activation in rheumatoid arthritis; T_{H2} pattern: predominantly humoral cell activation in SLE).

Various mechanisms facilitate the activation of autoreactive T helper cells:

1. A special subpopulation of T cells (NK cells) undermines T suppressor cell function and thereby induces T helper cell activation.
2. So called regulatory T-cells (T_{regs}, CD4+, CD25+) may show increased or reduced activation
3. Expression of an autoantigen together with an HLA antigen on monocytes and other antigen-presenting cells (i.e. dendritic cells) may lead to activation of T helper cells (e.g. as part of a viral infection).
4. Activation of membrane-bound toll-like receptors (TLRs) may trigger intracellular signal cascading with activation/proliferation of the cells involved, release of cytokines and onset of autoimmune reactions. Different TLR polymorphisms can be identified in various autoimmune diseases.
5. T helper cells may be activated by the alteration of a previously tolerated autoantigen through its conjugation with a viral or bacterial antigen or chemical substance (molecular mimicry, e.g. so-called arthrogenic peptide with HLA B 27).
6. Viruses can activate B cells and cytotoxic T cells without T helper cell intervention.

RHEUMATOID ARTHRITIS (RA) (RA) [M06.9]

Syn.:   Chronic polyarthritis (CP) [M06.9] (CP)

Def.:   RA is a systemic chronic inflammatory disorder resulting in synovitis, arthritis, bursitis, and tenosynovitis. Extra-articular manifestations are also possible. The intermittent, yet progressive course of disease can lead to joint destruction and disability.

Ep.:   Approx. 1% of adult population; at the age of > 55 years approx. 2%, with peak incidence between ages 55 and 75; f:m = 2-3:1; familial predisposition. Up to 70% of RA patients are carriers of HLA-DR4/DQB1 (compared with approx. 25% of healthy individuals); DR4 homozygous patients typically suffer a more serious (erosive) course of disease.

Aet.:   Unknown

Pg.:   In genetically predisposed individuals an unknown trigger mechanism (viral - e.g. EBV - or bacterial antigens?) induces an autoimmune disease with inflammatory infiltration of the joint synovial membrane by autoreactive T helper lymphocytes, B lymphocytes, plasma cells, and dendritic cells (which derive from monocytes/macrophages). At the centre of the immune response is the interaction between lymphocytes and monocytes/macrophages, leading to the production of pro-inflammatory cytokines (e.g. interleukin-1 [IL-1], IL-6, tumour necrosis factor-@ IL-15 etc.), immunoglobulins and autoantibodies against the Fc fragment of IgG (= rheumatoid factors [RF]). This results in complement activation and release of inflammatory mediators and cartilage-degrading enzymes (e.g. collagenase and elastase). Synovial membrane thickening (pannus)   is caused by invasion of macrophage-like cells (type A synoviocytes) and by proliferation of fibroblast-like cells (type B synoviocytes). The pannus proliferates and protrudes over the surface of the cartilage, often inexorably destroying it (tumour-like proliferation). Granulocytes from the synovial fluid that phagocytize immune complexes of IgG and RF are called rhagocytes. The systemic involvement of internal organs as a result of immune complex vasculitis is possible. Antibodies against cyclic citrullinated peptide (anti-CPP antibodies) may already be identifiable years before the onset of clinical manifestation of disease.

Note: Inflammation of the synovial membrane is really the villain in this drama

CI.: 1. Non-specific general symptoms

Fatigue, night sweats, possibly low-grade fever and myalgia; dull, brittle nails, pigmentation changes on the dorsal surface of the hands, palmar erythema (DD: liver disease)

2. Polyarthritis, possibly tendovaginitis and bursitis
   - Chronic synovitis with pannus formation and cartilage destruction are characteristic of RA. RA usually begins symmetrically in the small joints, esp. those of the fingers. Initially, pain on movement and swelling of metacarpophalangeal (MCP) and proximal interphalangeal (PIP) joints of the fingers are common: a firm handshake is painful = Gaenslen's sign; Normal wave-like appearance of PIPs is effaced by swelling. Arthritis of the wrist joint is associated with painful volar flexion of the wrist. Arthritis of the metatarsophalangeal joints can also cause compression pain in the toes. Joint effusions can occur during acute bouts of inflammation. Atrophy of the interosseal finger muscles leads to a sunken-in appearance of the dorsal aspect of the hand. Morning stiffness (>30 min) and poor circulation in any or all of the fingers are early symptoms. The distal interphalangeal joints (DIP) II–V, thoracic and lumbar spine are not affected.

Note: Rheumatoid arthritis patients can be recognized by looking at their hands
   - Carpal tunnel syndrome: synovitis of the tendon sheaths under the transverse carpal ligament can cause compression of the median nerve, leading to paresthesia of the thumb, index and middle fingers and possibly also thenar atrophy. Pain may also occur at the site of compression, especially at night, or on dorsiflexion of the hand. Dx: inspection and ultrasound
   - Sulcus ulnaris syndrome: occurs with elbow joint arthritis Dx: ultrasound
   - Baker's cyst in the knee (hernia of the knee joint capsule) - Dx: ultrasound! DD: rule out DVT of the lower leg!

3. Rheumatoid nodules (approx. 20%) are found in tendons and subcutaneously, especially over the extensor surfaces of joints (particularly the elbows).
   Hi: Wall of fibroblasts, epitheloid cells, and mononuclear cells arranged in the form of palisades around a fibrinoid focus.

4. Possible nail lesions: possibly reddish half-moons in the nail bed during acute attacks; subungual bleeding (vasculitis!); nail growth disorders, etc.

5. Extra-articular manifestations:
   - Heart: pericarditis and heart valve lesions (approx. 30%, mostly asymptomatic), granulomatous myocarditis (possibly with infarction-like ECG changes)
   - Lung: pleuritis (in autopsies 50%, mostly asymptomatic), pulmonary fibrosis (5% - DD: adverse reaction to methotrexate!), pulmonary nodules, bronchiolitis
   - Eyes: keratoconjunctivitis sicca (30%), scleritis, and episcleritis. Corneal ulcer as a rare manifestation with evidence suggesting increased mortality (vasculitis)
   - Blood vessels: digital vasculitis (rarely digital gangrene), vasculitis of the vasa nervorum (polyneuropathy), early-onset arteriosclerosis (possibly related to steroid therapy)

6. Sicca syndrome (secondary Sjögren's syndrome) in about 30% of patients

Co.: • Loss of joint function and joint deformity: I.e.. "swan-neck deformity" of the fingers due to hyperextension at the PIP joint and flexion at the DIP joint; boutonniere deformity of the fingers due to flexion at the PIP joint and hyperextension at the DIP joint caused by slippage of the extensor tendons towards the thumb; ulnar deviation of the fingers, possibly stiffening (ankylosis) of the joints, anterior atlantoaxial dislocation in the cervical spine (Dx: MRI) with cervical myelopathy and risk of sudden basilar cervical spine injury, i.e. related to neck hyperextension when patients are intubated (diagnosis: MRI) etc.
   • Side-effects of antirheumatic therapy (common!): e.g. gastric and/or duodenal ulcers and bleeding caused by NSAIDs; analgesic nephropathy; nephropathy caused by gold, D-penicillamine, etc.
   • Secondary (AA) amyloidosis (5%) with nephrotic syndrome and development of renal failure. An increase in serum amyloid A protein (SAA) is regarded as a risk factor.
   • Rarely, T-gamma-lymphoproliferative syndrome (0.5%) with lymphadenopathy, lymphocytosis (large granular lymphocytes) and granulocytopenia as subtype of Felty's syndrome (see below).

Special forms:
1. Caplan's syndrome: RA + silicosis (among miners)
2. Felty's syndrome: severe adult form of RA: hepatosplenomegaly, lymph node swelling, granulocytopenia. Characteristic lab findings are granulocyte-specific antinuclear antibodies (ANA) (85% of cases); 95% of cases are HLA-DR4-positive.
3. Late-onset rheumatoid arthritis (LORA):   begins after the age of 60, acute onset in 1/3 of cases; initial presentation often in one or several joints (misdiagnosis: inflammatory osteoarthritis!); often aggressive course with joint destruction, occasionally polymyalgic onset (→ DD).
4. Malignant form of RA:
   Rapid joint destruction + vasculitis-related extra-articular manifestations, massive change in inflammation parameters (ESR, CRP, RF, etc.)
5. **Juvenile chronic arthritis (JCA) = Juvenile idiopathic arthritis (JIA):** [M08.0-M09.0]
   Age of manifestation < 16 years

ILAR classification (International League Against Rheumatology) XE "ILAR classifikation"

a) Systemic JRA (Still's disease): [M08.2]
   Intermittently high fever, salmon-coloured maculopapulous exanthema, polyserositis of pleura and pericardium, hepatosplenomegaly, lymph node enlargement, leukocytosis, thrombocytosis, anaemia, high-grade inflammation. Joint involvement initially not predominant. Age of onset during early childhood (infancy). RF and ANA are negative. Complications: macrophage activation syndrome (MAS) with uncontrolled macrophage activation and marked cytokine release

b) Seronegative RA [M08.3]
   Symmetrical manifestation in > 4 large and small joints. ANA positive in 25%, age of onset early childhood, f>m

c) Seropositive RA [M08.0]
   Symmetrical manifestation in > 4 large and small joints. Age of onset: prepuberty. Joint destruction early on. 75% ANA positive, f>m

d) Oligoarthritis (type I): [M08.4]
   Asymmetrical manifestation. Persistent (i.e. 1-4 joints in chronic course > 6 months) or extended (i.e. > 4 joints in chronic course > 6 months). Age of onset: early childhood. In about 50% iridocyclitis in very severe forms, 80% ANA positive, f>m

e) Arthritis with enthesitis [M08.1]
   Oligoarticular onset. Bursitis, enthesitis, anterior uveitis. Onset during school age. Occasionally ANA positive, m>f

f) Psoriatic arthritis [M09.0]
   Arthritis and psoriasis or arthritis and 2 of three criteria (dactylitis, nail changes, psoriasis in 1st degree relatives). Positive RF is an exclusion criterion

g) Other types of arthritis [M08.9]
   Arthritis which does not meet criteria a) – f) or does not meet the criteria of several subtypes

6. RS3PE syndrome: ("remitting seronegative symmetric synovitis and pitting oedema): subtype of seronegative RA with symmetrical arthritis and soft swelling of the dorsal aspects of the hands. Responds well to steroid treatment.

DD:

- Mixed connective tissue disease (MCTD): systemic lupus erythematosus (SLE) (ANA), Sharp syndrome. Ribonucleoprotein antibodies (anti-RNP)
- Vasculitis: e.g. arteritis nodosa (biopsy)
- Spondylarthritis:
  - Association with HLA-B27
  - Ankylosing spondylitis (m>f, HLA-B27, sacroiliitis, vertebral / thoracic mobility ↓)
  - Reactive arthritis following enteritis or urethritis → Reiter's disease: arthritis, urethritis, conjunctivitis/iritis, possibly skin lesions
  - Enteropathic arthritis in the context of chronic inflammatory bowel disease, Whipple's disease
  - Psoriatic arthritis (skin and nail changes; joint symptoms may precede skin lesions)
- Rheumatic fever (ASL titer ↑, Jones criteria met)
- Lyme arthritis: asymmetric arthritis, knee joint often affected, History of tick bite and erythema migrans, antibody against Borrelia burgdorferi
- Brucella arthritis and parvovirus B19 arthritis: rare (pathogen and antibody detection)
- Infectious (purulent) arthritis (mostly monoarthritis with pathogen detection in joint tap)
- Arthritis (arthralgia) associated with viral infections
- Paraneoplastic syndromes in neoplastic diseases (carcinoma polyarthritis)
- Sarcoidosis with arthritis, e.g. Löfgren syndrome: bilateral hilar adenopathy, erythema nodosum, arthritis (ankle), elevation of ACE
- Behcet's disease: vasculitis of unknown etiology; Cl.: uveitis, oral and genital ulcers, arthritis
- Gouty arthritis (MTP joint of big toe = podagra, uric acid ↑, tophi)
- Active osteoarthritis (inflammatory flare of a degenerative joint disease)
- Polyarthrosis of the finger joints [M15.9] in some cases hereditary age-related lesions, DIP joints (Heberden's nodes) or PIP joints (with Bouchard's nodes); OA of the carpometacarpal joint of the thumb (rhizarthrosis); f > m
- Fibromyalgia syndrome: non-inflammatory disease of unknown origin; affects mainly middle-aged women; triad: diffuse pain, fatigue, sleep disturbances; typically presents with pressure-sensitive "tender points" on trunk and limbs.
- DD: carpal tunnel syndrome (compression of the median nerve → predominantly nocturnal pain + paresthesia especially of the first 3 fingers); symptoms triggered on maximum flexion or extension of wrist or percussion of the carpal tunnel (Hofmann-Tinel sign); NCV of median nerve ↓)
  - Chronic wrist strain through extreme flexion or extension (professional, recreational History)
  - Improper healing of a distal radius fracture, including trauma in the wrist area
  - Hormonal causes: hypothyroidism (fT4 ↓), acromegaly (GH ↑), diabetes mellitus, pregnancy
  - Rheumatoid arthritis and other rheumatic diseases
  - Gout (uric acid ↑)

- Amyloidosis in dialysis patients
- Idiopathic carpal tunnel syndrome (>40% of patients, mostly women)

Di.: ■ Laboratory:

a) Non-specific parameters of inflammation (= signs of activity):

ESR and CRP ↑, α/γ-globulins ↑, serum Fe ↓, serum ferritin ↑, normochromic/hypochromic inflammatory anaemia, mild thrombocytosis and leukocytosis

b) Immunological findings:

- Rheumatoid factors (RF) are initially positive in 40% of patients, and with disease progression in approx. 70–80% of cases = seropositive RA (other cases = seronegative RA)

  RF is not specific for RA. RF is also occasionally positive at low titers in the following:

  - approx. 5% of clinically healthy individuals (individuals aged 60 years or more >10%), low positive RF titer, mostly consisting of multi-reactive RF (M-type) capable of binding to various antigens, unlike specific RF against IgG in RA (S type).
  - Sjögren's syndrome (up to 50%), including connective tissue diseases in variable frequency
  - Occasionally liver disease, hepatitis C (resulting from cryoglobulin formation).

- Anti-CCP (cyclic citrullinated peptide)
- Sensitivity comparable to that of RF, high specificity (> 95%); anti-CCP has a high predictive value for the development of aggressive rheumatoid arthritis; level of titer increase correlates with the course of RA.
- Antinuclear antibodies (ANA) in approx. 35% of cases, titer often <1:160 (anti-dsDNA negative!)
- Circulating immune complexes in approx. 50% of cases

■ Detection of cartilage and joint manifestations:

- Arthrosonography: detection of synovitis, tendinitis, tendovaginitis, bursa effusions, Baker's cysts. A hyperperfusion of the synovial membrane detected in PW doppler suggest a high degree of activity. Using high-resolution arthrosonography, an experienced investigator can see signs of wear earlier than in an x-ray and just as early as by MRI.
- MRI shows inflammation, erosion of cartilage and bone up to 2 years earlier than in conventional x-ray
- X-ray: hands/hand joints, forefoot on both sides in 2 planes, cervical vertebrae in 2 planes, incl. x-rays in inclination/reclination. Frequent detection of atlanto-axial involvement, especially in active disease. ***Caution:*** compression in myeloma, MRI if necessary for further investigation

  Steinbrocker X-ray stages of RA:
  St. I: possible juxta-articular osteoporosis
  St. II: in addition, incipient destruction of cartilage and bone
  St. III: in addition, incipient subluxation/malalignment
  St. IV: joint destruction and deformation, joint luxation, ankylosis
  X-ray scores are important for studies (rating, Sharp/V.d. Heijde)

- Scintigraphic diagnostics:
  Two-phase scintigraphy with ^{99m}Tc phosphonate: the most sensitive diagnostic procedure for joint inflammation in the early phase = soft-tissue scintigraphy. Late phase = bone scintigraphy
  - scintigraphy with 99mTc human immunoglobulin: localization of synovitis
- Possible synovial fluid analysis: cell count ↑ (4000–50,000 white blood cells/μL depending on activity), detection of rhagocytes and RF, complement ↓; bacteriology: aseptic

Early-onset, oligosymptomatic, and seronegative cases (RF-negative) can be difficult to diagnose. The following diagnostic criteria may be helpful:

Diagnostic criteria of the American College of Rheumatology (ACR):

1. Morning stiffness for at least 1 hour
2. Arthritis in 3 or more joint areas: soft-tissue swelling or effusion in at least 3 joint areas simultaneously
3. Arthritis of the hand or finger joints: pain + swelling in the wrists, proximal interphalangeal (PIP) joints or metacarpophalangeal (MCP) joints
4. Symmetrical arthritis: simultaneous involvement of the same joint area on both sides of the body
5. Rheumatoid nodules: subcutaneous nodules over bony projections or extensor surfaces or in juxta-articular areas.
6. Detection of serum RF.
7. Characteristic X-ray changes in the hands: juxta-articular osteoporosis and/or erosions (OA lesions alone are not sufficient).

Four of seven criteria must be met to fulfill the diagnosis of RA.
Criteria 1–4 must have been present for at least 6 weeks.

Criteria for "early RA" (guidelines of the German Society of Rheumatology):

Cl.: Swellings in > 2 joints for ≥ 6 weeks and/or symmetrical pattern of distribution and/or morning stiffness ≥ 60 minutes
Lab.: In most cases elevated ESR and/or CRP

Diagnosis confirmed by detection of rheumatoid factor and anti-CCP antibodies

Th.: *__Remember:__ Effective disease-modifying therapy immediately after diagnosis of RA is crucial to the further course of the disease, since major joint damage already occurs in the initial stages of RA (window of opportunity, hit hard and early). A good therapeutic response early on may allow dose to be reduced later on (step-down therapy) depending on the clinical presentation. An initial combination regimen is recommended. The ACR criteria are often only met once irreversible damage has occurred and are therefore not suitable for early diagnosis. Therapeutic goals are not only symptom improvement and clinical remission, but also the inhibition of radiological disease progression.*

A) Physical: thermal therapy, cryotherapy, hydrotherapy, electrotherapy and massage therapy, therapeutic exercise. Cold compress for acutely inflamed joints - no heat! Whole-body cryotherapy in special cryo-chambers at -110°C (of which there are, however, relatively few) relieves pain for several hours.

__Note:__ The patient must regain activity as soon as possible; failure to do so may result in contracture of joint capsule and muscle atrophy.

B) Drug Therapy (adult dosage):

1. Corticosteroids

   Indication: • in <u>active RA</u> temporary use until disease-modifying antirheumatic drugs (DMARDs) start working

   • in <u>highly active RA</u> also longer term as "low dose" steroid therapy

   Side effects and contraindications: see chapter on corticosteroids

   Dosage: e.g. prednisolone initially about 20 mg/d, gradually reduced and discontinued once DMARD therapy is working. Longer-term <u>low-dose steroid therapy (2.5-5 mg prednisolone/d)</u> is very useful <u>in addition</u> to the DMARD regimen and delays onset of joint damage. In general, steroids should not be used long-term in dose regimens of more than 5 mg/d and are no substitute for DMARDs! Prophylaxis against osteoporosis with calcium and vitamin D3 is recommended!

2. Treatment with disease-modifying antirheumatic drugs (DMARDs):

   These drugs are effective in about 70% of cases. Onset of action occurs within about 2 months of treatment; methotrexate is often effective already after 4 weeks of treatment. Because of its <u>teratogenic</u> potential, patients must be advised to use reliable contraception while receiving methotrexate.

   <u>Regular clinical and laboratory tests are essential to record any side effects</u>

   DMARDs must be started early to avoid joint destruction, possibly also as part of a combination regimen.

2.1 Immunosuppressive drugs

   • Methotrexate (MTX)

   <u>Relatively well tolerated and effective substance!</u> May prolong life (especially by reducing the risk of MI)

   <u>Mechanism of action:</u> Folic acid antagonist (antidote: folinic acid = activated form of folic acid) with immunosuppressive effects

   <u>SE:</u> relatively frequent GI side effects, stomatitis and increase in hepatic enzymes. Bone marrow depression with cytopenia (5%/year), hair loss, occasionally arthralgia/myalgia within 24 hours of taking MTX. Less common are interstitial pneumonitis, hepatic fibrosis, acral nodulosis etc. Additional administration of folic acid (5 -10 mg/d following ingestion of MTX) reduces side effects and thereby premature discontinuation of treatment.

   <u>Ind: therapy of choice for medium to severe RA</u>

   <u>CI:</u> pregnancy, lactation, hepatic disorders, renal failure, hematological disorders, comedication with cotrimoxazole etc.

   <u>Dosing regimen: low-dose MTX therapy has proved successful:</u> 7.5 - 20 mg once a week as single oral or parenteral dose (parenteral administration markedly more effective with the same tolerability (remissions about 40% more frequent!). The patient can easily learn to administer MTX s.c.). On the day of taking MTX, do not take any NSAIDs, because they inhibit the elimination of MTX (risk of side effects increases)! MTX administration in women of child-bearing age taking contraception (risk of fetal malformations).

   Other immunosuppressants, e.g. azathioprine, ciclosporin A (Consider SE)

   <u>Leflunomide</u> (Arava®), a dihydroorotate dehydrogenase blocker with a half-life of several months! Fetotoxic metabolites persist substantially longer (for up to one year) → eliminating agent: cholestyramine. <u>Ind:</u> reserve medication if MTX does not help or is contraindicated

   <u>SE:</u> elevation of liver enzymes (about 5%), severe liver damage and hepatic failure (check liver enzymes every 2 weeks!), hypertension, interstitial pneumonia etc. Note CIs.

   <u>CIs:</u> pregnancy, women capable of child-bearing without contraception, liver damage etc.

2.2 Alkylating agents (e.g. cyclophosphamide)

   <u>SE:</u> see section on Cytostatic Agents

   <u>Ind:</u> Severe disease unresponsive to conventional therapy; vasculitis-related complications. *__Caution:__* cumulative dose! Risk of treatment-induced MDS, increased risk of urothelial carcinoma, in treatment of women who have not completed their family planning: if necessary ovarian protection (GnRh analogs)

2.3 Sulfasalazine (e.g. Azulfidine®):

   Only effective in mild, erosive disease

   <u>SE + CIs:</u> see section on ulcerative colitis

   <u>Dos:</u> 500–2000 mg/d; start with the lowest possible dose and slowly increase if well tolerated.

2.4 Hydroxychloroquine (Quensyl®):
   Only effective in mild RA in 50% of cases.
   SE + CIs: see section on Malaria; regular ophthalmologic check-ups due to the risk of retinopathy!
   Dos: hydroxychloroquine: 400 mg/d
2.5 Gold therapy:
   Hardly ever used now because of a number of serious AEs (especially with parenteral administration)
2.6 Biologicals:
   • Anti-TNFα therapy:
     Onset and mechanism of action: rapid (within 2 weeks), high response rate, anti-inflammatory + destruction-inhibitory affect: prevents radiological progression.
     SE: increased tendency to infection incl. septic infections (especially soft-tissue infections), reactivation of TB; worsening of heart failure and rarely damage to optic nerve during treatment with infliximab; increased risk of lymphoma. Other SE + CIs are to be considered (→ manufacturer's data). No long-term experience.
     Ind.: reserve medication if other therapy fails (high costs!)
   - Infliximab (Remicade®): monoclonal chimeric antibody against TNF-α; administration only i.v., half-life 26 days, only in combination with methotrexate owing to formation of neutralizing antibodies
   - Adalimumab (Humira®): Monoclonal completely humanized antibody against TNF-α; half-life: 14 days; s.c. administration
   - Etanercept (Enbrel®): fusion molecule from soluble TNF receptor and IgG1-Fc domain fragment. half-life almost 5 days, s.c. administration.
   • Interleukin-1 receptor antagonists (IL-1Ra): anakinra (Kineret®)
     As with anti-TNF-α therapy there is an increased risk for infections (high costs!)
   • Rituximab (MabThera®): anti-CD20 antibody
     Ind: refractory cases in which at least 2 TNF-α blockers have failed. Therapeutic principle: B-cell depletion, dose: 2 x 1000 mg i.v. every 2 weeks. Bear in mind SE + CIs (manufacturer's data), expensive, update prior vaccination status.
   • Abatacept (Orencia®): CTLA4-Ig, inhibition of T-cell co-stimulation. Indicated in cases where TNF-α blockers have failed. Long-term studies are still lacking.
3. Non-steroidal anti-inflammatory drugs (NSAIDs): symptomatic therapy, short duration of action, no effect on the course of disease.
   ■ Non-selective COX-1/2 inhibitors = cyclooxygenase inhibitors
     Mechanism of action: non-selective inhibition of cyclooxygenase 1 + 2 (COX 1 + 2) and subsequent prostaglandin synthesis. The anti-inflammatory effect is mediated via COX-2 inhibition. Inhibition of platelet aggregation and gastric SE result from COX-1 inhibition. The extent of COX-1 and/or COX-2 inhibition varies among individual substances: diclofenac is a more potent inhibitor of COX-2 than COX-1. Drugs with a shorter half-life (<5 h) are less problematic with regard to SE.
     - Propionic acid derivatives, e.g.:
       Ibuprofen: single dose: 400-800 mg, daily dose: up to 3200 mg
     - Arylacetic acid derivatives, e.g.:
       Diclofenac: single dose: 50 mg, daily dose: 50–150 mg
       Naproxen: (T_{50} up to 15 hours); daily dose: 250–1000 mg
     Note ASA is not used for treating RA (single dose with analgesic effect is too high and often causes SE)
     SE: - GI SE (approx. 30%):
             Gastric pain, possibly heartburn, peptic ulcers, possibly with (occult) bleeding or perforation. Risk of bleeding about 1 : 100/2 years. The ulcerogenic effect is due to an inhibition of prostaglandin synthesis with disturbed gastric mucus production. The following risk factors for severe GI SE should be considered:
             a) simultaneous corticosteroid therapy
             b) history (Hx) of ulcer
             c) high-dose NSAID therapy
             d) co-medication with anticoagulants
             e) age > 65 years

             In these cases, prophylactic use of proton pump inhibitors is indicated before long-term treatment with NSAIDs. In *Helicobacter pylori* infection, prophylactic Hp eradication therapy might be of benefit. To reduce gastric disturbance, take medicines after meals.

             Note: whereas glucocorticoids hardly increase the ulcer risk, NSAIDs are associated with a four-fold increase in the ulcer risk; the combination of these two substance classes, however, pushes up the risk by a factor of 15! Therefore, avoid combining these substances as far as possible. If the combination is unavoidable, it is absolutely essential to add a proton pump inhibitor. Do not prescribe several NSAIDs at the same time!

           - NSAID enteropathy: inflammation, haemorrhage, strictures, perforation of the small intestine; possible colitis with diarrhoea
           - NSAID cystitis

149

- pseudoallergic reactions: analgesic intolerance (with possible bronchospasm in asthma patients)
- allergic SE (urticaria, angioedema, bronchospasm)
- hematopoietic disorders
- CNS symptoms (e.g. headache, vertigo, confusion, hearing disturbances) - especially after high doses of indomethacin; tinnitus caused by ASA
- liver and kidney damage
- sodium and water retention, possible increase in blood pressure and worsening of heart failure.

Cls: - hepatic/renal dysfunction   - pregnancy and lactation
     - ulceration of the GI tract   - hematologic disorders, etc.
     - analgesic intolerance

■ Selective COX-2 inhibitors (Coxibe):

SE + Cls: similar to those for non-selective COX-1/2 inhibitors, but the frequency of GI AEs is said to be about 50% lower

Rofecoxib and valdecoxib have been withdrawn from the market because of an increase in cardiovascular events (MI, stroke). Etoricoxib has not yet been approved by the FDA. Generally, however, the long-term continuous use of NSAIDs (whether selective or non-selective) leads overall to increased cardiovascular complications. The indication should therefore be very strictly established in CHD, stroke or peripheral arterial occlusive disease. This also applies to patients at increased cardiovascular risk.

e.g.: celecoxib (Celebrex®) - dose regimen: 200 – 400 mg/d

Note: overall an elevated cardiovascular risk is seen with all NSAIDs except for naproxen. The advantage of Coxibe lies mainly in the lower rate of AEs in the GI tract. The increase rate of gastrointestinal AEs with non-selective COX inhibitors may be offset by the use of proton pump inhibitors.

4. Diet: some patients report alleviation of symptoms on a low-meat, low-fat diet ("red" meat in particular should be avoided). Fish should be preferred to meat, possibly with the addition of fish oil (omega-3 fatty acids). An unbalanced diet is to be avoided. No substitute for effective therapy with DMARDs!

C) Radiosynoviorthesis (RSO):

Internet information: *www.radiosynoviorthesis.com*

After localization of inflammatory joints by means of soft-tissue scintigraphy, radioactive substances (beta-emitters) are injected into painfully inflamed joints:

	90Yttrium for large joints	186Rhenium for middle joints	169Erbium for small joints
Half-life	2.7 days	3.7 days	9.5 days
Tissue area	11 mm	3.7 mm	1.0 mm

Assessment: effective supplement to DMARD therapy especially in individual joints; no systemic AEs; treated joints are free of pain in the long term; optimum effect after 3 - 6 months.

D) Synovectomy (by arthroscopy or surgery)

E) Treatment of complications: e.g. anaemia
- treatment of marked inflammatory anaemia with erythropoietin
- prophylaxis against and possible treatment of gastric ulcers (possibly with bleeding): proton pump inhibitor
- watch out early on for cardiovascular events or risk factors

F) Reconstructive surgery and prosthetic joint replacement (e.g. arthrodesis of the wrist, Hofmann-Tilmann operation of forefoot, TEP of knee joints etc.)

G) Rehabilitation, patient training and self-help groups (League against Rheumatic Disease)

Prg.: Prognostically unfavourable factors requiring early and consistent therapy are:
- Multiple joint involvement
- High RF titer and anti-CCP antibodies, high CRP, high ESR.
About 10% of cases show a mild course (type I RA), about 20% show variable disease activity (type II RA), and up to 70% show a severe progressive course (type III RA).
After 10 years, up to 50% of patients are disabled. Life expectancy is reduced by 7 years on average. Frequent causes of death are myocardial infarction (three-fold increase in risk). The possible fatal consequences of NSAIDs (gastric bleeding etc.) and DMARDs (e.g. infections as a result of biologicals) must also be borne in mind. There is an increase in solid tumours and malignant lymphomas.
Note: FFbH: Hannover Functional Ability Questionnaire; this serves as a means of recording the functional capacity of patients with rheumatic disorders. Functional capacity is taken to mean the possibility of coping with daily activities in the household, family, job and spare time. The FFbH is comparable with the HAQ (Health Assessment Questionnaire). The DAS 28 (disease activity score 28) is very well suited to assessing disease activity and should be used for documentation and assessment of efficacy especially in cases of intensive treatment/severe course of disease. See internet for details.

Syn: Spondyloarthropathies

Ep.: Prevalence 1-2% in Europe

Def.: Chronic inflammatory diseases, particularly of the axial skeleton, with a genetic predisposition (association with HLA-B27 up to 90% in ankylosing spondylitis) and absence of RF ("seronegative") with the following cardinal symptoms:

- Back pain due to sacroiliitis and vertebral involvement (spondylitis/-arthritis, syndesmophytes)
- Asymmetric oligoarthritis, often of the knee joints, radial involvement (dactylitis = sausage fingers)
- Inflammatory enthesopathy (inflammation of the tendon insertions, ligaments)
- Iritis or iridocyclitis (anterior uveitis) and other extra-articular manifestations

5 clinical entities:
- Ankylosing spondylitis (AS)
- Reactive arthritis and Reiter's syndrome
- Psoriatic arthritis
- Enteropathic arthritis with sacroiliitis in Crohn's disease, ulcerative colitis, Whipple's disease, and other intestinal diseases
- Undifferentiated spondylarthritis

Diagnostic criteria of the ESSG (European Spondyloarthropathy Study Group)
with a sensitivity and specificity of approx. 85%:

Vertebral pain	or	arthritis/synovitis
of an inflammatory type		asymmetric or predominantly of lower extremities

and one of the following criteria:

- Positive family history of AS, psoriasis, reactive arthritis, Crohn's disease, or ulcerative colitis
- Clinical finding or history of psoriasis
- Crohn's disease or ulcerative colitis
- Bilateral alternating gluteal pain
- Heel pain
- Sacroiliitis

Criteria of low-back pain of the inflammatory type	Sacroiliitis grading grade/findings
- Onset of disease before the age of 40	1 Faded articular cavity, pseudo-widening, moderate sclerosis
- Insidious onset of symptoms	
- Duration for at least 3 months	2 Irregular joint space widening, marked sclerosis, erosions, beaded appearance
- Morning stiffness	
- Improvement on movement	
At least four criteria must be met.	3 Joint space narrowing, erosions, sclerosis, partial ankylosis
	4 Total ankylosis

SPA is proven if there is X-ray evidence of bilateral sacroiliitis.

The existing criteria are not suitable for an early diagnosis, because sacroiliitis can only be detected after about 8 years on average using conventional x-ray techniques. Earlier diagnosis by means of MRI of ileosacral joints.

Criteria for early diagnosis of spondyloarthritis in patients with chronic back pain (according to Sieper and Rudwaleit 2006):
1. Sacroiliitis on x-ray or MRI + 1-2 clinical criteria   or
2. Inflammatory back pain + 2-4 further clinical criteria
Clinical criteria: inflammatory back pain, extraspinal criteria (enthesitis, psoriasis, colitis, uveitis), good response to NSAIDs, HLA-B27-positive, positive family history for spondyloarthritis.

ANKYLOSING SPONDYLITIS (AS) [M45]

Syn.: ankylosing spondylarthritis

Ep.: Prevalence: approx. 1% of the population, m:f = 3:1; familial accumulation
Age of manifestation: mostly between 20 and 40 years

Aet.: Genetic predisposition: 90% of patients are HLA-B27-positive (frequency in the normal population approx. 8%)
The chronic inflammatory process is induced by unknown factors

Cl.:
- ■ Sacroiliitis: esp. night/morning low back/gluteal pain stiffness, improvement with movement, possibly with radiation of pain to thighs, pain on percussion and displacement of the iliosacral joints (step-up test, Mennell's sign: low back pain when the lower leg of the patient in the lateral recumbent position is maximally flexed while the other leg is retroflexed).
- ■ Spondylitis: pain in the thoracolumbar junction region of the vertebral column (VC).
  Increasing loss of mobility of the VC and chest:
  Finger-floor distance (normally 0)
  Chest circumference difference on inhalation versus exhalation (normally >6 cm, less when older)
  Occiput-to-wall distance (normally > 15 cm)
  Chin-to-sternum distance (normally 0 cm)
  Schober's test: the 10 cm distance measured from L5 upwards with the patient standing upright must increase by at least 4 cm after maximum trunk flexion.
  Ott's test: the 30 cm distance measured from C7 down, with the patient standing upright, must increase by at least 2 cm after maximum trunk flexion.
- ■ Possible arthritis of peripheral joints (in 1/3 of patients)
- ■ Possible chest pains (synchondritis of the sternomanubrial synchondrosis, pubic bone pain (symphysitis)
- ■ Possible inflammatory enthesopathy: painful inflammation of the tendon insertions: Achilles tendon, plantar aponeurosis, trochanters, ischium, iliac crest
- ■ Possibly iritis (25%)
- ■ Relatively rare involvement of internal organs: e.g. cardiopathy (first-degree atrioventricular block), aortitis, possibly with aortic regurgitation, IgA nephritis, etc.

Co.: Ankylosis of vertebral column and chest, fixed kyphosis, rarely amyloidosis (1%)

Disease activity indices: e.g. BASDAI (Bath Ankylosing Spondylitis Disease Activity Index = 6 standardized questions on tiredness, pain sensation and morning stiffness) or BASFI = Bath Ankylosing Spondylitis Functional Index - see internet.

Lab.: depending on disease activity ESR and CRP ↑, HLA-B27-positive (90%)

Imaging techniques for visualizing of vertebral lesions and sacroiliitis:
- ▶ X-ray
  - Signs of sacroiliitis: blurred outline of joint with juxta-articular sclerosis and erosions (grading: see previous section on SPA)
  - Syndesmophytes: bone clasps bridging adjacent vertebrae (DD: ankylosing hyperostosis = diffuse idiopathic skeletal hyperostosis (DISH) or Forestier's disease, osteoproliferative degenerative vertebral lesions: spondylosis, osteochondrosis)
  - Spondylarthritis: osseous ankylosis of the facet joints, calcification of the vertebral ligaments, bamboo spine in end-stage disease, enthesopathy with ossifying periostitis, e.g. calcaneal spur
- ▶ MRI (see above)

DD:
- • Osteoporosis, slipped disc (hernia)
- • Tuberculous and bacterial spondylitis and spondylodiscitis (MRI, scintigraphy, pathogen identification)
- • Tumour-related vertebral symptoms
- • Other HLA-associated forms of spondylarthritis (see above)

Di.: History/clinical presentation, x-ray/MRI, diagnostic criteria (see above), HLA-B27

Th.: Systematic physiotherapy with guidance on regular independent exercises, training.

Drug therapy:
- • NSAIDs as needed or sometimes continuously
- • Corticosteroids if possible just temporarily during severe inflammatory flare-ups.
  In painful refractory sacroiliitis, possibly also intra-articular / peri-articular steroid injections under sterile conditions (bear in mind AEs + CIs, ideally CT or MRI-guided)
- • Salazosulfapyridine (e.g. Azulfidine RA®) in peripheral arthritis or early inflammatory back pain
- • Biologicals: reserve medication in severe, refractory pain or highly inflammatory activity: infliximab (Remicade®), etanercept: (Enbrel®) or adalimumab (Humira®) (see RA)
- • Bisphosphonates (e.g. pamidronate) to avoid microfractures, possibly with additional immunomodulation
- • Surgical treatment: joint replacement, if necessary straightening operation.

Prg.: Often runs an individually highly variable course that is characterized by flare-ups; a disability can be avoided in most cases if a systematic course of specific exercises is followed.
A mild course of the disease without severe ankylosis is mostly observed in women.

REACTIVE ARTHRITIS (REA) [M02.9] AND REITER'S SYNDROME [M02.3]

Def.: Inflammatory arthropathies which occur as a <u>secondary disease after GI or urogenital bacterial infections.</u>

Ep.: 2–3% of all patients with certain GI or urethral bacterial infections.

Aet.: 1. <u>Genetic predisposition:</u> 60-80% of patients are HLA-B27-positive
2. <u>Causative bacterial infection:</u>
   a) <u>Post-urethritis REA</u> after gonorrhoea and non-gonorrhoeal urethritis (NGU) due to *Chlamydia trachomatis* serovar D-K and mycoplasmas, most commonly *Ureaplasma urealyticum*
   b) <u>Post-enteritis REA</u> after infections caused by Yersiniae, Salmonellae, Shigellae, *Campylobacter jejuni,* and other enteric pathogens

Prg.: Chlamydiae and Yersiniae can lead <u>to persistent infections</u> which can trigger and maintain a reactive arthritis if there is a specific genetic predisposition. The arthritis is aseptic, i.e. no bacterial pathogens can be cultured from the joint tap. However, in some patients, non-replicative pathogenic components (e.g. bacterial antigens) can be detected in the joint tap.

Note.: HLA-B27 and some of the etiological bacteria (e.g. *Yersinia pseudotuberculosis*) share a common antigen pattern to some extent.

Cl.: After a <u>latency period of 2–6 weeks following enteric or urethral infection,</u> a secondary disease with arthritis and other symptoms develops. The full clinical picture of REA with 3 or 4 main symptoms is referred to as Reiter's syndrome and is found in one-third of patients.

Main symptoms of Reiter's syndrome:
1. <u>Arthritis:</u> often asymmetric, possibly migratory oligoarthritis with a predilection for the lower extremities, e.g. knee/ankle joints; occasionally also affecting the finger and toe joints (in the case of dactylitis = inflammation of all joints of a solitary finger or toe, a "sausage finger" and/or a "sausage toe" may be clinically observed).
2. <u>Urethritis</u>
3. <u>Conjunctivitis/iritis</u>
4. <u>Reiter's dermatitis:</u> erythematous psoriasiform dermatitis of the male genital mucosa (circinate balanitis); aphthoid lesions in the mouth; horny, partially pustular changes on palms of the hands and soles of the feet (keratoderma blenorrhagica); psoriasiform skin changes on the body
1–3 = Reiter's triad; 1–4 = Reiter's tetrad

Possible accessory symptoms:
■ fever
■ sacroiliitis
■ painful inflammation of the tendon insertions (enthesopathies)
■ rarely, visceral involvement: e.g. carditis, pleuritis

Lab.: ■ Non-specific inflammatory parameters: ESR and CRP ↑
■ Genetics: HLA-B27 (80% of cases)
■ <u>Identification of an enteric or urethral infection:</u>
   - pathogen detection: since the enteric or urethral infection has usually subsided by the time the REA occurs, the pathogen can occasionally be identified by means of PCR (in urethritis, cervicitis from morning urine and rarely also from the stool after enteritis); possibly detection from biopsies of the colon and ileum mucosa.
   - serological assays: infection cannot always be detected from the titer.
   - possible PCR detection of Chlamydiae from synovia or synovial membrane
■ <u>Exclusion of other rheumatoid diseases</u> (absence of RF, ANA, normal ASL titer, normal Borrelia titer, etc.)

DD: Other rheumatic diseases (see section on Rheumatoid Arthritis)

Di.: History (prior enteric or urethral infection) + clinical presentation + lab (HLA-B27, proof of infection) + ESSG criteria (see above)

Th.: 1. <u>Treatment of infection:</u>
In NGU caused by Chlamydiae or Ureaplasmae: doxycyclines or macrolides (e.g. erythromycin or clarithromycin). Since Chlamydiae can persist intracellularly in the form of metabolically inactive elementary particles, longer-term antibiotic treatment (up to 3 months) is recommended in REA! - Always think of treating the patient's partner as well! Treat gonorrhoea (see above) and/or enteritis according to the pathogen involved. The benefit of antibiotic therapy without detection of a florid infection has not been unequivocally demonstrated! In cases of REA following enteritis, an improvement resulting from the use of antibiotics has not been documented.
2. <u>Symptomatic treatment of REA:</u>
   - Physical therapy (in acute arthritis: cryotherapy)
   - NSAIDs
   - Possibly temporary corticosteroids in a hyperacute course and/or iridocyclitis
   - Possibly sulfasalazine in chronic course (see section on Rheumatoid Arthritis)

Prg.: Up to 80% of cases heal after 12 months. Oligosymptomatic course of REA has a more favourable prognosis than the full-blown clinical picture of Reiter's syndrome.

PSORIATIC ARTHRITIS [L40.5 + M07.3*]

Ep.: 10 - 20% of psoriasis patients (prevalence of psoriasis: up to 3% of the population in Europe)

Cl.: 5 forms (according to Moll and Wright 1973):
1. DIP and PIP joint involvement, such as Heberden's and Bouchard's osteoarthritis (5 %)
2. Deforming, mutilating rheumatoid arthritis = "telescope fingers" (5 %)
3. Symmetric arthritis such as RA (20%)
4. Asymmetric oligoarthritis, frequently HLA-B 27-positive (60%), possibly dactylitis with "sausage fingers" = involving all joints of a finger
5. Spondylarthritis with sacroiliitis, frequently HLA-B 27-positive (10%)
Variants: SAPHO syndrome (Synovitis - Acne - Pustular psoriasis - Hyperostosis - Osteitis): painful sternoclavicular hyperostosis, vertebral hyperostosis, peripheral arthritis, aseptic multifocal osteitis (bone scintigraphy!), pustular palmar/plantar psoriasis, if necessary bone sample in cases of suspected tumour
The course is often severe in HIV patients (skin and joints); *Caution:* HIV infection as possible trigger!

Di.: Clinical triad:
1. Red papules and plaques with silver-white desquamation particularly affecting the extensor sides of the limbs, the sacral and anal areas, and the hirsute head; possibly only discrete skin lesions behind the ears (search for sites of predilection)
2. Nail changes
Pitted nails → punctate dimples in the nail
"Oil drop" or "salmon patch" nails → yellowish-brown subungual spot
Onycholysis → nail that has lifted from its bed
Crumbly nails → thickened, crumbly nails
Missing cuticle → psoriatic focus at the nail wall
3. Arthritis; in spondylarthritis/sacroiliitis ESSG diagnostic criteria

Th.: Topical psoriasis therapy of the skin almost never improves the joints. Often no correlation between skin and joint involvement.
- Sulfasalazine (40% improvement of skin) in mild joint involvement (oligoarthritis):
- Immunosuppressants: e.g. methotrexate, leflunomide, ciclosporin - Ind: marked, erosive joint involvement (up to 60% of cases) → see RA.
- Caution when using steroids: improvement in arthritis; but risk of a dramatic worsening of the psoriasis (erythroderma!) even when the dose is reduced carefully.
- Biologicals: reserve medication in resistance to therapy: infliximab (Remicade®), etanercept (Enbrel®), adalimumab (Humira®) - see RA therapy: high response rates, high costs, bear AEs + CIs in mind! (See section on Rheumatoid Arthritis).

Prg.: Risk factors for an erosive course: polyarthritis, a high degree of inflammation, extensive involvement of the skin, HLA-DR3.

ENTEROPATHIC ARTHRITIS/SACROILIITIS

Syn: Intestinal arthropathies

Aet.: ■ Patients with chronic inflammatory bowel diseases (CIBD) – ulcerative colitis and Crohn's disease – develop arthritis in approx. 15% and sacroiliitis in approx. 25% of cases.
■ In Whipple's disease (see appropriate section) arthritis is found in 60% (often as first symptom!) and sacroiliitis in 40% of cases
■ Other chronic intestinal diseases
■ "Bypass arthritis" following gastrointestinal anastomosis with a blind loop: Migratory polyarthritis, often very painful - treatment: antibiotics

Di.: History/clinical findings; ESSG diagnostic criteria (see above) in sacroiliitis

Th.: 1. Treatment of underlying disease; if the CIBD improves, the arthritis also improves
2. Symptomatic therapy

CONNECTIVE TISSUE DISEASES

Connective tissues diseases (CTDs) comprise a group of diseases which preferentially affect the connective tissue and show similar morphologic changes (classical auto-immune diseases, involvement of internal organs).

Connective tissues diseases in the narrower sense of the term include
■ Systemic lupus erythematosus (SLE)
■ Polymyositis and dermatomyositis
■ Systemic sclerosis (scleroderma)
■ Sjögren's syndrome
■ Mixed connective tissue diseases (= MCTD, Sharp syndrome)

All CTDs occur more frequently in women. Abortive and mixed forms are possible.

Genetic predisposition seems to play a role in all CTDs (combination with particular HLA antigens). The underlying pathogenesis is of an auto-immune nature. The etiology is mostly unknown.

SYSTEMIC LUPUS ERYTHEMATOSUS (SLE)

Syn.: Lupus erythematosus disseminatus (LED)

Def.: Systemic disease of the skin and vascular connective tissue of numerous organs, with vasculitis/perivasculitis of the small arteries and arterioles associated with deposits of immune complexes consisting of DNA, anti-DNA, complement and fibrin.

Ep.: Prevalence: 50/100,000; incidence 5–10/100,000/yr (more common in the Afro-American US population, unknown in Central Africa); f:m = 10:1, mainly women of childbearing age; higher frequency of HLA-DR2 and DR3.
Late-onset SLE manifests itself after the age of 55 years (f:m = 2:1).

Aet.: Unknown

Pg.: Reduced activity and number of regulatory T cells, disturbed elimination of auto-reactive B and T cells. Formation and expansion of auto-reactive cell clones. Increased tendency toward apoptosis with reduced clearance of apoptotic material, triggering of immune reaction to DNA. Complement activation, partly congenital complement defects

Cl.:
- General discomfort (95%): fever, weakness, weight loss; more rarely lymphadenopathy
- Muscle/joint symptoms:
  Polyarthritis (>80%) without erosions, however, possibly with incomplete dislocation/malposition
  Myositis (40%)
- Skin changes (>70%): these give the name to the disease (wolf = lupus): skin disease.
  - Butterfly rash/erythema on the cheeks and bridge of the nose, avoiding the nasolabial fold
  - Discoid lupus: bright red papules with scaling and follicular hyperkeratosis
  - Photosensitivity of the skin with development of exanthema after exposure to light
  - Oronasal ulceration
  - Rarely secondary Raynaud's syndrome
  - Scalp involvement with cicatricial alopecia
  - Cutaneous vasculitis (livedo vasculitis, leukocytoclastic vasculitis, urticaria vasculitis)
  Confirm the diagnosis with skin biopsies of affected and unaffected skin using immunofluorescence: granular deposits of IgG, IgM or C3 along the basal membrane ("lupus band")

  Note: in SLE, lupus bands are also often found in macroscopically normal skin, which is not the case for the purely cutaneous form of lupus erythematodes!
- Organ system manifestations:
  - Cardiopulmonary changes (60–70%):
    Pleuritis and/or pericarditis with effusions
    Coronaritis, possibly MI, possibly lupus pneumonitis
    Libman-Sacks endocarditis, myocarditis, pulmonary infiltration
    Premature arteriosclerosis with infarction risk increased by a factor of up to 17
  - Renal changes (60–70%): see section on Lupus Nephritis (after SLE)
  - Neurological changes (60%): determine the prognosis together with renal involvement. Everything is possible, from vigilance deficits and depression to status epilepticus, apoplexy or an MS-like course. Mixed forms are also not uncommon:
    1. Focal form: microcirculation disorders, frequently positive for antiphospholipid antibodies, no lesions on MRI, SPECT sensitive, EEG with foci, symptoms tend to be neurological in nature.
    2. Diffuse form: MRI mostly without abnormal findings (possibly hippocampal atrophy), CSF shows slight increase in protein, possibly leukocytosis, anti-neuronal antibodies, symptoms tend to be psychiatric in nature.
    3. Peripheral nervous system: involvement in up to 15%

Lab.:
1. Non-specific signs:
   ESR ↑, CRP often normal, elevated values e.g. in infectious complications, α2/β-globulins ↑, complement (C3 and C4) ↓, hypochromic anaemia

2. Disease-specific immunologic findings:
   - Antinuclear antibodies (ANA) = high-titer ANF (antinuclear factors): 95%
   - Antibodies against double-stranded DNA (anti-dsDNA antibodies) are typical of SLE (60–90%)
   - Anti-Sm antibodies (25%)
   - Anti-Ro (SSA) (60%) and more rarely anti-La = SSB (20%) in subacute cutaneous LE

- Antiphospholipid antibodies (APA) (35%):
  - anti-cardiolipin antibodies (IgG, IgM) and anti-β2-glycoprotein 1 antibody (IgG, IgM)
  - lupus anticoagulant: patients with high-titer APA are at a higher risk for antiphospholipid syndrome with the triad arterial and/or venous thrombosis, abortion, and thrombocytopenia.
- Possibly antibodies against blood clotting factors (e.g. anti-F VIII-antibodies with possible inhibitor haemophilia)
- Circulating immune complexes

3. Often auto-antibody-induced cytopenia:
  - Coombs-positive haemolytic anaemia
    (DD: hypochromic anaemia in chronic inflammation)
  - Leukocytopenia (<4000/µL)
  - Lymphocytopenia (<1500/µL)
  - Thrombocytopenia (<100,000/µL)

Forms of the disease:

1. Cutaneous lupus erythematosus (CLE): [L93.0] usually only involves the skin; favourable prognosis.
   - Chronic discoid lupus erythematosus (CDLE):
     a) Localized form (in 90% on the head)
     b) Disseminated form (trunk, upper arms): tender plaques of hyperkeratosis with reddish inflammatory border + central atrophy. Favourable prognosis, since only the skin is involved in 95% of cases.

2. Subacute cutaneous lupus erythematosus (SCLE):
   This form takes a middle position with regard to the clinical picture and prognosis:
   - General malaise, arthralgia, myalgia
   - Skin changes
   - Possibly Sjögren's syndrome

   Renal involvement is rare. Most patients have the HLA-DR3 antigen and anti-Ro (SSA) (= antibody against cytoplasmic antigens). ANA are often found, although not usually anti-dsDNA antibodies.

3. Systemic lupus erythematosus (SLE):
   This form is defined by the ACR criteria (see below). Since there is often visceral involvement, the prognosis is serious and determined by the extent of the nephropathy, amongst other factors.

DD:
- Rheumatoid arthritis and other CTDs
- In the case of hematopoietic changes: hematological diseases
- In the case of renal manifestation: renal diseases of other pathogenesis
- In the case of neurologic manifestation: neurologic disorders of other pathogenesis
- Primary antiphospholipid syndrome without evidence of SLE
- Drug-induced lupus, e.g. by procainamide (20%), hydralazine (10%), methyldopa, phenytoin, neuroleptics, minocycline and etanercept
  The symptoms are usually limited to polyarthritis and pleuritis/pericarditis. In all patients, ANA and usually also anti-histone antibodies are found. No anti-dsDNA antibodies are found!

Idiopathic SLE	Drug-induced lupus
Anti-dsDNA antibodies	Anti-histone antibodies
Often HLA-DR2 and DR3	Often HLA-DR4
CNS affected in up to 60%	CNS rarely affected
Kidneys affected in >60%	Rare renal involvement, history of medicines! Reversibility after discontinuation of medicines.

Note: since neuroleptics and anticonvulsants can trigger a drug-induced lupus, one should consider this DD in cases where there are psychotic or epileptic symptoms!

Di.: SLE criteria of the American College of Rheumatology (ACR):
1. Butterfly erythema
2. Discoid lupus erythematosus
3. Photosensitivity
4. Oral or nasal mucosal ulcers
5. Non-erosive arthritis of 2 or more joints
6. Serositis (pleuritis, pericarditis)
7. Renal involvement (proteinuria >0.5 g/d or cylindruria)
8. CNS involvement
9. Haematologic findings: Coombs-positive haemolytic anaemia, thrombocytopenia, leukopenia
10. Immunologic findings: anti-dsDNA, anti-Sm, antiphospholipid antibodies
11. Antinuclear antibodies (ANA)

If at least 4 criteria are met, SLE is likely.

Note: These criteria are classification criteria for studies, but are not to be considered absolute for decisions on diagnosis and treatment.

Th.: Interdisciplinary therapy in keeping with the stage of the disease:
- "Drug-induced lupus": leave out the causative substances
- Cutaneous lupus: retinoids, light-screening ointment, topical steroids
- Systemic lupus:
  - Light protection (LSF 60), UV exposure even behind glass can lead to a flare-up.
  - Mild cases without visceral involvement: non-steroidal anti-inflammatory drugs (NSAIDs) + hydroxychloroquine; corticosteroids for inflammatory flare-ups
  - Serious cases with involvement of vital organs: high-dose prednisolone pulse therapy and/or immunosuppressants: in moderately severe cases, azathioprine; in severe organ system manifestations, cyclophosphamide bolus therapy (see therapy of lupus nephritis)
  - The most severe refractory cases of SLE:
    - possibly rituximab (see RA)
    - possibly autologous stem cell transplantation
  - Early treatment/correction of cardiovascular risk factors (smoking, hypertension, hyperlipoproteinemia)
  - Optimum antihypertensive therapy is essential for maintaining renal function!
  - Treatment of antiphospholipid syndrome: (see relevant section)
  - In long-term corticosteroid therapy: osteoporosis prophylaxis (calcium + vitamin D)

  Note: patients with active SLE should be warned not to become pregnant. Before a planned pregnancy, the SLE patient should be in stable remission without immunosuppressive therapy for > 6 months. The pregnancy of an SLE patient is always a high-risk pregnancy (close monitoring by obstetrics centre and rheumatologist).

  Neonatal lupus syndrome: passively acquired autoimmune disease in neonates whose mothers have anti-Ro (SSA) or anti-La(SSB) antibodies in the blood (→ placental transmission). Any skin changes spontaneously regress; and a possible congenital heart block may be irreversible → check fetal heart sounds from 16th week of gestation (bradycardia ?).

Prg.: As the disease may take a highly variable course, the 10-year survival rate in SLE today may be ≥ 90%; most frequent causes of death: cardiovascular diseases (especially CHD/MI), uraemia, cardiac and neurological complications and septic complications

LUPUS NEPHRITIS [N08.5, N16.4]

- Significant organ manifestation of SLE
- Major role in mortality and morbidity of SLE
- Typical example of immune complex glomerulonephritis
- Diverse glomerular lesions
- Variable clinical and laboratory medical signs

Classification of International Society of Nephrology/Renal Pathology Society (ISN-RPS) 2003:

Class I: Minimal mesangial lupus nephritis
Normal glomeruli under light microscopy, but mesangial immune deposits under immunofluorescence.

Class II: Mesangial proliferative lupus nephritis

Class III: Focal lupus nephritis
Active or inactive focal, segmental or global endocapillary or extracapillary glomerulonephritis involving < 50% of all glomeruli, typically with focal subendothelial immune deposits with or without mesangial lesions.
Class III (A): active lesions: focal proliferative lupus nephritis
Class III (A/C): active and chronic lesions: focal proliferative and sclerosing lupus nephritis.
Class III (C): chronic inactive lesions with glomerular scars: focal sclerosing lupus nephritis.

Class IV: Diffuse lupus nephritis
Active or inactive diffuse, segmental or global endocapillary or extracapillary glomerulonephritis involving ≥ 50% of all glomeruli, typically with diffuse subendothelial immune complexes, with or without mesangial lesions.
Class IV (S): diffuse segmental lupus nephritis: ≥ 50% of glomeruli involved have segmental lesions. Segmental lesions are defined as glomerular lesions involving less than half the glomerulum.
Class IV (G): Diffuse global lupus nephritis: ≥ 50% of glomeruli involved show global lesions.
Class IV-S (A): Active lesions: diffuse segmental proliferative lupus nephritis
Class IV-G (A): Active lesions: diffuse global proliferative lupus nephritis
Class IV-S (A/C): Active and chronic lesion: diffuse segmental proliferative and sclerosing lupus nephritis or diffuse global proliferative and sclerosing lupus nephritis
Class IV-S (C): Chronic inactive lesions with scars: diffuse segmental sclerosing lupus nephritis
Class IV-G (C): Chronic inactive lesions with scars: diffuse global sclerosing lupus nephritis

Class V: Membranous lupus nephritis
Class V lupus nephritis may occur in combination with Class III or IV and shows advanced sclerosis.

Class VI: Advanced sclerosing lupus nephritis
≥ 90% of glomeruli are globally sclerosed without residual activity.

<u>Cl.:</u> ■ Asymptomatic proteinuria and/or glomerular hematuria: granulated casts and/or acanthocytes
■ Acute nephritic syndrome
■ Nephrotic syndrome
■ Rapidly progressive glomerulonephritis
■ Chronic renal failure
■ Renal parenchymatous hypertension

<u>Di.:</u> Percutaneous renal biopsy should be performed in all patients with significant proteinuria (≥ 1 g/24 h) for the following reasons:
1. Differentiation between the different classes of lupus nephritis
   Note: an alternating course is possible within the different classes.
2. Exclusion of other disease manifestations such as thrombotic microangiopathy
3. Determination of activity and chronicity indices.

<u>Th.:</u> General measures:

Optimal antihypertensive therapy (ACE inhibitors have a renoprotective effect) and treatment of cardiovascular risk factors

Mesangioproliferative lupus glomerulonephritis (ISN/RPS Class II):
■ Steroid therapy alone is usually sufficient; favorable prognosis
■ If there is no clinical improvement, repeat biopsy

Focal and diffuse lupus nephritis (ISN/RPS Class III-IV):
Induction therapy
a) Corticosteroids in combination with i.v. cyclophosphamide bolus therapy over a period of 3-6 months.
b) Alternatively corticosteroids in combination with oral mycofenolate.
Maintenance therapy
a) To maintain remission, mycofenolate or azathioprine are suitable as an alternative in combination with corticosteroids:
b) If there is a remission, the patients should be checked for renal function, urinalysis and 24-hour proteinuria for at least 5 years after the diagnosis has been established.
If there is a relapse, renewed induction therapy with i.v. cyclophosphamide is justified. For maintenance therapy, the immunosuppressant on which the relapse occurred should not be given again, but treatment should be switched to the alternative described above.

Membranous lupus nephritis (ISN/RPS Class V)
1. Steroid monotherapy
2. In addition to the steroids:
   a) ciclosporin A or
   b) cyclophosphamide
3. Optimal adjustment of blood pressure (preferably with ACE inhibitors) - target value: 120/80 mmHg
4. Treatment of hypercholesterolaemia

Advanced sclerosing lupus glomerulonephritis (ISN/RPS Class VI)
No improvement in renal function to be expected as a result of therapy. Therefore no immunosuppressive therapy

<u>Prg.:</u> Remission criteria:
- Decrease in proteinuria <1 g/24 h
- Inactive sediment
- Normalization of complement factors
- Absence of extrarenal lupus activity

Unfavourable prognostic indices:
- Serum creatinine initially increased
- Nephrotic syndrome
- Renal hypertension
- Decrease in C3 complement
- Histologic ISN/RPS Class III or IV, high index of activity and chronicity

POLYMYOSITIS [M33.2] AND DERMATOMYOSITIS [M33.1]

<u>Def.:</u> Polymyositis (PM): systemic inflammatory disease of the skeletal muscles with lymphocytic, especially perivascular, infiltration.
Dermatomyositis (DM): polymyositis with skin involvement

<u>Ep.:</u> Rare disease, f:m = 2:1; high prevalence of HLA-B8 and HLA-DR3

<u>Aet.:</u> Unknown

Classification:
1. Idiopathic polymyositis (30%):
   Cardinal symptom: weakness of the proximal limb muscle
2. Idiopathic dermatomyositis (25%):
   Cardinal symptom: muscle weakness (as in 1) + skin lesions
3. Polymyositis/dermatomyositis in malignant tumours (10%)
4. Polymyositis/dermatomyositis with vasculitis in childhood (5%)
5. Polymyositis/dermatomyositis in connective tissue diseases ("overlap group") (30%)

Cl.:
1. Myositis of the proximal limb muscles with muscle weakness in the shoulder/pelvic girdle (100%) and cramp-like myalgia (60%). Difficulties in getting up and lifting the arms above the horizontal position. Possible fever.
2. Skin changes in dermatomyositis: livid red, oedematous erythema of the face, especially periorbital, troubled facial expression, lichenoid whitish to pale red papules on the extensor side of the fingers (Gottron's papules), nail fold hyperkeratosis (Keining's sign)
3. Involvement of internal organs:
   - Oesophagus: dysphagia (30%) → oesophageal manometry
   - Heart: interstitial myocarditis (30%), possibly tachycardia, ECG changes
   - Lung: (alveolitis, fibrosis) approx. 30 %

Special manifestations: Anti-Jo 1 syndrome: myositis, Raynaud's syndrome, often arthritis, fibrosing alveolitis, pulmonary fibrosis

Lab.:
- Non-specific inflammatory parameters, e.g. ESR ↑, possibly leukocytosis, etc.
- Muscle enzymes (CK, GOT, LDH, aldolase) ↑; possible detection of myoglobin in serum and urine
- Autoantibodies: ANA (50%), antibodies against histidyl tRNA synthetase (= anti-Jo 1) in 5% (dermatomyositis) up to 30% (polymyositis); anti-Mi2 (10%); anti-PmScl (10%), U1-RNP (15%), anti-SRP (5%, often with cardiac involvement)

DD:
- Drug-induced myopathy (CSE inhibitors and other lipid-lowering agents, corticosteroids = steroid myopathy, chloroquine/hydroxychloroquine) with elevation of CK
- Alcohol myopathy (up to 50% in patients with alcoholism), possibly with elevation of CK
- Inclusion body myositis: rare, painless myopathy with predominantly distal paresis; characteristic histology with inclusions of uncertain origin (viruses?)
- Polymyalgia rheumatica: pain and feeling of stiffness in the shoulder/pelvic girdle, occurs especially in older patients, in 50% of cases with simultaneous (giant cell) temporal arteritis (= Horton's arteritis), highly elevated ESR, (normal CK), promptly improving after corticoid therapy.
- Muscular dystrophy (muscle wasting, family history, electromyogram)
- Myasthenia gravis:
  Sx: Diplopia (double vision), ptosis, post-exertional muscle weakness, especially of the arms, thymus hyperplasia (65% of cases), thymoma (15% of cases).
  Dx: Detection of auto-antibodies against acetylcholine receptors (90% of cases); Tensilon® or Prostigmin® test: transient improvement after administration of acetylcholinesterase inhibitors; electrical stimulation (electromyogram): decrease in amplitude.
- Infectious myositis (Coxsackie viruses, trichinosis)

Di.:
- Clinical presentation (muscle weakness)
- CK elevation

Caution: Myositis is often overlooked and interpreted as a liver problem because only the transaminase values are measured and not the CK.

- Electromyogram
- MRI (oedema of affected muscles), possibly pyrophosphate scintigraphy
- Muscle biopsy with histology/immunohistology → DM: perivascular, perimysial CD4 T cell infiltrates; PM: CD8 T cell infiltrates in the muscle
- Search for tumour
- Echocardiography, chest x-ray, if necessary HR-CT scan for diagnosis of spread

Th.:
In the tumour-associated form, tumour resection can lead to an improvement. Otherwise, steroids and possibly immunosuppressants, e.g. azathioprine, MTX or ciclosporin A. Reserve medicines: cyclophosphamide, mycophenolate mofetil, sirolimus. In refractory cases, a temporary improvement can often be achieved by high-dose i.v. immunoglobulins. If necessary, high-dose chemotherapy with autologous stem cell transplantation.

Prg.: After 5 years of therapy: 50% complete remission (CR), 30% partial remission (PR), 20% progression

PROGRESSIVE SYSTEMIC SCLEROSIS [M34.0]

Syn.: Systemic scleroderma (SSc)

Def.: Systemic connective tissue disease with collagen buildup and fibrosis of the skin and internal organs + obliterating angiopathy with fibrosis and obliteration of the small vessels ("onion-skin" thickening of the intima) with infarction of the skin and internal organs.

Ep.: 1/100,000 of the population/year; usually in women in the third to fifth decade of life; the limited form is about 3 x more common than the diffuse form; f:m = 5:1.

Aet.: Unknown, frequent association with HLA-DR5 in the diffuse form and HLA-DR1, 4, and 8 in the limited form of the disease.

Pg.: Fibroblast dysregulation → overproduction of collagen + obliterating angiopathy. T cell dependent disease

Cl.: 1. Skin changes (100%) go through 3 stages:
oedema (e.g. of the hands: "puffy hands", "sausage fingers")- induration - atrophy
Disease onset mostly in the hands, followed later by centripetal progression (DD: dermatomyositis: centrifugal!)
Secondary Raynaud's syndrome (90%) with whitening of the fingers as a result of vasospasm, characteristic tricolour pattern: initial pallor, then cyanosis and finally reactive reddening/hyperperfusion; skin becomes tight and tense → sclerodactyly. Skin shrinkage results in painless contractures; occasionally ulceration, necrosis ("rat-bite necrosis"), and scarring of the fingertips. Rigidity of the face, reduction in the size of the mouth (microstomia), radiating furrows around the mouth ("tobacco pouch mouth"); little skin mobility over the sternum; pigment changes, telangiectasis.
Thibièrge-Weissenbach syndrome: a manifestation of SSc with microcalcifications in the subcutaneous tissue (subcutaneous calcinosis).

2. Arthralgia (20%), myalgia, possibly myositis

3. Organ system manifestations:
   - GI tract (80%):
   Frenulum sclerosis of the tongue; oesophageal motility disorder with dysphagia -> barium meal examination: wall rigidity, widening of the distal 2/3 of the oesophagus with impaired peristalsis (oesophageal manometry), reflux symptoms, "water-melon stomach"; intestinal pseudo-obstruction
   - Fibrosing alveolitis and pulmonary fibrosis (50%) with restrictive ventilation disorder (early CO_2 diffusion capacity ↓), fatal outcome from cor pulmonale and intercurrent pneumonia! Elevated risk for bronchial carcinoma
   - Cardiac involvement (up to 70%) with myocarditis and arrhythmias; possibly pericarditis
   - Renal involvement (20%) with multiple renal infarcts, nephrogenic hypertension microangiopathy with a risk of renal crisis. Renal involvement accounts for half of all deaths. ACE inhibitors improve the prognosis.

Two forms: (which illustrate the clinical spectrum)
   - Diffuse systemic sclerosis (dSSc): SSc with generalized oedema and sclerosis + involvement of visceral organs: detection of anti-Scl70 (= antitopoisomerase 1) in 40% of cases.
   - (Acral) limited form (lSSc): CREST syndrome: calcinosis of the skin, Raynaud's syndrome, esophageal involvement, sclerodactyly, telangiectasia; detection of anticentromere antibodies (ACA) in 70% of cases (however, no evidence of anti-Scl70)

Lab.: ANA (90%) are non-specific
Anti-Scl 70 found in 40% of cases of dSSc
ACA = anticentromere antibodies - found in 70% of cases of CREST syndrome or lSSc
Anti-RNA polymerase (20%): association with severe skin and kidney involvement
Anti-PM-Scl antibodies (5–10%) in dSSc with polymyositis (overlap syndrome), anti-fibrillarin antibodies (~ 5%), anti-Th(To) (~ 5%) association with pulmonary hypertension

DD:
   - Circumscribed scleroderma (localized scleroderma of the skin with lilac rings, never affects the hands; no visceral involvement)
   - Mixed connective tissue diseases (Sharp syndrome): anti-U1RNP
   - Scleroderma-like clinical pictures caused by chemical toxins (e.g. vinyl chloride, silicon dioxide)
   - Eosinophilic fasciitis (Shulman's syndrome): swelling of proximal limbs (excluding hands and feet), eosinophilia in blood and skin biopsy
   - Acrodermatitis chronica atrophicans in Lyme borreliosis
   - DD of Raynaud's syndrome (see appropriate section)
   - Nephrogenic fibrosing dermatopathy (NFD):
   Rare skin disease in patients with renal failure requiring dialysis: oedema, pruritus, pain, symmetric erythematous or hyperpigmented plaques in the region of the distal extremities (skin biopsy: Proliferating fibrocytes, thickened collagen fibres in epidermis and subcutis); uncertain pathogenesis (gadolinium SE?); Therapy: Phototherapy; Caution Gadolinium used as contrast medium in MRI

Di.:
   - Clinical picture (Raynaud's syndrome, skin changes)
   - Laboratory (ANA, possibly anti-Scl70, ACA)

■ Capillary microscopy (intravital microscopy of the nail-fold capillaries):
- "Slow pattern": dilatation of the giant capillaries, rarefaction of the capillaries
- "Active pattern": increase in avascular areas, etc.
■ A skin biopsy (if diagnosis is uncertain) shows typical changes only <u>before</u> the start of therapy
■ X-ray of the hands: possible calcifications (calcinosis cutis) and acro-osteolysis
<u>Diagnostics for determining complications:</u> Pulmonary and renal function diagnostics, renal diagnostics, lung function, echocardiography, osophageal diagnostics etc.

ACR criteria:	
Main criterion:	Scleroderma of proximal metacarpophalangeal (MCP) joints
Secondary criteria:	Sclerodactyly, pitted scars or loss of substance to soft tissue of distal fingers, bilateral basal pulmonary fibrosis
The main criterion or 2 secondary criteria must be met to establish the diagnosis.	

Th.: No causal therapy is known.
- Glucocorticoids in the early oedematous phase (max. 15 mg prednisolone/day); for as short a time as possible: Risk of renal crisis!
- <u>Immunosuppressants in severe forms:</u> e.g. methotrexate, cyclophosphamide
- <u>Symptomatic therapy</u> is important: physiotherapy to avoid contractures, warm oil/paraffin and mud baths + mild infrared A hyperthermia (= balneophototherapy); prophylaxis against Raynaud's symptoms: cold protection, calcium antagonists, ACE inhibitors.
- Prostaglandin analogs (iloprost, alprostadil), as well as endothelin receptor antagonists (bosentan), in trophic disorders of the extremities.
- ACE inhibitors have a nephroprotective effect.
- Treatment of pulmonary hypertension (see relevant section).
- No beta-blockers.
- Early high-dose chemotherapy + subsequent autologous stem cell transplantation where applicable in refractory cases and cases with an unfavourable prognosis (rapid progression, anti-Scl-70 antibodies, diffuse involvement of the skin, pulmonary involvement etc.).

Prg.: Depending on the extent of organ damage (heart, lung, kidneys), 10-year survival rate in the diffuse forms of disease approx. 70%.

SJÖGREN'S SYNDROME (SS) [M35.0]

Def.: Chronic inflammation of lachrymal and salivary glands and possibly other exocrine glands with <u>2 cardinal symptoms:</u> "dry eye, dry mouth" =
■ Keratoconjunctivitis sicca (KCS) with xerophthalmia (dry eye syndrome)
■ Decreased production of saliva with xerostomia (dry mouth)

Hi.: Lymphocytic infiltration of the salivary and lachrymal glands

Aet.:   1. <u>Primary form:</u> unknown aetiology

2. <u>Secondary forms:</u> Sicca syndrome in rheumatoid arthritis or other connective tissue diseases, also in hepatitis B or C and primary biliary cirrhosis

Ep.: Second most common rheumatic disease after rheumatoid arthritis; f:m = 9:1; frequent association with HLA-DR2 and -DR3

Cl.: <u>Sicca syndrome:</u>
Symptoms resulting from <u>drying of the eyes</u> (burning, foreign body sensation, etc.), the <u>mouth</u> and other mucous membranes; change in the composition of the saliva, caries (60%), swelling of the parotid gland (up to 50%), Raynaud's syndrome (40%), oesophagitis, possibly arthritis; lymphadenopathy (20%); also tendency to allergies and gluten-sensitive coeliac disease (10x more common)

Co.: ■ Corneal ulcerations
■ Rarely involvement of internal organs - e.g. lung [idiopathic interstitial pneumonia (10%): UIP or LIP)] and kidney (10 - 15%) - as well as vasculitis (10%)
■ Development of malignant lymphoma (MALT-NHL, approx. 5 %)
■ Neurologic symptoms: Peripheral neuropathy (5%), CNS involvement (up to 25%), inner ear hearing loss (approx. 25%)

Lab.: ESR ↑, leukopenia, anaemia, thrombocytopenia (facultative)
<u>Immunologic findings:</u>
- Hypergammaglobulinemia
- Rheumatoid factor (RF) (up to 50%)
- SS-B (= La) antibodies: } up to 70%

- SS-A (= Ro) antibodies:}
- Antibodies against epithelial cells in the excretory ducts of the salivary glands or muscarinic receptors (anti-M3)
- Cryoglobulins

DD:
- ■ Xerostomia of other origin: advanced age, cachexia, inflammation or tumours of salivary glands, effect of radiation, medicines with an anticholinergic action (atropine, antispasmodic agents, antihistamines, tricyclic antidepressants etc.); chronic graft-versus-host disease, etc.
- ■ Xerophthalmia of other origin (relatively common): medicines (see above), advanced age, vitamin A deficiency, dry air, etc.
- ■ Sarcoidosis, virus hepatitis, HIV, lymphoma, prior radiotherapy

Di.:
- ■ Clinical picture (sicca syndrome)
- ■ Ophthalmologic examination: slit-lamp (keratitis) + Schirmer test: detection of reduced lachrymation by inserting filter paper strip over the lower eyelid following anaesthesia of the cornea: wet zone in 5 minutes under 5 mm.
- ■ Saxon test: Measurement of saliva production by weighing a cotton swab that has been placed in the mouth for 2 minutes.
- ■ Possibly scintigraphic investigation of salivary gland secretion using $99mTc$ pertechnetate
- ■ Possibly biopsy from the inside of the lip or an enlarged salivary gland (sialadenitis with lymphocytic infiltration)

Th.:
- ■ Treatment of primary disorder in secondary forms of the disease
- ■ Symptomatic: providing a secretory stimulus for the salivary glands, e.g. with chewing gum etc., artificial saliva, eye drops, high fluid intake (2 l/d); ensure sufficient air humidity; protect eyes against wind with (sun) glasses; periodically close eyes for short periods, etc.; bromhexine can promote saliva and tear secretion. The pilocarpine derivative Salagen® promote salivary and lachrymal secretion. Hydroxychloroquine is useful in some patients (especially with arthralgia). Local administration of ciclosporin also promotes lachrymal production.
- ■ Good dental care, regular dental check-ups.
- ■ Hydroxychloroquine: In arthritis, reduction of hypergammaglobulinemia, saliva production improved
- ■ Immunosuppressive therapy should only be considered in cases of visceral involvement or vasculitis → consultation in centres.

Prg.: The course of primary SS is benign in most cases; increased mortality is a result of the increased incidence of lymphoma (5%). The prognosis of secondary SS is determined by the underlying cause of the disease.

SHARP SYNDROME [M35.1]

Syn.: Mixed connective tissue disease = MCTD

Relatively benign clinical course, with an overlapping symptomatology of SLE, scleroderma, polymyositis and rheumatoid arthritis. Cardiac, renal, or CNS involvement is rare. Raynaud's syndrome is always present, often in association with scleroderma-like skin changes, with swollen hands or sclerodactyly. The detection of ANA (antinuclear antibodies) is a characteristic finding in almost all patients. Closer differentiation reveals these antibodies to be antiribonucleoprotein (anti-U1 RNP). IgG deposits on the nuclei of the keratinocytes are often found in skin biopsies.

Th./Prg.: Treatment is determined by the organs involved and is similar to treatment in SLE. Immunosuppression with azathioprine, if necessary also ciclosporin A (CSA), is used in cases of high disease activity or visceral involvement.

VASCULITIS SYNDROMES

Def.: Inflammation of the blood vessels mediated by immunopathogenic mechanisms, leading to damage of the organs affected. The spectrum of clinical symptoms depends on the extent of the damage and the localization of the vessels and organs affected.

Classification:

- Secondary vasculitis syndromes: e.g. in
  - Rheumatoid arthritis, connective tissue diseases and other auto-immune diseases
  - Infectious diseases (e.g. HIV infection)
  - Ingestion of some medicines

- Primary vasculitis syndromes:

  1992 classification (Chapel Hill Consensus Conference):

  I   **Vasculitis of small vessels**
  1. Wegener's granulomatosis   ⎫
  2. Churg-Strauss disease   ⎬ ANCA-associated vasculitis syndromes of small vessels
  3. Microscopic panarteritis   ⎭
  4. Henoch-Schoenlein disease   ⎫
  5. Vasculitis in idiopathic cryoglobulinemia ⎬ Non-ANCA-associated vasculitis syndromes of small vessels
  6. Cutaneous leukocytoclastic vasculitis   ⎭

  II   **Vasculitis of medium-sized vessels**
  1. Classic panarteritis
  2. Kawasaki disease

  III **Vasculitis of large vessels**
  1. Giant cell (temporal) arteritis
  2. Takayasu disease (arteritis)

I   Vasculitis of small vessels

ANCA-ASSOCIATED VASCULITIS SYNDROMES OF THE SMALL VESSELS

Wegener's granulomatosis [M31.3]

Def.: Necrotising vasculitis predominantly of the small to medium-sized vessels with ulcerating, non-caseating granuloma in the respiratory tract (nose plus paranasal sinuses, middle ear, oropharynx, lung) and renal involvement in 80% of cases (glomerulonephritis, microaneurysms)

Ep.: Prevalence: 5/100,000

Aet.: Unknown, triggered in some cases by *Staphylococcus aureus*

Cl.: 2 stages:

1) An initial localised stage: Respiratory tract disease <u>without</u> glomerulonephritis and without systemic vasculitis
- Chronic rhinitis/sinusitis with possible haemorrhagic, scabby component, possible saddle nose, septum perforation, chronic otitis, possibly also mastoiditis
- Oropharyngeal ulcerations
- Pulmonary nodules ("coin lesions"), possibly with cystic changes (pseudo-cavitation), possibly subglottic stenosis of the larynx or bronchi

2) Generalized vasculitic stage with pulmonary-renal syndrome:
- Possibly alveolar haemorrhage with haemoptysis } pulmonary-renal syndrome
- (Rapidly progressive) glomerulonephritis   }
- Possibly episcleritis, arthralgia, myalgia, CNS symptoms
- Fever, weight loss, night sweats

DD:
- Infectious ENT and lung diseases (resistance to antibiotic therapy in Wegener's granulomatosis)
- Other vasculitis syndromes

Di.:
- Clinical picture (inflammation in nose or mouth, pulmonary findings, erythrocyturia)
- Biopsy from nasopharynx, lung and possibly kidneys → histologic triad: granuloma, vasculitis, glomerulonephritis
- ENT examination of paranasal sinuses with biopsies of mucosa (often non-specific)

Lab.:
- Often elevated ESR, erythrocyturia and increase in serum creatinine (glomerulonephritis), possibly leukocytosis, thrombocytosis, anaemia
- Detection of antineutrophilic cytoplasmic antibodies with cytoplasmic fluorescence pattern (cANCA) in most cases with target antigen proteinase 3 = anti-proteinase 3 antibodies (PR3-ANCA): 50% of cases in the initial localized stage, 95% of cases in the generalized stage.

X-ray: /CT
- Paranasal sinuses and chest: shadow of paranasal sinuses; infiltration, round foci, cystic changes in lung
- Cranial MRI or CT: Evidence of granuloma in the paranasal sinuses and possibly intracerebral lesions
- MRI/CT angiography: Detection of microaneurysms in the renal vessels (70% of cases)

ACR criteria (1990):
- Inflammation of the nose or mouth with ulcers and purulent nasal secretion
- Nodules, infiltration or cavitation in the chest x-ray
- Pathological urinary sediment with microhematuria or erythrocyte casts
- Granulomatous inflammation in arterial walls or perivascular regions confirmed by biopsy
At least 2 of 4 criteria must be met for the classification!

<u>Th.:</u> Stage-dependent:
- Initial localised stage: treatment with cotrimoxazole: 2 x 1 tablets of 160 mg trimethoprim and 800 mg sulfamethoxazole, possibly with additional low-dose prednisolone for a short time. Long-term remission in 2/3 of cases.
- Generalised stage: with life-threatening or organ-damaging course:
  1. Induction of remission:
    • Prednisolone 1 mg/kg b.w. daily + cyclophosphamide (CY) 2 mg/kg b.w. daily by mouth (maximum 200 mg) or CY i.v. as bolus injection every 14 days
    • In severe forms of disease 1 g methyl prednisolone i.v. daily for 3 days.
      Additional high-dose immunoglobulins i.v. in refractory cases
      Reserve options: Mycophenolate mofetil, 15-deoxyspergualine, infliximab, rituximab (anti-CD20 monoclonal antibody)
    • In mild disease, MTX 15 – 25 mg/week (CI: renal damage) + prednisolone 1 mg/kg b.w. daily may be sufficient to induce remission
  2. Maintenance therapy:
    After onset of remission, CY may be replaced with MTX or azathioprine (in the event of AE/CI for azathioprine switch to other immunosuppressants). The prednisolone dose is gradually reduced according to clinical picture and inflammation parameters. Six months after the start of therapy, the daily dose should not exceed 5 - 10 mg. Cotrimoxazole is given to treat/prevent a nasal infection with *Staphylococcus aureus*, as a result of which > 80% of patients remain in remission. here are currently no reliable recommendations for the duration of maintenance therapy.

<u>Prg.:</u> Without therapy, the prognosis is poor; with optimum treatment, the 5-year survival rate is about 85%, although organ damage (especially of the kidneys) worsens the prognosis.

CHURG-STRAUSS SYNDROME (ALLERGIC GRANULOMATOUS ANGIITIS) [M30.1]

<u>Def.:</u> Vasculitis predominantly of the small vessels with non-caseating granuloma and infiltration of eosinophilic granulocytes

<u>Aet.:</u> 1. Idiopathic
2. Drug-induced (e.g. montelukast)

<u>Cl.:</u> - Allergic asthma, possibly also allergic rhinitis
- Transitory pulmonary infiltrates, possibly fever
- Cardiac involvement (eosinophilic myocarditis, coronaritis)
- Mononeuropathy / polyneuropathy
- CNS vasculitis
- Frequent thromboembolism

<u>Lab.:</u> - Eosinophilia (in the blood and affected organs), total IgE ↑, possibly AP ↑
- pANCA in 40%, often MPO specificity

<u>Di.:</u> ACR criteria (> 4/6 criteria must be met)
  • Bronchial asthma
  • Transitory pulmonary infiltrates
  • Sinusitis
  • Eosinophilia > 10%
  • Polyneuropathy
  • Biopsy: Evidence of extravascular eosinophilia

<u>Th.:</u> As in Wegener's granulomatosis. In some cases, interferon alpha is effective.

<u>Prg.:</u> 5-survival rate approx. 60%; most frequent cause of death: myocardial infarction, heart failure

POLYARTERIITIS NODOSA [M30.0]

<u>Syn.:</u> Microscopic polyangiitis (mPAN)

<u>Def.:</u> An essential criterion of mPAN is that it affects small ("microscopic") vessels, but it may also involve larger vessels. Clinically, mPAN resembles Wegener's granulomatosis, but histologically the granulomatous inflammation is absent.

<u>Aet.:</u> Unknown

<u>Cl.:</u> - Renal involvement (70%): this essentially determines the prognosis: glomerulonephritis of variable histology, up to rapidly progressive GN with crescent formation. Development of nephrogenic hypertension, possibly with headache and development of renal failure

Urine: microscopic haematuria, proteinuria
- <u>Pulmonary vasculitis</u>, possibly with alveolar haemorrhage and blood in the sputum
- <u>Skin lesions</u> (40%): subcutaneous nodules, palpable purpura, predominantly of the lower extremities, possibly with necrosis (biopsy!)
- Further symptoms: polyneuritis, sinusitis, episcleritis, etc.

Note: A mixed picture of PAN with other vasculitis syndromes is referred to as polyangiitis overlap syndrome.

Lab.: ■ Anti<u>n</u>eutrophilic <u>c</u>ytoplasmic <u>a</u>ntibodies with perinuclear fluorescence pattern (<u>pANCA</u>), often with target antigen myeloperoxidase = anti-<u>myeloperox</u>idase antibodies (<u>MPO-ANCA</u>) in 60% of cases.

Note: pANCA are not specific to mPAN; they are also found in other vasculitis syndromes, such as ulcerative colitis, etc.

DD: ■ Mixed connective tissue diseases, especially SLE (leukopenia!)
　　　　 ■ Vasculitis of other origin

Di.: ■ Clinical picture, anti-MPO antibodies, biopsy / histology

Th.: As in Wegener's granulomatosis

Prg.: Steroids + cyclophosphamide bring about long-term remission in 90% of cases. With cotrimoxazole, more patients remain in remission.

NON-ANCA-ASSOCIATED VASCULITIS SYNDROMES OF THE SMALL VESSELS

Def.: Necrotizing vasculitis of the small vessels with leukocytic infiltration and decaying leukocytic nuclei. Deposits of immune complexes and complement are found in the inflammatory lesions. Skin manifestations predominate, although visceral organs may also be affected. Cutaneous leukocytoclastic vasculitis only affects the skin.

<u>Three clinical syndromes:</u>

● <u>Henoch-Schoenlein purpura (= allergic purpura) (HSP)</u> [D69.0]:]
　Occur.:　Mostly in preschool children
　Aet.:　Allergic vasculitis of the small blood vessels and capillaries associated with a prior infection of the respiratory tract (influenza A in 50% of cases)
　Pg.:　Type III hypersensitivity (Arthus reaction) with subendothelial deposits of immune complexes containing IgA in the small vessels and activation of the complement system.
　Cl.:　Fever + severe malaise - 5 common manifestations:
　　　　1. <u>Skin (100%):</u> Petechiae + rash ("palpable purpura"), especially on extensor surfaces of the legs + gluteal region
　　　　2. <u>Joints (65%):</u> Painful swelling of ankles and other joints ("Suddenly my child no longer wanted to walk.")
　　　　3. <u>GI tract (50%):</u> Colicky abdominal pain, vomiting, possibly GI bleeding with melena
　　　　4. <u>Kidneys</u> (clinical examination 30%, biopsy 80%): microhaematuria / macrohaematuria; <u>Hi.:</u> Mes-angioproliferative glomerulonephritis with mesangial IgA deposits
　　　　5. <u>CNS:</u> Headache, behavioural disorders, pathological EEG
　DD:　Purpura in meningococcal sepsis
　Di.:　- History / clinical picture: joint pain, abdominal pain, purpura with normal coagulation parameters
　　　　- Detection of circulating immune complexes, complement concentration often elevated initially; IgA ↑
　　　　- Biopsy/histology of skin lesions: Perivascular leucocyte infiltrations, vascular IgA deposits

● <u>Cryoglobulinemia</u> [D89.1]: Cryoglobulins are immunoglobulin complexes that precipitate in the cold; they mostly comprise IgM-IgG complexes (with monoclonal IgM as an autoantibody reacting with polyclonal IgG). Cryoglobulins often render patient rheumatoid factor positive (DD!). Detection of cryoglobulins (transport at 37°C in an insulated container). The cryo-precipitate is measured quantitatively either as cryocrit in % (reference value < 0.4 %) or by protein measurement of the washed precipitate (< 80 mg/L). The differentiation (monoclonal vs polyclonal) is carried out by means of immunofixation.

　Three types:　Type I:　Monoclonal cryoglobulin, mostly IgM (e.g. MM, Waldenström's disease etc.)
　　　　　　　　Type II:　Monoclonal and polyclonal immunoglobulins
　　　　　　　　Type III:　Polyclonal immunoglobulins (e.g. in rheumatic diseases)
　Aet.:1. Essential cryoglobulinemia
　　　　2. Secondary cryoglobulinemia: HCV infection, malignant lymphoma, connective tissue disease, plasmocytoma, Waldenström's disease etc.
　Cl.: Palpable purpura of predominantly acral distribution, arthralgia, glomerulonephritis with hematuria, proteinuria (50%), neuropathy, hypocomplementemia
　Dx: History/clinical picture/lab; possibly evidence of hepatitis C infection in cryoglobulinemia

● **Cutaneous leukocytoclastic vasculitis (CLV)** [D31.8]: Isolated angiitis of the skin without systemic manifestation

Th.: ▶ HSP: Glucocorticosteroids

Additionally cyclophosphamide in life-threatening or organ-damaging forms
▶ Cryoglobulinemia: Antiviral therapy in HCV-associated cryoglobulinemia (see relevant section); in essential cryoglobulinemia methotrexate, in progressive forms cyclophosphamide + corticosteroids; reserve treatment in refractory forms: rituximab
▶ CLV: symptomatic

Prg.: In HSP and CLV relatively good; after HSP, chronic renal failure may occur after years in some cases (→ long-term monitoring); the prognosis in essential cryoglobulinemia depends on the underlying disease

BEHCET'S DISEASE [M35.2.]

Def.: Multi-system disease with the histological picture of leukocytoclastic vasculitis; some systemic vasculitis with the involvement of arteries and veins

Aet.: Not known: Genetic + environmental factors (triggered by infection?)

Epid.: First manifestation between ages 20 and 40 years, m : w = 3 : 1. Common in Turkey and other countries bordering the former "silk route". The risk remains elevated among Turks living in Germany (prevalence approx. 20/100,000). This can be explained by the marked association with HLA-B51 (approx. 70%).

Cl.: - Skin/mucosa: Mouth ulcers (95%), genital ulcers (70%), pseudofolliculitis, papulopustules, erythema nodosum, vasculitis, pathergy phenomenon (pustule formation after intracutaneous puncture with 20 G needle at a 45° angle)
- Eye involvement (80%!): Uveitis anterior/posterior, panuveitis: risk of disease 25% within 5 years
- Arthritis (up to 70%)
- GI tract (up to 30%): Granuloma, ulcers, vasculitis, perforations (DD: Crohn's disease)
- CNS (up to 30%): CNS vasculitis, brainstem symptoms, sinus vein thrombosis
- Thromboembolism correlates with disease activity (arterial and venous!)

DD: Exclusion of hepatitis B, C, HIV, florid HSV infection

Di.: History, clinical picture

Exclusion of a viral infection (see DD), ophthalmologic investigation, pathergy test, detection of HLA-B51, focus search

Th.: Corticosteroids, colchicine in mild cases, if necessary immunosuppressants (azathioprine, CSA), also cyclophosphamide in life-threatening manifestations.

IFNα2a especially in cases with ocular involvement (long-lasting remissions described).
Reserve in severe refractory forms: TNF blocker

II. Vasculitis of medium-sized vessels

CLASSICAL PANARTERITIS NODOSA (cPAN) [M30.0]

Syn.: Classical polyarteritis nodosa

Def.: Vasculitis of medium-sized vessels without involvement of small vessels and without glomerulonephritis

Ep.: Incidence: 5/100,000 per annum; m : w = 3 : 1

Aet.: Hepatitis B infection; unknown causes

Cl.: General symptoms:
- Fever, weight loss (50%), night sweats
- Muscle and joint pain (65%)
- GI tract (50%): Colicky abdominal pain; possibly intestinal infarction
- Testicular pain
- Involvement of coronary arteries (80%): Angina pectoris, myocardial infarction in younger patients
- Involvement of cerebral vessels: Stroke in young patients
- Polyneuropathy (60%), mononeuritis multiplex, epilepsy, psychosis

Note: Since the disease can affect other organs, the clinical picture is diverse and the differential diagnosis difficult!

Lab.: Inflammation parameters:
ESR/CRP ↑, leukocytosis/granulocytosis, possibly thrombocytosis, possibly complement ↓
ANCA-negative, possibly detection of hepatitis B infection

<u>DD:</u>
- DD abdominal pain of uncertain origin
- DD of polyneuropathy
- DD fever of uncertain origin

<u>Di.:</u>
- Clinical picture / lab findings
- Arteriography of the splenic artery and coeliac tract with detection of microaneurysms
- Biopsy of clinically affected organs (e.g. muscle/skin biopsy): Vasculitis with granulomatous changes

<u>Th.:</u>
- <u>Hepatitis B-associated cPAN:</u> Antiviral therapy (see relevant section)
- <u>cPAN without hepatitis B:</u> Methotrexate, in progressive forms cyclophosphamide + corticosteroids

<u>Prg.:</u> Poor without treatment; the 5-year survival rate is about 90% with treatment

KAWASAKI DISEASE [M30.3]

<u>Syn.:</u> Mucocutaneous lymph node syndrome

<u>Ep.:</u> Most common vasculitis in small children; 80% of patients are <5 years old.

<u>Aet.:</u> Unknown

<u>Cl.:</u> Six main symptoms:
1. <u>Septic temperature</u> that does not respond to antibiotics
2. Mostly bilateral <u>conjunctivitis</u> with marked vascular congestion
3. <u>Stomatitis</u> with erythema of the posterior pharyngeal wall and a strawberry tongue, as in scarlet fever
4. Reddening of the hands (<u>palmar erythema</u>) and not infrequently also of the soles of the feet
5. <u>Rash:</u> in the second to third week, a mostly semilunar scaling of the skin starts at the fingertips
6. <u>Swelling of the cervical lymph nodes</u> in 50% of cases

<u>Lab.:</u> Parameters indicative of disease activity: ESR, CRP, α2-globulins, leukocytes ↑, platelets ↑
anti-endothelial cell antibodies = AECA

<u>Co.:</u> Aneurysm of the coronary vessels (20%) and, more rarely, of other arteries, coronaritis and myocardial infarction

<u>DD:</u> Scarlet fever, infectious mononucleosis (EBV infection)

<u>Di.:</u> To establish the diagnosis, 5 of the 6 main symptoms or 4 main symptoms + (MRI) evidence of coronary aneurysm must be present

<u>Th.:</u> High-dose immunoglobulins i.v. + oral ASA (no corticosteroids, which worsen the clinical picture!)

<u>Prg.:</u> Mortality approx. 1%; most frequent cause of death: myocardial infarction

III. Vasculitis of large vessels

GIANT CELL ARTERITIS (GCA)

Polymyalgia rheumatica = PMR [M35.3] and temporal arteritis [M31.6]

<u>Syn.:</u> Horton's arteriitis

<u>Def.:</u> Granulomatous arteritis of medium-sized and large arteries, in 50% of cases with giant cells: giant cell arteritis; <u>temporal arteritis:</u> mostly in the area supplied by the carotid artery (preferentially in the region of the external carotid branches and the arteries of the eye). <u>PMR:</u> Aortic arch/proximal extremity arteries. Clinical manifestation as temporal arteritis and/or polymyalgia rheumatica

<u>Ep.:</u> Most common vasculitis (GCA); mostly older women (75%); incidence (per 100,000/year): 5th decade <5; 6th decade >10; 7th decade 40; 8th decade almost 50

<u>Aet.:</u> Unknown (endogenous predisposition + exogenous trigger of infection?))

<u>Cl.:</u> ▶ **Temporal arteritis:** (approx. 60% also have PMR)
- <u>Thumping (temporal) headaches, possibly pain on chewing</u> (masseter pain, jaw claudication)
- <u>Ocular involvement about 30%:</u> Ocular pain, impaired vision, amaurosis fugax, risk of blindness! Ischaemic optic neuropathy; a cherry-red spot in the central fovea is observed when the central retinal artery is occluded
- <u>Prominent temporal artery</u> (indurated, tender and possibly pulseless)
- <u>More uncommon manifestations</u>, e.g. aortic arch syndrome with asymmetric blood pressure; mononeuritis multiplex, TIA/apoplexy, CHD
▶ **Polymyalgia rheumatica:** (approx. 20% also have temporal arteritis)
- <u>Severe</u> symmetric <u>pains in the shoulder and/or pelvic girdle</u> (especially at night), tenderness of affected muscles
- Morning stiffness (mostly > 30 min.)

▶ Bilateral subdeltoid or subacromial bursitis is a characteristic finding in PMR (sonogram, MRI).
▶ General symptoms: fatigue, possibly fever, loss of appetite/weight, night sweats, depression

Lab.: ESR ⇑ (mostly >50 mm in the 1st hour), CRP ↑, possibly slight leukocytosis and anaemia (CK normal, no autoantibodies)

DD: ▶ Temporal arteritis: Headache of other pathogenesis, amaurosis fugax in arterial occlusive disease of the carotid artery
▶ PMR: polymyositis/dermatomyositis (CK ↑), rheumatoid arthritis, especially late-onset RA (LORA); paraneoplastic myopathy

Di.: ▶ Temporal arteritis- diagnostic ACR criteria:
1. Age > 50 years
2. New or new kind of headaches
3. Abnormal temporal arteries (tenderness, weak pulse, swelling)
4. ESR > 50 mm in the first hour (sudden decline!)
5. Characteristic histologic changes in biopsy of temporal artery (caution: before biopsy, perform colour duplex sonography to exclude occlusion of the internal carotid artery with collateral circulation via the external carotid artery → CI for biopsy of the temporal artery. Since segmental involvement is possible, the biopsy tissue must be at least 20 mm long.)
   • Colour duplex of temporal arteries: Segmental "hour-glass" stenosis and hypoechoic halo, which probably corresponds to inflammatory wall oedema; this regresses after about 2 weeks with corticosteroid therapy. In cases with a characteristic halo and a corresponding clinical picture, many authors dispense with a biopsy of the temporal artery.
   • Ophthalmologic examination
▶ PMR: Characteristic clinical picture, positive therapeutic attempt: prompt response to corticosteroids (pain relief of PMR), CK and electromyography normal; arthrosonography: subdeltoid bursitis, trochanteric bursitis, biceps tendinitis

Th.: Glucocorticoids: e.g. prednisolone initially 40 - 60 mg/d according to acuteness of symptoms (in the case of amaurosis fugax, inpatient treatment with high-dose corticosteroids i.v.); after clinical improvement, gradually reduce dose by 5 mg/week; maintenance dose ≤ 7.5 mg/d for at least 2 years, otherwise risk of relapse. In PMR a dose of 20 mg/d is sufficient as initial therapy. If corticosteroids are poorly tolerated or the maintenance dose is too high, steroid use can be spared by combination with immunosuppressants (e.g. methotrexate).

Note: If temporal arteritis is strongly suspected, start steroid therapy immediately (owing to risk of blindness). The conclusiveness of the biopsy will not be altered during the first 14 days of corticosteroid therapy!
If there is not a dramatic improvement within a few days on corticosteroid therapy, the diagnosis must be reviewed (tumour, sepsis)!

Prg.: If untreated, loss of sight occurs in approx. 30%; with consistent therapy, the prognosis is relatively favorable: Often healing of polymyalgia (rarely also giant cell arteritis) after 1 - 2 years.

TAKAYASU DISEASE [M31.4]

Def.: Granulomatous inflammation of the aorta and its main branches; usually patients aged less than 40 years. It may result in occlusion of the vessels affected.

Criteria of the American College of Rheumatology (ACR) of 1990:
If 3 or more criteria are met: Takayasu with a specificity of 98%
• Age at start of disease < 40
• Intermittent claudication of upper/lower extremities
• Weak pulse/pulselessness of brachial artery ("pulseless disease")
• Blood pressure difference > 10 mm Hg between arms
• Vascular murmurs over the vessels (e.g. subclavian artery, aorta)
• Pathologic angiogram without any signs of arteriosclerosis or fibromuscular dysplasia

Ep.: Rare disease, incidence < 1/100,000/year in Europe and North America. Occurs especially in China, India, Japan, Korea, Thailand, Africa and South America. f > m

Cl.: 1. Prepulseless phase (pre-occlusive stage): insidious onset over years; fever, tiredness, weight loss, arthralgia
2. Pulseless phase (occlusive stage): panniculitis, erythema nodosum, Raynaud's symptoms. Intermittent claudication pain (mostly of arms). Involvement of cerebral arteries leads to impaired vision, visual field deficits, impaired concentration, stroke. Symptoms as in CHD and infarction of the relevant organs.

Lab.: ESR ↑↑ (often over 50 mm/h), anaemia, leukocytosis

DD: Arteriosclerotic diseases (CHD, peripheral artery disease, TIA); temporal arteritis and other vasculitis syndromes

Di.: Hx, doppler sonography, CT/MRI angiography, possibly PET with fluorine deoxyglucose for assessing disease activity

Th.: 1. Immunosuppression: glucocorticoids, MTX; reserve treatment: cyclophosphamide, anti-TNF therapy
2. ASA
3. Stenosis treatment: PTCA, stent, vascular surgery

Prg.: With treatment, the prognosis is favourable. Otherwise complications resulting from CHD and stroke.

APPENDIX

Fibromyalgia (FMS) [M79.0]

Def.: Multilocular pain syndrome with characteristic tender points associated with vegetative and functional symptoms.

Ep.: Approx. 2% of the population; f:m = 9:1, most common between the ages of 30 and 60

Aet.: Unknown

Classification:
- Primary FMS (with or without depression)
- Secondary FMS in:
  - Systemic rheumatic diseases (e.g. SLE, Sjögren's syndrome, RA etc.)
  - Infectious diseases (e.g. EBV infection, chronic hepatitis C etc.)

Cl.: ACR criteria of 1990 (American College of Rheumatology):
1. Pain in at least 3 areas of the body (left and/or right side, above or below the waist) for at least 3 months, with at least 11 painful areas out of 18 tender points tested.
   - Trapezius insertion at the base of the skull
   - Transverse ligament C4/C5
   - Trapezius muscle at the shoulders
   - Levator scapula muscle at the angle of the shoulder blade
   - Second costochondral junction
   - Lateral epicondyle (outer elbow)
   - Upper outer quadrant of buttocks
   - Major trochanter (lateral process of femur)
   - Medial fat pad of knee

   The diagnosis is confirmed by checking so-called control points, which are not usually tender in fibromyalgia. If more than 3 of 14 control points are tender, the diagnosis of fibromyalgia is doubtful:
   - Center of forehead 2 cm above the supraorbital ridge
   - Lateral/middle third of clavicle
   - Dorsal mid-forearm, between radius and ulna; 5 cm above wrist
   - Thumbnail
   - Thenar eminence (ball of thumb)
   - Biceps femoris muscle (mid-femur)
   - Calcaneal tuber (transition from heel to sole of foot)
2. Vegetative symptoms such as cold extremities, dry mouth, hyperhidrosis, trembling
3. Functional symptoms: insomnia, general exhaustion, paresthesia/dysesthesia, migraine, globus sensation, sense of swelling, feeling of stiffness, respiratory and cardiac symptoms, gastrointestinal symptoms, dysmenorrhea, dysuria.

Lab.: Normal

Imaging diagnostics: No specific findings

DD: - Tendopathy, inflammatory and degenerative disorders of the spine and joints
- Myofascial syndrome (MFS): Complex painful symptoms associated with trigger points and induced by excessive or inappropriate strain. The symptoms may regress when the causative factors are eliminated
- Polymyositis, polymyalgia rheumatica, and other diseases
- Psychoses; psychosomatic disorders
- Protracted viral infections

Di.: History/clinical picture, exclusion of other diseases, ACR criteria

Co.: Chronification, high degree of suffering because of resistance to therapy, disability

Th.: No causal therapy is known, numerous symptomatic approaches to treatment.
- Physical: heat/connective tissue massage, salt water bath / UVA phototherapy, whole body cryotherapy
- Relaxation techniques, physiotherapy, gentle endurance sports (e.g. swimming, walking, aqua-jogging etc.)
- Acupuncture, neural therapy, TENS (transcutaneous electrical nerve stimulation)

- Analgesics if required (e.g. acetaminophen), NSAIDs
- Antidepressants if required: e.g. low-dose amitriptyline
- In clinical trials (off-label use): anticonvulsants, e.g. pregabalin; serotonin/norepinephrine reuptake inhibitors: duloxetine; anti-Parkinson agent pramipexole – response rates overall 30 - 40%
- Self-help groups, if necessary psychosomatic therapy
- Patient training programs (outpatient or in a rehab clinic with experience of FMS)

Prg.: Decrease in symptoms after the age of 60 years. If the FMS is diagnosed and treated early on in the first 2 years of the disease, complete remission is seen in up to 50% of cases. In the later course of the disease, remission rates fall.

Chronic Fatigue Syndrome (CFS) [G93.3]

Ep.: Prevalence approx. 1%, f:m=2:1

Aet.: Unclear

Cl.: CDC Criteria for CFS (Centers for Disease Control, 1994)

Main criterion: Medically unexplained states of exhaustion lasting 6 months, which
- have not occurred before
- are not significantly improved by breaks / periods of rest
- lead to a marked decline in levels of activity
- are not the result of exertion

Secondary criteria:
- subjective perception of memory impairment
- painful lymph nodes
- muscle pain
- joint pain
- headache
- non-restorative sleep
- > 24 hour feeling of unwellness following exertion

Note: The diagnostic criteria according to Fukuda also list sore throat among the secondary criteria

Exclusion criteria
- Other diseases as a cause of exhaustion (anaemia, infectious diseases, sleep apnoea syndrome, Addison's disease, Cushing's disease, hypothyroidism, hypopituitarism, poorly controlled diabetes mellitus, malignant tumours, etc.)
- Psychosis, depression, dementia
- Alcohol or drug abuse
- Marked overweight; anorexia nervosa

Lab.: There are no lab findings characteristic of chronic fatigue syndrome, but lab results are used as to exclude other diagnoses.

DD: See exclusion criteria

Di.: History/clinical picture, exclusion diagnosis

Th.: Causal therapy is not known
Exercise therapy and behaviour therapy have been shown to have a favourable effect.

Prg.: 1/3 of patients show medium-term improvement; mortality is not increased.

Degenerative joint disease (osteoarthritis)

I am indebted to Prof. Dr. Hans-Peter Brezinschek (University of Graz, School of Medicine) for his significant contributions to this chapter.

Def.: Osteoarthritis (OA) is a slowly progressive, degenerative disease of joint cartilage and other joint tissues of multiple aetiology. It is primarily non-inflammatory. Inflammatory episodes are called active arthritis.

Ep.: Most common joint disease, increasing incidence in advanced age.
Approximately 20% of the general population will show signs of OA of the hip or knee between the ages of 50 and 60 years. Of these, approximately 50% will complain of symptoms (pain, impaired walking).

Aet.: 1. Primary (idiopathic) OA: no identifiable cause, genetic factors, age, i.e. Heberden's arthritis of the distal interphalangeal (DIP) joints with node formation, patients of advanced age (f:m=4:1)

2. <u>Secondary OA:</u> As a result of accidents, joint malposition, obesity, excessive/one-sided strain, rheumatic joint disorders etc.

Prg.: Cartilage damage → initially unmasking of collagen fibrils, followed in later stages by disintegration of cartilage fibres → Cartilage destruction (end-stage: eburnation), osteophyte formation at the joint margins, formation of pseudocysts from focal bone necrosis

Stages:
1. Asymptomatic
2. Active (=inflamed) stage of OA with acute pain
3. Clinically manifest OA with chronic pain and functional impairment

Cl.: A cardinal and early symptom is pain:
<u>Early triad:</u> Pain at start of movement, fatigue pain and exertional pain; radiating pain (i.e. knee pain in OA of the hip)
<u>Late triad:</u> Chronic pain, nocturnal pain, muscle pain; also impaired mobility, meteorosensitivity, crepitation
<u>Advanced stages</u> are associated with thickening of the joint contours, deformation, instability, muscle atrophy, joint displacement and muscle contractures.
In <u>active OA</u>, the joint is excessively warm and tender and may show effusion with swelling (sonography).

<u>Imaging procedures:</u> Sonography, x-ray, MRI
No radiological signs of OA are seen in the early stages of disease
Asymmetric joint space narrowing, subchondral sclerosis, pseudocysts and osteophytes. In severe cases, there may be major deformation of the joints, and secondary chondrocalcinosis may follow.
<u>Radiological signs of OA often do not correlate with the clinical presentation:</u> Only about 50% of patients with radiological signs of OA report symptoms (pain).

Lab.: No specific changes

DD: In active OA, rheumatic diseases of monarticular onset:
Lab findings (with CRP and rheumatoid serology), joint scintigraphy (see section on RA for further DD parameters)

Di.: History, clinical presentation, imaging procedures
The discrepancy between subjective symptoms (pain) and imaging results is a problem (see epidemiology). So-called OA scores are useful for quantifying OA-related symptoms (see also internet).

Th.: Therapeutic goals:
1. Stop or slow progression of OA
2. Reduce or eliminate OA pain and secondary inflammation
3. Improve / preserve functionality
A. <u>Causal therapy:</u> e.g.
• Minimally invasive treatment of accident-related damage to joints
• Early optimum therapy for underlying rheumatic disease, (e.g. RA)
B. Symptomatic therapy
1. <u>General measures:</u>
   • Weight loss in overweight patients
   • Reasonable alternation between load-bearing and load-relieving movements; avoidance of sports associated with joint loading, selection of suitable sports that do not impose excessive strain on the joints
   • Selection of shoes with soft soles (cushioned heels)
   • Keeping joints warm, avoidance of cold/wet conditions
   • Swimming in warm water (thermal baths), (aqua) gymnastics
2. <u>Physical therapy:</u>
   • <u>Medical physiotherapy</u>, aqua gymnastics
   • <u>Patient education</u>
   • Thermal therapy (ointments, patches, infrared therapy, mud etc.) in OA between inflammatory episodes
   • In active, painful OA: cold applications, electrical and ultrasound therapy
   • Isometric muscle activity training (muscle building and strengthening measures)
   • Walking exercises etc.
3. <u>Drug therapy:</u>
   Indication: Inflammatory active OA with pain
   • <u>Acetaminophen:</u> analgesic effect only
   • <u>NSAIDs:</u> analgesic and anti-inflammatory effects
     (see section on RA for details)
     Treatment recommendations: administration of NSAIDs in painful OA:
     − No chronic treatment, only for limited period during painful and inflammatory episodes
     − No combination of NSAIDS (only use one substance at a time)
     − Dose adjustment according to pain symptoms
     − Single doses as low as possible, but as high as necessary
     − Preference for substances with a short half-life

- Close monitoring of gastrointestinal and renal function
- Reduction of daily dose in elderly patients
- Ulcer prophylaxis with a proton pump inhibitor or misoprostol if long-term treatment is necessary
- Possibly intra-articular of glucocorticoid crystal suspension:

Only in inflammatory active OA which does not respond to alternative treatment options. Injection therapy is to be administered under strictly aseptic conditions for a limited period (no chronic therapy! No injection into the hip joint in view of a risk of bone necrosis!); also note contraindications and side effects, corticosteroids may damage the cartilage!

4. Orthopedic measures:

In OA of lower extremities: use firm footwear with cushioned heels, insoles, aids to the rolling motion of the foot and pronation support for varus OA of the knee. If surgical treatment is not possible, arthrodeses are recommended to stabilize the joints.

5. Surgery:
- Minimal invasive surgery; open surgery
- Joint replacement (establish indication in good time in elderly patients)

Note: Any alternative treatment measures not mentioned here are either not confirmed by clinical trial results or not supported by sufficient evidence.

172

Porphyrias

Internet-Info: *www.porphyrie.com, www.doss-porphyrie.de; www.porphyriafoundation.com; www.porphyria-europe.com*

Def.: Porphyrias are mostly hereditary disorders of haeme biosynthesis. Haeme is synthesized in 8 enzymatic steps from glycine and succinyl-CoA. Each of the synthesis steps can be affected by a partial genetic defect. Accumulation of porphyrines and/or their metabolites,before a particular enzymatic step associated with increased excretion via feces and urine are the consequence.
Typically erythropietic and hepatic poprhyrias are distinguished related to the predominant anatomic site of enzyme expression.
Based on course of disease, porphyrias are divided in acute and chronic forms of disease.

PPh.: The Human body contains two different and independent heme pools:

1. Erythropietic haeme pool (bone marrow reservoir of haeme required for haemoglobin synthesis, which is taken up by erythrocytes)
2. Hepatic Haeme pool (heme storage pool for synthesis of important heme containing enzyme systems such as cytochrome P450 mono-oxidase)

Disorders of heme metabolism can therefore present either as primarily of erythropietic or as primarily of hepatic origin, based on location of a particular enzymatic defect.

Biochemistry of enzymatic defects in porphyria

Enzyme defect	Genlocus	Associated porphyria	Type of porphyria	Inheritance	Symptoms
δ-aminolevulinate (ALA) synthase		X-linked sideroblastic anaemia (XLSA)	Erythropoietic	X-linked	
δ-aminolevulinate (ALA) dehydratase	9.q34	Doss porphyria/ALA dehydratase deficiency	Hepatic	Autosomal recessive	Abdominal pain, neuropathy
hydroxymethylbilane (HMB) synthase PBG deaminase)	(or 11q24.1-q24.2	acute intermittent porphyria (AIP)	Hepatic	Autosomal dominant	Periodic abdominal pain, peripheral neuropathy, psychiatric disorders, tachycardia
uroporphyrinogen (URO) synthase	10q25.3-q26.3	Congenital erythropoietic porphyria (CEP) Gunther's disease	Erythropoeitic	Autosomal recessive	Severe photosensitivity with erythema, swelling and blistering. Haemolytic anaemia, splenomegaly
uroporphyrinogen (URO) decarboxylase	1p34	Porphyria cutanea tarda (PCT)	Hepatic	Autosomal dominant	Photosensitivity with vesicles and bullae
coproporphyrinogen (COPRO) oxidase	3q12	Hereditary coproporphyria (HCP)	Hepatic	Autosomal dominant	Photosensitivity, neurologic symptoms, colic
protoporphyrinogen (PROTO) oxidase	1q23	Variegate porphyria (VP)	Mixed	Autosomal dominant	Photosensitivity, neurologic symptoms, developmental delay
Ferrochelatase	18q21.3	Erythropoietic protoporphyria (EPP)	Erythropoietic	Autosomal dominant	Photosensitivity with skin lesions. Gallstones, mild liver dysfunction

173

- Erythropoietic porphyria

- Congenital erythropietic porphyria (CEP)=M. Günther

Occ.: Extremely rare autosomal-recessive disorder, manifastation during infancy and early childhood (1-3 years)

Aet.: Reduced activity of uroporphyrinogen-III-synthetase, leading to excessive storage and excretion of uroporphyrine I.

Cl.: Severe photodermatosis (face, hands), red coloured urine (porphyra= purpuric), with positive fluorescence under UV light, red-brown discolouration of teeth, positive luminescence in long-wave UV light, haemolytic anaemia, splenomegaly

Th.: Strict protection from light, possibly allogeneic bone marrow transplantation

Prg.: Unfavorable

Erythropietic protoporphyria (EPP)

Syn.: Erythrohepatic protoporphyria

Occ.: Autosomal-dominant rare disorder

Aet.: Reduced activity of ferrochelatase > protoporphyrin ↑

Cl.: Photodermatosis: erythema, pruritus, urticaria following sun exposure; cholelithiasis, bilestones containing protoporphyrine, protoporphyrin crystal formation and deposition in hepatic parenchyma, approx. 10% of patients develop hepatic cirrhosis, 5% of patients develop and die from hepatic failure

Di.: Porphyrine detrmination in blood, faeces and urine

Th.: Betacarotin, light protection, ursodeoxic cholic acid, cholestyramine, hepatic transplantation, bone marrow/stem cell transplantation

- Hepatic Porphyrias

- Acute hepatic porphyria (AHP)

Four types of AHP can be distinguished

Three types are characterized by autosomal-dominant inheritance (enzymatic defect listed in brackets)
> Acute intermittent poprhyria = AIP (defect of porphobilinogen-desaminase)
> Hereditary coproporphyria = HCP (coproporphyrinogen-oxidase defect)
> Porphyria variegata = PV occurs in white South African Population – prevalence 3/1000 – protoporphyrine oxidase defect

One type is characterized by autosomal-recessive inheritance

δ-Aminolevulinicacid-dehydratase-defect-porphyria = Doss-porphyria (rarity)

Cl.: of all four types of porphyria is comparable

DD: Secondary porphyria as a result of primary hepatic parenchyma and blood cell disorders

Lead intoxication: lead serum level ↑, porpphyrine concentration in urine ↑

Acute intermittent porphyria (AIP)

Occ.: Second most common porphyria and most common acute hepatic porphyria; prevalence 10/100.000 (psychiatric patients 200/100.000). Only approximately 20 % of all AIP gene carriers will develop manifest disease with symptom onset after puberty.
Female-male: 3:1, peak of disease onset: third decade, no skin involvement

Aet.: autosomal-dominant disorder with reduced activity of hydroxymethylbilan-synthase=porphobilinogen-de(s)aminase (PBG-D) by about 50%. Approximately 200 different mutations of the PBG-D-gene can be identified on chromsome 11 (11q24).

Prg.: Approximately 2/3 of hepatic heme is used for cytochrome P450 synthesis. δ – Aminolevulinic acid synthase (δ - ALA) is the key enzyme required for heme biosynthesis.
Activity of this enzyme is regulated by heme via a negative feed-back mechanism.

Latent Phase
↓
Decline of hepatic heme pool (i.e. Medication related cytochrom P 450-induction)
↓
increased hepatic δ-ALA activity
↓
accumulation of haeme precursors and of PBG
↓
acute flare up of disease (bout)

Manifestation inducing factors:

1. all types of stress (including surgical interventions, infections), low caloric diet (fasting) and hypoglycemia
2. porphyrinogenic substances: alcohol, androgens/estrogens, a multitude of medical substances (barbiturates, metamizole, benzodiazepines, metoclopramide, enalapril, sulfonamide-antibiotics, etc.)
3. subtype: ovulocyclic premenstrual flares in women

Cl.: Manifold and confusing (misdiagnoses occcur frequently !

Possibly onset of crisis after medication intake !
1. Abdominal symptoms: Cramping, bloating and additional abdominal symptoms (appendectomy scars seen frequently)
2. Neurologic-psychiatric symptoms: adynamy, polyneuropathy with paresthesias, peripheral pareses (initially of hand and lower arm extensors), seizure activity, depression, psychiatric symptoms etc.
3. Cardiovascular symptoms: hypertonus, tachycardie etc.

Types of disease progression:

1. carrier states of enzymatic defects (asymptomatic clinic and blood chemistry)
2. latent disease with increased excretion of porphyrines incl. Increase of porphyrines in urine, otherwise asymptomatic)
3. clinical manifestation (20% of carrier states); predominant age of manifestation: 20-40 years of age

DD:

- various abdominal disease states, including acute abdomen
- neurologic and psychiatric disorders
- alcohol disease with related abdominal and neurological symptoms
- panarteriitis nodosa
- lead intoxication (lead serum level ↑)

Di.:

Trias: abdominal discomfort, palsy/pareses, psychosis, tachycardia

1. reddish coloured urin and dark discolouration of urine(dark coloured stains in underwear)
2. Proof of increased porphobilinogen in urine: Hoesch-Test (2 drops of fresh urine mixed with 2-3 ml Ehrlich Aldehyde reagent: red colouration). Schwartz-Watson-test: same procedure with additional shaking out of bile acid colors (not however of porphobilinogen).More complex however not more specific.
3. For proof of diagnosis and for follow up diagnosis:
4. Quantitative determination of δ -ALA, PBG and porphyrine concentration in 24 hour urine
5. Determination of PBG-D- activity in erythrocytes and analysis of genetic defects

Th.: Acute crisis situations require admission to an intensive care unit. Recommended is also getting in touch with a porphyria centre (see also *www.porphyria.com*)

- Discontinuation of any precipitating medication
- Haemarginate and Glucose IV inhibit hepatic δ -ALA synthase induction
- Haemarginate IV (Normosang) 3 mg/kg/d (for 4 days)
- Additionally glucose IV (4-6 g/kg/d) plus forced diuresis while monitoring water and electrolyte levels

1. symptomatic therapy with „safe drugs"
   - hypertension and tachycardia: use of beta-blockers
   - abdominal cramping: spasmolysis with atropin type drugs and paracetamol, possibly pethidine
   - sedation: use of chlorpromazine;nausea: ondansetron
2. patients suffering from repeated onset of acute crises require intermittent prophylaxis administration of 1 ampule Hämarginate/week.

Note: Ovulocyclic depnedent presentation of AIP : prophylactically trial with LHRH analoges possible

Preventive Measures:

- Patient Education; Issue porphyria patient-pass, avoid noxious triggers
- Family History and Examination in order to be able to identify latent carrier states.Hereditary disease counseling recommended
- Patients presenting with abdominal cramping: Think also of porphyria and look for one (in particular prior to any laparotomy of abdominal symptoms of unclear etiology !)

B) Chronic hepatic porphyria (Porphyria cutanea tarda)

Occ.: Most common type of porphyria: prevalence 15/100.000 population average;male:female=2-3:1, disease onset most commonly after age 40.

Aet.:
- type I PCT= acquired sporadic type of disease
- type II PCT=hereditary type (50%): autosomal-dominant inheritance of hepatic uroporphyrinogen-decarboxylase (Uro-D) deficiency.Mutation of uro-D-Gene locus on chromosome 1 p34.
Disease onset (manifestation) in conjunction with alcohol abuse (70%), estrogen intake (hormonal contraception), hepatitis C virus infection and various other hepatic disorders, HFE gene mutations, AIDS and haemodialysis

Cl.: - photodermatitis and photosensitivity with increased vulnerability, hyperpigmentation, bullous formation on light exposed areas of the skin, in particular on face and hands healing with residual scar formation.
1. occasionally dark urine, the acidified urine shows red fluorescence under UV light (uroporphyrine III is excreted in higher amounts than normal)
2. onset of hepatic damage with deposition of porphyrine in hepatic parenchyma (UV-fluorescence positive of hepatic biopsy material !). Occasionally, multiple rounded (1-3 cm diameter) well demarcated echo dense infiltrates of hepatic parenchyma without any impact on blood vessels (DD: hepatic metastases) can be found.
3. Frequently hepatic enzyme levels are increased (transaminases; gamma GT increase also observed in conjunction with alcohol abuse)

Clinical courses of disease presentation:

- carrier state (asymptomatic, enzyme defects)
- latent PCT (limited to porphyrinuria)
- manifest PCT (hepatic damage, photodermatitis)

Di.:
1. History of alcohol/medication intake plus clinical signs and symptoms (don't forget)
2. Determination of porphyrines in urine (UV-fluorescence, biochemical differentiation:Uro-,Hepta and Coproporphyrin)
3. URO-D activitiy in erythrocytes reduced
4. hepatic biopsy (Redfluorescence in UV light, histochemistry)

Th.:
1. avoid any type of triggering factors:alcohol, estrogens (cave oral contraceptives !)
2. therapy of hepatitis C if required (see there)
3. phlebotomies/blood-letting or isolated reduction of erythrocyte cell count with blood cell separator (=erythrocytaphereses)
4. Chloroquin: 2X 125 mg/week (watc for side effects !) - formation of chloroquin-porphyrin complexes, which are excreted renally
5. Avoid sunlight (protective clothing) use of sunscreen with high UV protection factor

Prg.: favorable outcome, if all triggers can be eliminated

Hyperuricemia and Gout

Def.: Hyperuricemia: serum uric acid levels > 6.4 mg/dl (380 umol/l). This is consistent with the solubility of uric acid in serum at 37 degree Celsius and pH 7.4

Ep.: Approximately 20% of the male population of the western civilized world suffer from hyperuricemia (>416 umol/l). The majority of the female population develops hyperuricemia after menopause when etsrogen levels are falling.
The risk of a gout attack increases with rising serum uric acid levels: Incidence rates at levels of > 9 mg/dl (>535 umol/l): Approx. 5%/ year.
The risk of developing nephrolithiasis approximates 0.2% / year in asymptomatic hyperuricemia and 0.8%/ year in patients who suffer from gout.

> **Note:** The incidence of onset of gout is positively correlated with metabolic system disorders and related types of endocrine dysfunction

Aet.: A) Primary hyperuricemia and gout:
1. disturbance of renal tubular uric acid secretion (>99% of cases): Reduced uric acid clearance → excretion of normal amounts of uric acid occurs only in patients with increased serum uric acid levels.
The disorder is characterized by a predominantly polygenetic inheritance mode and becomes manifest with a purine rich diet and obesity (prosperity syndrome). The majority of gout sufferers ahve a positive family history.
2. Excess production of uric acid (< 1%)
- Lack of Hypoxanthin-Guanin-Phosphoribosyltransferase (HG-PRT),
   Two types of HG-PRT deficiency are known:
- Lesch-Nyhan syndrome: X-chromosomal recessive inherited disease, which is characterized by extremely low levels of HG-PRT (< 1% of normal activity). Trias:hyperuricemia- progressive renal insufficiency- various neurologic deficits with inclination towards self mutilation
- Kelley-Seegmiller syndrome: activity of HG-PRT is reduced (1-20% of normal activity). Trias: hyperuricemia, nephrolithiasis, in up to 20% of patients presence of neurologic deficits, but without associated tendency towards selfmutilation
3. Extremely rare in creased activity of PRPP-synthetase (phosphoribosylpyrophosphatsynthetase)
B) Secondary hyperuricemia:
1. increased uric acid production: increased nucleic acid turnover in conjunction with leucemia, polycythemia, haemolytic anaemia, tumour lysis syndrome, cytostatiy or radiation tumour therapy with increased cell turnover
2. reduced renal uric acid excretion:- renal disease, -lactacidosis, - ketoacidosis (fasting metaolism,diabetes mellitus),- medication use (use of saluretics)

PPh.: Total body content of uric acid (uric acid pool)_ approximates 1g. In patients with gout uric acid levels may increase to up to 30 g and more.
Daily uric acid production averages about 350 mg of uric acid from endogenous synthesis and exogenic purin intake sources.
Two thirds are eliminated via the kidneys, one third via the intestines.
In humans uric acid is the end product of purine metabolism. However, many mammals possess the enzyme uricase which metabolises uric acid further, resulting in allantoin.
Serum Uric acid concentration can be determined with uricase.
Age, sex and type of nutrition are important influencing factors for serum uric acid levels.Taking solubility limits of serum sodium urate levels into consideration, hyperuricemia starts at serum uric acid levels of > 6.4 mg /dl (380 umol/l).

Prg.: Triggering factors are fast changes of serum uric acid levels, i.e. as a result of eating a purine rich meal in conjunction with or without alcohol, after fasting, or at the beginning of a uricosuric therapy.
Urate crystals are excreted into the synovial fluid, where granulocytes and macrophages take them up. This leads to release of inflammatory reaction mediators which cause a crystal-induced synovitis

Cl.: 4 stages:
o asymptomatic hyperuricemia (a lot more common than manifest gout)
o acute gouty attack
o asymptomatic interval between two gouty attacks
o chronic stage characterised by tophus formation and irreversible joint degeneration

- **acute gouty attack**
Triggers: Fasting and feasting (eating and/or drinking binges), stress, etc.
Sudden unexpected onset (frequently at night) of very painful monarthritis, 60% of cases affect the large toe joint (podagra- the bed sheet cannot be tolerated) , is associated with reddening, swelling and heat sensation in the affected joint.
Additional joint manifestations may occur in ankle and tarsal foot joints (approx.15%), knee joint (gonagra approx 10%), toe joints (5%), finger joints (5%), in particular first thumb joint (chiragra), wrist joint and elbow joint.

A typical gouty attack subsides after several days but symptoms may last up to 3 weeks.

An acute attack of gout is accompanied by signs and symptoms of a systemic inflammatory disorder namely by leucocytosis, temperature increase and increased erythrocyte sedimentation rate (ESR). Concomitant hyperuricemia is not obligatory.

- Chronic gout

Today observed less frequently, predominantly in patients who have not received any type of therapy.

- Urate deposition (tophi)

- soft tissue tophi (Diagnosis: murexid testing method as proof of presence of uric acid), i.e. Ear cartilage, large toe,calcaneus,olecranon, tendosynovitis as a result of urate crystal deposition (rarely carpal tunnel syndrome), bursitis

- - bone tophi > X-ray identification possible

- irregular or rounded bony defect caused by intraossary tophus formation in joint vicinitiy

- cup like joint mutilation near joint forming bones

- spiny osteophytes which extend into a tophus

- - periostal osteophytes surrounding tophi arroding the bone cortex

- Renal manifestations of hyperuricemia and gout

- urate nephrolithiasis

Beware: urate stones are X-ray negative (do not visualize) and predispose to urinary tract infections !

1. Urate nephropathy= primary abacterial interstitial nephritis

Early clinical symptom: albuminuria

Complications.: onset of hypertension, rarely chronic renal insufficiency,rarely onset of acute uric acid nephropathy and obstructive urate nephropathy:

2. excretion of large amounts of uric acid associated with high dose cytostatic therapy may lead to acute renal failure as a result of obstruction of renal tubuli and ureter

DD:

- Secondary hyperuricemia (History!)
- Acute monarthritis of different etiology

Beware: onset of acute monarthritis in a male patient: first think of gout, then think of reactive arthritis !

- Purulent arthritis caused by bacterial infections (ie. In conjunction with arthoscopies, arthrotomies joint injections and other related invasive joint surgery)
- Acute metacarpohalangeal artrhitis of the large toe
- Chondrocalcinosis=pyrophosphate crystal deposition=pseudogout

-deposition of calciumpyrophosphate dihydrate crystals in joint cartilage potentially causing acute crystal-induced synovitis in particular of the knee joint

- Aetiology: 1) idiopathic with advance age 2) hereditary 3) secondary related to different etiologies

X-ray: calcium deposits in tendons and joint cartilage

Diagnosis: proof of CPPD crystal deposition in joint punctates via polarisation microscope

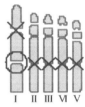

I II III VI V

DD: Pedal joint disease:
o Gout
x Rheumatoid arthritis

Di.:

1. Family History – Clinic- Laboratory

- Serum uric acid Early signs and symptoms:Early signs and symptoms:^

- Uric acid excretion in 24 hour urine

- Quoutient uric acid/creatinine (mg/dl) in spontaneously released urine (normal < 0.8)

2. Monarthritis of unclear etiology: prompt therapeutic response after administration of colchicine point to gout

3. X-ray of affected joints; determination of renal function

4. Special laboratory testing of enzymatic defects of purine metabolism if required

5. Synovial fluid analysis (polarisation microscopical assessment of bacterial contamination, leucocytoclastic uric acid crystal phagocytosis, etc.)

Th.: 1. Diet:

- ○ Normalization of body weight; increase fluid intake and diuresis (at least 1.5 l/d)
- ○ Beware with any type of cytostatic therapy and fasting: serum uric acid levels increase → increase fluid intake, uric acid neutralisation, administration of allopurinol
- ○ Low purine diet (< 300mg purine/d): reduced meat intake, no intake of animal inwards (like liver, kidney, brain) no intake of sardines, meat extract etc.
- ○ No alcohol intake: excess alcohol intake induces reactive lactic acidosis and can lead to temporary inhibition of renal urate excretion and thereby can induce a gout attack
- ○ Beware: no administration of diuretics which selectively reduce uric acid excretion !

2. Drug therapy:
a) acute attack:
- NSAIDS: drugs of first choice: i.e. Diclofenac, etc, in addition local cryotherapy
  Dose/side effects/contraindications: see chapter rheumatoid arthritis
- In the event of onset of untoward adverse events or if special contraindications require consideration administration of systemic steroids i.e. Prednisolone 20 mg/d once daily per os
- Colchicine is a second line treatment regimen related to its side effect profile
  mechanism of action: inhibition of phagocyte activity in diseased tissue
- - Side effects: gastrointestinal discomfort is observed in the majority of patients, dose-dependent onset of diarrhea, rarely agranulocytosis, myopathy etc.

- Contraindications: hepatic disorders, pregnancy and lactation (oral contraceptive use recommended until 6 months after discontinuation of treatment)
Dosing regimen: During the first 4 hours 1mg/h, followed by 0.5mg-1mg q 2 hours up to to maximum daily dose of 6mg. With onset of symptom improvement fast dose reduction is recommended.
Unfortunately, the majority of patients do not tolerate an optimized dosing regimen related to onset of GI side effects including diarrhea. Since colchicin works selectively best in acute gout for diagnostic reasons it should be the preferred initial treatment choice including in patients where signs and symptoms are not unequivocal.
b) Long-term therapy:
Patients with asymptomatic hyperuricemia and uric acid levels up until including 9 mg/dl (535 umol/l) are to follow a dietetic regimen. Drug therapy is indicated in patients suffering from clinically manifest gout and who consistently are suffering from increased serum uric acid levels > 9mg/dl (535 umol/l).
- Uricostatic treatment regimens:
  - Allopurinol: treatment of choice
  mode of action: inhibition of xanthinoxidase: reduced production of uric acid. Allopurinol inhibits progression of gout → goal: lowering of serum uric acid levels to levels of between 5,5 – 6,0 mg/dl (327 – 357 umol/l), in order to contribute to dissolution and excretion of uric acid. Initially, mobilisation of uric acid depots may induce onset of gout attacks in which temporary prophylaxis with NSAIDS is recommended.

SE: Rarely gastrointestinal disturbances, transaminase increase, leucocytopenia, rarely allopurinol hypersensitivity syndrome in 0.4 % of patients (=allopurinol-vasculitis associated with dermatitis, hepatitis and renal failure). In case of overdose or pre-existing renal insufficiency the risk of onset of serious side efects is increased !

Drug interactions: i.e. Inhibition of 6-mercaptopurine and of azathioprine, of theophyllin and phenprocoumon metabolism → dose reduction by up to 75% required.

Simultaneous administration of captopril is assocaited with an increased risk of leuopenia

Indications: 1. symptomatic hyperuricemia: arthritis urica, urate nephropathy, urate nephrolithiasis, tophi
2. Lack of clear cut clinical treatment recommendations for asymptomatic hyperuricemia. Generally, with serum uric acid levels > 9 mg/dl (535 umol/l) treatment is recommended by some specialists.

Contraindication: pregnancy and lactation

dosing: 100-300 mg/d; dose reduction in renal insufficiency patients.
- Rasburicase: recombinant urate oxidase, which catalyzes oxidation of uric acid to allantoin, which is water-soluble and therefore subject to renal elimination. Onset of action within hours (fast)
  Indication: tumour lysis syndrome: acute hyperuricemia during cytotoxic treatment of malignancies/leukemias.

Note: Serum samples from patients requiring rasburicase are to be sent to the lab cool,in a dry ice package in order to prevent continuation of enzymatic uric acid metabolism resulting in falsesl too low serum uric acid measurements.
- Uricosuric treatment:

Benzbromarone and probenecid

Mechanism of action: Increase of uric acid elimination via inhibition of tubular reabsorption of uric acid; initially, until normal serum uric acid levels are achieved the risk of tubular uric acid deposits and formation of urolithiasis is increase, thus tapering dose increase is recommended, patients are also to be encouraged to increase their fluid intake (2 l/d), also urine neutralisation to pH levels between 6,5 – 7,0 (i.e. with uralyte) is recommended.

Indication: Allopurinol incompatibility

Contraindication: Gout nephropathy (in particular nephrolithiasis and renal insufficiency), hyperuricemia

SE: Rarely allergic reactions, nephrolithiasis, gastrointestinal distumances, etc.

DISORDERS OF LIPID METABOLISM

(I hereby express my deep gratitude to Prof. Dr. Oette, former Head of the Department of Clinical Chemistry, University of Cologne, School of Medicine)

Internet-Info: *www.lipid-liga.de*

Synonyms: Hyperlipoproteinemias (HLP), Hyperlipidemias, Dyslipoproteinemias, Dislipidemias

Physiology: Plasma lipoproteins are composed of apolipoproteins and of lipids (triglycerides, cholesterin, phospholipids).

Density classes	*) in %	electrophoresis	Predominant function
chylomicrons	0	No migration	exogenousTriglyceride transport
VLDL	10	pre-β-lipoprotein	Endogenous triglyceride transpor
LDL	70	β-lipoproteins	Product of VLDL delipidation-cholesterol transport vehicle to extrahepatic cells-regulator of cellular cholesterolhomeostasis
HDL	20	α-lipoproteins	Transport of cholesterol to the liver-regulator of cellular cholesterolhomeostasis and lipolysis

*) approximate physiological levels in normal fasting serum (%)

Note: In hyperlipoproteinemia type III (HLP-type III) IDL (intermediate density lipoproteins)can be distinguished as separate lipoprotein serum fraction between pre-β and β-lipoprotein fractions. IDL are catabolites of VLDL, thus VLDL remnants and precursors of the LDL fraction. LDL subfractions with high atherogenicity (small dense LDL) and postalimentary presence of atherogenic chylomicron remnants shall be mentioned
Clinically signficant apolipoproteins (Apos) and occurrence: B-100 (VLDL;IDL;LDL);ApoE and Apo C-II (VLDL,IDL,HDL);Apo-A-I and Apo A-II (HDL).

Pathogenicity and atherogenicity:

VLDL,IDL and LDL increase and HDL decrease are linked with an increased atherosclerosis risk. Most significant risk increase is associated with extreme LDL and VLDL cholesterol serum elevations and HDL-cholesterol decreases. VLDL atherogenicity is low.
Prominent hypertriglyceridemias can trigger onset of acute pancreatitis in particular in the presence of chylomicrons (see refrigertor test). In case of predominant serum chylomicronemia, triglyceride serum levels of about 800 md/dl pose already an increased risk for onset of acute pancreatitis.

In case of predominant serum VLDL elevation triglyceride levels of about 2000 mg/dl are associated with an increased risk of pancreatitis and also increased arteriosclerosis risk (atherogenicity). However, even extreme increases of serum VLDL levels can occur without pancreatitis.

Classification:

▶ Based on assessment of serum cholesterol and serum triglyceride level assessment, 3 groups of hyperlipidemias can be distinguished:
- Hypertriglyceridemia (>200 mg/dl = > 2.3 mmol/l;more strict values > 150 mg/dl = >1.7 mmol/l)
- Hypercholesterolemia (>200 mg/dl = > 5.2 mmol/l)
- Combined hyperlipidemias (increase of both triglycerides and cholesterol)

Unit conversion for cholesterol: mg/dl = mmol/l X 38.6
Unit conversion for triglycerides: mg/dl=mmol/l X 88.5

Classification of lipoproteinconcentrations according to Frederickson (Lipoproteinelectrophoresis):

Type	I	II a	II b	III	IV	V
increase	chylomicrons	LDL	LDL &VLDL	IDL*	VLDL	VLDL&chylo microns
cholesterin	n	↑	↑	↑	n-↑	n-↑
triglyceride	↑	n	↑	↑	↑	↑
serum	lipemic	clear	Cloudy to lipemic	Cloudy to lipemic	Cloudy to lipemic	Cloudy to lipemic
Creamy surface layer	Yes & clear lower layer	-	-	-	-	Yes & cloudy lower level
distribution	rare	frequent	frequent	occasional	frequent	rare

* type III = dysbetalipoproteinemia with increase of cholesterol and Apo E rich IDL. These patients are Apo E 2-homocygotic

The original classification developed by Frederickson for phenotypisation of familial hyperlipoproteinemias is still useful as brief designation of the lipoproteinprofile, irrespective of the type of hyperlipoproteinemia (see below).
For typisation HDL and Lipoprotein (a)concentrations are not included.
Therefore, knowledge of HLP type is not suffificent for assessment of atherogenicity.
Noteworthy also, is that during the course of treatment the type of hyperlipoproteinemia can change from IV to IIB or V.

Ep.: In the age group of > 40 years more than 50% of the western nation population have cholesterol levels > 200 mg/dl (5.2. mmol/l). In addition, hypertriglyceridemias related to lifestyle and nutritional habits are common.

Note: Disorders of lipid metabolism are frequently associated with additional disorders linked with the metabolic syndrome: truncal obesity, pathological glucose tolerance testing and insulin resistance, hypertriglyceridemia,lowered HDL cholesterol levels and arterial hypertension.

Aet.:
Hyperlipoproteinemias are only symptoms:
From a pathogenetic point of view three classes can be distinguished:

I. Reactive types:
Metabolic overload.
Moderate HLP, predominantly related to poor nutritional and lifestyle habits
Hypertriglyceridemias, i.e. Related to increased alcohol consumption, high caloric and carbohydrate consumption
Hypercholesterolemias, i.e. Related to fat and cholesterol rich nutrition (animal fats, eggs)
Combined hyperlipoproteinemias are also bserved as a result of the described metabolic disturbances and nutritional stressors.

II. Secondary-symptomatic types:
Disease and medication induced metabolic system disorders
Common causes of hypertriglyceridemias: i.e. Poorly controlled diabetes mellitus,metabolic syndrome, obesity, pregnancy, high levels of alcohol consumption, renal insufficiency with haemodialysis, steroid therapy and occasional use of thiazid diuretcs, β-blocker therapy and use of hormonal contraceptives in women.
Common causes of hypercholesterolemias: i.e. Nephrotic syndroem, hypothyreosis, cholestasis (LpX ilevel ncrease), diabetes, pregnancy and medication related 8i.e. Steroids)
Causes of combined HLPs are identical with the ones described above.

III. Hereditary, familial (primary) hyperlipidemias:
Here it is important to obtain the patients respective family history

Note: Not necessarily all hereditary metabolic system disorders will become manifest during childhood or adolescence

1. Familial hypercholesterolemias (E78.0):
a) polygenic hypercholesterolemia
Unclear or very complex underlying moleculargenetic basis of disease. As a result of combined effects of endogenous (hereditary) and exogenous factors (nutritional habits, alcohol, obesity, life style) this manifests clinically most commonly as „common" hypercholesterolemia with total cholesterol levels between 250-400 mg/dl and a several fold increased risk of CAD.
b) monogenetic hypercholesterolemias
－ familial hypercholesterolemias (FH): functional mutation of the LDL receptor gene with autosomal dominant inheritance.

70 % of all LDL receptors are located in the liver which syntehsizes cholic acids from cholesterol (LDL= transport vehicle also for cholesterol). The hepatic ability to eliminate cholesterol from the blood is directl related to the number of LDL receptor on the hepatic cell surfaces (LDL receptor density). In heterozygote carriers of familial hypercholesterolemia the LDL receptor density is decreased, in homozygote carriers LDL receptors on hepatic cell surfaces are largely absent and either a total lack or minimal activity of residual LDL receptor activity on hepatic cell surfaces is observed.

- Frequency of heterocygote carrier state 1: 500, of homozygotic carrier state 1: 1 million of the average population.
- Heterozygote carriers typically have cholesterol levels between 300-500 mg/dl during adulthood and have a high likelihood of suffering an MI during their middle ages. The CAD manifestation risk in women is delayed by 7-10 years.
- *Note:* Women frequently underestimate the myocardial infarction risk.
- Homozygous patients have a cholesterol level of between 500-1200 mg/dl and suffer from CAD manifestations already during childhood and adolescence.
- Familial defect apolipoprotein B 100 (FDB):
- Functional mutation of the LDL receptor ligand gene. Apolipoprotein B 100, only LDL protein, represents the ligand of the LDL receptor. Frequency: approx. 1:600 : 1: 1000; autosomal dominant inheritance, until now only heterocygous forms have been described.LDL-cholesterol levels and CAD risk is comparable with a mild type of familial hypercholesterolemia.
- Apolipoprotein E variants:

Patients who carry the Epsilon 4-allellic variation of apolipoprotein E and the phenotypic variations E3/4or E4/4 show a reduced LDL receptor activity and as a consequence suffer from mild to moderate serum LDL-cholesterol increase. Untreated this is associated with an increased CAD risk. Carriers of apolipoprotein E4 also are at increased risk for development of Alzheimers Disease.

2. Familial combined (mixed) hyperlipidemia (E78.2)

Relative frequency: 1-2:100, autosomal dominant inheritance mode. Molecular genetically not understood. APO B 100 is produced in excess. Serum cholsterol levels increase to up to 350 mg/dl und triglycerid levels vary between 200-400 mg/dl. CAD risk increases with LDL cholesterol levels. *Beware:* Don't mix up with type III hyperlipoproteinemia

3. Familial hypertriglyceridemia (E78.1)

Frequency approx. 1:300 (?), occasionally occurrence in conjunction with metabolic syndrome, HDL-cholesterol ↓; triglyceride levels 200- >100 mg/dl, risk of onset of pancreatitis with high levels. Arteriosclerosis risk elevated in particular when HDL cholesterol levels are reduced. Disease as such most likely characterized by different moleculargenetic thus polygenetic traits, gene mutations of the enzyme lipoprotein lipase may play a role.

4. Familial dysbetalipoproteinemia (E78.2)
Syn: VLDL-remnant or type III hyperlipoproteinemia

Even though this genetic variant (apolipoprotein phenotype E 2/2= apo E 2-homozygosity) is observed quite commonly, manifestation of this metabolic disorder is quite rare (1:5000-1:10.000). Cholesterol levels 300-800mg/dl, serum triglyceride level 400 - 1000 mg/dl. IDL is increased. With high levels yellowish palmar xanthoma development is typical, early onset of arteriosclerosis. Milder forms are difficult to assess.

5. Chylomicronemia syndrome (E78.3)
   ○ occasionally found as part of a distinctive hypertriglyceridemia or found in conjunction with the rare familail type V-HLP. Latter type is complex from a moleculargenetic point of view.
   ○ Fat induced HLP (type I) is associated with lipoproteinlipase deficiency or apolipoprotein C II-deficiency (rarely observed).

Note: Presence of chylomicrons increase the risk of pancreatitis

6. Lipoprotein (a)-hyperlipoproteinemia=Lp (a)-elevation (E78.4)

Lp (a)contains an apolipoprotein, which competes against plasminogen for endothelial cell binding sites (antiplasminogenic effect). In addition, Lp (a) increases expression of plasminogen-activator-inhibitor 1 (PAI-1).
With increased lipoprotein (a) levels most likely local thrombolysis at the endothelial cell level is inibited, which favors plaque formation. However, alternative pathogenetic mechanism are being discussed. Lp (a)-levels of > 30 mg /dl are an independent risk factor for development of arteriosclerosis. In particular with LDL cholesterol level increases Lp (a) level elevations are to be monitored closely and a more stringent therapeutic reduction of LDL-cholesterol levels is recommended.

Note: in particular in association with serum IDL cholesterol level increases simultaneously increase lipoprotein (a) level increases are associated with increased risk of arteriosclerosis development including related complications like CAD.

7. Familial hypoalphalipoproteinemia (E 78.6)
- HDL-cholesterol lowering < 40 mg/dl (1,0 mmol/l). Atherogenicity is being underestimated

A high number of CAD patients has lower amounts of serum HDL cholesterol levels which are not only hereditary. Secondary HDL serum level reductions are to be found in numerous different conditions: i.e. Obesity, meatbolic syndrome, type II diabetes, hypertriglyceridemia, cigarette smoking and intake of anabolics.
HDL cholesterol increase associated with levels > 65 mg/dl, which occur more common in females, do not present a risk for development of arteriosclerosis and do not require treatment. Large scale epidemiological studies (i.e. Framingham study) even point toi the protective effect of HDL, which is why for overall assessment of the

atherogenic risk potential of IDL cholesterol HDL levels should be taken into consideration as well. Diagnostic significance increases when assessing the quotients total cholesterol/HDL-cholesterol or LDL-cholesterol/HDL-cholesterol.

HDL serum levels are dependent on various factors amongst which iare the activity of the cholesterol transfer protein (CETP) and the serum level of sex hormones.

<u>Cl.:</u>

- <u>Arteriosclerosis and related diseases</u> (see also pathogenicity and atherogenicity):
- CAD and MI
- Peripheral arterial vascular occlusive disease
- Arterial occlusive disease of cerebral arteries and stroke

Hypercholesterolemia and CAD:
In the presence of normal HDL cholesterol levels infarct morbidity rises rapidly whith total serum cholesterol level increases (related to LDL cholesterol) above 200 mg /dl (5,2 mmol/l) doubles at levels of 250 mg /dl (6,5 mmol/l), and quadruples at 300 mg/dl (7,8 mmol/l) compared to the risk at 200 mg /dl (= achievable serum limit which should not be exceeded). However, even at this level the risk of infarction is increased, if the cholesterol fractions show an atherogenic constellation:HDL- cholesterol < 35 mg/dl (0,91 mmol/l) or LDL-cholesterol > 150 mg/dl (3,9 mmol/l). Two thirds of all infarction patients have HDL levels < 35 mg /dl (PROCAM-study !).) <u>Even increased triglyceridelevels, in particular in association with HDL level decreases are associated with an increased risk of myocardial infarction.</u>

Long-term therapeutic LDL cholesterol serum level reduction is generally associated with a reduction of myocardial infarction risk by approximately 30% and reduction of total mortality by approx. 25%: This observation holds true for means of primary prevention (WOS study) as well as for means of secondary prevention of CAD patients (i.e. 4-S study). Opitmal lowering of serum LDL cholesterol levels in particular in younger patients can induce partial regression of athersclerotic plaque formation and can contribute to a conversion of instable (danger of plaque rupture) to stable plaques.

- Pancreatitis
occurrence: together with pronounced hypertriglyceridemias (see also patogenicity and atherogenicity)
- Xanthomata
- Tendon xanthoma (Achilles tendon, finger extensors)
- Plane xanthoma (interdigital creases), tuberous xanthoma (knee, elbows), xanthelasma (eye lids)
Occurrence: hypercholesterolemia (FH,FDB), phytosterolemia (very rare) and less pronounced with type III-HLP.
- Eruptive xanthomata (buttocks, lower arm extensor surfaces)
Occurrence: distinctive hypertriglyceridemias
- Palmar crease xanthoma (type III-HLP)
- Arcus cornea
Occurrence: FH, FDB and possibly type III as well as with extremely low HDL cholesterol levels.
- Fatty liver disease
Occurrence: hypertriglyceridemias

<u>Di.:</u> Laboratory:
- Triglyceride
- Total serum cholesterol, LDL and HDL cholesterol
- Total cholesterol/HDL-quotient; LDL/HDL quotient
Lipoprotein (a)
- Can be used as diagnostic indicator for HLP-type II determination of apolipoprotein E-genotype
- Serum lipemia check for chylomicrons (fridge test). After 24 hours chylomicrons will have formed a creamy top layer
LDL cholesterol level determination (according to Friedewald): LDL-cholesterol=total cholesterol – (0,2 X triglyceride) – HDL-cholsterol. However, with clearly elevated triglyceride and Lp(a)-levels as well as in type III hyperlipoproteinemias these calculations and the resulting values are of no use.

<u>Note:</u> For accurate serum triglyceride level determination blood withdrawal after a 12 hour fasting period is recommended, because serum triglyceride levels fluctuate meal dependent. However, other serum parameters like total cholesterol, Ldl-cholesterol, Hdl-cholesterol and lipoprotein (a) are not or only very little influenced by the last meals.

- <u>Differentiation between reactive-physiological, secondary and hereditary/familial disorders of lipid metabolism.</u> Differential-diagnostic examination to rule out metabolic syndrome, diabetes mellitus,hepatic and gall duct diseases, pancreatitis,thyroiddisorders, renal disorders, hyperhomocysteinemia and hyperuricemia;assessment of nutritional and lifestyle habits, history of alcohol and medication intake are recommended. Also bodymass-index and body fat distribution (waist/hip circumference) assessment are indicated. Hyperthyreosis is associated with decreased serum LDL levels, hypothyreosis may be associated with decreased LDL levels.
- <u>Determination of arteriosclerosis risk profile:</u>atherogenic cost, surplus weight, disorders of lipid metabolism, smoking, arterial hypertension, metabolic syndrome, diabetes mellitus, hyperhomocysteinemia, hyperfibrinogenemia, postive family history of myocardial infarction, inactivity etc. (see also coronary artery disease =CAD and atherogenicity of lipoproteins)

- Special diagnostic measures are required for suspected familial disorders of lipid metabolism: i.e. DNA analysis (LDL-receptor, ApoB100, Apo E et al.). At present genetic LDL-receptoranalysis is quite laborious and unable to identify all mutations. By now more than 500 mutations are known. So far, only few relevant mutations of the Apo B 100-gene were identified.
- Family screening in case of familial hypercholesterolemia: DNA-test for heterogeneous gene carriers.

Note: Hyperlipoproteinemia is a symptom and not a diagnosis! Diagnostic examples: secondary diabetic hypertriglyceridemia; familial hypercholesterolemia, type IIa, heterozygous type with LDL-receptormutation.

Th.:   Therapeutic goals:
▶ Triglyceride < 200 mg/dl (<2.3 mmol/l), increased risk (i.e. Diabetes mellitus)< 150 mg/dl (<1,7 mmol/l)
▶ LDL-cholesterol adjustment under consideration of arteriosclerosis risk category

Risk*)		LDL-cholesterol-goal value
High risk	CAD or CAD-equivalents	< 100 mg/dl(<2,6mmol/l)***)
Average risk	2 or more risk factors	< 130 mg/dl(<3,4mmol/l)
Low risk	0 or 1 risk factor	< 160 mg/dl (<4,1mmol/l)

*) risk definition and calculation: see chapter CAD
**)HK-aequivalent:ie.diabetes mellitus, ischaemic cerebral strioke, peripheral arterial occlusive disease
***) in high risk patients LDL goal levels should e <70 mg/dl (< 1.8 mmol/l)

▶ HDL-cholesterol > 40 mg/dl (> 1,0 mmol/l) in men, > 50 mg/dl (1.3 mmol/l in women)
in particular secondary HDL-level decrease can be infuenced therapeutically (s.a. and also section medication).

Target value for LDL-cholesterol-/HDL-Cholesterol-quotient (approximation):
High risk < 2,0
Average risk <3,0
Low risk <4,0

Beneficial in addition to LDL-cholesterol adjustment in particular when HDL cholesterol levels are low. Rounded values.

▶ Additional therapeutic goals:
Avoidance and treatment of pancreatitis;prevention of development and elimination of xanthomas and of fatty liver disease; improvement of serum lipid level in particular in the presence of extreme hypertriglyceridemias

Therapeutic approach:

1. Improvement of nutrtional habits and of lifestyle
2. Elimination of causative factors in secondary types, i.e. Improved diabetes mellitus treatment regimmen resulting in optimized serum glucose levels, treatment of hypothyreosis, weight normalization, discontinuation of alcohol intake (possibly alcohol tolerance testing (alcohol loading and discontinuation)
3. Elimination and/or treatment of additional risk factors, i.e. Diabtes mellitus, hypertonus, cigarette smoking,hyperhomocysteinemia, obesity, physical inactivity, etc.
4. Nutritional therapy:
a) Cholesterol lowering diet:
- Reduction of fat intake < 25% of total calories =cal%
- Fat exchange: avoid saturated animal fats, prefer mono and polyunsaturated fatty acids.
- Carbohydrates: 50-60% cal%, preference of complex carbohydrates, increased fruit and vegetable intake;mediterranean diet
- Protein:up to 15% cal%
- Fiber: 20-30 g/d
- Cholesterol intake restriction < 300mg/d, <200 mg/d in patients with already increased LDL cholesterol levels.
Noteworthy: 1 egg yolk ≈ 270 MG Cholesterol. Do not use more than 2-3 eggs per week!
- Normalization of body weight, caloric well-balanced diet
- Irrespective of HLP-type regular consumption of sea food (i.e. 1-2 X per week) with high content of omega 3-fatty acids (in particular eicosapentanoic acid and docosahexaenic acid). Also important is daily intake of sufficient amounts of iodine (150.200 ug/day).
Following consequently a healthy diet protects against development of arteriosclerosis, lowers LDL-cholesterol levels by about 20-60 mg/dl and improves drug therapeutic responsiveness.
Consequent physical training (averaging caloric consumption of about 2000 kcal/week) contributes further to serum LDL cholesterol level lowering , HDL cholesterol level increase and lowering of serum triglyceride levels.
Increased activity is also beneficial for heart and circulation, body weight, and diabetes meelitus.
b) Triglyceride lowering diet:
- Discontinuation of alcohol intake
- Body weight normalization

► Without chylomicronemia:
- Reduction of total fat intake <30cal%
- Fat exchange (preference of vegetable fats (linoleic acid and/or oilrich fatty acid), sea food consumption)
- Avoid mono and disaccharides, prefered intake of complex carbohydrates
- Oncrease number of daily meals (i.e. 5)
- In case of poor response and excessive triglyceride increase regularly low calorie days

Noteworthy: Triglyceride lowering therapies commonly are associated with distinctive serum HDL cholesterol level increases

► With chylomicronemia: type I (very rare), type V (rare) or temporary in case of presence of uncontrolled severe hypertriglyceridemia, type IV:
As above, however daily fat consumption needs to be maximally restricted (below 10%). Use of medium chain triglycerides.
With onset of pancreatitis (upper abdominal pain etc.) several days of fasting (NPO (nothing per mouth), in patients with serious disease immediate plasma-exchange therapy. In case of frequently recurring excessive serum hypertriglyceridemia routinely low caloric days (i.e. 1X week) are highly commendable

5. Lipid- owering therapy:

► Drug therapy:

• Statins=Cholesterol-synthesis-enzymeinhibitor =CSE-inhibitor=HMG-CoA reductase inhibitor (Hydroxymethylglutaryl-Coenzyme A-reductase):

Mode of action: Inhibition of key enzyme of cholesterol synthesis → lowering of intracellular cholesterolconcentration in hepatocytes → counterregulatory increase of IDL receptor numbers on hepatocytic cell surfaces→lowering of serum LDL cholesterol levels by 20-60%. Additionally, atherogenicity prevention effects, i.e. Cia improved endothelial cell function.
Note: Stains are the most eficacious LDL choleserol lowering drugs known and can also be used for the treatment of type III hyperlipoproteinemia. They have been shown to reduce risk of myocardial infarction and of total mortailty when used as primary preventive therapy (i.e. West of Scotland study) as well as second line preventive treatment regimen of coronary artery disease (i.e. 4S study, HPS study) !

Generic name	Trade name	Dose (mg/dl)
Atorvastatin	Sortis	10-80
fluvastatin	Cranoc, Locol	20-80
lovastatin	as generic	10-80
pravastatin	Pravachol	5-40
rosuvastatin	Crestor	10-40
simvastatin	Zocor	5-80

Statins do not only lower serum LDL-choleserol levels but also moderately serum VLDL and triglyceride levles. Mild increases of serum HDL cholesterol levels are common. Rosuvastatin and atorvastatin are the most potent statins. Proper dosing and selection of the most appropriate statin rarely leads to discontinuation of therapy. Maximum therapeutic response per mg statin is achieved at low doses.

Subsequent dose doublings lower serum LDL cholesterol levels only by about 6% (6% rule). During the initial dose adjustment stage, therapeutic response control is recommended for each dose adjustment after 3 weeks. Statins can be administered together with cholic acid complex forming agents or with ezetimib.

Adverse events:occasionally myopathy with muscle weakness and/or muscle pain, occasionally with CK- increase. Rarely, extreme CK increase = life-threatening rhabdomyolysis. Occasionally GI discomfort, transaminase increase. Additional side effects occur less frequently (see product package insert):

Measure of risk minimization:

- Optimal dose adjustment, possibly prefer combination regimen (see below)
- Avoid concomitant use of dangerous drugs (see also drug interactions)
- Discontinuation if muscle pain and/or red stained urine are observed (myoglogin related, urinary strip testing)
- Monitoring of toal serum CK levels.
- Avoid strenuous activity (i.e. Marathon running)
- Cave in patients with hypothyreotic disease and in patients with limited renal functional

Drug interactions: Since statins are metabolised by the hepatic cytochrome P450 enzyme system, drug interactions with different medications which are metabolized by this enzyme system cannot be entirely excluded → it is important therefore to check the respective company product incormation! The risk for rhabdomyolysis is increased i.e with simultaneous administration of ciclosporine, fibrates, nicotinic acid derivatives and erythromycin.

Contrainidications: Hepatic disorders, skeletal muscle disorders, pregnancy, lactation, use in children. A combination regimen consisting of statins and fibrates increases the the risk for rhabdomyolysis by a factor of 10. In patients with serious metabolic system disorders or in the event of occurrence of dose related side effects, a combiation regimen of statin use itogehter with an ion exchange resin or cholesterol absorption inhibitor has been recommended.

- Anion exchange resins or bile acid binders:

Are mostly used in combination with statins.

Mechanism of action: Nonresorbable, basic anion exchange resins, bind ileal bile acids and thereby remove them from enterohepatic circulation. This induces hepatic LDL receptor stimulation and thereby lowering of serum LDL cholesterol levels. Decrease of coronary morbidity and letality was proven. However evidence of reduction of total mortality is lacking.

- Colestyramine (i.e. Quantalan): 3X4-8g/d before meals with plenty of fluid, tapered dose administration over a period of 1-3 weeks. In combination with statins 2-3 X 4 g/d, possibly 2X4 g before breakfast is enough.
- Colestipol: 3X5 – 10 g/- in combination with statins 2-3 x 5g/d, possibly 2X5 g before breakfast. Intake just loke colestyramine.

Adverse events:Frequently, gastrointestinal side effects: gastric fullness sensation,bloating,obstipation,reduced resorption of fat soluble vitamins (A,D,E,K), which should be substituted in the event that high dose anion exchange resins are administered.

Note: Since drugs with acid groups are bound to exchange resins, these substances are to be administered 3h or earliest 4h after intake of the exchange resins. These substances are i.e. aspirin, certain statins,thyroxine, coumadin, digitalis preparations and vitamin C.

Important:in order to reduce obstipation increase fluid intake !

- Cholesterolabsorption inhibitors

Ezetimib (Ezetrol®)

Ind: Hyperchoesterolemia. In combination with statins (dual principle)highly recommendable

SE: Rarely abdominal discomfort, increase of transaminases, myopathy with total serum CK increase, myalgias, etc.

Dosage: 10 mg/d (LDL cholesterol lowering effect up to 18%)

Note: No long term study results available !

- Fibrates =clofibrate acid derivatives:

Mechanism of action: complex mode of action, amongst them :increased catabolism of triglyceride rich lipoproteins.
Intake of fibrates will lower serum VLDL/LDL-cholesterol levels and increase serum HDL levels. Latter effect is important to lower CAD risk.

Products:
- Fenofibrate: 250 mg (retard)/d
- Bezafibrate: 400 mg (retard)/d

SE: Occasionally GI disturbances, transaminase increase, hair loss, negatiive impact on sexual potency, myopathy with muscle pain and CK increase, rhabdomyolysis very rarely, increased risk of gallstone formation, rarely other SE → see also company product information.

Drug Interactions: Increase of therapeutic efficacy of sulfonyl urea containing antidiabetics, and of coumadin type anticoagulants.

Contraindications: Simulltaneous therapy with CSE inhibitors represents a relative contraindication (see also drug interactions of statins); renal insufficiency, hepatic disorders, pregnancy, lactation, etc.

- Nicotinic acid:

Well-known active ingredient with positive lowering effect of all lipoprotein fractions, including Lp(a). An improved galenic formualtion with retarded release mechanism of the active ingredient is associated with improved tolerability (Niaspan ®)

Administration requires special measures of precaution (→ see product information)

Combination with statins restricted possible
Dosing: up to 2000 mg/dl
SE: most common SE=flush (other side effects and contraindications → see product information)

Ind: combined hyperlipoproteinemia and primary hypercholesterolemia, in particular with low HDL-cholesterol, provided therapeutic management with recommended drugs is not possible.

► Extracorporal LDL-elimination (LDL-apheresis):

1. Elimination of LDL from plasma; methods:
- Immunoadsorption of IDL-and Lp(a) (LDL-apheresis method developed at the University of Cologne, Germany)
- Adsorption of IDL and Lp (a) to dextransulfatecolumns (developed in Japan)
- Heparin induced extracorporeal LDL precipitation = H.E.L.P.: Elimination of LDL , fibrinogen and of Lp(a) (developed at the university of Goettingen; Germany)
2. Adsorption of LDL and Lp (a) from blood via the DALI-method (Fresenius Inc.)= Direct Adsorption of Lipoproteins (Method for whole bood adsorption without plasma separation)

Ind.: Severe familial hypercholesterolemia with unsatisfatory LDL lowering drug treatment and high arteriosclerosis risk as well as extreme Lp(a)- increase.
LDL elimination is performed once a week, additionally patients are treated with a cholesterol lowering drug treatment regimen.

Summary of therapeutic options

Hypercholesterolemia:

Weight reduction, dietary adjustment, lifestyle improvement and treatment of any underlying causes. Thereafter, initiation of drug therapy.

Substance Class	Maximal lowering of LDL-cholesterol
-statins	Up tp 60%
-anion exchange resins	up to 30% combined up to and >60%
-ezetimib	up to 20%
Extracorporeal LDL elimination	60-80%

Note: Individual response to statins is highly variable. The table shows the maximally achievable lowering. Ezetimib and anion exchange resins should not be combined. Statins and ezetimib as mon or combination product (ie.e Simvastatin plus Ezetimib = Inegy®) will induce a more pronounced serum LDL cholesterol level lowering.

- **Hypertriglyceridemia:**

Weight reduction, dietary adjustment, lifestyle improvement and treatment of any underlying causes are successful in the majority of patients
Additionally required drug therapy i.e with fibrates. In patients with low HDL-cholesterol levels and unfavorable LDL/HDL cholesterol ration therapeutic administration of statins is recommended.. Even careful combination of fenofibrate i.e. With pravastatin or fluvastatin is associated with a distinct increase of risk of rhabdomyolysis (relative contraindication). Lowering of triglycerides commonly is associated with serum HDL-cholesterol increase.
Important: With serious lipaemia induced pancreatitis frequently caused by excess alcohol consumption or poorly controlled diabetes mellitus, immediate plasma exchange therapy is indicated.

- **Type III-Hyperlipoproteinemia**

Diet and lifestyle change as with hypertriglyceridemia. Check fibrates and statins for efficacy.
- **Combined Hyperlipidemia:**
Therapeutic approach as with hypercholesterolemia. High trglyceride levels can be treated just as like hypertriglyceridemia
- **Low HDL or high Lp(a)**
Adjustment of serum LDL-cholesterol levels to low levels, possibly cautious combination with nicotinic acid (see above)

Obesity (E 66.9)

Internet-Infos: i.e. *www.medizin.uni-koeln.de* (→ Leitlinien)
www.adipositas-gesellschaft.de

Definition: Obesity is defined as excess of adipose tissue increase by more than 30% in women and by more than 20% in men. Body mass index (BMI) determination allows indirect assessment of total adipose tissue mass.

Body mass index = BMI = body weight/height (m)^2

Weight Classification (Europe, USA)	BMI (kg/m²)
Normal weight	18,5 – 24,9
Surplus weight (pre-obesity stage)	25,0 – 29,9
Obesity (grade I)	30,0 – 34,9
Obesity (grade II)	35,0 – 39,9
Obesity (grade III)	40 or more

Epidemiology: Age-dependent increase of prevalence in western industrialized nations. Germany is the first-ranking country in Germany ! In Europe already 25% of school children are obese; Approx. 1/3 of the adult population are overweight ! Approx. 1% suffer from extreme obesity with BMI >40. Asian populations are assessed according to lower BMI standards. Low prevalence during war times.

Note: The three most important causes of avoidable diseases and death cases are:
1. Smoking 2. Alcoholism 3. Obesity

Aet.: 1. primary obesity (approx.95%) → causative factors:
- genetic factors: in approx. 5% of all extremely obese patients a monogenetic mutation of the Melanocortin 4-receptor (MC4R) was identified, most commonly these patients are binge eaters. The ob-gene codes for leptin synthesis, a hormone,which down regulates the hypothalamic appetite sensation via receptors. Since all obese patients have increased serum leptin levels, it is assumed that these patients suffer from leptin resistence.- A subgroup of patients with metabolic syndrome was identified as carriers of the gene mutation GNB3-825T. In approx. 20% of patients a mutation of the FTO gene is supposed to be causative for the obesity. Additionally, rare obesity syndromes have been described.
- Excess calorie consumption (supernutrition) , lifestyle, physical inactivity
- Psychological factors (stress, frustration, loneliness → eating as reward, as instrument of consolation, as form of dependency („binge eater"); loss ofnormal hunger and satiety sensation, nicotin renunciation)
  2. secondary obesity (approx. 5%)
- endocrine disorders: Cushing Disease, hypothyroidism, insulinoma, testosterone deficiency in men etc.)
- Central-nervous system disorders related obesity: brain tumours (hypothalamus, hypophysis) and status post surgery or radiation of this type of disorders.

Prg.: Total daily caloric intake (hypercaloric food, in particular fat loaded food) exceeds daily requirements ladditionally physical activity level is reduced)
Increase of body weight occurs beyound the normal weight limits as a result of increase of adipose tissue mass by approx. 75% and increase of fat free mas by approx. 25%. The caloric equivalent of one kg body weight appropriates about 7000 kcal.

Obesity is not a disease as such, it develops disease state qualities related to its associated morbditiy and mortality. Being overweight by 20% or more beyond the normal weight limits increases the risk of development of health problems (Framingham-study). Mortality of obese patients with a BMI > 35 kg/m² doubles compared with persons within normal weight limits.

CI.:
- reduced physical exertion levels with possible onset of exertion dyspnoea and rapid fatigue onset
- possibly discomfort and pain in strained joints (hip and knee joints) and of the spine
- increased sweating
- diminished feelings of self worth

Co.:
- Metabolic syndrome (wealth syndrome)
commonly joint presence of truncal obesity, dyslipoproteinemia (triglycerides ↑, HDL-cholesterol ↓), hyperuricemia, essential hypertension, pathological glucose tolerance test, type II diabetes. Obesity is the manifestationfactor for these disease types.
- Nonalcoholic fatty liver disease
- Obesity is a risk factor for:
  - arterial hypertension
  - coronary artery disease and stroke
  - deep venous thrombosis and thrombo-embolic complications (in particular after surgical procedures)
  - narcolepsy and sleep-apnoea syndrome
  - cholecystolithiasis
  - EPH-gestosis
  - cancer (i.e. colon/rectal cancer, endometrial cancer, breast, prostate etc.)
  - arthritis (spine, hip and knee joints)
- Hormonal disturbances

- Men: increase aromatase activity of fat cells: estrogens ↑, testosteron ↓ with possible development of sexual dysfunction
-women: androgens ↑ → possibly hirsutism, hair loss, seborrhea, acne secindary amenorrhea, infertility, polycystic ovary syndrome (PCOS)
- intertriginal infections, striae
- obesity has a negative impact on congestive heart failure
- possible onset of reactive depression and social problems.

Memo: The three most important internal complications of obesity are:
- cardiovascular diseases
- type 2-Diabetes mellitus (Diadipositas)
- tumour

Di.:
- Determination of body weight with BMI (accurate) or Broca-formula (orientation)
- Assessment of fat distribution type by measuring waist size between rib circumference and anterior iliac spine (undressed) after normal expiration. Measuring results of > 94 cm (males) and > 80 cm (females) the risk for obesity assocated metabolic disorders is clearly increased (IDF-definition 2005).
- Android (proximal, truncal or abdominal) adipose tissue distribution type: Truncal or abdominal „apple type": Health risks are particularly high with this distribution type.
- Gynoid (distal, hip or gluteofemoral adipose tissue distribution): Hip or femoral dominant „pear- type" (health risks are lower compared with the android type of fatty tissue distribution)
- Localized type of adipose tissue distribution: i.e. „breech-type" distribution
- Assessment of posible presence of additional coronary risk factors or disease states associated with the meatbolic syndrome: serum lipid levels, serum uric acid levels, fasting blood glucose levels, blodd pressure monitoring results
- Assessment of nutritonal status and history:Eating habits, physical activity and sense of health and well-being
- Exclusion of any type of endocrine disorder: determination of basal TSH levels, dexamthason suppression testing, oral glucose tolerance testing
- Exclusion of bulemia: disturbed eating habits with binge eating attacks, followed by self-induced vomiting, etc.

Th.:   Indication: BMI > 30 kg/m² and/or disease which are caused or aggravated by obesity: psychosocial stress levels.

3 Pillars:
1. **change of nutritional habits, reduction of daily caloric consumption levels.**
2. **Physical therapy/ persistence training**
3. **behaviour therapy/group-dynamic therapy**

Therapy of obesity is an active task requiring lifelong change of nutriotnal and life style habits. Prerequisite to success is the patients insight and recognition that he/she has a weight problem and the motivation to overcome this problem. Most successful are long-term group dynamic treatment option (i.e. Weight watchers) with diet counseling and behavoir therapy (prevention of eating disorders, rellearning of natural hunger and satiety sensations, stress reduction, self assurance training, acquiring of coping mechanisms without „ a grab into the fridge" etc, as well as regular physical activity/endurance training

ad 1: calorie reduction
Ind: Aiming at weight reduction by reaching an energy deficit level
Note: Not the weight reduction is the problem but rather maintenance of the achieved goal weight. All reduction diets make sense only as part of a long-term therapy concept and when the patient is able to maintain the achieved goal weight after the diet.
Goal should be slow and persistent weight reduction rather than fast weight loss of many kilograms.

CI.:
- Normal weight
- Children and adolescents
- Pregnancy and lactation
- Patients with eating disorders
- Cardiac disease and serious eating disturbances
► Calorie reduced balance diet with approx 1200 kcal/d containing at least 50g protein/day. Since fat has the highest caloric value, however a low satiety level, the diet should be low-fat.
► Diets with very low daily caloric intake levels (< 1000 kcal/d) will lead to a fast weight reduction, however are not ideal for long-term weight maintenance and controlled
Note: Generous fluid intake is recommended with all reduction diets (at least 2,5 l/day.
Very low calory diets should only be employed for a limited time period (6 weeks) under a physicians supervision, followed by adjustment ot a healthy balanced diet.

So called fasting regimens lasting for a limited total time period of 1.2 weeks which are characterized by several
days of intake of vegetable juice or rice or fruit or raw vegetables are to be assessed according to the same criteria
as any other diet therapy. The same contraindications and evaluations apply.

After a normal weight has been reached readjustment from a hypocaloric to an isocaloric eating regimen is
recommeded.. This should be low fat, rich in fibre and salt restricted (5g/d), alcoholic beverages should be
consumed sparingly.

In patients with increased cardiovascular risks saturated fat intake should be restricted to < 10% and cholesterol
intake < 150 mg/1000 kcal; polyunsaturated fatty acid intake should be increased to up to 10% of the daily caloric
intake levels.

Additional therapeutic options:

- Drugs for weight reduction (anti obesity drugs):
- Orlistat (Xenical)
Mode of action: non-resorbable lipase inhibitor → reduction of total fat absorption by about 30% (at a dosing level of 3x
120 mg/d)
SE: Steatorrhea and spoiling of underwear, bloating, malabsoprtion of fat soluble vitamins, in some patients onset of
hypertension. Side effect related rate of treatment discontinuation > 25%.
Drug interactions: possibly reduced resorption of fat soluble substances (i.e. Ciclosporine)
Indication: Supplemental therapeutic option in patients with a BMI > 30 kg/m², provided baseline therapy is
unsuccessful.

Appetite suppressants: can be associated with serious side effects such as cardiac valve disorders, pulmonary
hypertension, and therefore are not recommendable.
- Surgical treatment options (i.e. Gastric banding, laparoscopic stomach bypass etc.) should only be performed in
special therapy centres where the required experience was obtained. Asociated mortality should be < 0.5%.
- Complications with gastric banding: band leakage, slipping,-migration etc.
- Complication with stomach bypass surgery: leakage, ileus etc.
Weight reduction of more than 50% is possible.!
Indication: BMI > 40, in particular in the presence of psychosocial and organic disease comorbidity. Complications and
contrainidcations are to be considered !

Prg.:   With good motivation and consequent therapy about 25% of obese patients are able to achieve long-term weight
reduction.
Weight reduction by about 10 kg is already associated with lowering of total mortality by > 20%, diabetes associated
mortailty is reduced by > 30%, obesity associated cancer risk is reduced by > 40% !

Note: Short-term crash diets are not successful in the long run. Often times a short-term diet success is followed by an even
more pronounced weight gain with associated increaed frustration levels (yo-Yo effect): Long-term success rates are
dependent on the patients motivation and willingness to change nutrtional and life style habits for good and for the better !

Pro.:   Preventive medicine programs aimed at improvement of nutritional, dietary and health habits in the general
population.

Eating disorders
Anorexia nervosa (F50.0) and bulimia nervosa (F 50.2)

Internet-infos: *www.bzga-esstörungen.de*

Def.:   Anorexia nervosa is characterized by intentional diet restriction aimed at achieving and maintianing a below normal
weight (under weight).
In comparison, bulimia nervosa is characterized by binge eating episodes and self induced vomiting. However most
bulimia patients maintain a normal weight level.
An atypical eating disorder, psychogenic hyperphagia (binge eating disorder) is characterized by binge eating
episodes without self-induced vomiting or any other control mechanism, therefore resulting in surplus weight gain.

Ep.: Increasing prevalence of eating disorders predominantly in adolescent females and women between the ages of 15-35. Disease onset during adolescence or young adulthood is most common. Sex distribution F:m= 10:1. Prevalence of anorexia nervosa approx. 150/100.000, double the amount for bulimia nervosa

Aet.: Distorted relationship between Ideal and actual body perception, in particular during puberty. Rejection of female role functioning models leads to development of various compensation and regression mechanisms leading to a shift of sexual impulse to an oral level. Predisposing factors are of genetic, socio-cultural, familial and intrapsychic origin. They are relevant for the development of typical psychological problem areas such as low self esteem, identitiy and autonomy conflicts as well as stress intolerance. The diet is a sign of particular symptom development.

Cl.:

Anorexia nervosa:
- Decresed body weight of at least 15% or BMI< 17.5 kg/m²
- Self induced weight loss through low calorie eating habits as well as one of the following behaviour mechanisms
- self induced vomiting
- self induced laxative use
- excessive physical activity
- diuretic/appetite suppressant abuse
- disturbed body image
- endocrine disorder (in women: amenorrhea, in men: libido and potency loss)

Bulimia nervosa
- persistent thoughts about eating
- on average at least 2 binge eating episodes per week for at least 3 months
- different mechanisms aimed at weight controlled
- self induced vomiting
- laxative abuse
- temporary fasting
- use of diuretics, appetite suppressants or thyroid hormones
- pathological fear of becoming overweight
- commonly previous anorexia nervosa

Co.:

With poor nutrition and states of underweight:
- Bradycardia, hypotonia, hypothermia, delayed stomach emptying, amenorrhea, osteoporosis, pseudocerebral atrophy, bone marrow hypoplasia

With vomiting:
Cardiac rhythm disturbances, renal insufficiency, pseudo-Bartter syndrome with edmea formation, poor dentition (dental enamel loss), reflux oesophagitis, metabolic alcalosis etc.

Lab.: Lowering of blood glucose levels.potassium,chloride, magnesium, sodium, Triiodinethyronine, LH, FSH,estrogen, vitamin D, increase of amylase and cortisol levels.

DD:

Weight loss of different etiology:
- Addisonse disease
- Hyperthyroidism, pheochromocytoma
- Adrenal gland insufficiency
- Uncontrolled diabetes mellitus
- Malassimilation syndrome
- Gastrointestinal disorders with insufficnet food intake, reduced resorption, increased losses
- Chronic infectious disease states, in particular intestinal parasitic disorders
- Neoplasms (tumour cachexia)
- Age related cachexia/ poor eating habits of the elderly
- Poverty/food deprivation

Psychiatric disorders:
- Anorexia, bulimia
- Psychogenic vomiting
- Depressive syndrome of different etiology
- Schizophrenia
- Toxicomania

<u>Th.:</u>

General Measures:
- Outpatient or inpatient
- Inpatient treatment with high physical risks, increased self mutilation and/or increased suicide risk
- Education
- Nutritional counseling
- Consideration of possible consequences like fluid and electrolyte disorders, amenorrhea, osteoporosis, depression
- Cognitive behavioural therapy
- Alternative psychotherapy
- Possibly focal psychodynamic therapy, family therapy
- Psychological intervention linked with regular control of bodily status

Specific Measures:
- Anorexia nervosa:
  - Refeeding until goal weight is reached (premorbid weight or BMI 20 kg/m^2 in < 170 cm, 85% of Broca reference weight in > 170 cm)
  - Envisioned weight gain per week on an outpatient basis: 0.5 kg, in patient 0.5 - 1 kg /week)-
  Required daily calories 1500-3000 kcal (calculated based on daily individual requiremnts plus predetermined daily weight gain)
- NG tube nutritional supplementation on an inpatient basis in high risk patients
- Parenteral nutrition (TPN) in patients with serious gastrointestinal dysfunction
- In some patients psycopharmacologically active drugs may be required

- Bulimia nervosa

  Antidepressive drug treatment regimen, preferentially SSRI because of the more favorable side effect profile (i.e. Fluoxetin)

Prg.: Mortality rate 0.5 - 1 % per year of treatment, high rate of chronicity, increased suicide rate (in particular in bulimia nervosa patients)

VI. ENDOCRINOLOGY

Internet infos: www.dgae-info.de; www.endokrinologie.net; www.diabetes.cme.de
www.aace.com; www.endosociety.org

DIABETES MELLITUS ("sweetened with honey") [E14.9]

Internet infos: www.diabetes-deutschland.de; www.diabetes-webring.de; www.diabetes-world.net;
www.diabetes.ca; www.deutsche-diabetes-gesellschaft.de; www.diabetes.org

Def.: In most cases diabetes mellitus is a hereditary chronic metabolic disorder, which is characterized by an absolute or relative deficiency of insulin. After a longer duration of the disease, damage occurs to blood vessels and to the nervous system.

Ep.: Prevalence of manifest diabetes is age dependent: Age < 50 y. 1 - 2 %, age > 60 y. circa 10 %, age > 70 y. up to 20 %. > 90 % are type 2 diabetics and circa 5 % are type 1 diabetics. In the USA 4 % of the obese juveniles are type 2 diabetics! The number of type 2 diabetics in a population increases with the extent of overnutrition.
In the native Inuit of Greenland diabetes is relatively rare, whereas it is relatively frequently found in American Pima Indians.

Aetiologic classification : (WHO and ADA = American Diabetes Association, 1997)

I. Type 1 diabetes: cell destruction, which leads to absolute insulin deficiency
   A) Immune mediated
      Special form: LADA (latent autoimmune diabetes (with onset) in adults): Type 1 diabetes with late manifestation, in which insulin deficiency develops rel. slowly, so that it is often   confused with type 2 diabetes. (The majority of these patients are aged 25 – 40 at diagnosis).
   B) Idiopathic (infrequent in Europe)

II. Type 2 diabetes: This can range from a predominantly insulin resistance with a relative insulin deficiency to a predominantly secretory defect with insulin resistance.

III. Other specific types:

   A) Genetic defects of β cell function (autosomal dominant pathway):
      „Maturity-onset Diabetes of the Young (MODY) without auto-Ab detection and without adiposity: Manifestation is before 25 years of age; circa 1 % of all diabetics:

MODY type	Gene	Abbreviation	Chromo-some	PPh	Notes
MODY 1	Hepatocyte nuclear factor 4 alpha	HNF-4alpha	20q	Reduced insulin secretion, decreased glycogen synthesis	Low triglycerides
MODY 2 (15 %)	Glucokinase	GK	7p	Reduced insulin secretion	Mild course, mostly without late complications
MODY 3 (65 %)	Hepatocyte nuclear factor 1 alpha	HNF-1 alpha	12q	Reduced insulin secretion	Renal glycosuria
MODY 4	Insulin promotor factor-1	IPF-1	13q	Reduced insulin secretion, defective receptor for sulfonylureas	
	Pancreatic duodenum homebox-1	PDX-1			
MODY 5	Hepatocyte nuclear factor 1 beta	HNF-1beta	17q	Reduced insulin secretion	Renal cysts, malformations of genitalia
MODY 6	Neuro D1 or BETA2	Neuro D1	2q	Abnormal trans-scription regulation of the Beta cells	

B) Genetic defects in insulin action

C) Chronic pancreatitis

D) Endocrinopathies: Acromegaly, Cushing´s syndrome, phaeochromocytoma, hyperthyroidism, so-
matostatinoma, glucagonoma, aldosteronoma

E) Drug induced, e.g. glucocorticoids, thyroid hormones, diazoxide, beta-adrenergic agonists, thiazides

F) Infections, e.g. congenital rubella, CMV

G) Uncommon forms of immune-mediated diabetes, e.g. anti-insulin receptor antibodies

H) Genetic syndromes, which are sometimes associated with diabetes, e.g.
Down´s syndrome, Klinefelter´s syndrome, Turner´s syndrome etc.

IV. Gestational diabetes (GDM)

Classification according to the clinical severity (WHO, 2000):

- IGT: Impaired glucose tolerance
- NIR: Noninsulin requiring (type 2 diabetics)
- IRC: Insulin required for control (type 2 diabetics, who need oral antidiabetic agents + insulin)
- IRS: Insulin required for survival (type 1 diabetics and type 2 diabetics without own insulin production)

Aet.: ■ **Type 1 diabetes** (< 10 %):

Destruction of beta cells in the islets of Langerhans caused by autoimmune insulitis with absolute insulin
deficiency. Blood glucose levels increase, when circa 80 % of all beta cells are destroyed. Genetic factors play a
predisposing role: 20 % of type 1 diabetics have a positive family Hx (for type 1 diabetes) and HLA types DR 3
and/or DR 4 are present in > 90 % of patients. The following findings point to an autoimmune insulitis in the
case of a newly diagnosed type 1 diabetic:

■ Infiltration of the islets of Langerhans by autoreactive T lymphocytes

■ Detection of autoantibodies:
- cytoplasmic islet cell Ab (ICA):   antigen: gangliosides
- anti-GAD-Ab (GADA):   antigen: glutamic acid decarboxylase
- anti-IA-2-Ab: antigen: tyrosine phosphatase 2
- insulin auto-Ab (IAA):   antigen: (pro)insulin

Detection of the ICA by immune fluorescence very costly, therefore most commonly replaced by anti-GAD-Ab
and anti-IA-2-Ab.

Detection of ICA in type 1 diabetes 80 %, GADA and IA-2A together > 90 %, IAA age dependent 20 – 90 %
(diagnostically not relevant)

■ Temporary remissions with immunosuppressive therapy

If GADA and IA-2-Ab are present in a healthy individual, the risk to develop diabetes type 1 within the next 5
years lies around 20 %.

■ **Type 2 diabetes** (> 90 %):

Pathogenetic 2 defects play a role:

■ Impaired insulin secretion

In type 2 diabetics the early postprandial insulin secretion is defective; this leads to postprandial
hyperglycaemia.

It remains unclear whether the deposition of insular area amyloid polypeptide (IAPP) seen in B cells of type 2
diabetics is linked to defective insulin secretion.

■ Decreased insulin action (insulin resistance)

Cause: Pre-receptor defect, receptor defect with down regulation, post-receptor defect = impairment of the
signal transduction, e.g. of the tyrosine kinases

Note: Healthy individuals express only one iso form of the Human-Insulin-Receptor on the cell membrane: HIR-
A. Type 2 diabetics express both iso forms: HIR-A and HIR-B.

Remember: The majority of diseas develops because of a metabolic syndrome (= prosperity syndrome):
Frequent concurrence of the 4 risk factors: Obesity with emphasis on the trunk (abdominal),
dyslipoproteinaemia (triglycerides , HDL cholesterol), essential hypertension and impaired glucose
tolerance or type 2 diabetes mellitus. At the onset of the metabolic syndrome an insulin resistance of the
insulin dependent tissues (e.g. skeletal muscle cells) is found, so that increased insulin levels are required for
the cellular utilization of glucose. The hyperinsulinaemia increases the feeling of hunger, leads to obesity, and
favours the development of early arteriosclerosis.

■ Abdominal obesity with a waist circumference ≥ 94 cm (m) or ≥ 80 cm (f) in Caucasians (different values apply for other ethnic groups)

■ Plus two of the following factors:
- Triglycerides > 150 mg/dl (1,7 mmol/l)[*)]
- HDL cholesterol   < 50 mg/dl (1,29 mmol/l)[*)] f
                       < 40 mg/dl (1,04 mmol/l)[*)] m
- Blood pressure > 130/85 mm Hg[*)]
- Fasting plasma glucose > 100 mg/dl (5,6 mmol/l) or type 2 diabetic
[*)] or previous therapy of one of these impairments

Note: There are definitions of the metabolic syndrome, that differ from this (WHO, NCEP-ATP III).

Remember: Overnutrition combined with obesity are the determining factors for the manifestation of type 2 diabetes mellitus! Circa 80 % of type 2 diabetics are overweight.
High insulin levels reduce the sensitivity and density of insulin receptors (= down regulation) and thus the effect of insulin. This results in a further increase of insulin levels (vicious circle). Therapeutic principle is the elimination of overeating and obesity   resulting in a reduction of insulin levels again, and restoration of the sensitivity and density of insulin receptors!

Other manifestations of type 2 diabetes:
- Stress factors: Infections, trauma, operations, stroke, myocardial infarction etc.
- Endocrinopathies and drugs are considered separately in the diabetes classification.

	Type 1 diabetes	Type 2 diabetes
Pathogenesis	insulin deficiency	insulin resistance
Anatomy	asthenic	mostly pyknic/obese
Onset	often rapidly	slowly
Predominant age of manifestation	12. - 24th year	> 40th year
B cells	reduced to < 10 %	reduced only moderately
Plasma insulin / C peptide	low or lacking	initially increased
Autoantibodies (IAA, GADA, IA-2A)	+	
Metabolic status	unstable	stable
Ketosis tendency	strong	low
Response to sulfonyl ureas	lacking	good
Insulin therapy	necessary	only when insulin reserve is exhausted

■ **Gestational diabetes (GDM):** [O24.4]
Def.: Impairment of carbohydrate metabolism recognized for the first time during pregnancy. In most cases disappears after completion of pregnancy, but an increased risk of 50 % for further GDM exists for a subsequent pregnancy. The risk of developing permanent diabetes mellitus is up to 45 %/10 years.
Occ: 3 % of all pregnancies!
Co.: 1. For the mother: Increased risk of preclampsia, infections of the urinary tract, polyhydramnios and caesarian section.
      2. For the child: Diabetes is the most common cause of increased intrauterine death and perinatal morbidity of the child: Embryofoetopathy diabetica with increased birthweight > 4.500 g and macrosomia; increased risk of respiratory distress syndrome, postpartal hypoglycaemia, hyperbilirubinaemia, hypocalcaemia, hyperglobulia etc.

Genetics and inheritance:
Polygenic-multifactorial inheritance; variable penetrance of the diabetogenic genes.
Genetic mutations in type 2 diabetes: ATP-sensitive potassium channel; protein PC-1; PTPN1; GNB3-825T; TCF7L2; SLC30A8 etc.
- Type 1 diabetes:
The risk of the child developing type 1 diabetes is approx. 5% if father is affected, 2.5% if mother is affected, 20 % if both affected.
The disease risk for siblings of a type 1 diabetic is high in identical twins (circa 35 %) and depends, in the remaining cases, on the extent of the HLA identity: HLA identical siblings have a risk of circa 18 %, HLA haplotype identical siblings have a risk of circa 6 %; siblings differing in their HLA types have little risk of developing type 1 diabetes.

- Type 2 diabetes:
For children with one type 2 diabetic parent the probabiltiy developing type 2 diabetes is up to 50 %. The risk for identical twins is 100 %.

<u>Cl.:</u>   Manifestation of diabetes mellitus:
While the development of manifest type 1 diabetes is relatively rapid, type 2 diabetes manifests slowly and often unnoticed, so that diagnosis is often only made on routine examination of blood or urine.
- ■ <u>Nonspecific general symptoms:</u>
  Tiredness, lack of energy etc.
- ■ Symptoms caused by hyperinsulinism and <u>temporary hypoglycaemia</u> (initial status of type 2 diabetes): High urge to eat, sweating, headaches etc.
- ■ Symptoms due to <u>hyperglycaemia and glycosuria with osmotic diuresis:</u> <u>Polyuria, thirst, polydipsia, weight loss</u>
- ■ Symptoms due to <u>impairment of electrolytes and fluid balance:</u> Nocturnal cramps of the calf muscles, visual disturbances (changing turgor of the crystalline lense)
- ■ <u>Skin appearances:</u>
  - Pruritus (often anogenital localization)
  - Bacterial / fungal skin infections (e.g. furunculosis!, candidiasis!)
  - Rubeosis diabetica (diabetic facial blush)
  - Necrobiosis lipoidica (mostly on both lower legs, brownish red areas, ulcerations possible)
- • <u>Impotence,</u> amenorrhoea

<u>Co.:</u>   1. <u>Macro-/microangiopathy:</u>
Diabetic vessel damage is subdivided into <u>non-specific macroangiopathy</u> and a <u>diabetes-specific microangiopathy</u> with thickening of capillary basement membranes. The nonenzymatic glycosylation of proteins in the basement membrane caused by hyperglycaemia seems to play a role in the formation of the microangiopathy. The thickness of the basement membrane correlates with the duration of the diabetes. The microangiopathy can lead to acral necrosis even though the peripheral pulses may still be palpable.

   1.1. <u>Macroangiopathy with early arteriosclerosis:</u>
   - <u>Coronary heart disease:</u> Stenosing arteriosclerosis of the large epicardial coronary arteries: <u>55 % of diabetics die of myocardial infarction!</u>
     <u>Specific features of CHD in diabetes:</u>
     ▪ Diffuse distribution pattern of CHD with predominant involvement of the distal coronary arteries and of the main trunk
     ▪ Impaired angina perception threshold caused by ADN (see below) with poss. painless infarctions and silent ischaemia
     ▪ Infavourable prognosis
   - <u>Peripheral arterial occlusive disease</u>
   - <u>Arterial occlusive disease of cerebral arteries and ischaemic cerebral infarction</u>

   Remember: Diabetics who also suffer from hypertension have a 20 – 30 % probability of a cardiovascular event (myocardial infarct, cerebral infarct) within the next 10 years (high risk group. If in addition a diabetic nephropathy develops, the cardiovascular risk increases to > 30 %/10 years!
   Pain as a premonitory symptom (angina pectoris, pain on exercise in intermittent claudicatio) is often absent due to concominant neuropathy!

   1.2. <u>Microangiopathy:</u>
   - Glomerulosclerosis (Kimmelstiel-Wilson disease)
   - Retinopathy
   - Neuropathy
   - Microangiopathy of intramural small coronary arteries (small vessel disease)

   1.2.1. <u>Diabetic nephropathy (DN)</u> [E14.2]
   <u>Def.:</u>- Persistent (micro-)albuminuria (> 20 mg/l)
      - Arterial hypertension
      - Decreasing glomerular filtration rate
      - Increased cardiovascular risk
   <u>Ep.:</u> The average progression to DN is circa 2,5 % per year in type 2 diabetes, therefore after 10 years circa 25 % (similar figures are valid for type 1 diabetes). In patients with an increased serum creatinine the mortality rate is around 20 %/year (mainly due to cardiovascular mortality). In manifest DN 75 % of type 1 diabetics and 20 % of type 2 diabetics develop end-stage renal failure within 20 years. In Europa and USA up to 50 % of all dialysis patients are diabetics.
   <u>Pa.:</u>   Type 1 diabetes: <u>Glomerulosclerosis</u> (Kimmelstiel-Wilson disease)
        Type 2 diabetes: <u>Nonspecific</u> vascular and tubulointerstitial renal changes as a consequence of complex risk factors of the metabolic syndrome.
   <u>Pg.:</u> Hyperglycaemia → activation of growth factors in the kidneys (TGF-β and angiotensin II)

   - ■ Renal hypertrophy with increase in size of the glomeruli and thickening of the basal membrane
   - ■ Increased glomerular permeability with microalbuminuria
   - ■ Glomerulosclerosis, interstitial fibrosis
   - ■ Renal failure

Risk factors for accelerated progress of DN:
- Arterial hypertension
- Extent of the albuminuria
- Level of diabetic control (HbA1c)
- Hypercholesterinaemia
- Smoking
- possible high protein intake

Remember: Early symptom is a microalbuminuria of 30 - 300 mg/24 h or 20 - 200 mg/l in a random urine (because the microalbuminuria has a range of variation of up to 40 % laboratory testing should be repeated). The risk of renal and cardiovascular complications rises linearly with increasing albuminuria! Temporary/reversible increases in albumin excretion are seen in urinary infections, febrile illnesses, physical effort, uncontrolled blood pressure or blood sugar etc
Frequency and severity of diabetic nephropathy correlates with the duration of the diabetes and the quality of metabolic control. Early antihypertensive therapy (also of a borderline hypertension!), especially with ACE inhibitors delays the progressiom of diabetic nephropathy to end-stage renal failure, and reduces the cardiovascular + total mortality!

Stages of diabetic nephropathy:

Stage	Albumin excretion (mg/l)	Creatinine clearance (ml/min)	Remarks
1. Renal damage with normal renal function a) Microalbuminuria b) Macroalbuminuria	20 - 200 > 200	> 90	Serum creatinine in the normal range. Blood pressure increasing in the normal range or hypertension, dyslipidaemia, rapid progress of coronary heart disease, AVD, retinopathy and neuropathy
2. Renal damage with renal failure a) Low grade b) Moderate grade c) High grade d) Terminal	> 200 decreasing	60 - 89 30 - 59 15 - 29 < 15	Serum creatinine borderline or increased hypertension, dyslipidaemia, tendency to hypoglycaemia, rapid progression of coronary heart disease, AVD, retinopathy and neuropathy, development of anaemia, impairment of the bone metabolism

1.2.2. Diabetic retinopathy : [E14.3+H36.0*]
    Occ.: Type 1 diabetes: 90 % after 15 years
          Type 2 diabetes: 25 % after 15 years
          30 % of all blindness in Europe is due to diabetes!
    Pg.: Microangiopathy; neovascularisation is triggered by an angiogenic growth factor. Poor diabetic control, hypertension and smoking lead to a faster progression of diabetic retinopathy.

    ■ Nonproliferative retinopathy (background retinopathy):
     - Mild: Only microaneurysms
     - Moderate: In addition single intraretinal haemorrhages, venous calibre changes with pearlstring-like veins
     - Severe: Microaneurysms and intraretinal haemorrhages in all 4 quadrants or veins with pearlstring appearance in at least 2 quadrants or intraretinal microvascular anomalies (IRMA) in at least 1 quadrant (4-2-1-rule)
      ■ Proliferative retinopathy:
    Neovascularisations in vicinity of the optic papilla = NVD (neovascularization disk) or in other parts of the retina = NVE (neovascularization elsewhere) with or without vitreal or epiretinal haemorrhages.
    Co.: Retinal detachment and secondary neovascular glaucoma
    ■ Diabetic maculopathy: a) focal b) diffuse c) ischaemic
    Macular oedema, hard exudate, intraretinal haemorrhages; central vision endangered!

1.2.3. Diabetic neuropathy: [E14.4] Dependent on duration of diabetes and quality of metabolic control. After 10 years of disease circa 50 % of patients have a neuropathy.
    Pg.: Unclear; possible impairment of the microcirculation of the vasa nervorum + metabolic impairment (e.g. nonenzymatic glycosylation of structural proteins etc)

    ■ Peripheral sensorimotor polyneuropathy (80 %): More pronounced distally, symmetrical, with reduced sensitivity to stimuli and loss of power, especially feet/lower legs (paraesthesia, "burning feet"), areflexia (achilles reflex missing bilaterally), decreased thermosensitivity and algesthesia, later possibly also motor impairment. Determination of superficial sensitivity using Semmes-Weinstein´s monofilament, which is applied onto the sole with a pressure of 10 g. Determination of temperature sensation e.g. with a "tip-therm tube".

Early symptom: Decreased vibration sense   measuring with 64 Hz tuning fork (128 Hz) according to Rydel-Seiffer with a graduation scale of 0 - 8. The vibrating tuning fork is applied e.g. onto the the medial malleolus and the patient indicates with eyes closed, how long he or she can feel the vibration. A graduation value of < 6 out of a total of 8 is pathological.

Special diagnostic:
- Pedography (= measuring of the dynamic pressure distribution pattern of the plantar soles while walking): Decreased strain on the toes during increased pressure load of the ball of the forefoot.
- Measuring of nerve conduction velocity: In polyneuropathy
  DD: Polyneuropathies of other origin: alcohol abuse, neurotoxic drugs (nitrofurantoin, barbiturates, cytotoxic drugs etc), chemicals (solvents, heavy metals, insecticides etc), paraneoplastic syndrome, malabsorption syndrome, polyarteritis nodosa etc

- ■ Rarer manifestations of diabetic neuropathy: e.g.
  - Pronounced diabetic polyneuropathy:
    Asymmetric proximal diabetic neuropathy with pains in the hip region and at the ventral thigh, reduction of the ipsilateral patellar reflex, poss. paresis of the quadriceps muscle.
  - Peripheral facialis paresis; pareses of the ocular muscles (diplopia)
  - Diabetic radiculopathy mainly with unilateral segmental pains and sensoric impairments in a region of the trunk.

- ■ Autonomic neuropathy (AN): (second most common!)
  Def.: Neuropathy of the autonomic nervous system (sympathetic and parasympathetic nervous system)

  - Cardiovascular AN:
    Occ.: 15 % of diabetics at time of diagnosis, > 50 % of diabetics after 20 years of disease duration; mortality increased circa 4fold due to ventricular arrhythmias leading to ventricular fibrillation (sudden cardiac death).
    - ■ Silent myocardial ischaemia and painless myocardial infarctions with increased mortality
    - ■ Reduced variability of heart rate with fixed frequency
      a) In an ECG at rest and in a 24 h-ECG
      b) During maximum inspiration and expiration (difference of the heart rate < 9/min)
      c) During a Valsalva manoeuvre
      d) During an orthostatic test
    - ■ Resting tachycardia (vagus damage)
    - ■ Asympathetic postural hypotension (sympathic damage): Reduced systolic/diastolic blood pressure and absent reflex tachycardia when moving to a standing position.
    - ■ Poss. flattened or inverted circadian blood pressure curve with elevated nocturnal blood pressure readings
      Special diagnostic: Detection of a cardial sympathic dysinnervation (predominantly of the posterior cardiac wall) by ^{123}J-MIBG scan.

  - AN of the gastrointestinal tract (parasympathetic damage)
    - ■ Impairment of oesophageal motility, poss. with dysphagia
    - ■ Gastroparesis with epigastric fullness/pressure feeling, poss. postprandial hypoglycaemia
      Di.: Ultrasound (detection of reduced peristaltic and delayed gastric emptying), poss. special diagnostic: C^{13}-octan acid breath test or scan of gastric emptying
    - ■ AN of the intestines with postprandial diarrhoea alternating with constipation
    - ■ Anorectal dysfunction (incontinence)

  - AN of the urogenital tract (damage to the parasympathetic system)
    - ■ Atonic bladder and disturbance of micturation poss. with residual urine and predisposition to urinary tract infections
    - ■ Erectile dysfunction and absence of the spontaneous nocturnal tumescence (circa 30 % of all diabetics)

  - AN of the neuroendocrine system:
    Reduction/absence of the hormonal contraregulation in hypoglycaemia (decreased perception of hypoglycaemia !)
    Reduced release of catecholamines during orthostatic and physical strain

  - AN of thermoregulation:
    Reduced sweating, vasodilatation (warm and dry diabetic foot!)

  - AN of the pupils: Decreased pupillary reflexes (special diagnostic by means of pupillometry: slow mydriasis)

1.2.4. Diabetic foot syndrome (DFS, 25 % of older diabetics): [E14.7]

Def.: Syndrome of variable clinical picture with differing aetiologies, which in case of injury to the foot can result in infected ulcers and complications as serious as amputation of the limb.

Grades of severity of the foot lesions according to Wagner:
Grade 0   Foot at risk without any lesion
Grade 1   Superficial wound
Grade 2   Wound reaches to tendon or capsule
Grade 3   Wound reaches to bone or joint
Grade 4   Necrosis limited to forefoot or heel
Grad 5   Necrosis of the whole foot

The classification according to Wagner is completed by information regarding the presence of infection or ischaemia:
A: No infection/no ischaemia
B: With infection (most common pathogens: staphylococci, enterococci, pseudomonas aeruginosa; often mixed infections)
C: With ischaemia
D: With infections and ischaemia

- ■ Neuropathic diabetic foot (50 % of all DFS):
  - Warm foot with very dry skin (no foot odour !)
  - Impaired sensitivity (sensitivity to vibration and/or touch ↓), reduced sensation of pain and temperature (with danger of unnoticed trauma!)
  - Palpable foot pulses
  - Doppler index (ankle pressure/arm pressure) normal (impairment through media sclerosis)
  - Transcutaneous pO₂ normal
  - Impairment of the unrolling-motion of the foot with increased pressure load beneath the head of the metatarsal bones and the big toe
  - Co.: Infections; painless neuropathic ulcers (= malum perforans) at pressure points (heel, ball of foot), often triggered by missing or incorrect foot care, tight foot wear, (micro-)trauma.
  In the advanced state poss. development of a diabetic-neuropathic osteoarthropathy (DNOAP)with necroses in the region of the metatarsophalangeal joints, tarsometatarsal joints or other foot joints (Charcot´s foot).

- ■ Ischaemic foot in peripheral vascular disease (PVD), esp. of the arteries of the lower legs (circa 50 % of all DFS):
  - Medical Hx: - Diabetes mellitus   - Hypercholesterinaemia
                       - Arterial hypertension   - Smoking
  - Intermittent claudication
  - Cool, pale foot with poss. livid discolouration
  - No palpable foot pulses
  - Doppler index (RR ankle : RR arm) < 0,9; transcutaneous pO₂ ↓
  - Sensation intact
  - Acral necrosis/gangrene, threat of amputation (Germany: circa 15.000 major amputations per year = amputations above the ankle
  Di.: Pulse status, ankle-arm pressure index, coloured duplex ultrasonography, MR angiography; consultation with vascular surgeon

- • Combined form of neuropathic and ischaemic diabetic foot
  (circa 35 % of all DFS)
  Claudication and pain at rest can be reduced or absent because of the neuropathy! Prognosis unfavourable!

  Remember: As long as a PVD is missing, the foot pulses are well palpable !

2. Diabetic cardiomyopathy

Remember: CHD, arterial hypertension and diabetic cardiomyopathy are the 3 risk factors for the development of heart failure in diabetics. The mortality of diabetics with heart failure is circirca 15 %/year.

3. Reduction of immune resistance with tendency to bacterial skin and urinary tract infections

4. Deficiency of lipid metabolism: Triglycerides ↑, HDL cholesterin ↓

5. Fatty liver

6. Diabetic coma, hypoglycaemic shock

7. Hyporeninaemic hypoaldosteronism with hyperkalaemia, hyponatraemia, hyperchloraemic metabolic acidosis and poss. hypotension (see there for details)

Di.:   ▶ **Medical history** (family history, pregnancy complications etc)

        ▶ **Clinical picture and laboratory investigations:**

          ✦ **Determination of blood glucose:**
          Specific according to the hexokinase ferment method:

The largely identical intermediate stages of impaired glucose homeostasis (impaired fasting glucose = IFG) and the pathological glucose tolerance are risk factors for future diabetes mellitus and cardiovascular disease.

Diagnostic reference values for the determination of diabetes mellitus (American Diabetes Association, 2003):

Stage	Fasting venous plasma glucose	Random blood sugar	Oral glucose tolerance test (OGTT)
Diabetes	≥ 126 mg/dl (≥ 7,0 mmol/l) *)	≥ 200 mg/dl (≥ 11,1 mmol/l) and symptoms of a diabetes	2 h value ≥ 200 mg/dl (≥ 11,1 mmol/l)
Impaired fasting glucose = IFG	≥ 100 - 125 mg/dl (≥ 5,6 – 6,9 mmol/l)		Pathologic glucose tolerance („impaired glucose tolerance = IGT") 2 h value ≥ 140 - 199 mg/dl (≥ 7,8 < 11,1 mmol/l)
Normal	< 100 mg/dl (< 5,6 mmol/l)		2 h value < 140 mg/dl (< 7,8 mmol/l)

*) For glucose determination of capillary blood a fasting blood glucose ≥ 110 mg/dl (≥ 6,1 mmol/l)is considered as diabetic, because glucose levels in capillary blood are circa 10 % lower than in the plasma.

Capillary blood levels between 90 mg/dl (5,0 mmol/l) and < 110 mg/dl (6,1 mmol/l) are within the range of impaired fasting blood glucose levels.

Explanations:
Fasting blood sugar (= FBS) is the decisive test for the diagnosis of a diabetes mellitus and for the monitoring of its therapy. It is just as accurate as the OGTT´s 2 hour value in predicting the risk of developing a microangiopathy. It is simple, efficient and cheap. The reading should be ensured by a repeated value.
Fasting is defined as a period of 8 hours without nutrition.
Random blood sugar = at any time of the day, without relation to meals; symptoms are diabetes associated symptoms such as polyuria, polydipsia and weight loss.
Blood glucose levels measured by instruments using strips can vary up to 15 % from the actual value.

Note: • Normal blood glucose values at the moment are defined as ≤ 100 mg/dl (≤ 5,6 mmol/l) in the venous plasma.
• The possibility of incorrect low blood sugar levels has to be taken into consideration when measuring serum glucose, because of the in vitro glycolysis (break down circa 10 % per hour!). Serum samplings without addition of glycolysis inhibiting agents (e.g. sodium fluoride) must not be used for glucose testing.

DD: Temporary hyperglycaemia after myocardial infarction, stroke, increased intracranial pressure, acute intoxications (CO), after ingestion of thiazide diuretics etc

■ **Determination of glucose in the urine** (in the morning urine, in day portions and in the 24 h urine):
Repeated findings of glucose in the urine indicate apart from few exceptions (s.b.), the existence of diabetes mellitus. Every diabetic should analyze his or her individual renal threshold (blood glucose level, at which glyco-suria appears for the first time).
The normal renal threshold level for glucose is around 180 mg/dl glucose in the blood (lower in pregnancy around < 150 mg/dl → poss. physiological glycosuria in pregnancy). The physiological glucosuria can be up to 15 mg/dl. The lower detection level of the test strips is circa 30 mg/dl.

Consider: In cases of diabetic nephropathy the renal threshold for glucose can be increased (to up to 300 mg/dl), meaning that in these cases no glycosuria is observed despite hyperglycaemias of e.g. 200 mg/dl. Therefore the absence of glucose in the urine does not exclude a manifest diabetes (early diagnosis of diabetes with FBS)! Therefore the self monitoring of glucose in the urine is not a suitable way to achieve a normoglycaemic therapy target.
The exceptional case of glycosuria despite normoglycaemia is found in renal diabetes due to tubular partial functional deficiency. Hereditary defects of the glycometabolism (pentosuria, lactosuria, galactosuria, fructosuria) are excluded by the specificity of the enzymatic determination method.

■ **Determination of ketones** (β-hydroxybutyrate, acetoacetate, aceton) in the blood. Fast testing instruments detect the leading agent β-hydroxybutyrate. In diabetic ketoacidosis (DKA) levels of > 3,0 mmol/l β-hydroxybutyrate are found.

■ **Oral glucose tolerance test (OGTT):**
Ind: The OGTT is not recommended for routine clinical examination (significance in unclear cases).
Conditions:
- Avoidance of a state of hunger (at least 3 days ≥ 150 g CH/d)
- Need to remain nil by mouth 10 h before the test
- No febrile illnesses
- In women not during menstruation

Causes of abnormal OGTT: Various physical factors (e.g. myocardial infarction, long confinement in bed etc) and also drugs (diuretics, corticosteroids, oestrogens etc) lead to elevated blood sugar levels. Therefore the OGTT should not be performed in these circumstances.

In patients after a partial gastric or upper intestinal resection and also in patients with the malabsorption syndrome, the intravenous glucose tolerance test should be performed.

Performing the test: After determinating the FBS, adults drink a solution of 75 g of glucose. Blood glucose dtermination 120 minutes after glucose uptake. The OGTT is contraindicated if the fasting blood glucose levels are already clearly pathological (see above).

■ **Continuous BS measuring over 24 h** (sensor method or microdialysis): Special diagnostic test in specific circumstances (e.g. clarification of unexplained hypo- or hyperglycaemias)

■ **Diagnosis of gestational diabetes:**
Because of the absence of clinical symptoms, screening of pregnant women with increased diabetic risk at 24 - 28 weeks gestation. Determination of blood glucose 60 min. after 50 g glucose (or oligosaccharide mixture; no preparation required, not fasting). A blood sugar > 140 mg/dl (> 7,8 mmol/l) is suspicious of gestational diabetes; further clarification by means of standardized OGTT.

■ **Screening examination for diabetes mellitus:**
Fasting BS for individuals > 45 y. every 3 y. For risk groups earlier:
- Overweight, hypertension, dyslipidaemia
- Positive family history (relative of 1. degree)
- Members of ethnic groups with high diabetic risk (e.g. Pima indians)
- After delivery of a child with a birth weight of > 4.500 g
- History of gestational diabetes
- History of pathological glucose tolerance or impaired glucose homeostasis

■ **Screening for further risk factors** such as premature arteriosclerosis (hypertension, hyperlipoproteinaemia, smoking etc)

■ **Test for microalbuminuria** (at least 1 x/year in diabetics)

■ **Examination of risk groups for type 1 diabetes for anti-GAD-AB and anti-IA2-AB only in clinical trials**, because at the moment without therapeutic significance.

Th.: 1. Diet, normalisation of weight
2. Physical activity increases the sensitivity of the muscles to insulin!
3. Drugs: a) oral antidiabetics , b) insulin
4. Patient education and monitoring
5. Exclusion of or therapy for poss. further risk factors for premature arteriosclerosis
6. Prophylaxis of, and therapy for, complications

To 1 - DIET:
In type 2 diabetes therapy must always start at the onset of impaired glucose tolerance, to avoid vascular complications! In this respect normalizing weight is of highest priority (target value: BMI < 25). If this is successfull drug therapy is often unnecessary and progression to diabetes can be avoided or delayed. In times of hunger the number of manifest type 2 diabetics is at its lowest.

In the mostly normal weight type 1 diabetic, food and insulin supply must be optimally coordinated, in order to achieve a normoglycaemic metabolic state: In conventional insulin therapy meals must be adapted to a strict pre-set insulin therapy system. In the intensified insulin therapy the insulin supply is adapted to a relatively free calculable nutrition uptake as required! (see below)

Daily energy requirement (in kcal):
Normal weight x 32 in mild physical activity (most frequent case)
Normal weight x 40 in moderate physical activity
Normal weight x 48 in heavy physical activity
Normal weight (according to Broca) in kg: Height in cm - 100 (women - 10 %)
Body mass-Index:

$$\frac{\text{Body weight (kg)}}{\text{Height (m)}^2} \rightarrow \text{Normal index: } 18,5 - 24,9 \text{ kg/m}^2$$

1 kcal = 4,2 kilojoules	
1 g carbohydrate	= 4,1 kcal = 17,2 kJ
1 g protein	= 4,1 kcal = 17,2 kJ
1 g fat	= 9,3 kcal = 38,9 kJ
1 g alcohol	= 7,1 kcal = 30 kJ

- No large meals, several small ones instead.

- Composition of the diet:
  - Proteins 10 - 15 % of total calories (low fat meat, fish, vegetable proteins). In diabetic nephropathy low protein diet.
  - Fat: 30 % of total calories, possibly with a high proportion of unsaturated fatty acids. If additional disorder of the lipid metabolism is present, the proportion of fat of the total calories should be reduced to< 25 %.
  - Carbohydrates: According to the remaining calorie requirement, 50 - 60 % → calculation according to bread units (BU) 1 BU = 10 g CH (corresponds to approx. half a roll). The Langerhans' islets excrete for every BU circa 1 IU insulin. The amount of BU can be determined from exchange tables. In conventional insulin therapy the BU are distributed in a ratio of 2 : 1 with meals and between meals, to avoid a hypoglycaemia between 2 main meals. This is not valid for the intensified insulin therapy.

    Fast resorbable monosaccharides (glucose) and disaccharides are unfavourable (saccharose = cane sugar, lactose = milk sugar). Allowed sweeteners are saccharine, cyclamate, aspartame (note: aspartame is carcinogenic in animal studies). Sugar exchange products (fructose, xylitol) undergo glycolysis independently of insulin and therefore have only a minor glycaemic effect; however, in a bad metabolic status the impact can be more significant.

    The maximal permissible amount of carbohydrates (carbohydrate tolerance) corresponds to the daily amount of carbohydrates at which no essential glucosuria occurs and at which the blood glucose levels after a meal remain < 150 mg/dl.
  - Large amounts of high-fibre roughage lead to a delay of the carbohydrate absorption and a decrease of the blood glucose levels in type 2 diabetics.
  - Alcohol only occasionally up to a max. of 30 g, always together with carbohydrates (alcohol inhibits gluconeogenesis in the liver → danger of hypoglycaemia).

to 3.:
- ### TREATMENT OF TYPE 1 DIABETES:
  INSULIN SUPPLY - diet - physical activity - education
- ### STEPWISE TREATMENT OF TYPE 2 DIABETES ACCORDING TO PHASES

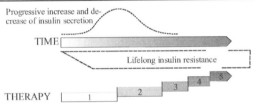

  1. **WEIGHT NORMALIZATION** – diabetic diet – physical activity – education

     Remember: Trials show, that the progression of type 2 diabetes can be halted by normalization of weight and regular physical activity!
  2. In addition metformin in overweight type 2 diabetics. Sulphonylureas are recommended in normal weight.
  3. If control is inadaequate addition of a 2. oral antidiabetic agent: Acarbose, insulinsecretagogues or glitazone
  4. Secondary failure of SU therapy (= exhaustion of the B cells) is observed on average after 10 y., when combination of SU and injection of an intermediate insulin in the evening (bedtime insulin).
  5. With exhaustion of endogenous insulin production, SU lose their effect → conventional or intensified insulin therapy is commenced.

Drugs:
Oral antidiabetics:
Insulinotropic and noninsulinotropic drugs:

Insulinotropic = β-cytotropic	Noninsulinotropic = nonβ-cytotropic
Sulphonylureas, meglitinides, exenatide sitagliptin	Biguanides (metformin), α-glucosidase inhibitors, glitazones
Effect on the β-cell	Peripheral effect
Treatment of the secretion deficit	Treatment of the insulin resistance
Effect also in later stages of disease	Effect mainly in earlier stages of disease
Danger of hypoglycaemia	No danger of hypoglycaemia
Danger of weight gain	Suitable for obese patients

- **Biguanides:** Metformin
  - Ef.: - Delayed intestinal glucose absorption
    - Inhibition of hepatic gluconeogenesis   } extrapancreatic effects
    - Increased glucose uptake in muscles
    - Appetite lowering effect (→ poss. weight rediction)

202

Because of their effect biguanides are desirable substances for the treatment of the insulin resistance found in overweight type 2 diabetics, since they neither induce hypoglycaemia or reinforce the hyperinsulinaemia.

Remember: In the UKPD study metformin performed in all areas (micro- and macroangiopathy, lethal outcome) better than all other forms of therapy and is therapy of 1. choice in overweight type 2 diabetics with respect to their CI.

SE: Very rarely lactic acidotic coma (with high mortality) only if disregarding the contraindications; often gastrointestinal complaints etc

CI: Renal insufficiency!, decompensated cardiac failure, respiratory impairment, severe hepatic impairment, conditions which predispose to hypoxia of tissue, wasting, slimming diet, fasting, acute concommitant illness, pregnancy, before and after surgery, 48 h before and after intravenous urography (danger of lactic acidosis!), alcoholism; old age is a relative CI and others

Dose: 1 - 2 x 500 – 1.000 mg/d after the meals; to start with smallest dose.

- α-glucosidase inhibitors: Acarbose (glucobay®), miglitol (Diastabol®)

Eff.: Competitive inhibition of glucoamylase, saccharase, maltase in the intestinal mucosa. Postprandial blood glucose peaks are flattened. The unfissured carbohydrates stimulate the enterohormone GLP-1 (glucagon like peptide) in the lower small intestine. This sensitizes the β-cells to glucose stimulus.

Similar effects of α-glucosidase inhibitors can be achieved with a high-fibre diet.

SE: At higher doses symptoms of carbohydrate malabsorption can occur (flatulence, bloating, abdominal pain, diarrhoea), increase in liver enzymes etc

CI: Pregnancy, age < 18 y., severe renal insufficiency, chronic intestinal disease

Dose: Careful adjustment of dose: Initially 50 mg/d at beginning of a meal. If tolerated slowly increase the dose to 3 x 50 mg/d (higher doses cause more side effects.

- **Glitazones (thiazolidinediones):** In the ProActive study mild decrease of clinical end points with pioglitazone. On the basis of a meta analysis the suspicion was raised of a slight increased cardiovascular risk with rosiglitazone.
  - Rosiglitazone (Avandia®); rosiglitazone + metformin (Avandamet®)
  - Pioglitazone (Actos®)

Eff.: „Insulin-sensitizer", improves the sensitivity of the peripheral cells for insulin (decrease of the insulin resistance).

SE: e.g. weight gain, oedema → poss. deterioration of cardiac impairment; rarely hepatic damage. Isolated cases of macular oedema with rosiglitazone; suspicion of increased risk of fractures in women.

Ind: For combination therapy in type 2 diabetes with metformin or SU

CI: Liver diseases, cardiac impairment, renal impairment, pregnancy, lactation etc.

Dose: e.g. rosiglitazone 4 - 8 mg/d or pioglitazone 15 - 45 mg/d; monitoring of liver enzymes

- **Insulinsecretagogues (insulinotropic substances):**

1. Sulphonylureas (SU):

Eff.: Stimulation of insulin secretion by increasing sensitivity of B-cells to glucose. Proof of efficacy for risk reduction of clinical end points exist for glibenclamide (UKPDS).

Ind: Type 2 diabetes with sufficient remaining intrinsic insulin production - provided that dietetic therapy alone (weight normalization!) is not enough (step 3 of the graduated therapy plan).

Note: At the time, when diabetes manifests itself, the majority of diabetics still have exaggerated levels of insulin in the blood. Here sulphonylureas lead to normoglycaemia, but the metabolic syndrome deteriorates through the additional hyperinsulinaemia! Therefore weight normalization and physical activity are an indispensable part of therapy! Overweight type 2 diabetics are treated initially with metformin.

CI: - Type 1 diabetes
- Pregnancy (change to insulin)
- Severe renal impairment, hepatic impairment
- Diabetic metabolic imbalances (precoma/coma, acidosis/ketoacidosis)
- Reduced cognitive states (e.g. accidents, operations)
- Diabetic gangrene
- Allergy to sulphonylureas

SE: - Hypoglycaemia - causes:
- Overdosage
- Reduced nutritional intake
- Physical exertion
- Alcohol consumption
- Renal impairment (delayed renal elimination)

- Gastro-intestinal irritation
- Allergic reactions (sulphonamide allergy)
- Rarely blood disorders (agranulocytosis, haemolytic anaemia)

Interaction of sulphonylureas with other substances:

(which can predispose to hypoglycaemia or to inhibition of its efficacy), e.g.:

Increase (risk of hypoglycaemia)	Reduction	Risk factors for the occurrence of severe hypoglycaemias under SU therapy
Beta-adrenoreceptor blockers	Glucagon	Age > 70 years
ACE inhibitors	Öestrogens, gestagens	Cerebrovascular or cardiac diseases
Coumarin derivates	Corticosteroids	Renal or liver impairment
ASA	Phenothiazine derivates	Alcohol
Nonsteroidal antiinflammatories	Saluretics	Erratic food intake
Sulphonamides	Thyroid hormones	Diarrhoea
Clarithromycin	Sympathomimetics	Physical exertion
Gatifloxacin	Diazoxide	
Alcohol (be careful !)	Nicotinic acid derivates	

Examples of preparations:

■ Glibenclamide has the greatest hypoglycaemic effect. Decreased risk of microvascular complications shown (UKPD study)
Dose: 1,75 - 10,5 mg/d

■ Glimepiride (Amaryl®): Is given as a single dose immediately before breakfast.
Dose: 1 - 3 (6) mg/d

All SU are commenced at the smallest dose, slowly increasing. Attention has to be paid to the danger of nocturnal hypoglycaemia! In the first 4 weeks tight blood glucose monitoring is necessary, as after 2 - 3 weeks the metabolic status often improves and then poss. dose reductions can be indicated! At maximum dose of glibenclamide 2/3 are given in the morning and 1/3 of the dose in the evening. A dose at midday is unnecessary, as the islet cells remain stimulated by the morning dose. Regular blood glucose measurement is necessary; in weight reduction dose reduction. Caution in older patients with irregular mealtimes! Mild renal impairment can prolong the duration of action.

2. **Sulphonylurea analogues: Meglitinides:** Evidence of efficiency regarding the risk reduction of clinical endpoints is not yet available.
Repaglinide (Prandin®), nateglinide (Starlix®)
Eff: Meglitinides are so-called postprandial glucose regulators. The effect on FBG is less noticable. They lead to short term insulin secretion from the B-cells via a blockage of the ATP-sensitive potassium channels. Similar to an intensified insulin therapy they are taken with meals. The risk of hypoglycaemia is thought to be less than with SU. Precondition: Good patient education + compliance
Ind: Type 2 diabetes, step 3: Alternative to SU
CI: Similar to SU
SE: Hypoglycaemias, gastrointestinal SE, rarely increase in liver enzymes, visual disturbances, allergy
Interactions: No combination with drugs, which interfere with CYP3A4 (e.g. gemfibrozil)
Dose: e.g. repaglinide (Prandin®) 0,5 - 2,0 mg before meals; start with lowest dose!

Failure of the SU therapy:

Primary failure of the SU:
Relatively rare occurrence in IDD = late manifestation type 1 diabetics

Secondary failure of the SU:
a) Supposed (reversible) secondary failures:
- "Diet failure": Overweight type 2 diabetics, in which the possibility of dietary manipulation is not yet exhausted.
- Transient deterioration of glucose homeostasis through stress situations or infections
b) Genuine secondary failure in optimal diet and weight normalization:
A secondary failure rate of circa 5 % annually is estimated. Secondary failure occurs after an average of 10 years of diabetes duration and is a consequence of exhaustion of the B-cells causing insulin deficiency . As a result the genetically pre-set insulin resistance can no longer be compensated. The cardinal symptom is hyperglycaemia despite maximal therapy with SU.
In order to find out whether insulin is needed the determination of the C-peptide is recommended in relation to the fasting blood glucose (FBG):

$$\frac{\text{C-peptide (ng/ml)} \times 1000}{\text{FBG (mg/dl)}} \quad \rightarrow \text{Values} < 11,7 \text{ indicate a need for insulin}$$

Th.: Combined therapy with SU + insulin:

■ **Basal insulin supported oral therapy (BOT)**
- Continuation with SU and metformin therapy.
- Additional dose of an intermediate (NPH)-insulin in lowest dose possible before bedtime. Alternative: Bedtime dose of long-acting analogue insulin glargine (Lantus®): Starting with a small dose (6 IU) and increasing if required very slowly and in small steps. Usually 8 - 16 IU is sufficient! FBG should be in the normal range.

Advantages:
- Only 1/3 of the insulin dose is needed compared to monotherapy with insulin alone.
- A relatively good adjustment is achieved with <u>one</u> insulin.

■ **Prandial or supplementary insulin therapy (SIT):**
Precondition: Insulin production is still sufficient for the basic supply; only increased insulin requirement at meals is supplemented.
Continuing oral antidiabetics + additional injection of a small preprandial dose of normal insulin. Starting dose: FBG (mg/dl) x 0,2 = daily insulin dose. Dividing of this dose in the ratio 2 : 1 : 1 (breakfast/lunch/dinner).

■ **Sitagliptin** (Januvia®), a DPP-4-inhibitor: Inhibition of dipeptidyl peptidase 4, which is responsible for the breakdown of the glucagon-like peptide 1. GLP1 stimulates insulin secretion and inhibits glucagon secretion. Long-term data are still awaited.
Ind: Type 2 diabetes, in addition to metformin and/or SU, if these alone have an inadequate effect.
SE (e.g. gastrointestinal disturbances etc) and <u>CI</u> (e.g. renal insufficiency etc) have to be considered!
Dose: 2 x 50 mg/d

■ **Exenatide** (Byetta®)
Eff: Incretin-mimetic (like GLP-1 = glucagon-like peptide), which increases insulin secretion and inhibits glucagon secretion. No hypoglycaemias occur, because insulin secretion only is increased in the presence of an elevated glucose level. Long-term data still are awaited.
Ind: Type 2 diabetes in combination with metformin and/or SU, if they alone have an inadequate effect.
SE: Frequently nausea, occasionally diarrhoea; AB formation against exenatide with occcassional reduction of its effect etc.
Dose: Initially 2 x 5 µg/d s.c. circa 30 minutes before the main meals. Increase of dose after 4 weeks possible to 2 x 10 µg/d s.c.

INSULIN

Insulin is synthesized in the B-cells of the Langerhans islets from the precursors pre-proinsulin and proinsulin; the <u>C-peptide</u> (connecting peptide) is segregated from the centre of the molecule chain of the <u>proinsulin</u>. Since insulin and <u>C-peptide</u> are secreted equimolarly and simultaneously into blood, measuring C-peptide allows an assessment of B-cell pancreatic function. Compared to insulin determination in serum, C-peptide measurement in serum has the advantage that no cross reaction is possible with insulin antibodies, and measurement of the exogenic insulin dose remains unaffected. Furthermore the determination of C-peptide is less likely to be influenced by short-term fluctuations of insulin synthesis, due to its longer biological half-life (circa 25 minutes).
In patients with insulin dependent diabetes, C-peptide is lowered.
The secretion of insulin, which is stored in the granula of the B-cells, is proportional to the blood glucose level. Due to rapid inactivation of circulating insulin through <u>insulinases</u>, the plasma half-life is short (5 minutes). An excess of <u>contrainsulinergic hormones</u> (STH, ACTH, corticosteroids, glucagon, adrenalin, thyroxine) can lead to a diabetic metabolic status (WHO classification 3D).

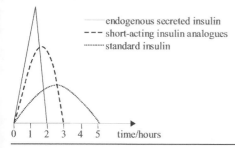

———— endogenous secreted insulin
- - - short-acting insulin analogues
·········· standard insulin

Action mechanisms of the insulin:

1. <u>Membrane effect:</u> Stimulation of the transport of <u>glucose</u>, amino acids and potassium into the muscle and fat cells.

2. Metabolic effects: Stimulation of the anabolic metabolic processes (glycogen synthesis, lipid synthesis, protein synthesis) and inhibition of the catabolic processes (glycogenolysis, lipolysis, proteolysis).

In diabetics the uptake of glucose into the cells is more difficult (insulin resistance a./o. insulin deficiency). Only higher blood glucose levels can sufficient glucose get into the cells. The deficiency in glucose in the cells of the adipose tissue storage leads to a reduced production of glycerin phosphate, so that fatty acids are not synthesized to triglycerides and leave the fat tissue. In the liver they are broken down via acetyl-CoA to ketone bodies (diacetic acid, β-hydroxybutyric acid, acetone). The ketone bodies, which are also used by the muscle cells as a source of energy, reduce the cell´s permeability for glucose, so that the situation is worsened further (insulin antagonistic effect of the ketone bodies).

- The daily insulin requirement for a 'normal individual' is 40 IU of insulin (more in the obese):
  20 IU of insulin for the ingestion and 20 IU of insulin for the basal metabolism.
- 1 IU of insulin lowers the blood glucose level by 30 - 40 mg/dl.
- 1 BU (bread unit) increases the blood glucose level by 30 - 40 mg/dl.
- Conclusion: 1 IU of insulin neutralizes on average 1 BU.

Insulin secretion can be divided into two components:
1. Basal insulin secretion
2. Insulin secretion dependent on meals.

The basal insulin secretion with low insulin concentrations in the blood (5 - 25 µU/ml) suppresses the release of glucose from the liver. Therefore in insulin deficiency increased blood glucose levels occur as well in the fasting period. On the other hand, meal dependent insulin secretion is required for the utilization and storage of glucose from the nutrition.

Indications for an insulin therapy:
1. Type 1 (insulin dependent) diabetes
2. Type 2 diabetes: Insulin therapy in time, if diet + oral antidiabetics no longer lead to good control.
3. Pregnancy, if diet alone does not lead to normoglycaemia.
4. Diabetic complications (microangiopathies, [pre-]coma diabeticum)
5. Poss. perioperative conditions or those diabetics requiring intensive medical care

Insulin preparations:
Human insulins and insulin analogues are in current use.

Insulin is available in Germany in two different concentrations. As **U40** (40 IU/ml) for the conventional injection with insulin injection devices, and as **U100** (100 IU/ml) in cartridges for insulin pumps and injection aids (pen).

1. Short-acting insulins:

a) Standard insulin: Onset of action after 30 - 60 min.; duration of action circa 5 h
- Insuman Rapid®
- Actrapid® (Novo Nordisk)
- Huminsulin® Normal (Lilly)

  Ind: - In metabolic imbalances and when used for first time
  - For intermittent therapy (e.g. perioperatively)
  - For intensified conventional therapy (ICT) and for insulin pump therapy

Administration: Subcutaneous; in patients in coma intravenous
In s.c. administration duration between injection and meal circa 15 - 20 min.

b) Short-acting insulin analogues: Onset of action after 5 - 10 min., duration of action circa 3,5 h
Insulin Lispro (Humalog®)
Insulin Aspart (NovoRapid®)
Insulin Glulisin (Apidra®)
Variation of the amino-acid sequence avoids subcutaneous hexamer synthesis, thus more rapid resorption; onset of action circa 15 min., duration of action circa 3 h.
Advantage: No delay of meal after injection, fewer postprandial hypoglycaemias; insulin between meals may possibly no longer be necessary. Possible postprandial injection.
Disadvantage: Its effects may be too short for acting on slowly resorbable carbohydrates; basal insulin supply must be dosed exactly; not mixable with conventional NPH insulin
CI: Disorders affecting gastric emptying, e.g. in AN with gastroparesis (→ danger of hypoglycaemia!) etc

2. Depot insulins:
By combining insulin with protamine, zinc, insulin preparations with longer duration of effect are achieved.
Administration: Subcutaneous; depot insulins must not be given intravenously!

a) Intermediate insulins:
NPH insulins (neutral protamine hagedorn insulins)
Principle: Insulin protamine cristals; onset of action after circa 60 min., duration of action 9 - 18 h
Examples of human insulins:
- Insuman® Basal(Hoechst)
- Insulin Protaphan HM® (Novo-Nordisk)
- Huminsulin Basal® (Lilly)
Ind: Combination therapy insulin + oral antidiabetics (SU, metformin); conventional and intensified conventional insulin therapy (ICT)
Intermediate-acting insulin analogue: Insulin Detemir (Levemir®) – not mixable with standard insulin!

b) Long-acting insulins:
Onset of action after circa 60 min., duration of action up to 24 h
Ind.: Coverage of basal insulin demand in the intensified conventional insulin therapy (ICT).
Example of human insulin: Ultratard® HM (insulin zinc cristals)
Steady-state conditions are only achieved after 3 - 5 days, due to their long duration of effect.

Long-acting insulin analogues:
Insulin Glargine (Lantus®)
Insulin Detemir (Levemir®)

3. Insulin mixtures from standard insulin + NPH insulin:

NPH insulins can be freely mixed with standard insulin (not possible with zinc insulins). For virtually all needs corresponding trade preparations are available.

Ind: Conventional insulin therapy with 2 (- 3) daily injections, division of doses: 2/3 in the morning, 1/3 in the evening
Time between injection and meal with standard insulin approx. 30 minutes (insulin analogues without delay between injection and meal); administration s.c.

Examples:
- Actraphane (Novo-Nordisk) 10/-20/-30/-40/-50: The numbers indicate the proportion in % of standard insulin (the rest if NPH insulin).
- Huminsulin Profil II, III (Lilly) - 20, 30 % standard insulin
- Insuman Comb 15, 25, 50 Hoechst®
- Humalog® Mix 25, 50 (Lilly) = combination of Lispro® and Lispro NPH
- NovoMix 30 (combination of 30 % Insulin Aspart + 70 % Aspart NPH)

4. Inhalative insulins (discontinued in the UK) : e.g.

Exubera®: A rapid acting pulverized insulin - bioavailability circa 10 %; onset of action after 5 - 10 min., duration of action circa 5 h. 1 mg corresponds to circa 3 IU of s.c. injected insulin. Long term studies are not available. Smokers, as well patients with lung diseases (e.g. asthma or COPD) should not use it. Indication: e.g. rare problems with s.c. administration.

Complications of insulin therapy:

1. Hypoglycaemia

Cause: Overdosage (rare due to suicidal attempts), missing or inadequate food intake, increased physical activity, weight reduction, interaction with drugs (e.g. beta-blockers) and alcohol (life-threatening hypoglycaemias in drying-out cells!).

Note: When change from animal to human insulin a 10 - 20 % decreased insulin demand has to be expected (danger of hypoglycaemia !).

2. Lipodystrophy of the fat tissue at the injection sites

Pro.: Systematic alteration of injection sites (whole abdominal area and lateral thigh - not on upper arm)

3. AB related complications are extremely rare when human insulin is used.

4. Insulin resistance:

Additional requirement of insulin due to impairment of the interaction between insulin and its receptor on the cell surface a./o. due to glucose utilization in the cell → Causes:
- Overweight (most common)
- Hypertriglyceridaemia
- Infections (common)
- Increase of insulin antagonizing hormones (s.a.)
- Stress / trauma
- Ketoacidosis (Pre-/coma diabeticum)
- AB against insulin (extremely rare for human insulin)

Note: Pseudoresistance occurs in hyperinsulism (due to too high insulin doses: Hypoglycaemias and afterwards reactive hyperglycaemias): in this circumstance only a gradual reduction (!) of the insulin dose helps (see below: Somogyi effect).

A) Conventional insulin therapy (CT)

with intermediate insulin or insulin mixtures with intermediate insulin + standard insulin: Satisfactory stabilization can only be achieved with at least 2 injections/d. 2/3 to 3/4 of the daily dose is injected before breakfast, the rest before dinner (time between injection and meal = 30 minutes with standard insulin - with insulin analogoues no time difference). Better stabilization is possible with 3 injections

In the morning: mixed insulin – at lunch time: standard insulin – in the evening: mixed insulin

Disadvantage: A fixed pre-set dose of a depot insulin without addition of standard insulin is not sufficient to neutralize the blood glucose increase after nutrition. On the other hand, the insulin level is unphysiological high between meals, so that inbetween meals are necessary to avoid hypoglycaemia: As a result the patient must obey a fixed meal schedule: If the patient does not eat enough his insulin dose will be too excessive and vice versa.

Remember: Conventional insulin therapy = The patient must eat, because he injected insulin!

Note: Morning hyperglycaemia can have 3 causes:
1. The duration of action of a single morning dose of a depot insulin is too short, so that the blood glucose level increases during the night and esp. in the morning.
   Th.: A second dose of insulin in the evening (morning/evening ratio: 2 - 3 to 1).
2. Somogyi effect: The patient receives an insulin dose in the evening, which is too excessive: This causes nocturnal hypoglycaemia (nocturnal blood glucose testing around 3 - 4 o'clock) and reactive hyperglycaemia in the morning.

Patients who tend to have nocturnal hypoglycaemias should not go to bed with a blood glucose level < 120 mg/dl, because by 3 am it will have decreased by 30 - 40 mg/dl, only increasing again after 3 am through to morning time. If patients are below 120 mg/dl at 23.00 they should eat one or two BU. Nocturnal testing around 3 h is only necessary if hypoglycaemia is suspected, for example if the patient undertook a lot of sport, or after alcohol consumption.
Th.: Decrease of the insulin dose in the evening!
3. Dawn phenomenon: Despite constant insulin supply hyperglycaemia occurs in some patients in the early morning (after 6 o´clock). The cause is increased insulin demand in the second half of the night, due to increased growth hormone (GH) secretion (esp. type 1 diabetes).
Di./Th.: Blood glucose testing in the night (e.g. 22 / 2 / 4 o´clock) and adaption of the insulin dose in the evening(intermediate- or long-acting insulin), or use of an insulin pump → supply of an elevated basal rate in the early morning hours.

Additional remarks
• Readjustment of poorly controlled diabetes:
  Never change diet and insulin simultaneously, otherwise the overview is lost. The patient remains on his old routine system for 2 days, frequent blood glucose measurements are performed, and only then is the insulin therapy changed.
• If adjustment carries a danger of hypoglycaemia (which the patient must be educated about) dextrose should be at hand. Relatives should be familiar with the emergency treatment of hypoglycaemia in case of hypoglycaemic shock (1 mg glucagon i.m. or s.c.).
• During longer lasting unusual muscle exercise (e.g. sport on weekends) the insulin demand decreases, so that on that corresponding (and poss. also following) day only a reduced insulin dose (e.g. 50 %) should be injected.

B) Intensified insulin therapy

Basis /bolus concept:
Dividing the insulin level of a healthy individual into a basal rate and additional meal dependent (prandial) rate, insulin peaks can be imitated in an insulin dependent diabetic in 2 ways:

a) Intensified conventional insulin therapy (ICT):
  The basal insulin demand is covered by at least two injections of an intermediate-acting insulin (poss. single dose of a long-acting insulin). In cases of normal daily routine, sometimes a single dose of a depot insulin in the evening is sufficient. The injection in the evening is based on the daily routine of the patient and on the nocturnal blood glucose curve. Usually it is applied between 22 - 24 o´clock.
  Around 40 - 50 % of the total daily insulin dose falls into the basal insulin supply. The other 50 - 60 % of the daily dose is divided between the meal dependent (prandial) bolus doses of standard insulin or short-acting insulin analogues. The individual dose depends on the size of the meal, the preprandial tested blood glucose level, the time of the day and on any planned physical activity. An interval between injection and meal is not absolutely necessary in this case, but circa 15 min. is desirable.
  There is a circadian insulin sensitivity, and therefore the insulin demand per carbohydrate unit (CU) varies at different times:
  Insulin demand per CU: In the morning circa 2 IU, at noon 1,0 IU, in the evening 1,5 IU
  The adjustment of the dose of the standard insulin in levels which differ from the target blood glucose (90 - 120 mg/dl) is based on the experience, that 1 IU of standard insulin lowers blood glucose by circa 30 mg/dl (in blood glucose levels ≤ 300 mg/dl). In blood glucose levels > 300 mg/dl 1 IU of standard insulin lowers the blood glucose by circa 60 mg/dl.

  Examples of injection patterns of the ICT (S = standard insulin, D = depot insulin)

Breakfast	Lunch	Dinner	At night (23 h)
S	S	S	D
S + D	S	S	D
S + D		S	D
S	S + D	S	D

  In patients with sufficient basal insulin secretion a trial can be done with a supplementary insulin therapy (bolus of a rapidly acting insulin at the main meals without basal insulin).

b) Insulin pump therapy:
  Here standard insulin is used exclusively. Continuous subcutaneous insulin infusion (CSII) is achieved by means of an external pump. In modern instruments the basal rate can be programmed separately for each hour, so that e.g. a Dawn phenomenon can be counteracted in an optimum way. At mealtimes the patient releases – dependent on the preprandial blood glucose level and the desired amount of food – bolus insulin doses via the insulin dosing instrument. In this case they are insulin pumps without an automatic glucose sensor (the blood glucose testing is done manually by the patient) = "open-loop-system". The ideal instrument would be an insulin pump with continuously working glucose sensor (e.g. the Ulm "glucose clock"), in which the insulin supply is regulated = feedback-regulated pumps = "closed-loop-system". With insulin pump therapy the insulin demand is usually lower!

  Co.: 1. Local infections
       2. Decompensation into coma if insulin flow is blocked
       3. Danger of hypoglycaemia in case of insufficient self control of blood glucose

<u>Ind.:</u> - Pregnancy
  - Severe Dawn phenomenon
  - Threatened late complications of diabetes etc

<u>Preconditions for an intensified insulin therapy:</u>
- Cooperative patients with the ability to decide therapeutic options
- Intensive diabetes education
- Daily self monitoring of the metabolic status (at least 4 self testings of blood glucose)
- Supervision of patients by experienced diabetic physicians

<u>Advantages of the therapy:</u>
- Ideal control of metabolism
- Individual synchronized arrangement of food intake (the patient injects insulin, when he wants to eat).
  The results of the <u>Diabetes Control and Complication Trial (DCCT)</u> in type 1 diabetics show, that the rate of diabetic late complications (retinopathy, nephropathy, neuropathy) is reduced by 50 % with intensified insulin therapy. Optimising the metabolic state prevents further progress of existing complications. However, this therapy causes a threefold increased risk of hypoglycaemia.

Therapeutic target:

<u>Prevention of diabetic late complications by targeting an almost normoglycaemic metabolic status:</u>
1. • <u>Fasting blood glucose and preprandial 80 - 110 mg/dl (4,4 - 6,1 mmol/l)</u>
   Blood glucose postprandial ≤ 140 mg/dl (≤ 7,8 mmol/l)
   Blood glucose self monitoring by the educated patient
   • Urine free of glucose
   • Negative for ketones
   • Albuminuria < 20 mg/l
2. <u>Avoidance of hypoglycaemic reactions</u>
3. <u>Normalization of body weight and blood lipids</u> • Target values:
   LDL cholesterol < 100 mg/dl (< 2,6 mmol/l)
   HDL cholesterol > 45 mg/dl (> 1,1 mmol/l)
   Triglycerides < 150 mg/dl (< 1,7 mmol/l)
4. <u>Normalization of the glycosylation long-term parameter HBA1c (estimation of control every 3 months):</u>
   Through glycosylation of haemoglobin via the instable aldimine form (labile HbA1) the stable ketoamin form (stable HbA1) is formed, consisting of 3 sub-fractions a, b and c. Since the essential c-fraction (HbA1c) corresponds to 70 % of the HbA1, both parameters have the same informative value. HbA1c reflects as "<u>blood glucose memory</u>" the metabolic blood glucose status of the patient in the preceding last 8 weeks. <u>In good metabolic control the HbA1c fraction is normal (< 6,5 % - depending on the laboratory´s method).</u>
   False low concentrations are seen in reduced erythrocyte lifespan (e.g. haemolytic anaemia) and in the first half of pregnancy.
   False high concentrations can occur in renal insufficiency, hyperlipoproteinaemia, chronic alcohol abuse, second half of pregnancy and lactation period, high dose salicylate therapy.

   Remember: If the HbA1c value increases to 7 %, the risk of myocardial infarction increases by 40 %; for values around 8 % the risk of infarction increases by 80 % (UKPD study).
   A 1 % point reduction of the HbA1c value decreases diabetic complications by 20 % (UKPD study). The risk of hypoglycaemia increases 3fold!

   <u>Interpretation of blood glucose and HbA1c:</u>
   • Normal blood glucose, high HbA1c:
     - Pretending of having a good diet discipline only for a short period before the ambulant testing is performed
     - In unstable metabolic status high HbA1c values indicate metabolic decompensation in the preceding weeks despite normal blood glucose.
   • Elevated blood glucose, satisfactory HbA1c values: Only temporary increase of blood glucose (e.g. stress related high glucose during visit at the doctor) in otherwise satisfactory stabilization
   • Normal blood glucose and HbA1c values: Good metabolic status in the last 4 - 8 weeks
   • Elevated blood glucose and HbA1c values: Bad metabolic status in the last 4 - 8 weeks

   • HbA1c determination alone is not recommended for the <u>diagnosis</u> of a diabetes due to lack of sensitivity. As <u>long-term parameter</u> however HBA1c is absolutely essential for <u>metabolic monitoring</u>.

5. <u>Eliminating of poss. further risk factors for premature arteriosclerosis:</u>
   • Smoking cessation
   • BP < 130/< 80 mm Hg
   • With proteinuria ≥ 1 g/d the target RR is < 125/< 75 mm Hg

   Remember: For every 10 mm Hg lowering of systolic BP, diabetic complications decrease by 12 % (UKPD study).

6. <u>Regular examination to detect poss. late complications (documenting in the health record)</u>
   - Screening for (micro-)albuminuria, urea, creatinine i.s.
   - Inspection of the feet by the doctor

- Patient education concerning prevention of foot complications (self inspection of the feet, professional foot care and adaequate shoes, protection from injuries etc)
- Pulse status, neurological status
- Ophtalmological examination with ophthalmoscopy, poss. fluorescence angiography

7. Since diabetics are cardiovascular high risk patients, primary prevention with ASA (100 mg/d) is considered.

8. Early prevention and treatment of complications:

   Basis: Best possible blood glucose control and treatment/elimination of other vascular risk factors

   ▶ Diabetic foot syndrome (DFS)

   Precondition: Interdisciplinary collaboration in diabetic foot clinics: Differentiating between neuropathic foot (neurologic diagnostic) and/ or peripheral vascular disease (PVD) (vascular investigation)

   Therapy points:
   Foot care (patient education !) - relief of pressure – diabetes shoes for relief – avoidance of trauma and infections – cleaning of wounds/débridement of necrotic coating + treatment of infections – revascularisation therapy in PVD. Bypass grafting of peripheral vascular occlusion, as well as interventional measures in patients with DFS prevent, in the majority of cases, amputation of the endangered leg. Poss. additional hyperbar oxygenation (HBO).

   Frequency of bacterial infections in chronic wounds in the DFS: Staph. aureus (50 %) alone or combined with enterobacter (40 %), streptococci (30 %), staph. epidermidis (25 %) etc. After taking culture swabs, initial treatment with broad-spectrum antibiotics, changing according to culture results

   Remember: Unsuitable footwear is the most common cause of pressure sores/ulcerations/necrosis. The target of the previous 10 years, the so-called St. Vincent declaration, to reduce amputations in diabetics by 50 %, was not achieved! No amputations before vascular + diabetic specialist consultation! Revascularisation therapy in vascular surgical centres can reduce the high number of major amputations!

   Prg.: Without good prevention and therapy high risk of amputation. After amputation 50 % of the patients die within 3 years (caused by further complications of diabetes).

   ▶ Diabetic retinopathy - annual ophtalmological examination !

Nonproliferative (background-) retinopathy		Proliferative retinopathy	
Microaneurysms intraretinal haemorrhages	IRMA `pearlstring` veins	Proliferations of vessels	Preretinal haemorrh. retinal detachment
↑	↑	↑	↑
Panretinal laser coagulation			Surgery of vitreous body (vitrectomy)

   Remember: Intensified insulin therapy in type 1 diabetes can reduce the risk of diabetic retinopathy by circa 75 % (DCCT study).

   ▶ Diabetic polyneuropathy:
   • Only optimum blood glucose stabilization can lead to improvement!
   • The opinions concerning α-lipoic acid are controversial.
   • In case of failure poss. attempt with amitriptyline in low dose (consider SE + CI)

   ▶ ADN with gastroparesis: Metoclopramide often loses its action after few weeks. Attention has to be paid to postprandial hypoglycaemias in insulin therapy → poss. adjustment of injection-meal interval! In therapy resistent gastroparesis poss. implantation of a gastric pacemaker.

   ▶ Diabetic nephropathy (DN): Annual screening for micro- and\or macroalbuminuria!
   • Aim for BP readings in the normal lower range! (see above), preferably with ACE inhibitors or AT1 blockers, which act renoprotectively.
   • Avoidance of nephrotoxic substances
   • In persisting proteinuria: Protein restriction (0,8 g/kg BW/d), preferably fish and vegetarian proteins, diet low in sodium chloride (limit NaCl supply to 6 g/d).

   ▶ Erectile dysfunction:
   • Urological history + diagnostic (exclusion of a deficiency in testosterone and of a hyperprolactinaemia; drug history; SKAT test, diagnostic of arterial + venous vessels)
   • Therapeutic options:
   - Phosphodiesterase-5-inhibitors: Sildenafil (Viagra®), vardenafil (Levitra®), tadalafil (Cialis®)
     SE: e.g. headaches, facial flushing, drop in BP, esp. In combination with nitrates, molsidomin or alpha-blockers; rare visual disturbances or loss of eyesight
     CI: CHD, after myocardial infarction or stroke; simultaneous therapy with nitrates or molsidomin; arterial hypotonia, cardiac insufficiency among others
   - MUSE (medicamentous urethral system for the erection) with prostaglandin E1 analogues, e.g. alprostadil is more comfortable for the patient than the selfinjection therapy into the cavernous body of the penis.
   - Vacuum pump if venous flow is too rapid.

▶ **Diabetes therapy in pregnancy (incl. gestational diabetes):**
  - Close cooperation between physician and obstetrician
  - Extensive patient education
  - If therapy with diet alone not possible, intensified conventional insulin therapy or insulin pump. Oral antidiabetics contraindicated. In known diabetes optimization of the metabolic control preconception.

  Therapeutic targets: Preprandial blood glucose 60 - 90 mg/dl, 1 h postprandial < 140 mg/dl, 2 h postprandial < 120 mg/dl, median blood glucose < 100 mg/dl, normal HbA1c. If control is optimal, infant mortality is comparable with nondiabetics (< 1 %).
  Gestational diabetes usually resolves post partum, but a high risk remains for later development of diabetes.

  Consider an alteration of the insulin sensitivity in pregnancy:
  1. Increasing insulin sensitivity with higher risk of hypoglycaemia in the 8. - 12. week of gestation
  2. Decreasing insulin sensitivity during the 2. half of pregnancy → increase dose.
  3. Returning insulin sensitivity immediately after delivery → reduce dose.
  4. Breast-feeding reduces the insulin requirements by circa 5 IU.

▶ **Diabetes and surgical procedures:**
  - Circumstances: Insulin patient:
  Preoperative minimum requirement: circulation stable, normal water and electrolyte balance, constant pH, blood glucose < 200 mg/dl.
  Perioperative separate infusion of glucose 5 % plus necessary electrolytes (100 - 200 ml/h) + standard insulin i.v. via pump. Regulate insulin supply depending on blood glucose level (hourly controls). Check serum potassium every 4 h.
  Alternative: appropriate insulin supply by application of insulin pumps.

Current blood glucose (mg/dl)	Insulin dose (IU/h)
120 - 180	1,0 if preoperative daily requirement < 40 IU
	1,5 if preoperative daily requirement 40 - 80 IU
	2,0 if preoperative daily requirement > 80 IU
> 180	each time 0,5 IU more
< 120	each time 0,5 IU less
≤ 100	reduce insulin supply or stop, increase glucose supply, blood glucose controls every 15 - 30 min.

  In the following procedures a reduction in insulin requirement has to be taken into account postoperatively, with the danger of hypoglycaemias:
  • Amputation of an extremity because of gangrene
  • Excision of an infected organ (e.g. gallbladder)
  • Drainage of an abscess or of a phlegmon
  • Hypophysectomy, adrenalectomy, pheochromocytoma surgery
  • Delivery via caesarian section

  - Circumstances: Type 2 diabetes/patient on oral antidiabetics:
  Metformin is to be stopped 48 h before surgery, no sulphonylureas on the day of surgery!
  Minor and medium surgery: Infusion with 5 % of glucose (adding necessary electrolytes), to check blood glucose hourly.
  Blood glucose < 200 mg/dl → surgery.
  Blood glucose > 200 mg/dl → Insulin supply (see above)
  SU with first postoperative meal, blood glucose monitoring
  Major surgery: Preoperative change to insulin

  Remember: Blood sugar normalization through insulin therapy can reduce the mortality in surgical intensive care patients by 30 % and septic complications by nearly 50 %!

▶ **Combined transplantation of pancreas and kidney:**
  Ind: Type 1 diabetes with end stage renal failure
  Co.: Rejection reactions, infections, transplant pancreatitis and   thrombosis, pancreatic fistulas, intraabdominal abscesses
  Results: 10 y survival rates circa 70 % (best results if HLA compatibility between donor + recipient)

Islet cell transplantation:
From the donor pancreas an islet cell suspension is processed through treatment with collagenase and cell separator and injected into the portal vein of the recipient; subsequent immunosuppressive therapy of the recipient.
Ind: Type 1 diabetic with end stage renal failure, who also receive a kidney transplantation.
Success rate: In the majority of cases the transplanted islet cells are rejected within 5 years.

▶ **Therapy forms/diagnostic under clinical trials:**
  • Development of an artificial endocrine pancreas = "Closed-Loop-System", consisting of continuous working glucose sensor, micro computer and insulin pump (e.g. `Ulm glucose clock´). Here the insulin supply is regulated via glucose controls (feedback regulated).

- Development of noninvasive blood glucose machines
- Gene therapy of the type 1 diabetes (reprogramming of liver cells for insulin production)

Prg.: While the mortality from coma in diabetes has dropped from > 60 % (around 1900) to approx. 1 % (insulin, oral antidiabetics), nowadays diabetic mortality is determined by the extent of vascular damage: Vascular related causes of death in diabetes mellitus nowadays approach 80 %! Therefore every diabetic with a vascular risk factor (e.g. hypertension) should receive prophylactically ASA (100 mg/d). Reduce LDL cholesterol with statins to 100 mg/dl (in very high risk to 70 mg/dl).

With early aggressive treatment of diabetes and hypertension the prognosis is good; in cases of unsatisfactory control of diabetes life expectancy and quality are reduced.

The prognosis of type 2 diabetes can be improved substantially, by early normalisation of weight and central obesity!

Most common cause of death: Myocardial infarction (55 %) and/or renal failure (> 40 %). In former times almost 10 % of type 1 diabetics became blind due to retinopathy!

Poor metabolic control further increases the danger of late complications such as autonomic diabetic neuropathy and diabetic foot syndrome.

Pro.: Various intervention trials for prophylaxis of type 1 diabetes (e.g. the risk for type 1 diabetes could be reduced by 80 % in Finland due to vitamin D prophylaxis).

COMA DIABETICUM [E14.0]

Causative factors:
Absolute or relative insulin deficiency
- Absent exogenic insulin supply:
    - First manifestation of previously unrecognized diabetes
    - Omitted injection; interrupted insulin supply in insulin pumps
    - Oral medication instead of insulin (in insulin dependence)
- Insufficient exogenic insulin supply:
    - Insufficient dose prescribed
    - Technical error in the calculation of dose and injection
- Increased insulin demand:
    - Infection (pneumonia, UTI etc)   - Myocardial infarction
    - Dietary error   - Hyperthyreosis
    - Surgery, accident, pregnancy   - Therapy with saluretics,
    - Gastrointestinal disorders   corticosteroids

In 25 % of cases it is a so-called manifestation coma, meaning that diabetes mellitus is diagnosed for the first time in the state of the coma. Infections represent the most common triggering cause (circa 40 %)!

Pg.: The ketoacidotic coma if typical of type 1 diabetes, the hyperosmolar coma of type 2 diabetes.

Remember: Absence of a diabetic ketoacidosis (DKA)does not exclude a coma diabeticum!

Pathogenesis of ketoacidotic coma:

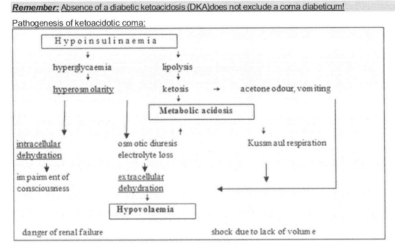

- Pathogenesis of hyperosmolar coma:
  A relative insulin deficiency leads to decreased peripheral glucose utilization, while at the same time there is an increased hepatic release of glucose. In this case small amounts of insulin prevent the ketosis through inhibition of the lipolysis in the fat tissue.

Cl.: 3 clinical forms of the diabetic decompensation:
- Cardio vascular form (lack of volume, shock)
- Renal form (acute renal failure)
- Pseudoperitonitic form: Signs of peritoneal irritation, atonia of stomach and intestines, esp. gastric wind (→ aspiration tube!)
  DD: Acute abdomen

Assessment of the severity of an impairment of consciousness using the Glasgow coma scale:

		Points
eye opening	spontaneous	4
	in response to speech	3
	to response to pain	2
	no eye opening	1
verbal reaction	oriented	5
	confused	4
	random words	3
	moaning, no words	2
	none	1
motor response	carrying out requests	6
	localizing response to pain	5
	withdraws to pain	4
	flexor response to pain	3
	extensor posturing to pain	2
	no response to pain	1
maximum score		15
minimum score		3

	Hyperosmolar coma (typical of type 2 diabetes)	Ketoacidotic coma (typical of type 1 diabetes)
Precoma	– loss of appetite, vomiting – thirst, polydipsia, polyuria – weakness, tachypnoea – signs of the dehydration with tendency to collapse (strongest in the hyperosmolar coma)	
Coma	Insidious onset !	poss. pseudoperitonitis (abdominal pain) poss. acidotic (deep) respiration
	2. Dehydration and development of shock (pulse ↑, RR and CVP ↓) 3. Oligo-anuria, reduced reflexes 4. ECG: signs of hypokalaemia and poss. arrhythmias laboratory: 5. Hyperglycaemia 6. Glucosuria 7. Serum Na$^+$ normal or mild decreased 8. Serum K$^+$ variable: despite loss of potassium the K$^+$ levels in the serum can be normal or elevated due to acidosis before the start of the insulin therapy. 9. HCT + Hb ↑, leucocytosis	
	hyperglycaemia > 600 mg/dl hyperosmolality > 300 mosmol/kg H2O minimal acetonuria	hyperglycaemia > 350 mg/dl ketonuria: in urine stix acetone +++ ketonaemia: β-hydroxybutyrate > 5 mmol/l metabolic acidosis with standard bicarbonate 8 - 10 mmol/l anion gap increased through ketone bodies

Serum osmolality (in mosmol/kg H_2O) = 1,86 x Na$^+$ + glucose + urea + 9
(all in mmol/l; if quoted in mg/dl → to divide glucose by factor 18 and urea by factor 6)
Gap of anions (in mmol/l) = (Na$^+$) – (Cl$^-$) – (HCO3$^-$)
Reference value: 8 - 16 mmol/l

<u>DD:</u>   **Causes of a loss of consciousness:**
    1. <u>Toxic:</u>
       - Exogenous intoxication (esp. <u>alcohol</u>, heroin, sedatives, psychopharmaceutical drugs)
       - Endogenous intoxication (uraemia, coma hepaticum)
    2. <u>Cardiovascular:</u>
       - Collapse
       - Shock
       - Stokes-Adams syncope, circulatory arrest
    3. <u>Endocrine disorders:</u>
       - <u>Hypoglycaemic shock, coma diabeticum</u>
       - Addison´s disease
       - Thyrotoxic crisis and myxoedematous coma
       - Hypophyseal coma
       - Hypercalcaemic crisis
       - Diabetes insipidus
    4) <u>Cerebral diseases</u> (**_Note:_** Often with reactive hyperglycaemia!)
       Hypertonic massive haemorrhage, encephalomalacia, subarachnoid haemorrhage, sub-/epidural haematoma, head-brain injury, epilepsy, meningitis, encephalitis, sinus thrombosis, generalised seizure etc
    5) <u>Psychological:</u> Hysteria
    6) <u>Anoxaemic:</u> Suffocation, hypercapnia in respiratory failure
    7) <u>Lactoacidotic coma</u>
       <u>Cause:</u> Severe hypoxia, following fructose infusion in fructose intolerance, very rarely SE of biguanide therapy (di.: blood lactate ↑)

DD	Coma diabeticum	Hypoglycaemic shock [E15]
development	slowly, days	sudden, minutes
Hunger		+ + +
Thirst	+ + +	
muscular system	hypotonic **never convulsions**	hypertonic, tremor
Skin	dry !!!	Sweating
breathing	Deep respiration,* ketotic odour	Normal
Eye balls	soft	Normal
	fever, abdominal pain	delirium prodrome (incorrect diagnosis: alcohol abuse!); poss. picture of a cerebral infarction with neurological deficit; Babinski positive, poss. epileptic seizures

* In hyperosmolar coma normal breathing, because no ketoacidosis (→ and no ketotic odour!).

The <u>DD</u> between coma diabeticum and hypoglycaemia is easily made with capillary finger glucose testing.
Should there be the slightest doubt in the differential diagnosis (emergency service, blood glucose meter not available), insulin should never be administered, as it can quickly cause death!

<u>Di.:</u>   History/clinical picture - laboratory (blood glucose ↑, in diabetic ketoacidosis (DKA) β-hydroxybutyrate ↑)

<u>Th.:</u>   Intensive therapy unit
    A) <u>General measures:</u>
       • Control of respiration, circulation, fluid/electrolyte balance
       • Insertion of urinary catheter for fluid balance monitoring (+ antibiotic prophylaxis)
       • Central venous catheter for measurement of CVP
       • Gastric tube (because of gastric atonia and pyloric spasm with nausea)
       • Close laboratory monitoring (blood glucose hourly, potassium + blood gases every 2 h)
       • Ulcer and thromboembolic prophylaxis (low-dose heparin)

    B) <u>Specific therapy:</u>
       1. <u>Treatment of dehydration and hyperosmolarity:</u>
       In untreated diabetic coma hypernatraemia due to dehydration occurs, but there is also a concomitant sodium renal loss. With normal urine output and a moderate hypernatraemia (< 155 mmol/l) rehydration with normal saline (0,9 % NaCl) is indicated. The use of hypotonic solutions (e.g. 0,45 % NaCl) increases the risk of rapid fluid shift into the liquor cerebrospinalis, and is therefore only used with caution, even in severe hypernatraemia.
       Dose in relation to time: In the 1. hour 1000 ml, after that depending on the CVP: 0 cm → 1.000 ml/h, 1 - 3 cm → 500 ml/h, 4 - 8 cm → 250 ml/h, 9 – 12 cm → 100 ml/h. In the first 8 hours the average fluid requirement totals 5 - 6 l. After the 8. hour 250 ml/h is often sufficient.

214

If cardiovascular collapse is imminent, additional albumin solution or human plasma protein solution.
Dose adjustment depends on diuresis and clinical picture (in patients with cardiac impairment avoid too rapid infusion → danger of pulmonary oedema!).

2. Insulin therapy:
Use only standard insulin! Plasma half-life of insulin circa 5 minutes. Various dose patterns are recommended. The „low-dose" insulin therapy with an initial bolus of 20 U i.v., followed by 5 - 10 U of standard insulin/h i.v. via dosage pump has proved successful in most patients.
If hypokalaemia exists before insulin therapy (which rarely is the case), it should be balanced first. No insulin therapy without accompanying volume supply (point 1).
The blood glucose should not be lowered faster than 100 mg/dl per hour and initially not < 250 mg/dl (too rapid lowering of the blood glucose can lead to damages to the retina).
Advantage of the „low-dose" insulin therapy: Smaller incidence of hypokalaemia and hypoglycaemia in the course of the therapy as well as lower risk of cerebral oedema.
Disadvantage: Some patients require higher doses: If blood glucose does not come down within 2 h with the initial dose, the dose must be doubled (in rare cases far higher amounts of insulin are necessary in order to break an insulin resistance). When the blood glucose is reduced to 250 - 300 mg/dl, the supply of standard insulin is reduced to 2 U/h with simultaneous infusion of 5 % glucose solution.

3. Correction of acidosis:
Through the effects of insulin, the acidosis is counteracted efficiently by inhibition of the lipolysis. A mild acidosis therefore does not require correction! Cautious administration of bicarbonate only if the pH drops to < 7,1, and give only 1/3 of the calculated requirement to avoid provoking a dangerous hypokalaemia!

4. Electrolyte balance:
- Sodium replacement with the fluid substitution
- Potassium replacement:
  Ind: After start of the insulin therapy, as soon as the blood glucose decreases
  CI: Anuria, hyperkalaemia
  Dose: Depending on the serum K^+ level and on the pH. In pH > 7,1 the following standard values are to be used:

Serum K^+ (mmol/l)	K^+ substitution (mmol/h)
< 3	20 - 25
3 - 4	15 - 20
> 4 - 5	10 - 15

In this phase avoid cardiac glycosides (danger of digoxin toxity!). In cases of significant hypokalaemia (< 3 mmol/l) poss. interruption of the insulin supply.
- Phosphate replacement:
  Ind: Poss. in serum phosphate < 0,5 mmol/l
  CI: Renal insufficiency
  Dose: Circa 50 mmol/24 h

Remember: Low dose insulin therapy and slow correction of the metabolic imbalance reduces complication rate! Fluid shifts in the CNS during diabetic coma take time to normalise. It is to be expected that the patient may not regain consciousness immediately, following correction of blood glucose, pH and volume/electrolyte imbalances.

Transition from coma therapy to oral nutrition:
Build up with light diet, e.g. initial gruel diet, while before every meal a small dose of standard insulin is given s.c.. After that readjustment of the diabetes.

HYPOGLYCAEMIA [E 16.2] and HYPOGLYCAEMIC SHOCK [E 15]

Def.:   Blood glucose < 40 mg/dl (< 2,2 mmol/l) or
Whipple Trias: BG < 45 mg/dl (< 2,5 mmol/l) + hypoglycaemic symptoms + disappearance of these symptoms after administration of glucose

Aet.:   A) Fasting hypoglycaemia:
- Insulinomas: See chapter
- Extrapancreatic tumours (e.g. liver cell carcinoma)
- Very rarely paraneoplastic secretion of insulin-like peptides (e.g. IGF II)
- Severe liver disease (reduced gluconeogenesis and glucose release), uraemia (substrate deficiency for gluconeogenesis)
- Insufficiency of the adrenal cortex or anterior pituitary (failure of contrainsulin working hormones)
- Very rare β-cell hyperplasia in the first years of life (nesidioblastosis) through mutation of the sulfonylurea receptor
- Glycogenoses

- Renal hypoglycaemia (renal diabetes mellitus)
- Hypoglycaemia of the newborn of a diabetic mother

B) Reactive (postprandial) hypoglycaemia:
- Early stages of diabetes mellitus
- Gastric emptying disorder due to autonomous neuropathy (diabetic gastroparesis)
- Dumping syndrome after gastric resection
- Reactive hypoglycaemia in vegetative instability (increased vagal tone)
- Rare hereditary defects (e.g. leucin hypersensitivity, fructose intolerance)

C) Exogenous hypoglycaemia:
- Overdosage of insulin or sulphonylureas (most common cause)
- Hypoglycaemia factitia: Nonaccidental due to insulin injections or ingestion of sulfonylureas (psychotic, suicidal or criminal)
  Characteristics: Hypoglycaemia occurs erratically and independently of meals. Affected individuals are often health professionals or relatives of diabetics.
- Excessive alcohol consumption while fasting
- Interactions of drugs with antidiabetic agents (e.g. sulfonamides, nonsteroidal antirheumatics, beta-blockers, ACE inhibitors)

Causes of a hypoglycaemia in diabetes mellitus:
1. Most common, relative overdosage of insulin or sulphonylureas, e.g. if a patient omits usual food intake during an intercurrent illness, while continuing to take the hypoglycaemic agent in the usual dose! Following new adjustment of sulphonylureas the metabolic state can improve after approx. 3 weeks, when hypoglycaemia can occur unless dose reduction is instituted. During intensified insulin therapy with optimum blood glucose and HBA1c levels, hypoglycaemia can easily be induced. In the event of frequent hypoglycaemias, the perception of hypoglycaemia isreduced, so that autonomous warning symptoms are frequently not perceived in time.
2. Interference with blood glucose lowering drugs
3. Absolute overdose (accidental, suicidal, criminal)
4. Heavy physical strain
5. Alcohol consumption (alcohol inhibits gluconeogenesis)

Cl.:

Phases	Symptoms and clinical signs
C) Autonomous symptoms:	
a) Parasympathetic reactions	Very intensive hunger, nausea, vomiting, weakness
b) Sympathic reactions	Agitation, sweating, tachycardia, tremor, mydriasis, hypertension
4. Central nervous = neuroglucopaenic symptoms	Headaches, endocrine mental syndrome (mood swings, irritability, poor concentration, confusion), impaired coordination, primitive automatisms (grimacing, grasping, smacking), convulsions, focal signs (hemiplegia, aphasia, diplopia), somnolence. Hypoglycaemic shock = hypoglycaemic coma, central impairment of respiration and circulation

In severe autonomic neuropathy the symptoms under 1 can be reduced or absent!
Glucose is the only source of energy used in cerebral metabolism → high sensitivity of the brain to hypoglycaemia.

DD: Psychoses, epilepsy, CVA etc

Remember: In suddenly occurring, aetiologically unclear neurological or psychiatric symptoms, always think of hypoglycaemia and check blood glucose!

Di.: Determination of blood glucose concentration: Hypoglycaemic symptoms occur in most cases only with values < 50 mg/dl (in diabetics they can occur above that value).

In spontaneous hypoglycaemia in nondiabetics the cause must be clarified by further investigations:
Determination of blood glucose, serum insulin and C-peptide during a spontaneous hypoglycaemia or in the 72 h fasting test with determination of the insulin/glucose ratio during a hypoglycaemia (see chap. insulinoma).
Insulin and C-peptide show a parallel increase in endogenous secretion; in hypoglycaemia due to exogenous insulin supply (hypoglycaemia factitia) the C-peptide is reduced! After taking sulfonyl-ureas (e.g. in a suicidal attempt) insulin and C-peptide are increased. Detection of glibenclamide i.s. or proinsulin i.s. (high in insulinoma, normal after having taken sulfonylureas) are in this instance of diagnostic help.
Delayed hypoglycaemia can be assessed by an OGTT over 5 h.

Th.: A) Causal:
As far as possible elimination of the triggering cause, poss. preserving a blood sample for further diagnostic evaluation

B) Symptomatic:
Mild hypoglycaemia (still conscious): 5 - 20 g of glucose = dextrose = grape sugar (poss. as well saccharose = cane and beet sugar) orally. Oligosaccharide drink (fruit juices, cola) are also suitable, as long as no therapy with acarbose (α-glucosidase inhibitor) took place.

<u>Severe hypoglycaemia:</u> 25 - 100 ml of 40 % glucose i.v., poss. repeated after 20 min. or subsequently 5 % glucose per infusion (until blood glucose circa 200 mg/dl).
<u>Glucagon:</u>
If no venous access possible, the patient is aggressive or first aid has been unprofessional: 1 mg of glucagon i.m. or s.c. (e.g. GlucaGen Hypokit®): Increase of endogenous glucose production. Glucagon has no effect if glycogen reserve is exhausted.
After regaining consciousness immediately continue administration of glucose orally or i.v. while monitoring the blood glucose.

Therapy of reactive hypoglycaemias in vegetative instability: Diet poor in carbohydrates, rich in fat and proteins in form of many <u>small</u> meals, application of parasympatholytics.

Therapy of dumping-syndrome: See chapter

Therapy of insulinoma: See there

Pro.: Education of diabetics to enable them to be aware of early symptoms of a hypoglycaemia (increase of the "hypoglycaemia awareness").

THYROID GLAND

Internet infos: *www.schilddruesenliga.de; www.thyroidmanager.org; www.infoline-schilddruese.de; www.schilddruese.de; www.forum-schilddruese.de*

PPh.: Daily iodine turnover (= demand): 200 µg iodine. Germany is an iodine deficiency area (average iodine supply in Germany: < 100 µg/d)   supplementation recommended for all pregnant women with 200 µg of iodine/d. This supply can not be guaranteed in a typical diet, which in Germany is not supplemented with iodine! Iodized table salt would only ensure a sufficient supply of iodine, if also the industrial salt would be iodized.
Synthesis of the thyroid hormones:

1. Iodination:
   Active transport of iodide from the blood into thyroid gland cells mediated through the sodium-iodine symporter (NIS). Oxidation from J⁻ to J2.

2. Iodisation:
   Iodisation of tyrosine to 3-monoiodotyrosine (MIT) and MIT to 3,5-diiodotyrosine (DIT).

3. Coupling:
   From each one molecule of MIT and molecule DIT, L-triiodothyronine (T3) is formed and from two molecules DIT L-tetraiodothyronine = L-thyroxine (T4) is formed.

4. Storage:
   T3 and T4 are synthesized and stored within thyroglobulin (Tg).

5. Hormone incretion:
   After proteolysis of thyroglobulin, T3 and T4 are released into the blood where they are binded mainly to transport proteins: TBG (thyroxine binding globulin), TBPA (thyroxine binding prealbumin = transthyretine) and albumin. The ratio of free to protein binded hormone is less than 1 : 1.000. Only free hormone is biologically active. Conversion from T4 to T3 takes place extrathyroidally. The T3 is synthesized to approx. 80 % from peripherally monodeiodized T4 (at the same time the same amount of hormonally inactive rT3 = reverse T3 is formed).

   Biological half-life: • T3: circa 19 h.
                                    • T4: circa 190 h.

Ef.: Effects of the thryroid hormones:
   • Increase of basal and total metabolism
   • Stimulation of growth and development (in antenatal thyroid hormone deficiency there is impairment of brain maturation, delay of bone growth and of closure of epiphyses).
   • Effect on the nervous system:
     Hypothyroidism: Apathy
     Hyperthyroidism: Hyperexcitability
   • Effect on the muscle:
     Hypothyroidism: Decreased tendon reflexes
     Hyperthyroidism: Poss. myopathy
   • Stimulation of calcium and phosphate turnover
   • Inhibition of glycogen and protein synthesis
   • Increase of cardial sensitiveness for catecholamines: ➜ Tachycardia can occur in hyperthyroidism

Feedback system of the thyroid gland:
TRH (thyrotropin releasing hormone) is released from the hypothalamus which leads to synthesis and release of TSH (thyroid-stimulating hormone) from the anterior lobe of the hypophysis. The effect of TSH is due to a stimulation of the adenylate cyclase in the membrane of the thyrocytes. TSH increases enteral iodine resorption, thyroid hormone production and emptying of T3 and T4 from the thyroglubulin depots. Persistent release of TSH leads to a hypertrophy of the thyroid gland.
The blood level of circulating, free and therefore not protein binded hormones (FT3/FT4) depend on a feed-back system. Falling levels stimulate the production of thyroid hormones by the thyroid gland via higher centres, whereas when hormone levels are high the central stimulus decreases (TSH/TRH) and the hormone production is reduced (negative feed-back).

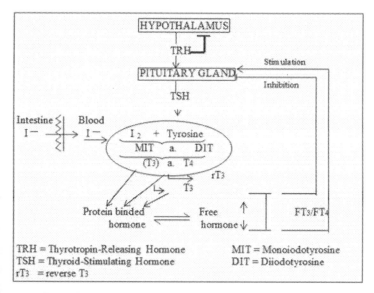

TRH = Thyrotropin-Releasing Hormone   MIT = Monoiodotyrosine
TSH = Thyroid-Stimulating Hormone   DIT = Diiodotyrosine
rT3 = reverse T3

DIAGNOSTIC

- History/clinical picture:
  - Palpation of the thyroid gland: Enlargement (= goiter)?, consistency?, tenderness?, node? thrill? murmur?
  - Pulse, ocular findings etc
- In vitro and in vivo diagnostic and additional examinations

IN VITRO DIAGNOSTIC

1. Thyroid-stimulating hormone (TSH)
Due to the negative feed-back mechanism between the blood level of thyroid hormones and the secretion of TSH the determination of basal TSH is one of the most sensitive parameters for the assessment of the thyroid function; basal TSH is usually sufficient as a screening test for the exclusion of a thyroid function disorder.
Normal range: 0,4 – 2,5 mU/l = euthyroidism; 2,5 - 4,0 mU/l border range
TSH ↑ : Primary (thyrogenic) hypothyroidism (or extremely rare: secondary hypophyseal hyperthyroidism)
TSH ↓ : Found in hyperthyroidism, autonomous thyroid gland, thyroxine therapy and rarely in hypophyseal hypothyroidism

2. Thyroid gland hormones (note reference range of each examining laboratory)
Difficulties in interpretation arise with the measurement of the total hormone concentration in the serum (= protein binded + free hormones), because changes of the protein binding relations influence the total hormone determination. These can be measured by determining the thyroxine binding globulin (TBG), but measurement of the free hormone concentration is more secure.

- **Free triiodothyronine (FT3)**
  Reference range: 2,2 - 5,5 pg/ml in serum

- **Free thyroxine (FT4)**
  Reference range: 0,6 - 1,8 ng/dl in serum

3) Determination of iodine in the urine → Ind.:
1. Clarification whether a hyperthyroidism was induced by contamination with iodine.
2. Epidemiologic statement about the iodine supply of a population: If sufficient iodine supply (200 µg/d) at least 140 µg/d is excreted in the urine.

4) Thyroid gland autoantibodies
- **Thyroglobulin antibody (TgAB or Anti-TG)**
  Increased titres in 70 % cases of autoimmune thyroiditis Hashimoto; in immunogenic hyperthyroidism (Basedow type), endocrine orbitopathy and sometimes in healthy individuals.

- **Antibodies against thyroidal peroxidase (anti-TPO-AB) = microsomal AB (MAB)**
  Interpretation of increased titres similar to TgAB (increased titres in 90 % cases of Hashimoto disease)
- **TSH receptor autoantibodies (= TRAB)**
  are found in 80 % of cases in immunogenic hyperthyroidism (Basedow type)
  The bioluminescence assay for TRAB is called TSAB (thyroid stimulating antibodies) and has a sensitivity of 98 %.

5) Tumor markers
- **Serum thyroglobulin (Tg)**
  Reference range: up to 50 ng/ml
  Thyroglobulin does not only exist in the follicles of the thyroid gland, but also in traces in the serum of healthy individuals. Tg is below the detectable limit (< 1 ng/ml) in:
  - Agenesis of the thyroid gland
  - Hyperthyroidism factitia
  - Following thyroidectomy of a differenciated carcinoma of the thyroid   An increase of Tg points to metastases.
- **Serum calcitonin**
  Medullary thyroid carcinoma (= C-cell carcinoma) produces increased calcitonin, which can be detected in the serum.

IN VIVO DIAGNOSTIC

1) Ultrasonography:
- Position, shape, size of the thyroid;
  Volume of one thyroid lobe = length x width x depth x 0,5
  Upper reference limit of the total thyroid volume: 18 ml (female) and 25 ml (male)
- Echo structure: Anechoic cysts; hypoechoic or hyperechoic thyroid nodule: Frequency circa 20 % in Germany (f > m; increase with age). Smaller nodules < 1 cm $\varnothing$, which are not palpable, are usually monitored with ultrasound + TSH determination. Nodules > 1 cm $\varnothing$ need further investigation (scintigraphy → in cold nodules: Aspiration cytology). Calcifications can point to carcinoma. But: The sonography just indicates a pfobable benign diagnosis of a defined finding. Histology alone (with restriction cytology) is needed for a definite diagnosis.
- Relation of the thyroid to neighbouring organs (trachea etc), circulation (colour doppler)
- Detection of suppressed thyroid areas, which are not visible in the scintigraphy.

2) Scintigraphy:
Standard examination is the quantitative scintigram of the thyroid with ^{99m}technetium-pertechnetate (^{99m}Tc04, a gamma radiation source with T50 of 6 h) and determination of the radionuclide uptake into the thyroid: TcTU = Technetium Thyroidal Uptake.
An examination with iodine radioisotopes is performed only for specific indications: Dose calculation before radioiodine therapy, carcinoma of the thyroid, detection of ectopic parts of the thyroid. 2 isotopes are in use:
- ^{123}J : Gamma emitter; T50 = 13,3 h
- ^{131}J : Gamma/beta emitter; T50 = 8 days (rel. high exposure to radiation)
Scintiscanning of the thyroid is not necessary in younger patients with diffuse struma, homogeneous echo pattern and normal thyroid function parameters (FT3/4, basal TSH).
Scintigraphic validity:
- Position, shape and size of thyroid, detection of ectopic thyroid tissue (e.g. lingual goiter)
- Functional topography: Presentation of thyroid gland areas with increased or reduced metabolic activity, which are distinguished by the extent of the nuclide accumulation:
  - Cold nodule: Does not or hardly accumulate.
  - Warm nodule: Accumulates slightly more than the rest of the thyroid tissue.
  - Hot nodule: Accumulates intensively, while the residual thyroid tissue accumulates less or not at all. This could be the finding of an unifocal autonomy (synonym: autonomous adenoma); for details see hyperthyroidism.

Remember: Focal findings in scintiscanning must be correlated with sonographic findings.

A cold nodule which appears anechoic in sonography corresponds with a cyst.
A cold nodule which is not anechoic is suspicious of a carcinoma (frequency of carcinoma circa 4 %) and therefore requires definite clarification:
Repeated fine needle aspiration with cytology, Tc-99m-MIBI scintigraphy (accumulation points to carcinoma), if the slightest suspicion of carcinoma or diagnostic uncertainty: Surgery with histologic clarification.
Risk criteria for cold nodules:
- Exposition to radiation in the past
- Sex (4fold increased probability for malignancy in males)
- Juvenile age
- Local findings: Solitary nodule, rapid development, not moveable

3) Suppression test
Ind: Detection of autonomous thyroid gland

The amount of autonomous thyroid tissue is measured by the determination of the thyroidal ^{99m}Tc-pertechnetate uptake = TcTU under conditions of suppression, which means after taking LT4 in suppressive dose (e.g. 150 µg LT4/d over 2 weeks). In a TcTUsupp > 1,5 % it has to be expected, that exogenic iodine supply triggers a hyperthyroidism!

Through comparison of basic scintigram (without taking thyroid hormone) and suppression scintigram (after taking thyroid hormone) autonomous thyroid areas can be detected, which became independent from the regulating control of TSH and therefore present themselves isolated in the suppression scintigram.

4) Fine needle aspiration biopsy
Simple and safe method. Only contraindication: Haemorrhagic diathesis.
Main indication: Selection of cold and hypoechoic/anechoic nodules, which require surgical clarification. In circa 30 % of the biopsies however no sufficient cytologic assessment is possible (research from the Mayo clinic).

Additional examinations:
• Radiography
Trachea target view and oesophageal barium swallow document displacement and stricture in a large (retrosternal) goiter. Carefully compare radiologic space-occupying findings with scintigram (congruency ?).
Valsalva manoeuvre during transillumination: Instability of the tracheal wall (tracheomalacia) in a large goiter?
Chest x-ray in two planes: Detection of intrathoracic struma and metastases (follow-up of thyroid gland carcinoma).
CT of thyroid carcinomas (expansion? relapse? infiltration? lymphomas? Non^{131}I storing tumour tissue?) - Beware iodine containing contrast medium!

• Ultrasonography and MRI of the orbitae: In unilateral exophthalmus a tumour must be excluded in the differential diagnosis.

Remember: If even slight suspicion of functional thyroid disorder do not use contrast medium before clarification of thyroid function. Reason: After exogenous iodine thyroid scanning is impossible for a long period and a latent hyperthyroidism (in case of a thyroid autonomy) may become manifest!

• Whole body bone scintigram with ^{99m}Tc-MDP
Search for bone metastases (follow-up) in thyroid gland carcinoma. Only 60 % of the bone metastases of differenciated thyroid gland carcinomas show a pathologic increase of activity.

• Tc-99m-MIBI scintigram
Tc-99m-MIBI accumulates (non-specific) in malignant thyroid tissue and can detect metastases, which do not store ^{131}I. Should accumulation be found in a cold nodule, suspect thyroid gland carcinoma. During follow-up of patients with thyroid gland carcinoma no discontinuation of thyroid hormones is necessary.

• PET: Search for thyroid gland carcinoma metastases

EUTHYROID GOITER [E04.9]

Def.: Enlargement of the thyroid gland with normal hormone production; noninflammatory, nonmalignant.

Ep.: More than 90 % of all thyroid disorders are euthyroid goiters, most common endicrine disorder: In iodine deficient areas like Germany circa 30 % of the adults! f : m = 1 : 1

Aet.: a) Endemic: Circa 30 % of the German population
    Genetic defect of follicle epithelial cells + iodine deficiency as a factor of manifestation
    b) Sporadically: f : m = 4 : 1; endocrine strains with increased thyroid hormone requirement (pubercy, pregnancy, climacterium), lithium and other noxious substances which can lead to goiter
    Seldom: Pendred syndrome (bilateral sensorineural hearing impairment + eu- or hypothyroid goiter, autosomal recessive inheritance, mutation of the SLC26A4/PDS gene

Pg.: • Intrathyroidal iodine deficiency is the essential factor in the pathogenesis of the endemic goiter! It causes activation of local intrathyroidal growth factors: e.g. "epidermal growth factor" (EGF), "insulin-like growth factor I" (IGF I) etc → resulting in hyperplasia of the thyrocytes.
    • Thyroid hormone deficiency → TSH secretion → hypertrophy of the thyrocytes
    Iodide therapeutically inhibits cell hyperplasia and LT4 inhibits cell hypertrophy (indirect via inhibition of TSH secretion).

Morphogenesis: Hyperplastic diffuse goiter → colloid goiter → nodular goiter

Cl.: Staging according to WHO classification:
    Grade 0: Mild thyroid enlargement but the enlargement is neither palpable nor visible. Grade 0 can only be detected by ultrasonography: In females a goiter is diagnosed if thyroid volume of > 18 ml, in males if > 25 ml.
    Grade 1: Enlargement is palpable, but not visible on inspection of the neck.
    Grade 2: Enlargement of the gland visible and palpable.
    Grade 3: Goiter is visible from behind the patient.

1. Tracheal complications: 3 degrees:
   - Displacement of trachea without narrowing
   - Signs of compression → poss. stridor, engorgement of the upper blood vessels
   - Tracheomalacia ('scabbard' trachea)
2. Development of autonomous thyroid gland:
   Depending on goiter age, size and nodular transformation the iodine deficiency goiter shows an increasing tendency of developing a TSH independent functional autonomy. Older patients with large nodular goiter show in > 50 % a functional autonomy! In this case a latent hyperthyroidism can be present, in which the autonomous thyroid mass exceeds a critical limit (TcTUsupp > 1,5 %) while the thyroid hormone concentration in the blood is still normal. Increased iodine supply through iodic drugs or contrast medium can trigger a hyperthyroidism in these cases! (for further details see chap. hyperthyroidism)
3. Development of cold nodules (risk of carcinoma 4 %)

DD: 1. In retrosternal goiter (as the most frequent cause of a widening of the upper mediastinum):
   Other causes of a widening of the upper mediastinum (bronchial carcinoma, malignant lymphomas, teratoma, thymoma, aortic aneurysma etc) → scintiscan of the thyroid to project onto the x-ray; CT, MRI.
2. Carcinoma of the thyroid gland:
   Nodular thyroid changes are common in iodine deficient areas such as Germany, but only 4 % of cold thyroid nodules are a carcinoma.
   Risk factors for carcinoma of the thyroid are history of exposure to radiation (esp. of head and neck region), male sex, < 20 and > 60 years of age, positive family history (of carcinomas of the thyroid), scintigraphically cold nodules → clarification through fine needle aspiration cytology, completed through immunocytology in follicular neoplasias (with monoclonal AB against thyroperoxidase, galectin 3 and CD44v6).

   Remember: Cold solitary nodules in the scintiscan (85 % of all nodules of the thyroid), which are not anechoic in the ultrasonography, must be investigated with fine needle aspiration cytology!

Di.: 1. Basic diagnostic tests:
   - TSH basal (= screening test): Normal values in euthyroid patients
   - Ultrasonography of the thyroid
2. Additional diagnostic tests if conspicuous basic results:
   - FT3 and FT4: Normal
   - Calcitonin in hypoechoic nodules
   - Scintigraphy of the thyroid
   - Poss. chest x-ray (retrosternal goiter ?)
   - Fine needle aspiration cytology in cold nodules:
     A negative finding does not exclude a carcinoma, since small carcinomas < 1 cm $\varnothing$ are often not hit in the biopsy. The findings of the cytology should be divided in:
     - Negative (no suspicion of malignancy)
     - To be clarified further histologically. (The differential diagnosis of follicular adenoma or carcinoma can not be clarified via cytology !)
     - Positive (= malignancy)
   - Diagnosis of a latent hyperthyroidism:
     - Clinically mostly euthyroid
     - FT3 and FT4 normal
     - TSH basal ↓
     - In the scintigram autonomous areas become visible.

Th.: **a) Conservative:**
   1. Iodide substitution:
      Ind: Therapy of choice in euthyroid goiter without autonomy
      Eff.: Elimination of the intrathyroidal iodine deficiency → regression of hyperplasia of the thyrocytes (reduction effect as seen in LT4 substitution: Circa 30 - 40 % of the original volume).
      CI: Thyroid gland autonomy with latent hyperthyroidism, iodine allergy
      As the proportion of autonomous thyroid parts increases in older patients with long standing goiter, autonomous thyroid must be excluded via suppression test before the start of a iodide therapy.
      Preparations: e.g. Jodetten® Henning
      Dose: Adults 200 µg/d (children 100 µg/d)
   2. Combination therapy with iodide + LT4:
      According to evidence long term monotherapy with LT4 of the euthyroid goiter is not indicated. A combination of LT4 + iodide is well justified pathogenetically, in gravidas it is the therapy of choice.
      Eff.: Administering thyroid hormones reduces TSH production, which is the reason for the hypertrophy (reduction effect circa 30 - 40 % of the original volume). After stopping LT4 however the thyroid volume increases again.
      - LT4 preparations: e.g. L-thyroxine Henning®, Euthyrox®
        Dose: Slowly increasing dose: Initially 50 µg LT4/d; at intervals of 1 - 2 weeks increase dose to 75 and finally to 100 µg LT4/d = optimum dose for 75 % of the patients.

- Combination preparations with iodide + LT4: e.g. Thyronajod® Henning
  Dose: As in single components

Regular control examinations:
- The individually correct dose is determined by basal TSH and control of the thyroid hormones: A low normal TSH (0,5 - 0,8 µU/ml) with normal FT3 and FT4 values is targeted.
- Monitoring of neck circumference, palpatory findings and ultrasonography of the thyroid, weight and questions about the state of health.
  In too high dose: Hyperthyroidal symptoms (= hyperthyroidism factitia).
  In too low dose: "Therapy resistent" goiter and time delay.
  SE: of a LT4 therapy:
  - Hyperthyroidism factitia in excessive doses, osteoporosis in long-term therapy with TSH suppressing doses
  - Interaction with other drugs:
    LT4 decreases the effect of insulin
    LT4 increases the effect of anticoagulants
  CI: Fresh myocardial infarction, angina pectoris, acute myocarditis, adrenal insufficiency (untreated). Eliminate cardiac arrhythmias and cardiac insufficiency before LT4 therapy. Goiters with autonomy.

b) Surgical therapy:
Differs according to findings from partial resection to total resection
Ind: Large nodular goiters, esp. If the neck organs are impaired; goiters with autonomy. In cold nodules with the slightest suspicion of malignancy a hemithyroidectomy is performed.
Co.: - Recurrent nerve palsy: Lowest risk (< 1 %) with intraoperative neuromonitoring of the n. recurrens. Laryngoscopy before and after surgery.
- Parathyroid tetany (< 1 %)
Postoperative relapse prophylaxis depending on the residual volume of the thyroid:
> 10 ml: Only iodide (100 - 200 µg/d)
3 - 10 ml: Iodide + LT4 (75 - 125 µg/d)
< 3 ml: Only LT4 (see above)
Monitoring of therapy: To keep TSH basal in low normal range with normal FT3, FT4.

c) Radioiodine therapy:
Reduction effect up to 50 % of original volume
Ind: Recurrence of goiter, increased surgical risk, if surgery is refused, goiters in older age, multifocal autonomy of the thyroid gland
CI: Age of growth, pregnancy, suspicion of malignancy
In careful dosage a hypothyroidism as a late result can be observed in < 10 % of cases; this can be treated easily.

Pro.: Supplementing the population with iodine: In Switzerland the frequency of goiter could be reduced from 30 % to 3 % within 60 years through iodizing of the table salt.

Remember: Goiter prophylaxis with iodide is the best way to reduce the frequency of functional autonomy and the iodine induced hyperthyroidism!

Goiter prophylaxis of all pregnant women with iodide (200 µg/d): In the 12th week of pregnancy the fetal thyroid gland begins hormone synthesis and requires this amount of iodine!

Notice: Use of iodized table salt only in private households and salt-water fish consumption are mostly insufficient in Germany to cover this iodine demand and therefore are no replacement for a iodide substitution!

HYPOTHYROIDISM   [E03.9]

Congenital hypothyroidism:

Occ.: 1 in 5.000 newborns

Aet.: Athyroidism, thyroid gland dysplasia or ectopy; rarer: Defect in hormone biosynthesis or incretion; extremely rare: Hormone resistence (T3 receptor defect).

CI.:   At birth:
- icterus neonatorum prolongatus   - hypokinesia
- poor drinking   - poss. reduced muscle reflexes
- constipation   - poss. umbilical hernia
Later:
- growth underdevelopment (height)
- underdevelopment in maturation (age of bone and teeth)
- mental and psychological retardation, low intelligence

- hearing difficulties, speech disturbance

The untreated full picture (cretinism) is extremely rare in areas with good medical care.

Di.: Early diagnosis determines the prognosis! By law prescribed screening for hypothyroidism of the newborns: On the 3rd day of life 1 - 2 drops of blood from the heel are dropped on a filter paper: TSH determination. In congenital hypothyroidism: Increased basal TSH.

Th.: Life-long substitution therapy with T4 as early as possible! Regular controls of the hormone status.

Remember: The microsomia can still be influenced in delayed T4 substitution; the cerebral damages however are irreversible! If a congenital hypothyroidism is treated 3 weeks too late, A-levels in later life will no longer be possible.

Acquired hypothyroidism

Aet.: 1. Primary (thyrogenic) hypothyroidism:
- - Result of autoimmune disorder (Hashimoto thyroiditis), sometimes in connection with a polyglandular autoimmune syndrome (see there)
- - Iatrogenic: After thyroidectomy, after radioiodine therapy, drug related (e.g. thyrostatic drugs, lithium, sunitinib, amiodarone etc)

Remember: Amiodarone can induce hypo- but also hyperthyroidism.

2. Secondary hypophyseal hypothyroidism: (very rare)
Simmonds´syndrome (see chap. hypopituitarism)
3. Tertiary hypothalamic hypothyroidism (very rare)

Cl.: - Physical and mental decline in performance, poor motivation, tiredness, slowing down, disinterest (facial expression !), prolonged Achilles tendon reflex time
- Increased sensitiveness to cold
- Skin: Dry, cool, pasty, pale-yellow, scaly
- Poss. weight gain due to generalized myxedema
- Dry, brittle hair
- Constipation
- Raw, hoarse voice (incorrect diagnosis: disorders of the larynx)
- "Myxedema heart": Bradycardia, cardiac enlargement with poss. cardiac insufficiency resistent to digitalis therapy, low voltage ECG
- Early arteriosclerosis due to hypercholesterinaemia
- Poss. myopathy with CK elevation
- Poss. disturbed menstrual cycles, defects in spermatogenesis, infertility, increased abortions

Hypothyroidism in older individuals: Often oligosymptomatic or uncharacteristic: Cold intolerance; motor and mental retardation, which does not correspond to age; disturbances of memory, depression (consider age-linked hypothyroidism and check TSH!); poss. periocular mucinous oedemas.

Incorrect diagnoses in age-linked hypothyroidism: "early aged - sclerotic - depressive - immobile - apathetic"

Myxedema coma: Nowadays extremely rare, high mortality; manifestation factors: Infections, operations, traumas etc.

Cardinal symptoms:
1. Hypothermia (rectal temperature often not measurable!)
2. Hypoventilation with hypoxia/hypercapnia and poss. CO_2 anaesthesia
3. Bradycardia and hypotonia
4. Myxedematous aspect (often women of higher age)

Di.: • Latent (subclinical) hypothyroidism:
- - FT3 and FT4: Normal
- - TSH basal: Depending on the TSH value 3 degrees of severity:
  Grade 1: TSH 2,6 – 4,0 mU/l – Grade 2: TSH 4,1 – 10 mU/l – Grade 3: TSH > 10 mIU/l
• Manifest hypothyroidism:
Clinical picture + laboratory:

DD	FT$_4$	TSH basal	
Primary = thyrogenic hypothyroidism (mostly)		↑	Goiter: +/-
Secondary = hypophyseal hypothyroidism (very rare)	↓	↓	Goiter: never Gonadotropins ↓ poss. ACTH ↓

In secondary hypophyseal hypothyroidism additional diagnostic tests (see chap. hypopituitarism)

- In Hashimoto thyroiditis
  Detection of AB against thyroidal peroxidase (TPO-AB) in 95 % and thyroglobulin AB in 70 % of cases
- Ultrasonography and poss. fine needle biopsy (e.g. in lymphocytic Hashimoto thyroiditis)
- Scintigram:
  Heavily reduced or absent storage of radionuclide in the thyroid.

DD: **Low T3-/Low T4 syndrome**
In seriously ill patients on ICU FT3 and FT4 can be decreased. Unlike in hypothyroidism the concentration of reverse T3 (r-T3) is elevated in the low T3 syndrome. The patients are considered as euthyroid, a substitution therapy is refused in most cases.

Th.: • Manifest hypothyroidism:
Permanent substitution with LT4 + life-long control examinations. The more distinct the hypothyroidism, the lower and slowlier the substitution therapy must be started! Danger: Angina pectoris attacks, cardiac arrhythmias!
Initially: 25 - 50 µg LT4/d; monthly dose increase by 25 µg LT4/d; maintenance dose: 1,5 - 2,0 µg per kg of body weight/d.
The individual optimum LT4 dose is established on the basis on the well-being of the patient and the basal TSH, which should be between 0,5 – 2,0 mU/l (normalizing of TSH lasts 6 – 8 weeks).

• Latent (subclinical) hypothyroidism:
Should be treated with LT4 because of increased risk of early arteriosclerosis and poss. cause of infertility.

• Myxedema coma: Intensive care unit!
- Ventilatory support, securing vital functions
- Supply of glucocorticosteroids, glucose infusion, regulation of electrolyte and water balance (often hyponatraemia)
- Thyroxine i.v. (initially 100 - 200 µg)
- Poss. slow rewarming

HYPERTHYROIDISM [E.05.9]

Ep.: Incidence of the autoimmune hyperthyroidism (Basedow disease): circa 40/100.000/year

Aet.: 1. Immunogenic hyperthyroidism (Basedow disease, Graves disease [E05.0])
2/3 of cases manifest after the 35th year of life; f : m = 5 : 1
a) Hyperthyroidism without goiter
b) Hyperthyroidism with diffuse goiter
c) Hyperthyroidism with nodular goiter
2. Hyperthyroidism with autonomy of the thyroid gland
The majority of cases manifest in older age !
According to the distribution of the autonomous thyroid tissue in the scintiscan three forms are distinguished:
- Unifocal autonomy (synonym: autonomous adenoma)[E05.1]
Causes of unifocal thyroid autonomy are constitutive activating mutations in the gene of the TSH receptor (80 %) or of the Gs-α-protein (up to 35 %).
- Multifocal autonomy [E05.2]
- Disseminated autonomy [E05.0]

Note: Marine-Lenhart syndrome = combination of Basedow disease + autonomy of the thyroid (Occ.: Up to 10 % of patients with Basedow disease in iodine deficient areas)

3. Rarer forms of hyperthyroidism:
- Temporary in subacute thyroiditis
- In carcinoma of the thyroid
- Iatrogenic: 1. Exogenous supply of thyroid hormones (hyperthyroidism factitia [E05.4])
2. Amiodarone induced hyperthyroidism
- Very rare central hyperthyroidism, e.g. TSH overproduction due to pituitary adenoma
- Very rare paraneoplastic TSH production

Pg.: to 1.: **Immunogenic hyperthyroidism = immunehyperthyroidism** (Basedow disease):
Genetic disposition (familial accumulation, increased occurrence of HLA-DQA1*0501 and -DR3) + unknown triggering agent (infections ?).
The hyperthyroidism is caused by TSH receptor autoantibodies (TSH-R-AB = TRAB), which have a thyroid stimulating effect.

to 2.: **Thyroidal autonomy**:
Most common cause of autonomy of the thyroid are goiters due to iodine deficiency.
In every normal thyroid autonomous areas exist, which resist the feed-back regulation of the hypothalamus/hypophysis = physiologic, basal autonomy (therefore there can never be a complete suppression in

the suppression test). A latent hyperthyroidism is present, if in euthyroid patients the autonomous thyroid mass exceeds a critical limit (TcTUsupp > 1,5 - 3 %).

The amount of the autonomously produced thyroid hormones depends on two factors:
- Mass of the autonomous thyroid tissue
- Amount of iodine supply

In iodine deficient areas the autonomous thyroid portion can become relatively large without exceedind the euthyroidism. Exogenous iodine supply (e.g. iodine containing contrast medium and drugs such as amiodarone) however then triggers a hyperthyroidism. This explains why in iodine deficient areas (e.g. Germany) circa 80 % of the nonimmunogenic hyperthyroidisms develop due to exogenic iodine supply! Nutritional iodine (iodine in table salt, salt-water fish) is usually not the triggering cause of a hyperthyroidism.

Cl.:   1. of the hyperthyroidism:
- Goiter (70 - 90 % of the patients); with high vascularization of the struma an auscultatoric thrill can be heard above the thyroid gland.
- Psychomotoric agitation: Fine tremor [R25.1] of extended fingers, increased nervousness, irritability, sleeplessness
- Sinus tachycardia, poss. arrhythmias (extrasystoles, atrial fibrillation), increased blood pressure amplitude, (systolic) hypertonus
- Weight loss (despite ravenous hunger), poss. hyperglycaemia (due to increased metabolism with mobilizing of the fat and glycogen depot)
  DD: Untreated diabetes mellitus
- Warm moist skin, soft thin hair
- Intolerance to warmth, sweats, poss. subfebrile temperatures
- Increased bowel frequency, poss. diarrhoeas (constipation however does not exclude hyperthyroidism!)
- Myopathy: Weakness of thigh muscles, asthenia
- Poss. osteoporosis due to negative calcium balance: In 15 - 20 % of the cases hypercalcaemia, hypercalciuria, increased alkaline phosphatase
- Pathologic glucose tolerance (50 %)
- Poss. fatty liver
- Poss. disturbance of the menstrual cycle, infertility (less common than in hypothyroidism)

2. Additional symptoms in immunogenic hyperthyroidism (Basedow disease):
- Endocrine ophthalmopathy/orbitopathy in circa 60 % of cases (for details see there)
- Basedow´s triad (50 % of cases): Goiter, exophthalmus, tachycardia
- Pretibial myxedema is rare (< 5 % of cases): Nonpitting oedema.
  Similar to in endocrine orbitopathy there are deposits of glycosaminoglycans (GAG) in the subcutaneous pretibial tissue, rarely as well in the forearm or shoulder areas. Spontaneous regression is possible.
- Seldom acropachy (digital clubbing)

Clinical picture of specific forms of hyperthyroidism:

• Age-linked hyperthyroidism (above 60 years): "Masked" mono- or oligosymptomatic hyperthyroidism:
- Weight loss, loss of strength (incorrect diagnosis: "Tumor cachexia")
  Caution in this case with a rushed "tumour search" with iodous contrast medium before measuring of thyroid function. If a hyperthyroid patient is given iodine containing contrast medium, he is brought into a thyrotoxic crisis, which is a life-threatening critical condition!
- Depression of "older age"
- Heart insufficiency ("high-output failure")
- Arrhythmias (extrasystoles, atrial fibrillation)

• Thyrotoxic crisis/coma [E05.5]
  Spontaneous in hyperthyroidism; often after iodine uptake (contrast medium, drugs) in patients with autonomy of the thyroid gland, after stopping an antthyroid therapy; after thyroidectomy, if surgery did not take place during an euthyroid state. Operations or additional severe secondary diseases during florid hyperthyroidism.

  3 Stages (according to Hermann):

  St. I: - Tachycardia (> 150/min.) or tachyarrhythmia e.g. atrial fibrillation
             - Fever up to 41 °C, sweating, dehydration
             - Psychomotoric agitation, tremor, anxiety
             - Vomiting, diarrhoeas
             - Muscle weakness, asthenia
  St. II: Additionally disturbance of consciousness, somnolence, psychotic states, disorientation
  St. III: Additionally coma with poss. adrenocortical insufficiency and circulatory collapse.

DD: Psychosis, other febrile illness e.g. infections, cocaine or amphetamine abuse, tachycardia of other origin; subacute thyroiditis (BSG ↑) → always check TSH !
Vegetative dystonia: The hyperthyroid patient often has increased bowel frequency, warm hands and does not talk about his complaints. The patient with vegetative dystonia has the tendency to constipation, often cold hands and emphasizes his complaints.

226

Hyperhidrosis (excessive sweating – in contrast to the physiological sweating during physical strain, fever, heat, stress)
　　1. Secondary hyperhidrosis:
　　　- Endocrine causes: Menopause, pregnancy, hyperthyroidism, phaeochromocytoma, carcinoid, diabetes mellitus, male hypogonadism
　　　- Neurological disorders, malignancies
　　　- Drugs (opioids, neuroleptics, drugs for Parkinson´s disease etc), vitamin D deficiency
　　2. Primary hyperhidrosis (genetic, psychological):
　　　Typical distribution pattern (palms, feet, axillae, face, scalp); triggering cause: emotional stress.
　　　Therapy options in primary hyperhidrosis:
　　　- Autogenic training, bio-feedback, psychotherapy
　　　- Sage preparations, clonidine, Bornaprin (Sormodren®)
　　　- Aluminium chloride containing antiperspirants (e.g. as roll-on applicator)
　　　- Iontophoresis, botulinum toxin injections
　　　- Ultima ratio: Poss. endoscopic thoracic sympathectomy (ETS)

Di.: ▶ Diagnosis of a manifest hyperthyroidism
　　1. History (iodous drugs, topical agents such as povidone-iodine, contrast media)

　　2. Clinical picture (symptoms of hyperthyroidism)

　　3. Laboratory:
　　　• Basal TSH reduced (= screening test)
　　　• FT3 nearly always elevated
　　　• FT4 elevated in 90 %
　　　• In immunogenic hyperthyroidism detection of TSH receptor autoantibodies (= TRAB) in the new bioassay in > 95 % and anti-TPO-AB (70 %).
　　　• Detection of iodine in the urine if iodine contamination triggered off hyperthyroidism

> ***Remember:*** In reduced basal TSH a FT4 determination alone is not sufficient, because there are isolated T3 hyperthyroidisms (e.g. in the early stage of a hyperthyroidism).
> In the differential diagnosis the possibility of extreme iodine deficiency must be taken into account: FT3 ↑, FT4 ↓, normal TSH basal = euthyroid functional status. Thyrostatic therapy contraindicated.

> ***Note:*** Diagnosis of rare central hyperthyroidism (e.g. due to a TSH producing pituitary adenoma): Thyroid hormones elevated + TSH basal not suppressed, but poss. even increased! → DD: The same results are found in the very rare thyroid hormone resistance: Congenital defect of the β-thyroid hormone receptor, 1 : 50.000.

　　4. Imaging procedures:
　　　• Ultrasonography: Defined or diffuse hypoechoic picture + hypervascularization in the colour duplex.
　　　• Scintigraphy: TcTU ↑
　　　　- Homogeneous intensive radionuclide accumulation in immunogenic hyperthyroidism
　　　　- Unifocal, multifocal or disseminated radionuclide accumulation in the 3 forms of hyperthyroid autonomy

　▶ Diagnosis of hyperthyroidism factitia (= exogenous supply of thyroid hormones):
　　a) Unintentionally in the context of a substitution therapy
　　b) Deliberate abuse of the patient, mostly for the purpose of weight reduction (e.g. anorexia nervosa)
　　　- Discrepancy between excessive elevation of FT3/FT4 and clinical condition of the patient
　　　- Crucial: Total suppressed iodine uptake (TcTU ↓) in the thyroid gland (intact control mechanism !)
　　　- TRAB and Tg not measurable.

　▶ Diagnosis of a latent hyperthyroidism: TSH basal ↓, FT3+4 normal

Th.: Therapy of hyperthyroidism
　　No causal therapy known. Therapy choice depends amongst other factors on patient´s age and the form of hyperthyroidism.
　　a) Antithyroid drugs
　　b) Surgical therapy
　　c) Radioiodine therapy

A) Antithyroid drug therapy
　　Antithyroid drugs → Ind: Every hyperthyroidism is treated with antithyroid drugs until euthyroidism is achieved.
　　▶ Sulphur containing antithyroid drugs inhibit the synthesis of MIT and DIT, but not the incretion of the already synthesized hormones (T3, T4), therefore circa 6 – 8 days of latency before the effect takes place.
　　　• Propylthiouracil (Propycil®)
　　　• Thiamazol (Favistan®, Thiamazol®)
　　　• Carbimazole
　　▶ Perchlorates inhibit uptake of iodide into the thyroid gland (e.g. Irenat®)
　　　Ind: e.g. rapid blockage of thyroid in thyroid autonomy and necessary administration of iodous contrast media.

- Allergic reactions with exanthema, fever, joint/muscle pains etc
        - Thrombo-/leucocytopenia; seldom allergic agranulocytosis (monitoring of leucocytes!)
        - Liver enzyme changes, cholestasis etc

Carbimazole is a prodrug, which is transformed into thiamazol. Dose ratio of carbimazole : thiamazol = 1,5 : 1.
Dose: Carbimazole initially 15 - 30 mg/d, maintenance dose: 2,5 - 15 mg/d

Hyperthyroidism due to thyroid gland autonomy should be treated with definite therapy after achieving an euthyroid state, since in most cases a relapse follows after stopping the thyroid antagonists → radioiodine therapy or surgery; indications see below

In immunogenic hyperthyroidism the antithyroid drug therapy is continued for circa 1 year. After stopping of the therapy a relapse occurs in circa 50 % of cases. Smoking increases the risk of relapse! The TRAB level 6 months after diagnosis/therapy of the immune hyperthyroidism is of prognostic relevance: In values > 10 IU/l a remission is unlikely and a definitive cure is necessary through radioiodine therapy or surgery. In high risk patients use definitive therapy in an early stage.

In immunogenic hyperthyroidism attention has to be paid that a hypothyroidism is avoided by all means, because in such a case a poss. existing endocrine orbitopathy would deteriorate!

Additional drug therapy: In tachycardia beta-blockers, e.g. with propranolol, which inhibits the conversion T4 → T3.

B) Surgical therapy

Always achieve euthyroidism first with antithyroid drugs! Only then operate: In Basedow disease subtotal resection of the thyroid gland (< 2 ml residual thyroid gland). Total thyroidectomy in suspicion of malignancy.

Ind: - Large goiter
        - Displacement of adjacent structure
        - Suspicion of malignancy (e.g. cold nodules)
        - Thyrotoxic crisis
CI: Small diffuse goiter; inoperability; florid hyperthyroidism (untreated)
Co.: - Postoperatively substitution required hypothyroidism (up to 100 %)
        - Recurrent nerve palsy: Lowest risk (< 1 %) in intraoperative neuromonitoring of the n. recurrens. Before and after surgery laryngoscopy!
        - Parathyroid tetany (< 1 %)
        - Mortality: very rare

C) Radioiodine therapy:

Since the radioiodine therapy takes weeks to take effect, treatment with antithyroid drugs must be done before and afterwards. ^{131}I dose: 200 - 2.000 MBq, depending on the therapy plan:

5. Function optimized dose plan with low rate of hypothyroidism
6. Ablative dose plan with periodical hypothyroidism
Lifelong LT4-Substitution is necessary in both cases.

Ind: - Immunogenic (Basedow) hyperthyroidism
        - Thyroidal autonomy
        - Relapse of hyperthyroidism after thyroidectomy
        - Smaller goiters
        - Contraindications for surgery
        - Progressive endocrine orbitopathy
CI.: - Growth age, fertile women without secure contraception for at least 6 months
        - Pregnancy and lactation period
        - Florid (untreated) hyperthyroidism
        - Suspicion of malignancy (→ surgery !)
Co.: - Temporary, harmless radiation thyroiditis
        - Hypothyroidism (see above)
        - Poss. deterioration of an endocrine orbitopathy (→ prophylactic administration of steroids)
        No genetic radiation risk known! A slightly increased risk of a myelodysplastic syndrome however is suspected.

Therapy of the latent hyperthyroidism:

- Contraindication for iodous drugs (e.g. amiodarone) and contrast media, because those trigger a manifest hyperthyroidism!
  If in exceptional circumstances an investigation with iodine containing contrast medium must be performed, a prophylactic administration of perchlorate and possibly additional thiamazol for two weeks is recommended.
- In order to protect the patient from the risk of hyperthyroidism from an uncontrolled iodine supply a relative indication for a prophylactic therapy is given: Autonomous areas can be selectively eliminated by radioiodine therapy under suppressive conditions. If in case of artrial fibrillation the causal connection with a latent hyperthyroidism is suspected, a temporary therapy trial with antithyroid drugs can be tried. If a sinus rhythm is achieved, the association is probable and radioiodine therapy is recommended.

Management of the thyrotoxic crisis: Always on ICU!

a) Causal therapy:

    1. Inhibition of hormone synthesis: Thiamazol 80 mg i.v. every 8 h. In addition potassium perchlorate is recommended (1.500 mg/d).

    2. In life-threatening iodine induced thyrotoxic crisis plasmaphoresis and nearly total thyroidectomy are the most effective causal measures.

b) Symptomatic therapy:

- Fluid, electrolyte, calory substitution parenterally (common mistake: Missing a dehydration!): 3 - 4 l of fluids/d; 3.000 Kcal/d
- Beta blockers with caution regarding to SEs + CIs
- Glucocorticosteroids are recommended because of relative adrenocortical insufficiency; furthermore they suppose to inhibit the conversion from T4 to T3.
- Physical lowering of temperature
- Poss. sedatives
- Thromboembolic prophylaxis

Mortality of the thyrotoxic crisis: > 20 %

ENDOCRINE ORBITOPATHY (EO) [E05.0]

Syn: Endocrine ophthalmopathy

Occ.: EO is associated with immunogenic hyperthyroidism (Basedow disease) in > 90 % of cases. EO is regarded as an extrathyroidal manifestation of Basedow disease (like pretibial myxedema and acropachy). > 90 % sufferers are hyperthyroid, some are euthyroid, rarely hypothyroid. EO can precede hyperthyroidism, appear at the same time or follow it. No correlation between degree of severity of the EO and current thyroid function exists.

Aet.: Unknown; probably genetically related autoimmune disorder (autoantibodies against TSH receptor = TSH-R-AB = TRAB). TSH receptors are also found in the orbital tissue. 8fold increased risk of EO in Basedow patients, who are smoker!

PPh.: In EO there is an infiltration with autoreactive T lymphocytes, fibroblast proliferation with collagen increase and deposition of glycosaminoglycanes (GAG) into periorbital tissue and lateral bulbi muscles. The results are exophthalmos and decreased motility of the bulbi with double vision.

Cl.: Eye symptoms as a result of the exophthalmos:
- Infrequent blinking (Stellwag sign)
- Visible sclera stripe above the cornea when looking straight (Dalrymple sign)
- Lag of the upper eyelid when looking downward (Graefe sign)
- Impairment of ocular convergence (Moebius sign)
- Earliest sign: (Often) Swelling of the lateral parts of the eyebrows
- Photophobia, foreign body sensation, painful pressure behind the eyes, double vision, deterioration of vision

Further:
- Symptoms of hyperthyroidism (Basedow disease) in > 90 % of cases
  But: Occurrence of the EO as well in euthyroidism!
- Rarer pretibial myxedema (dermatopathy): Large porous skin of slushy consistence, due to deposits of glycosaminoglycanes in the subcutaneous pretibial tissue, rarely also affects the forearm or shoulder area. Spontaneous remission possible (< 50 % of cases)
- Seldom acropachy (clubbing of the distal phalanges of the fingers and toes)

7 symptoms are regarded as parameters of disease activity:
- Spontaneous retrobulbar pain   - Chemosis (oedema of the conjunctiva)
- Pain during eye movements   - Redness of the eyelid
- Swelling of the caruncles   - Oedema of the eyelid
- Conjunctival injection

Severity degrees: 6 stages (according to Grußendorf and Horster):

    I   Anamnestic complaints: Foreign body sensation, lacrimation, light sensitivity, feeling of retrobulbar pressure

    II   Lid retraction and changes of the coonective tissue: Conjunctivitis, chemosis (= oedema of the conjunctiva), periorbital swelling, thickening of lacrimal glands

    III Protrusio bulbi: a) mild, b) marked, c) very distinct
         Measured by the distance between the anterior surface of the cornea and the orbital lateral edge using an ophthalmometer according to Hertel or Gwinup

    IV Eye muscle paralysis causing diplopia

    V   Corneal ulcerations due to lagophthalmos

    VI Loss of vision, possible blindness

TRAB positive. The higher the TRAB level, the more active the EO progression.

Lab.: TRAB positive. The higher the TRAB level, the more active the EO progression.
Often hyperthyroid (basal TSH ↓, FT3/4 ↑); poss. also euthyroid, seldom hypothyroid (in inhibiting TRAB)

Imagining diagnostic: U/scan, MRI, photographic documentation of the course

DD: In unilateral eye findings: Retrobulbar tumour, cavernous sinus thrombosis, abscess, mucocele etc. - exophthalmos most commonly bilateral!

Di.: Clinical picture – thyroid gland tests - ophthalmic examination

Th.: No causal therapy known. Interdisciplinary collaboration between thyroid gland specialist/endocrinologist, ophthalmologist, poss. radiotherapist and surgeon.
- Achieve euthyroid thyroid gland function (thyroid depressants), whereas hypothyroidism is to be avoided by all means (deterioration of the EO!).
- No smoking!
- Local measures: Shaded glasses, dexpanthenol eye ointment at night, sleep with elevated head end of the bed (→ reduces eyelid oedema of the eyelid in the morning); artifical tear drops if required.
- Corticosteroids
- Retrobulbar radiotherapy of the orbita missing the lenses
- Surgical decompression of the orbita: Various procedures, e.g. removal of fat tissue (method according to Prof. Olivari, Wesseling).
- In clinical trials: Octreotid therapy

Prg.: 30 % improvement, 60 % no change, 10 % deterioration

INFLAMMATION OF THE THYROID GLAND

Acute thyroiditis [E06.0]

Occ.: Very rare

Aet.: Bacterias, viruses, effects of radiation, traumata

Cl.: Acute onset with fever, local pain, poss. swelling of the regional lymphnodes

Lab.: CRP, ESR ↑, poss. leukocytosis, euthyroidism

U/scan: In colliquations poss. anechoic areas

Fine-needle aspiration biopsy: Detection of granulocytes in bacterial inflammation, poss. identification of the organism

Th.: In case of bacterial origin antibiotics, in case of an abscess pus is to be aspirated (culture, cytology), poss. Incision

Subacute granulomatous De Quervain thyroiditis (dĕ-kār-van[h]´) [E06.1]

Occ.: f : m = 5 : 1 (predominantly women in the 3. - 5. decade of life)

Aet.: Unclear, often following a viral infection of the airways; genetic disposition (increased occurrence of HLA-B 35).

Cl.: - Exhaustion and malaise, poss. fever; no swelling of lymph nodes
- Thyroid gland often tender, but not always.

Lab.: - Extreme ESR elevation ! CRP ↑; normal white blood cell count
- Thyroid function: Initially often hyperthyroid, later euthyroid, poss. temporary hypothyroidism

U/scan: Hypoechoic, partly confluent areas of the thyroid gland

Scintiscan: Strongly decreased radionuclide uptake of the thyroid (TcTU ↓) or cold nodule

Fine-needle biopsy + histology: Granulomatous thyroiditis with epitheloid and Langhans giant cells

Th.: Spontaneous cure in circa 80 %; no thyroid depressants, poss. NSAIDs; in case of localized pain corticosteroids (1 mg prednisolone/kg BW) → remission of complaints after administration of prednisolone within 24 h! If there is no clinical improvement with steroids reconsider the diagnosis!

Internet infos: *www.hashimotothyreoiditis.de*

Occ.: Most common form of thyroiditis: Prevalence 5 - 10 %, most common cause of a hypothyroidism: f : m = 9 : 1; predominantly women between 30 - 50 y.; increased association with other autoimmune disorders (see chap. "polyendocrine autoimmune syndromes")

Aet.: • Familial disposition: 50 % of the relatives of the patients also have auto-antibodies; increased association with HLA markers (HLA-DR3, -DR4, -DR5), often vitiligo, alopecia
• Hepatitis C

Cl.: Onset mostly unnoticed, the majority of patients are diagnosed in the late stage, when the lymphocytic inflammatory destruction process has resulted in a hypothyroidism.

U/scan: Inhomogenous hypoechoic sound pattern of the frequently small thyroid

Scintiscan: Reduced radionuclide uptake of the thyroid (TcTU ↓)

DD: Autoimmune thyroid gland disorders:
1. Autoimmune thyroiditis (Hashimoto): TPO-AB positive (circa 70 %)
2. Immune hyperthyroidism (Basedow disease): TRAB positive (> 95 %) and TPO-AB positive (circa 70 %)
3. Variants of the AIT (see below)

Di.: 1. Detection of increased titres of:
- anti-TPO-AB: 95 % of cases
- Thyroglobulin antibodies (TgAB): 70 % of cases
2. Fine needle biopsy + histology: Lymphocytic thyroiditis, in the late stage fibrosis/atrophy
3. Thyroid function: Development of hypothyroidism

Th.: LT4 substitution in hypothyroidism is obligatory (optimal dose determined by measuring TSH). Some authors recommend LT4 administration for euthyroid autoimmune thyroiditis (immunosuppressants and steroids are without benefit).

Variants of AIT

• „Silent" thyroiditis: Variant of AIT with mild course; poss. only temporary.
• Post-partum lymphocytic thyroiditis: [O90.5] In circa 4 % of pregnant women a temporary and usually clinically latent thyroid function disorder occurs during the post-partum period with an often positive finding for TPO-AB.
• Iatrogenic autoimmune thyroiditis
- Cytokine induced thyroiditis (due to therapy with interferon or IL-2)
- Amiodarone induced thyroiditis

Note: Chronic fibrosing thyroiditis is an extremely rare form of chronic thyroiditis = Riedel goiter [E06.5], which causes stony-hard infiltrating and immuring.

THYROID MALIGNANCIES [C73]

Ep.: Most common endocrine neoplasm. - Incidence: circa 4/100.000 diseases per year; f : m = 3 : 1 for differentiated carcinoma; balanced sex ratio for C cell and anaplastic carcinoma.

Aet.: - Genetic factors (medullary carcinoma)
- Ionizing radiation: In the survivors of the atomic bomb in Japan as well as after atomic bomb tests on the Marshall islands an increased relative risk of 1,1/Gray of thyroid radiation dose was shown. Following the Tschernobyl reactor accident circa 1.500 children in Belorussia, Ukraine and Russia developed (especially papillary) thyroid carcinoma; The incidence of thyroid carcinoma in adults also increased in these 3 countries.
- Unknown factors

Classification:
A) Thyroid carcinoma (TH-Ca.):
1. Differentiated carcinomas: a) Papillary carcinoma circa 60 %
                                   b) Follicular carcinoma circa 30 %
2. Undifferentiated (anaplastic) carcinoma circa 5 %
3. Medullary (C cell) carcinoma circa 5 %
B) Rare malignancies of the thyroid (malignant lymphoma, sarcomas etc.)
C) Metastases of extrathyroidal tumours

Differentiated thyroid carcinomas

Follicular carcinoma predominantly metastasizes haematogenic into lungs and bones, the papillary carcinoma mainly lymphogenic. Regional lymph node metastases of papillary carcinoma often present before the primary tumour, which in many cases is of a minimum size, however it can occur multicentrically (normal scintiscan of the thyroid). The primary tumour and its metastases can be treated successfully with 131J.

Ad 2.: **Anaplastic (undifferentiated) thyroid carcinomas**
do not participate in iodine turnover. Therefore radioiodine therapy impossible.

Ad 3.: **Medullary (C cell) thyroid carcinomas (MTC):**
- No participation in iodine metabolism, radioiodine therapy therefore not successfull.
- C cells produce calcitonin: Increased levels in metastases (tumour follow-up: Relapse indicator). 1/3 of patients suffer diarrhoea.
  a) Sporadic MTC (85 %): Age peak 50. - 60. y.
  b) Familial FMTC (15 %):

	MEN 2a (70 %)	MEN 2b (10 %)	NonMEN (20 %)
Age peak	20. - 30. y.	10. - 20. y.	40. - 50. y.
Inheritance	a u t o s o m a l	d o m i n a n t	
RET-protooncogene	EXON 10,11	EXON 16	EXON 10,11

Multiple endocrine neoplasms (MEN): [D44.8]
MEN 2a = Sipple's syndrome: C cell carcinoma + phaeochromocytoma (50 % of cases) + primary hyperparathyroidism (20 % of cases)
MEN 2b = Gorlin syndrome: Similar to MEN 2a, in addition ganglioneuromatosis (e.g. tongue) + Marfanoid habitus.
NonMEN: Only C cell carcinoma (FMTC only)
- Genetic screening of the family members of patients with MTC: If genetic test positive: Prophylactic thyroidectomy in pre-school age (6 years). In the rare MEN 2b the prophylactic thyroidectomy should be performed after diagnosis of the gene mutation. Postoperatively regular preventive examinations for phaeochromocytoma (determination of the catecholamines in the serum and in the urine) and for primary hyperparathyroidism in MEN 2a (serum calcium + parathormone).

History:　　- Radiation of neck or thymus region 10 - 30 years ago? Other radiation exposure?
　　　　　　　- MEN patients in the family (C cell carcinoma)?
　　　　　　　- Growth of goiter inspite of sufficient substitution?

Cl.:　　Only circa 25 % of all sonographic detected thyroid gland carcinomas present clinical signs
　　　　- Hard goiter nodules
　　　　- Local late symptoms: Solid, bumpy goiter, fixed skin, cervical and/or supraclavicular lymph nodes, hoarseness (recurrent nerve paresis), Horner's symptom complex (miosis, ptosis, apparent enophthalmos), neck, ear and occipital pains, stridor, dysphagia and upper vein engorgement

Di.:　　• U/scan: Irregularly demarcated hypoechoic areas are suspicious of malignancy.
　　　　• Scintigraphy: Cold nodes, which do not accumulate.

　　　　Remember: A cold node in the scintiscan, which is not anechoic on the u/scan, is always suspicious of a malignancy (esp. in younger males) and must definitely be investigated:
　　　　- Fine needle aspiration biopsy with cytology (90 % accuracy)
　　　　- In persisting suspicion of malignancy (also in negative cytology) surgery followed by histology (preparation with capsule)

　　　　• Calcitonin determination in the serum
　　　　• CT, MRI of the neck region
　　　　• Search for metastases: Chest X-ray, CT, bone scan, PET
　　　　• In C cell carcinoma genetic analysis for point mutation in the RET protooncogene
　　　　• If familial C cell carcinoma is suspected to offer genetic counselling and family examination.

Th.:　　Always combination of surgery, radiotherapy, nuclear medicine:
　　　　1. Surgery: Radical thyroidectomy + removal of the cervical lymph nodes. Postoperatively the endogenous TSH production increases markedly due to decline of the thyroid hormone concentration. Furthermore rh-TSH can be given to improve radioiodine accumulation as precondition for the subsequent
　　　　2. Ablative radioiodine therapy: 3 -4 weeks postoperatively ^{131}I scan of the whole body in order to detect iodine storing residual thyroid gland tissue and metastases. After that high dosed therapy with ^{131}I in several fractions until ^{131}I storing tissue is no longer detectable. ^{131}I dose: 1.000 up to > 10.000 MBq.
　　　　SE: Temporary radiation thyroiditis, gastritis and sialadenitis. The risk of a later acute leukaemia is 1 % (in very high repeated radioiodine therapies).

　　　　Memo: In radioiodine therapy of benign thyroid gland disorders no increased leukaemia risk has been detected.

3. Suppressive hormonal treatment of the thyroid gland with LT4: As high as possible up to the tolerable limit, in order to prevent an increased TSH stimulation of the poss. still existing metastases (TSH target range: < 0,1 mU/l).

4. External radiotion therapy: In undifferentiated tumours (undifferentiated tumours are more radiosensitive, no radioiodine accumulation).
   C cell carcinomas are radioresistant, radical surgery is crucial.

5. Palliative chemotherapies in the context of research trials in inoperable, nonradioiodine accumulating thyroid carcinoma and rapidly progressing medullary thyroid carcinoma.

Follow-up:
Control examinations in intervals of 6 months.
- History, palpation findings, U/scan, laboratory   } Prognosis in age < 45 y. is better than in older patients.
- Detection of carcinoma metastases and relapse: }
  For routine follow-up examinations determination of Tg
  If suspicion of relapse or metastases the scintiscan with 131I is still necessary. In this case recombinant human (rh)TSH can be given i.m. before the examination, so that T4 does not have to be stopped.
- Detection of lung/bone metastases:
  - Chest X-ray, CT
  - Tc-99m-MIBI-SPECT
  - ^{18}F-FDG-PET: Diagnostic of relapse/metastases of thyroid carcinomas independent of the iodine accumulation ability
  - Whole body bone scan with ^{99m}Tc-MDP, which has a strong bone affinity
- Tumor markers:
  - Thyroglobulin (Tg) is formed both by the normal thyroid gland and by differentiated cells of a thyroid gland carcinoma. Therefore Tg does not help in the diagnosis of thyroid gland carcinoma. After radical thyroidectomy however the Tg production ceases. A reincrease of the Tg level after radical surgery of a differentiated thyroid gland carcinoma indicates a tumour relapse and/or metastases! Poss. Tg determination after administration of rhTSH.
  - Calcitonin in the serum in C cell carcinoma (see above); in family screening examination calcitonin is determined after pentagastrin stimulation.

Prg.: 10 year survival rates of all cases:
- Papillary thyroid Ca.: > 90 %
- Follicular thyroid Ca.: > 75 %
- Medullary thyroid Ca.: circa 50 %
- Anaplastic thyroid Ca.: Mean survival time 6 months

Pro.: In radioactive fallout (e.g. nuclear power plant accidents) blockage of the iodine uptake into the thyroid gland through single administration of potassium iodide (e.g. Thyprotect®); Dose age dependent (in 13 - 45 years of age 100 mg); in age > 45 no administration of iodide is recommended because of increased risk of an induction of a hyperthyroidism in autonomy of the thyroid gland.

PARATHYROID GLAND, VITAMIN D METABOLISM AND HOMEOSTASIS OF CALCIUM/PHOSPHATE

CALCIUM

99 % of the body´s calcium is found in the bone as hydroxylphosphatite (= mixed salt consisting of calcium carbonate and phosphate). Only 1 % of the calcium is present extracellularly.

Daily requirement of calcium:

Adults 1.000 mg/d

Juveniles, pregnancy, lactation period, postmenopausal and males > 65 y.: 1.500 mg/d

Normal range of the calcium levels in the serum: Total calcium: 2,2 - 2,7 mmol/l
                                                    Ionized calcium: 1,1 - 1,3 mmol/l

The serum calcium level is subject to a marked circadian rhythm with highest levels in the morning.
- Circa 45 % of the serum Ca^{++} are bound to proteins (4/5 to albumin and 1/5 to globulin).
- Circa 5 % of the serum Ca^{++} are complexed bound to bicarbonate, citrate and phosphate
- Circa 50 % of the serum Ca^{++} are present as free ions = biological active fraction

The quantity of the ionized calcium fraction (Ca^{++}) depends on:
1. Protein concentration in the serum: high protein concentration → low Ca^{++} (and vice versa)
   Should it be only possible to measure the total Ca, the albumin corrected Ca can be calculated if the albumin level differs from the reference value of 4 g/dl:
   Corrected Ca (mmol/l) = measured Ca (mmol/l) - 0,025 x albumin (g/l) + 1,0
   Rule of thumb: A decline of the albumin of 1 g/dl results in a decrease of the total Ca of circa 1 mg/dl (0,25 mmol/l).
2. Blood pH: Alkalosis → low Ca^{++} (the cause of tetany in hyperventilation)
                  Acidosis → elevated Ca^{++}

In the routine serum calcium determination the total calcium is measured, which is equivalent to the ionized calcium concerning the diagnostic value. In case of aberrations from the normal serum protein concentration however a correction to normal serum protein values must be performed (via formula or nomogram).

Calcium homeostasis: If the serum Ca^{++} level decreases, parathormone causes an increase, should the serum Ca^{++} level increase, calcitonin causes a decrease of the Ca^{++} concentration in the serum.

Note: Defects in the ionic balance can lead to impairments of the neuromuscular excitability. This finds expression in the serum electrolyte formula according to Gyoergy:

$$K = \frac{(K^+)*(HCO_3-)*(HPO4^{--})}{(Ca^{++})*(Mg^{++})*(H^+)}$$

Increased potassium leads to an increase in neuromuscular excitability, and vice versa.

PHOSPHATE

The normal serum phosphate concentration ranges between 0,8 - 1,6 mmol/l.
20 % of the serum phosphate is protein-bound.
The homeostasis of calcium phosphate metabolism is regulated by parathormone and vitamin D.

PARATHORMONE

Parathormone (PTH) is synthesized in the 4 epithelial bodies as pre-pro-parathormone, a peptide consisting of 115 amino acids. A peptide consisting of 84 amino acids leaves the gland; the N-terminal peptide with the amino acid sequence 1-34 carries the biological activity; half-life of the hormone circa 3 minutes. In the blood intact PTH and PTH fragments are circulating. The determination of intact PTH is independent of the renal function.

An ionized calcium < 1,25 mmol/l stimulates the parathormone secretion. In calcitriol deficiency PTH issecreted relatively more compared to the serum calcium. A mild hypomagnesaemia stimulates – similar to the hypocalcaemia – the PTH secretion, in severe hypomagnesaemia the PTH secretion is reduced. High phosphate concentration in uraemia stimulates directly the PTH secretion.

Between calcium and parathormone a negative feedback regulation exists physiologically (inverted change of both quantities). The negative feedback regulation continues to exist in hypercalcaemia due to tumours, vitamin D intoxication and sarcoidosis: In these patients suppressed PTH levels are found. Whereas in hypoparathyroidism both parameters are decreased and in primary hyperparathyroidism both are elevated.

The effect of parathormone occurs via stimulation of the adenylate cyclase in kidneys and bones: In the kidneys phosphate is increasingly excreted and calcium reabsorbed. Through the dropping phosphate level the renal 1α-hydroxylase is stimulated with increased synthesis of 1,25(OH)$_2$vitamin D$_3$, by which the enteral calcium absorption is accelerated. A negative calcium balance of the bone occurs in <u>pathologically</u> increased PTH concentrations (however not in physiological PTH concentrations).

<u>Reference range:</u> 1,5 - 6,0 pmol/l. There is a discrete circadian rhythm with mild elevated levels towards the evening, furthermore minimum pulsations (mostly less than 0,5 pmol/l) are observed.

CALCITONIN

Calcitonin is synthesized in the C-cells of the thyroid gland; it consists of 32 amino acids and causes an inhibition of the osteoclastic activity. The secretion of calcitonin is regulated via the Ca^{++} concentration in the blood: High Ca^{++} levels increase the secretion of calcitonin, low Ca^{++} levels inhibit the secretion.

In the medullary C cell carcinoma of the thyroid calcitonin is elevated (tumour marker !). There are no clinical symptoms of decreased calcitonin levels.

<u>Procalcitonin</u> is regarded as a sensitive marker for sepsis.

VITAMIN D METABOLISM and PARATHYROID FUNCTION

<u>Vitamin D3</u> is either synthesized from 7-dehydrocholesterol under the influence of UV light in the skin or it must be supplied with the diet. Vitamin D3 synthesized in the skin or vitamin D3 taken up from the food together with vitamin D2 is bound to <u>vitamin D binding protein</u> (DBP, transcalciferin) in the plasma and transported to the liver.

In the liver vitmin D3 is transformed to 25-OH-D3 (= calcifediol). From that the biologically very active 1α-25(OH)2-D3 = calcitriol is synthesized in the kidneys. This transformation is regulated by the phosphate concentration: Low phosphate levels increase the calcitriol formation and vice versa. Calcitriol increases the enteral resorption of calcium and phosphate.

Furthermore calcitriol influences the <u>immunoregulation</u>, e.g. stimulation of the interleukin 1 production by monocytes and inhibition of the interleukin 2 production by T lymphocytes.

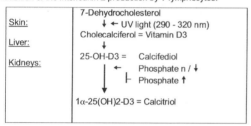

REGULATION OF THE SERUM CALCIUM CONCENTRATION

1. Decrease of serum Ca^{++} →
2. Secretion of <u>parathormone</u> →
3. Stimulation of phosphate excretion in the kidneys →
4. Decrease of serum $HPO4^-$ →
5. Stimulation of formation of calcitriol in the kidneys →
6. Enteral reabsorption and osseous mobilisation of Ca^{++} and $HPO4^-$ →
7. Normalization of serum Ca^{++}

PRIMARY HYPERPARATHYROIDISM [E21.0]

Def.: Primary disease of the parathyroid gland causing increased production of parathormone (PTH).

Ep.: Incidence up to 45/100.000/year

Aet.: 1. <u>Solitary adenomas</u> (80 %), multiple adenomas (5 %) of the parathyroid gland; 80 % of parathyroid adenomas are situated behind the thyroid, circa 10 % are situated in the anterior mediastinum
2. Hyperplasia of the epithelial bodies (15 %) - histology: Hyperplasia of the water clear cells or main cells
3. Rarely carcinoma of the epithelial bodies (< 1 %)

Rarely <u>multiple endocrine neoplasms</u> (<u>MEN</u>) are observed (see there).

Pg.:   In the pHPT a hypercalcaemia results from the three points of action of the PTH:
- Increased osseous calcium mobilisation (elevated pyridinolines in the urine)
- Increased intestinal calcium absorption (calcitriol mediated)
- Increased tubular calcium reabsorption

Cl.:   > 50 % of the patients have no or just non-specific complaints (coincidental diagnosis of a hypercalcaemia)

1. Renal manifestations (40 - 50 %):
   - Frequently nephrolithiasis (calcium phosphate or oxalate)
   - Rarer (unfavourable for the prognosis) nephrocalcinosis

   Typically a restriction of the concentrating ability, which is ADH refractory and which leads to polyuria with polydipsia. Advanced cases can lead to renal insufficiency.

   Note: Renal stones are frequent in pHPT, but only a small proportion of patients with calcium containing stones have a pHPT (circa 2 %).

2. Bone manifestations (circa 50 %):

   The increased parathormone activity leads to an increase of the osteoclasts, reactively also of the osteoblasts, while a negative bone balance results. The osteoclastic activity leads in marked cases to subperiostal resorption lacunae and acro-osteolyses of hands/feet. The formerly observed brown tumours (osteodystrophia cystica generalisata of Recklinghausen) are hardly observed nowadays. Radiologically the most common sign often is only a diffuse osteopenia, which is seen in views of the hands in 40 %, in views of the vertebral column in 20 %; the skull – second most commonly affected – appears radiologically in a "frosted glass" effect; the lamina dura of the dental alveoli shows erosions. Pains in the vertebral column and extremeties are symptoms of a skeletal manifestation. In the laboratory the alkaline phosphatase and the hydroxyprolin excretion are increased if bones are affected.

3. Gastrointestinal manifestations (circa 50 %):
   - Loss of appetite, nausea, constipation, meteorism, weight loss
   - Rarer gastric/duodenal ulcers (hypercalcaemia → hypergastrinaemia → HCl)
   - Rarer pancreatitis (circa 10 %; the pancreatitis can lower the calcium level and thus can mask a pHPT!)

   Note: Also gall stones occur twice as frequently as in the standard population.

   Beware: From the classic symptoms "stones,bones,groans, moans" the nephrolithiasis plays the dominating role nowadays !

4. • Neuromuscular symptoms: Rapid exhaustion, muscular weakness and atrophy, QT shortening in the ECG
   • Psychiatric symptoms: Depressive mood swings

5. Hypercalcaemic crisis (< 5 %):

   The pHPT can exacerbate at any time without prodromal symptoms into a hypercalcaemic crisis, especially if a factor is added which favours the hypercalcaemia (e.g. confinement to bed, therapy with calcium, vitamin D, thiazides).

   Sy.: • Polyuria, polydipsia
        • Vomiting, dehydration, adynamia
        • Psychotic features, somnolence, coma

   Due to rapid development of a renal insufficiency with increase of the serum phosphate calcifications can occur in various organs (e.g. cornea, tunica media of the arteries). Cardiac
   arrhythmias can lead to a sudden death. Mortality up to 50 %!

Lab :

	Serum	Urine
Calcium	↑	↑
Phosphate	↓	↑
Intact PTH	↑	
Alkaline phosphatase	↑	

DD: of the hypercalcaemia: See there

Di.:   **in hypercalcaemia syndrome:**
1. Intact parathormone (intact PTH) in pHPT ↑, in tumour hypercalcaemia ↓
2. Parathormone related peptide (PTHrP): ↑ in tumour hypercalcaemia
3. 1,25(OH)$_2$-vitamin D3: ↑ in sarcoidosis and vitamin D intoxication
4. Tumor screening and X-ray/scintiscanning of the bones

Di.:   **of the pHPT:**

Calcium	Intact PTH	Diagnosis
↑	↑	primary HPT (+ tertiary)
↓	↑	secondary HPT
↑	↓	Tumor hypercalcaemia; sarcoidosis: 1,25(OH)2 vit.

An increase of <u>serum calcium</u> (> 2,6 mmol/l = 5,2 mval/l = 10,5 mg/dl in normal renal function and normal total protein), in at least three determinations on different days, and of <u>intact PTH</u> points to a pHPT with a probability of over 95 %.

Note: In rare cases a normocalcaemic pHPT can be present. This is the case, if there are simultaneously a vitamin D deficiency, a renal insufficiency or an albumin deficiency.

- Localization diagnostic:
  - Ultrasound (adenomas are hypoechoic with a hyperechoic edge)
  - Spiral CT and MRI with 3 D reconstruction
  - 99mTc-MIBI (Metoxyisobutylisonitril) scintiscanning, poss. combination with SPECT
  - Intraoperative presentation by experienced surgeon

Th.: A) <u>Surgery in centres:</u>

Ind:1. Symptomatic pHPT
  2. Indications for surgery in the asymptomatic pHPT:
      - Serum calcium > 0,25 mmol/l above the upper normal limit
      - Impairment of the creatinine clearance
      - Decline of the bone density (T score < -2)
      - Co-factors, which can favour a hypercalcaemic crisis (see above).
      - Age < 50 years

The disease is curable through surgical removal of the enlarged epithelial bodies in time. Intraoperatively all existing epithelial bodies must be presented. In secure preoperative localization of the adenoma poss. as well endoscopic surgery.

- Isolated removal of adenomatously enlarged (weighing > 50 mg) epithelial bodies
- In hyperplasia of all epithelial bodies: Total parathyroidectomy with simultaneous autologous transplantation of epithelial body residuals, e.g. into the M. brachioradialis or M. sternocleidomastoideus (by that way it can be operated again without any difficulties in case of reoccurring hyperplasia).

In case of successful surgery the intraoperative PTH determination shows a decline of circa 50 %.

Removed epithelial bodies are usually cryoconserved, in order to be able to perform an autologous transplantation in the (rare) case of a <u>definitive</u> postoperative hypofunction. <u>Temporary</u> postoperative hypofunctions are observed especially in patients with increased alkaline phosphatase as sign of a bone manifestation (recalcification tetany). Postoperatively the calcium level must be monitored tightly in order to substitute calcium in time if necessary.

B) <u>Conservative:</u>
If there are no indications for surgery the following recommendations apply:
- Drink plenty; no application of thiazides or digitalis preparations
- Osteoporosis prophylaxis in postmenopausal women with oestrogen/gestagen combination
- Monitoring (every 3 months)

Therapy of a hypercalcaemic crisis:
See chapter hypercalcaemia

Prg.: Good in case of early diagnosis + surgery as long as there is no renal insufficiency present.

SECONDARY HYPERPARATHYROIDISM [E21.1]

If a nonparathyrogenic disorder leads to a decrease of the serum calcium, the parathyroid gland reacts secondarily with an increased secretion of PTH.

1. Renal sHPT:

See renal osteopathy in chap. chronic renal insufficiency (phosphate, creatinine ↑)

2. sHPT in normal renal function:

a) <u>Enteral causes:</u> Malabsorption syndrome with decreased resorption of calcium

b) Seldom <u>hepatic causes:</u>
  - Liver cirrhosis (impaired transformation of D3 to 25-OH-D3)
  - Cholestasis (impaired resorption of vitamin D3)

c) Deficient exposure to sunlight

Cl.: • Symptoms of the underlying disease
      • Poss. bone pains

Di.: Calcium in the serum ↓, phosphate in the serum normal, PTH intact ↑

Th.: • Treatment of the underlying disease (see also chap. renal osteopathy)
  • Substitution of vitamin D3 and poss. calcium

TERTIARY HYPERPARATHYROIDISM [E21.2]

A "tertiary hyperparathyroidism" is present, if during the course of a sHPT a hypercalcaemia develops. This is not due to a newly occurring autonomy (as in the pHPT), but rather an imbalance between PTH secretion and demand, if for example after a kidney transplantation suddenly the PTH demand becomes so low, that even the basal secretion (in the context of a sHPT) of the hyperplastic epithelial bodies is too high, a hypercalcaemia develops.

Th.: Poss. surgical

HYPOPARATHYROIDISM [E20.9]

Def.: Hypofunction of the parathyroid gland (epithelial bodies) with deficiency in parathormone. Cardinal symptom is hypocalcaemic tetany.

Aet.: 1. Most commonly postoperative after neck surgeries (e.g. strumectomy)
  2. Seldom idiopathic (autoimmune genesis?)
  3. Very rarely aplasia of parathyroid gland and thymus (DiGeorge syndrome)

Cl.: 1. Functional symptoms:
  - Hypocalcaemic tetany [E83.5]
    (in 75 % of the cases in the idiopathic, in 40 % of the cases in the postoperative hypoparathyroidism): Convulsions with preserved consciousness, often linked with paraesthesias, *main d´accoucheur* posture (carpopedal spasms), spasm of the laryngeal muscles
  - Chvostek's sign:
    If positive twitching of the muscles of the angle of the mouth is produced while tapping the facial nerve around the cheek
  - Trousseau' sign:
    If positive *main d´accoucheur* posture occur by inflating a blood pressure cuff for some minutes at 20 mmHG above systolic pressure on the arm.
  - ECG: Prolongation of QT interval

  2. Organic changes:
    Impaired growth of hair and nails, cataract formation, calcification of basal ganglia, osteosclerosis

  3. Psychological changes (irritability, depressive mood swings)

DD: 1. Hypocalcaemias of other origin (intact PTH ↑):
    Acute pancreatitis, malabsorption syndrome, peritonitis, healing process of a rachitis or osteomalacia (vitamin D deficiency), renal insufficiency, infusion of EDTA or citrated blood, rarer causes.

  2. Normocalcaemic tetany: [R29.0] Most common!
    Decrease of ionized calcium because of alkaloses (mostly respiratoric alkalosis due to psychogenic hyperventilation).

  3. Pseudohypoparathyroidism (very rare):
    As seen in the genuine hypoparathyroidism calcium is decreased and phosphate is elevated in the serum: The parathormone (intact PTH) however is not decreased as in the genuine hypoparathyroidism, but even increased as seen in secondary hyperparathyroidism. 4 types:
    Type Ia: Reduction of the Gs portion in the adenyl cyclase receptor complex
    Type Ib: Normal Gs activity, probably PTH receptor defect
    Type Ic: Gs activity normal, defect in the catalytic unit
    Type II: PTH receptor adenyl cyclase complex functionally normal, the intracellular response mediated by cAMP is blunt.

    Note: Gs = Guanine nucleotide binding protein, is activated by the PTH receptor type I after coupling of PTH and mediates inthat way the PTH effect.

    The pseudohypoparathyroidism has an increased familial tendency and comes along with organic stigmata (e.g. in type Ia shortening of metacarpal or metatarsal bones, short stature, heterotopic calcifications). In certain individuals in families of these patients the phenomenon is called pseudo-pseudo-hypoparathyroidism, if they show the typical stigmata without any calcium impairment.

Di.: 1. Typical serum constellation: Calcium + magnesium ↓, phosphate ↑
2. Urine status: Excretion of calcium, phosphate and cAMP ↓
3. Parathormone (intact PTH) ↓
4. Ellsworth-Howard test: If pseudohypoparathyroidism is suspected: After administration of PTH there is a more than twofold increase of phosphate in the urine compared to the original level in healthy individuals; the increase is less in patients with pseudohypoparathyroidism.

The detection of a hypocalcaemia and hyperphosphataemia and at the same time normal creatinine (exclusion of a renal insufficiency) and normal albumin level (exclusion of a malassimilation syndrome) renders the diagnosis of a primary hypoparathyroidism very probable, the detection of a decreased parathormone level confirms the diagnosis.

Th.: In tetany: Slow i.v. injection of 20 ml of calcium gluconate solution 10 %

Beware: Calcium and digitalis have a synergistic effect! Therefore never give calcium i.v. to a digitalized patient!

In aetiologically unclear tetanic seizures taking of blood samples before the symptomatic administration of calcium for determination of calcium and phosphate levels.

Long term therapy:
Vitamin D (high dosed 40.000 IE cholecalciferol/d or 0,5 - 1,5 g 1,25-(OH)2 vitamin D3 = calcitriol (Rocaltrol®) + calcium orally (1 - 3 g/d) under monitoring of the serum calcium and the calcium excretion in the urine (patient card). Target range: To maintain serum calcium in the lower normal range. If the serum phosphate does not drop during therapy, poss. additionally phosphate binding substances.
In overdose danger of hypercalcaemia, nephrocalcinosis, nephrolithiasis and deterioration of renal function.

OSTEOMALACIA [M83.9] **AND RACHITIS** [E55.0]

Def.: Rachitis („English disease", „rickets") is the terminology for deficient mineralization and a disorganisation of the epiphyseal plates, osteomalacia is a deficient mineralization of spongiosa and compacta. Therefore in children both defects occur simultaneously, while in adults after closure of the epiphyseal plates only an osteomalacia can occur.

Aet.: 1. Vitamin D deficiency: Malabsorption syndrome, deficient vitamin D supply, lack of UV light (e.g. in older people)
2. Defects of the vitamin D metabolism:
   - Related to the liver (liver cirrhosis, drug interactions, e.g. hydantoins) with deficient synthesis of $25\text{-}OH\text{-}D_3$ = calcifediol.
   - Related to the kidneys with deficient synthesis of $1,25\text{-}(OH)_2\text{-}D_3$ = calcitriol:
     In most cases chronic renal insufficiency
   - Seldom vitamin D dependent rachitis (Vit. D dependent rickets = VDDR):
     VDDR1: Genetically conditioned 1α-hydroxlase deficiency
     VDDR2: Genetically conditioned defect of the intracellular Vit. D receptor
3. Vitamin D independent osteomalacias are rare in renal tubular function defects (phosphate diabetes, renal tubular acidosis), phosphate deficiency etc.

Lab.: In vitamin D dependent osteomalacia:
- Hypocalcaemia + increased alkaline phosphatase
- Malabsorption syndrome: Hypophosphataemia and decrease of $25(OH)\text{-}D_3$
- Renal insufficiency: Hyperphosphataemia and decrease of $1,25(OH)_2\text{-}D_3$

Cl.: Skeletal pain, bowing of bones (poss. bow legs), adynamia, defects of gait (varus form of the femoral necks, weakness of the gluteal muscles with waddling gait); Rachitic rosary = defined swelling of the ribs at the junction with their cartilages; risk of tetany etc.

X-ray: Looser's zones (radiolucent bands running transverse to the longitudinal axis of the bones =

noncalcified osteoid), deformities of bones and vertebras

Bone biopsy with histology (poss. after preceding tetracycline labeling of the mineralization front)

Di.: History (underlying disorder !) + clinical picture + X-ray + laboratory

Th.: ● In vitamin D deficient rachitis: Substitution of vitamin D_3
● In vitamin D metabolic disorders: Therapy of the underlying disorder and substitution of metabolic active vitamin D metabolites, e.g. $1,25(OH)_2\text{-}D_3$
Therapy control through monitoring of the serum calcium and of the calcium excretion in the urine (patient card)
→ danger of the hypercalcaemia with all its consequences.

Internet infos: *www.osteoporose.org; www.osteoporose.com; www.osteofound.org; www.lutherhaus.de/dvo-leitlinien*

Def.:   Osteoporosis is a skeletal disorder characterized by decreased bone stability, which predisposes to an increased risk of fractures.
The clinically most relevant consequence of osteoporosis is the increase in fractures of the femoral neck, forearm and vertebras. From an age of 75 y. on the 10 year fracture risk for a neck of femur is even without additional risk factors > 20 %. Neck of femur fractures have a 1 year mortality rate of 10 - 20 %.

Ep.:   Most common bone disease in older age. Primary osteoporosis is the most frequent (95 %) . 80 % of all osteoporoses affect postmenopausal women. 30 % of all women develop a clinically relevant osteoporosis after the menopause, in the age > 70 years senile osteoporosis increases gradually in both sexes.
Secondary osteoporoses are rarer (5 %), mainly due to treatment with glucocorticosteroids and immobilisation.

Aet.:   1. Primary osteoporosis (95 %)
- Idiopathic osteoporosis of young people (rare)
- Postmenopausal osteoporosis (= Type I osteoporosis)
- Senile osteoporosis (= Type II osteoporosis)

2. Secondary osteoporosis (5 %)
- Endocrine causes: Hypercortisolism, hypogonadism, hyperthyroidism etc.
- Malabsorption syndrome with decreased supply of calcium and/or vitamin D
- Immobilisation
- Iatrogenic/drug related: Long-term therapy with corticosteroids (≥ 5 mg prednisolone/d) or heparin

3. Disorders which can be associated woth osteoporosis: e.g. rheumatoid arthritis, Crohn´s disease

4. Hereditary disorders:
Osteogenesis imperfecta, Ehlers-Danlos syndrome, Marfan syndrome, homocysteinuria

Risk factors for the development of a primary osteoporosis:
■ Factors that cannot be influenced therapeutically:
- Age: Bone mass reduces with increasing age.
- Sex: Women have a lower bone mass, which further clearly decreases during menopause.
- Genetic factors: Osteoporosis in the family history
■ Most important factors which can be influenced therapeutically:
- Deficiency in sex hormones or reduced time of exposure to oestrogens < 30 years (late menarche, early menopause)
- Physical inactivity
- Nutritional factors: Cachexia; deficiency in calcium and vitamin D
- Heavy consumption of cigarettes and/or alcohol

Metabolic characteristics:
- "Fast loss" of bone density: Bone mass loss due to increased turnover ("high turnover"): Loss of trabecular bone density > 3,5 % per year: Typical for the early postmenopausal osteoporosis within the first 10 years after the menopause.
- "Slow loss" of bone density:Bone mass loss due to reduced turnover ("low turnover"): Loss of trabecular bone density < 3,5 % per year: Typical for the late postmenopausal osteoporosis > 10 years after the menopause.

Types of distribution:
- Generalized osteoporosis
  - Postmenopausal osteoporosis (type I): Spongiosa pronounced bone mass loss
  - Senile osteoporosis (type II): Spongiosa plus compacta affecting bone mass loss
- Localized osteoporosis (e.g. Sudeck syndrome, joint adjacent osteoporosis in rheumatoid arthritis)

Characteristic	Type of osteoporosis	
	I (postmenopausal)	II (senile)
Age (years)	50 – 70	> 70
Sex (f : m)	Women	2 : 1
Bone loss	More trabecular than cortical	Equally trabecular and cortical
Most frequent fractures	Vertebral body	Femoral neck, humerus, radius, vertebral body
Aetiological factors	Oestrogen deficiency	Ageing process, decreased mobility, poss. deficiency in calcium and/or vitamin D

<u>Stages:</u>

Clinical stage	Criteria (DXA measuring)
0 Osteopenia [M81.9] (Preclinical osteoporosis)	• T score: - 1,0 up to - 2,5 SD • No fractures
1 Osteoporosis (without fractures)	• T score: < -2,5 SD • No fractures
2 Manifest osteoporosis (with fractures)	• BMD reduced • 1 - 3 vertebral fractures without adequate trauma
3 Advanced osteoporosis	• BMD reduced • Multiple vertebral fractures • Often as well extraspinal fractures

<u>T score:</u> Standard deviation (SD) below the mean value of the bone density of healthy individuals aged 30 y. (with „peak bone mass")
<u>BMD</u> = Bone mineral density
<u>Z score:</u> Standard deviation below the mean value of the bone density of a reference group of the <u>same age</u>.
During childhood and adolescence the bone mass builds up under the influence of sex hormones and reaches the maximum value around the 20th year of age („peak bone mass"). Men have a 30 % higher „peak bone mass" than women. After the 40th year of life there is a slow reduction of bone mass in all individuals (around 0,5 %/year is considered physiologic). Women have a higher bone loss during the first 10 years following the menopause (2 % per year and more).

<u>Cl.:</u>   • <u>Bone pains</u> (esp. in the spine)

      • <u>Low trauma fractures</u> or fractures without apparent cause (<u>spontaneous fracture</u>)

      • Collapse of vertebral bodies result in a <u>round back, gibbus formation and reduction of body height</u> (> 4 cm) with fir tree-like cutaneous folds at the back ("Tannenbaum phenomenon"): Monitoring of body height!

<u>X-ray:</u> of thoracic and lumbar spine in 2 views if fracture of the vertebral body is suspected.
A reduction of the bone mass of less than 30 % is not detectable in the standard X-ray.

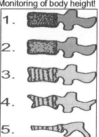

1. The homogenous structure of the normal vertebral body resembles fine tissue.
2. In beginning osteoporosis cover plates and vertical trabeculae protrude.
3. In marked osteoporosis horizontal trabeculae are hardly detectable, the vertical trabeculae are rare and strongly accentuated.
4. Cod fish vertebra
5. Wedge shaped vertebra and "crush fracture"

<u>Lab.:</u>   • <u>Basic investigation:</u>
      - ESR/CRP, differential blood count
      - Ca, PO4, γ-GT, AP, creatinine, electrophoresis, basal TSH

      • <u>Poss. further investigations for clarification of a secondary osteoporosis</u> (testosteron, oestrogens etc.)

      • <u>For monitoring of therapy determination of a bone-forming and a bone-resorption parameter with high specifity:</u>
        ▪ <u>Marker of the bone-resorption (osteoclast activity)</u> must decline during successfull therapy
          - Pyridinium cross-links (pyridinolin and desoxypyridinolin, measured in total concentration and in free fraction)
          - Carboxy-terminal type I collagen telopeptide
          - Carboxy-terminal octapeptide
          - Tartrate resistant acid phosphatase
          - Aminoterminale collagen type I Telopeptid
          - Bone sialoprotein (BSP)
        ▪ <u>Marker of the bone-formation (of the osteoblast activity), e.g.:</u>
          - Alkaline phosphatase (AP isoenzymes, bone specific AP)
          - Osteocalcin (very instable → freeze serum sample instantly)
          - Carboxy-terminal propeptide of the type I procollagen (PICP)
          - Amino-terminal propeptide of the type I procollagen (PINP)

<u>DD:</u>   1. <u>Malignant tumours:</u> e.g. plasmocytoma, Waldenström macroglobulinaemia, bone metastases of carcinomas
     2. <u>Primary hyperparathyroidism</u> (pHPT)
     3. <u>Osteomalacia</u> (see above)

DD	Serum calcium	Serum phosphate	Alkaline phosphatase	Additional diagnostic
Osteoporosis	n	n	n / ↑	X-ray, osteodensitometry
Osteomalacia	↓	n / ↓	↑	Vitamin D ↓, X-ray, poss. histology
Malignant tumours	↑	n	↑	Tumor search, ESR, blood count, electrophoresis, PTHrP
pHPT	↑	↓	↑	Intact parathormone ↑

Di.: Clinical picture - densitometry – special diagnostic:

Measuring bone density (densitometry): The DXA (dual energy X-ray absorptiometry) is the method of choice: Measuring the area density of the bone mineral concentration (g/cm^2).

The densitometry shows a decreased amount of minerals in the bone in osteoporosis and in long-term controls an increased loss of bone mass.

Remember: The diagnosis of a primary osteoporosis is a diagnosis of exclusion!

Th.: A therapy with drugs is recommended, if the 10 year risk of vertebra and proximal femur fractures is > 30 % and if the T score of the DXA bone density measuring of the lumbar spine or of the whole proximal femur is < -2,0. The threshold value for the recommendation of a drug therapy can be taken from relevant reference charts (see internet infos).

1. Causal: e.g. testosterone substitution in hypogonadism, reduce/replace corticosteroids

2. Symptomatic:

General measures for the prohpylaxis of osteoporosis and fracture, such as improving the coordination and muscular power, avoiding drugs which stimulate osteoporosis or increase falls

- Calcium rich diet (milk/-products), mobilization, physical and physiotherapeutic treatment; rehabilitation sport, self-support groups. If in pain poss. temporaray analgesics.

- Supplementation of calcium + vitamin D:
  - Sufficient calcium supply: Prophylaxis and basic therapy of the osteoporosis;
    Dose: 1.000 mg/d for adults, 1.500 mg/d from the 65th year of age on

    Note: Avoid calciuretic diuretics (e.g. furosemide), rather use calcium saving drugs (e.g. hydrochlorothiazide)

  - Vitamin D: In senile osteoporosis there can mostly be found a (latent) calcium and vitamin D deficiency. For older people the supply of vitamin D3 = cholecalciferol (600 - 800 IU/d) is recommended. This way the risk of femoral neck fractures a´can be decreased!

3. Drugs:

- Drug class A (research results consistently positive):
  - Bisphosphonates: Reduction of the incidence of vertebral + nonvertebral fractures by 50 % proven for alendronic acid (Fosamax®) 10 mg/d or 70 mg/1 x per week, risedronate sodium (Actonel®) 5 mg/d or 35 mg/1 x per week. For ibandronic acid (Bonviva®) 150 mg 1 x per month orally or 3 mg every 3 months i.v. only a fracture lowering effect for vertebral fractures is proven.
    Ef.: Antiresorptive effect through inhibition of the osteoclasts. Bone density ↑
    SE: Irritation of the oesophageal mucosa, gastro-intestinal disturbances, rarely scleritis, osteonecroses of the upper and lower jaw are observed in high dose long-term therapy (e.g. in tumour patients) ; among other things consider SE/CI
  - Strontium ranelate (Protelos®) reduces in postmenopausal osteoporosis the risk of spinal and hip fractures.
    Ef.: The substance has a calcimimetic effect on the cation sensing receptor and thus shows a stimulating effect on osteoblasts while at the same time inhibiting effect on osteoclasts.
    Dose: 2 g granulate 1 x daily 2 h separated from calcium/magnesium supply
    SE: Nausea, diarrhoea (gradually increasing dose !), cephalgias, dermatitis and eczemas, temporarily asymptomatic increase of CK, mildly elevated incidence of thromboembolic complications
    CI: Thrombophilia, TVT in the history etc
  - Raloxifene (Evista®), a selective oestrogen receptor modulator (SERM), has an effect on bones and lipid metabolism similar to that of oestrogens, there is however no oestrogen effect on breast and uterus. No gynaecological bleedings are observed. The risk of vertebral fractures and breast carcinomas is reduced. Neck of femur fractures however were not significantly reduced in the trials. Climacteric complaints can be triggered. Coordinate question of indication between endocrinologist and gynaecologist, e.g. osteoporosis patients with increased risk of breast carcinoma.
    Dose: 60 mg/d
  - Therapy with parathormone (PTH):
    - Teriparatide (Forsteo®): PTH 1-34
      Ind.: Complicated course of the osteoporosis; maximum duration of therapy 18 months (20 µg/d s.c.)
      SE: Nausea, arthralgia, headaches, vertigo etc.

- Recombinant parathormone (Preotact®): PTH 1-84
  Decrease of the incidence of spinal fractures
  Ind.: Postmenopausal osteoporosis in high risk patients
  SE/CI to be considered (e.g. hypercalcaemia)
- Oestrogens in postmenopausal osteoporosis of the woman have a anti-resorptive and a fracture lowering effect. Due to increased risk of myocardial infarction, cerebrovascular accident, TVT and breast cancer an oestrogen/gestagen therapy is not advised.

Endpoint	Incidents under oestrogen/gestagen therapy per 1.000 women	
	After 2 years	After 5,2 years
CHD	3 additional	4 additional
CVA	1 additional	4 additional
Venous thrombosis	6 additional	9 additional
Breast cancer	0 additional	4 additional

Ind.: Oestrogen deficiency of various genesis (e.g. oestrogen substitution in premature menopause)
- Drug class B (research results controversial)
  - Fluorides:
    Ef.: Stimulate the osteoblasts and increase the bone mass. A reduction of the fracture rate however could only be shown in the spinal region, whilst not in other bones.
    SE are to be considered: Gastrointestinal disturbances, ankle joint pains, development of a fluorosis in overdose (narrow therapeutic window)
  - Calcitonin (parenteral or nasal)
    Ef.: Shows an anti-resorptive effect through inhibition of the osteoclasts and has in addition an analgetic effect.
    Ind.: Poss. in osteoporotic related bone pains
- Prophylaxis of femoral neck fractures: Wearing of hip protectors (= elastic cotton trousers with lateral protection pads)

PAGET DISEASE [M88.9]

Syn.: Ostitis deformans

Def.: Localized mono- or polyostotic, progressive skeletal disorder of unclear origin. Increased bone turnover processes are characteristically with the risk of deformations, chronic pains, fractures, articular and neurologic complications.

Occ.: After osteoporosis second most common disorder of the bone. Familial clusters. The disease is most frequent in England, very rare in Asians and Africans.
Prevalence in Western Europe: 1 - 2 % of the > 40 year old individuals. High number of unreported oligo- or asymptomatic cases. Only 30 % of the cases are diagnosed before death.
m > f, age peak > 40 y.

Aet.: Unknown (viral origin ?; in 30 % mutation of the RANK gene)

Pg.: In the beginning there is an uncontrolled stimulation of the osteoclastic bone resorption (early phase). Secondarily there follows an exaggerated unruly bone build-up (late phase). The result is a bone, which is distended, mechanically less stable with the tendency to fractures and deformations.

Pat.: Most commonly the pelvis is affected, followed by femur, tibia, skull, lumbar vertebra. Thickening + bowing/deforming of the long bones.

Cl.: 1/3 of the patients are asymptomatic (incidental finding)
- Local bone pains, poss. increased skin temperature above affected bone
- Poss. bowing and shortening of the legs (e.g. "sabre sheath" tibia)
- Poss. increase of the head circumference (hat does not fit any more)

Lab.: Alkaline phosphatase (AP) ↑ (osteoblast isoenzyme) = good activity parameter!
Excretion of pyridinium cross-links in the urine (marker of the bone resorption)

Co.: Arthrosis due to malposition of the joints, fractures, poss. radicular compression syndrome in spinal involvement; hearing difficulties in skull involvement (circa 40 %): Conduction deafness due to ankylosed ear bones or inner ear deafness due to compression of the VIII. cranial nerve; hypercalciuria + nephrocalcinosis; volume load of the heart due to increased bone perfusion; seldom (< 1 %) osteosarcoma as a late complication

DD: Bone tumours, osteomyelitis, hyperparathyroidism

Di.: • Clinical picture / AP
• X-ray: 3 phases (which might be as well detectable in a patient at the same time)

243

- 1. phase: Osteolyses are early manifestations (esp. skull and long bones)
- 2. phase: Mixed picture of osteolytic and osteosclerotic areas
- 3. phase: Predominantly scleroses with thickening and deformations of the involved bones; ungainly and coarse spongiosa with isolated osteolyses
- Bone scan: Screening test for further bone lesions; increased technetium uptake in involved bone. Every increase in storage must be clarified radiologically.
- Poss. bone biopsy (mosaic structure, increase of multinuclear giant osteoclasts and osteoblasts)

Th.: Symptomatic:
- Inhibition of the pathologically increased osteoclast activity through:
  - Bisphosphonates are the drugs of choice, e.g. disodium pamidronate i.v. (Aredia®), zoledronic acid (Aclasta®) or oral preparations: Tiludronic acid (Skelid®), risedronate sodium (Actonel®). In early and consequent therapy bone deformations can be prevented. The therapy is applied in cycles. The aim is to normalize the activity parameters (esp. AP).
    (SE + CI see chap. osteoporosis)
  - Calcitonin is less effective than bisphosphonates
- Analgesia if required
- Treatment of fractures, malpositions of bones and joint damages
- Adequate supply of calcium (when taking at least 2 h interval to bisphosphonates, which otherwise are less resorbed) and vitamin D
- Physiotherapy, physical therapy (no thermotherapy, since the bones are already too warm)

ADRENAL CORTEX

Synthesis of the adrenocortical steroids:

	Zona glomerulosa	Zona fasciculata	Zona reticularis
Group	Mineralocorticoids	Glucocorticoids	Androgens
Main representative	Aldosterone	Cortisol	Dehydroepiandrosterone
Main effect	Na⁺ retention, K⁺ release from the cell Fluid retention	Gluconeogenesis with hyperglycaemia and protein break down, prevention of water entry into the cell	Protein synthesis Virilization
Secretion rate	50 - 250 µg/24 h	20 - 30 mg/24 h	m: 3,0 mg/24 h f: 0,7 mg/24 h
Plasma concen-tration	2 - 15 µg/100 ml	6 - 25 µg/100 ml	m: 0,3 - 0,85 µg/dl f: 0,2 - 0,6 µg/dl

Aldosterone, cortisol and to a lesser extent corticosterone are the most important corticosteroids. The adrenal androgens (dehydroepiandrosterone, androstenedione) are in a man insignificant, in the woman they stimulate (together with the androgens of ovarian origin) the secondary sexual hair growth. Women transform circa 60 % of the androstenedione to testosterone in the peripheral tissue. Under the influence of the 11-dehydrogenase (esp. in the liver) cortisol can be found partly as cortisone.
Cortisol is found in the blood in:
- 75 % bound to the transport protein transcortin (= CBG = corticosteroid-binding globulin).
- 15 % bound to albumin
- 10 % in free form.
Normally the transport capacity of the CBG is circa 20 µg cortisol/100 ml of plasma; an increase of the plasma cortisol level above this limit (normal cortisol level, dependent on the time of day: 6 - 25 µg/100 ml) leads to a disproportionally high increase of the free portion of cortisol in the plasma. Synthetic glucocorticoids are not bound to transcortin; this explains the stronger inhibiting effect of synthetic glucocorticosteroid on the ACTH production (see below). The plasma half-life of cortisol is circa 90 min., for the synthetic glucocorticosteroids it is partly several times longer. After having been metabolized in the liver the cortisol metabolites are excreted via the kidneys as glucuronides.

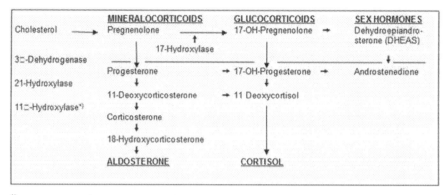

*) Inhibition of the 11β-hydroxylase by metyrapone

RENIN – ANGIOTENSIN – ALDOSTERONE - SYSTEM (RAAS)

The RAAS exists as a circulating system in the tissues of myocardium, vessel walls, kidneys and other organs.

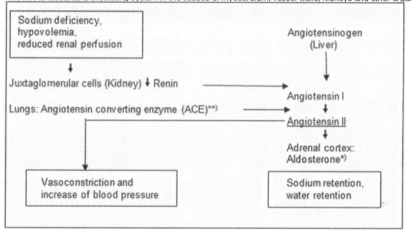

*) = Inhibition of the aldosterone effect through spironolactone
**) = Location of effect of the ACE inhibitors

The mineralocorticoid effectiveness of aldosterone : corticosterone : cortisol is 1.000 : 1,4 : 0,8.

The aldosterone effect works on the renal tubules, the intestinal epithelium, the saliva and sweat glands, resulting in a decline of the Na^+ concentration and an increase of the K^+ concentration in urine, saliva, sweat and intestinal secretion (vice versa in the serum !).

Regulation of the aldosterone secretion:
a) Stimulating:
   • The renin-angiotensin system is most important (preservation of constancy of the circulating blood volume)
   • Increase of the serum potassium
   • ACTH (less important)
b) Inhibiting:
   Atrial natriuretic peptide (ANP)

[E26.0]

Ep.: Prevalence of the <u>normokalaemic</u> Conn`s syndrome: 5 - 10 % of the hypertensive patients → as a result <u>most common cause of a secondary hypertension!</u>
Prevalence oc the classic <u>hypokalaemic</u> Conn´s syndrome: < 0,5 % of the hypertensive patients

Aet.:
- <u>2/3 of cases: Idiopathic hyperaldosteronism (IHA) through bilateral (rarely unilateral) hyperplasia of the zona glomerulosa</u> (often milder clinical picture with normal potassium)
- <u>1/3 of cases: Aldosterone producing adenomas (APA) of the adrenal cortex:</u> Often markedclinical picture with hypokalaemia
- Rarer causes:
  - Familial hyperaldosteronism type I and type II:
    - Type I = <u>G</u>lucocorticoid <u>r</u>emediable <u>a</u>ldosteronism (GRA): Fusion between the ACTH-dependent expressed gene of the 11β-hydroxylase and the gene of the aldosterone synthase, resulting in a ACTH-dependent secretion of aldosterone + hybrid steroids
    - Type II: Presented as adenoma or hyperplasia
  - Aldosterone producing carcinoma: rare

Cl.:
- Mostly only hypertension
- Less than a third of the patients shows the „classic" <u>clinical triad</u>:
  - <u>Hypertension</u> (cardinal symptom), with poss. headaches and poss. organ damages
  - <u>Hypokalaemia</u> with poss. muscle weahness, constipation, ECG changes, polyuria, polydipsia
  - <u>Metabolic alkalosis</u>

Most patients are normokalaemic!

Lab.:
- Plasma aldosterone ↑, plasma renin activity and concentration ↓ aldosterone/renin quotient ↑
- Poss. hypokalaemia and metabolic alkalosis
- Increased aldosterone and aldosterone metabolites (tetrahydroaldosterone and aldosterone-18-glucuronide) in the 24 h urine: Rather low sensitivity
- **No** hypernatraemia due to escape phenomenon of Na+ saving effect of aldosterone!

DD:

RR	Aldo-steron	Re-nin	K⁺	Syndrome	Cause
	↑	↑	n-↑	Essential hypertension + Diuretics	Secondary hyperaldosteronism through Na+ deficiency
				Renovascular: Renal artery stenosis	Secondary hyperaldosteronism due to renal ischaemia
				Renoparenchymatous	
	↑	↓	n-↓	Primary hyperaldosteronism	See above
	N	↓	n	Low-renin essential hypertension	Essential hypertension
↑	↓	↓	n-↓	Liddle´s syndrome	Constitutively active Na+ channel → increased Na+reabsorption
				Hypertensive form of the AGS (11β-hydroxylase deficiency)	Increased production of deoxycorticosterone
				Apparent mineralocorticoid excess	Renal 11 β-hydroxysteroid dehydrogenase deficiency (which inactivates cortisol at the mineralocorticoid receptor)
				Pseudohyperaldosteronism due to chronic licorice ingestion	Inhibition of the renale 11β-hy-droxysteroid dehydrogenase
				Cushing`s syndrome	Mineralocorticoid effect of cortisol
			↑	Gordon syndrome	Increased activity of Na⁺Cl⁻ co-transporter: Na⁺ reabsorption ↑
n	↑	↑	n	Functional	Hyponatriaemia, hypovolaemia
				Restricted liver function	Decreased hepatic metabolism of aldosterone
n-↓	↑	↑	↓	Bartter syndrome type I, II, III	Mutation of renal transport channels → decreased Na⁺ reabsorption
				Gitleman syndrome	→ secondary hyperaldosteronism

Di.: 1. <u>Suspicion of Conn´s syndrome</u>: especially in: Patients with spontaneous or diuretic induced hypokalaemia, hypertension difficult to be controlled (≥ 3 antihypertensives), hypertension grade III, patients with serendipitously discovered adrenocortical tumour (adrenal incidentaloma)

2. Screening test:
   Simultaneous measuring of plasma aldosterone and plasma renin activity (PRA, decrease on ice) or plasma renin concentration (PRC) for determination of the aldosterone/renin quotient (ARQ) → Increase of ARQ + aldosterone indicate a Conn syndrome

   Beware: Falsely high aldosterone/renin quotient under beta-blocker therapy, falsely low under therapy with ACE inhibitors, angiotensin receptor blockers and aldosterone antagonists.

3. Confirmation of the diagnosis: Absent/decreased aldosterone suppression in the
   - Saline infusion test: 2.000 ml 0,9% NaCl infusion over 4 hours i.v.
   - Fludrocortisone suppression test (high sensitivity and specifity, but complicated):
     0,1 mg fludrocortisone every 6 hours over 4 days.

     Beware: Danger of hypokalaemia (regular potassium substitution and monitoring of potassium) and of hypertensive crisis

   - Captopril test: Determination of aldosterone before and 2 h after administration of 25 mg of captopril

4. Determination of the underlying subtype by means of further diagnostic:

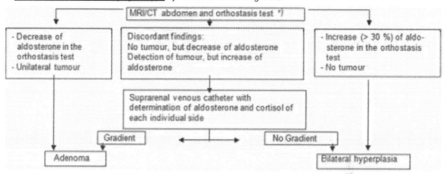

*) Orthostasis test: In IHA an increase (> 30 %) of aldosterone occurs after 2 h of orthostasis, in APA aldosterone decreases. Simultaneous determination of cortisol, in order to exclude ACTH induced acute increase of aldosteroneanstieg.

In suspicion of glucocorticoid remediable aldosteronism (GRA): Poss. determination of 18 hydroxycortisol +18 oxocortisol (no routine diagnostic), genetic testing

Th.: - IHA: Spironolactone (50 - 100 mg/d) + poss. K+ saving diuretics and further antihypertensives
- APA: Laparoscopic adrenalectomy after 4 weeks of preceding therapy with spironolactone
- GRA: Administration of low dose dexamethasone + family screening for the corresponding mutation
- Aldosterone producing carcinoma: Surgery + chemotherapy with mitotane; unfavourable prognosis

---

HYPOALDOSTERONISM [E27.4]

Aet.: 1. Primary hypoaldosteronism with increased renin levels:
   Addison´s disease, defect aldosterone synthesis; temporary after removal of an aldosterone producing adenoma with suppression of the contralateral adrenal gland.
2. Secondary hypoaldosteronism with decreased renin levels = hyporeninaemic hypoaldosteronism (= renal tubular acidosis (RTA), type IV):
   - In patients with diabetes mellitus (frequently)
   - Drug induced: Therapy with mineralocorticoids, ACE inhibitors, prostaglandin synthesis inhibitors; heparin long-term therapy

Cl.: Poss. hypotension with corresponding symptoms

Lab.: Hyponatraemia, hyperkalaemia (poss. dangerous), metabolic acidosis

DD: Pseudohypoaldosteronism (= receptor defect in the distal tubulus)

Di.: Plasma aldosterone ↓; plasma renin ↑ (primary) / ↓ (secondary)

Th.: In the primary form (e.g. Addison´s disease) therapy with mineralocorticoids (fludrocortisone); in secondary forms poss. omitting of triggering drugs, otherwise if clinically relevant also administration of mineralocorticoids.
Therapy adjustment through monitoring of electrolytes and plasma renin

GLUCOCORTICOSTEROIDS

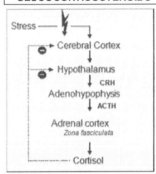

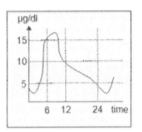

Diurnal rhythm of the
cortisol concentration in
the blood

The hormone production of the adrenocortical gland follows a circadian (day-night) rhythm. The minimum production is around midnight, the maximum production in the mornings between 6 - 8 o´clock. This physiological rhythm is abolished in Cushing´s disease! 70 % of the daily cortisol production are released in the morning hours. In this time the organism tolerates best small therapeutic doses of corticosteroids without disturbance of the feedback mechanism. Therefore the total daily dose should be given in one go in the mornings before 8 o´clock in a long-term therapy with corticosteroids!

The hypothalamus controls via the corticotropin-releasing hormone (CRH) the production of adrenocorticotropic hormone (ACTH) in the anterior pituitary. ACTH stimulates the adrenal cortex to synthesize cortisol. A decline of the cortisol level in the blood is the physiologic stimulus for the secretion of CRH and ACTH, which results in the stimulation of the cortisol production. High cortisol levels inhibit the secretion of CRH and ACTH (feedback regulation).

If long-term corticosteroid therapy interferes with the feedback mechanism, it can lead to the atrophy of the adrenal cortex. Sudden discontinuation of the hormone supply the results in an acute insufficiency of the adrenal cortex → therefore gradual tapering downward of a long-term corticosteroid therapy!

Ef.: Pharmacologic effects of the glucocorticoids:
7. Glucocorticoid effect of all adrenal cortex hormones with an oxygen function at C11: Gluconeogenesis: Glucose production from amino acids and intermediary products (lactate, pyruvate, glycine). Result: Catabolism with muscular atrophy and osteoporosis. The glucocorticoid effect is counteracting the effects of insulin: Contribution to a diabetic metabolic state.

2. Effects on lipid metabolism:
Hyperlipidaemia, increased lipolysis, fat mobilisation from the periphery, fat deposition in the liver, redistribution of fat characterized by truncal obesity.

3. Effects on the haematopoietic and lymphatic tissue:
d) Leukocytes ↑, eosinophiles and lymphocytes ↓, reduction in lymphatic tissue and inhibition of the B and esp. T lymphocyte activity; result: susceptibility to infections, anti-allergic and immunosuppressive effect.
e) Increase of the erythrocytes and platelets with thrombosis favouring effect (decrease also of the anti-thrombins)

4. Inhibition of inflammation, exsudation and proliferation in connective tissue, epithelium and mesenchyme; result:
– Antiphlogistic effect
– Delayed wound healing and ulcerogenic effect

5. Hypocalcaemic effect: Inhibition of the enteral calcium resorption + stimulation of the renal calcium excretion

6. Mineralocorticoid effect of cortisol in relation to aldosterone equals 1 : 1000   sodium retention, potassium excretion, shift of potassium from the intra- to extracellular compartment in exchange for sodium and hydrogen ions.

Beware: The undesirable effects mentioned above only occur in hypercortisolism or pharmacologic therapy with glucocorticoids in unphysiologically high doses.

Synthetic glucocorticosteroids:
Since the naturally existing glucocorticoids cortisol and corticosterone also influence the electrolyte balance (Na+ and water retention → possibly oedemas, hypertension; potassium loss), there has been a search for other therapeutic steroids. Prednisolone and prednisone only still have a weak mineralocorticoid effect compared to cortisol, in their other effect they are 4 - 5 times more potent than cortisol.
By introduction of atoms in the prednisolone molecule in position 6, 9 or 16 further synthetic steroids were created, which hardly have a mineralocorticoid effect, but whose antiphlogistic efficacy is by far higher than that of prednisolone.

Equivalent doses of glucocorticosteroids (oral administration):

	Dose equivalence (mg)	Biological half-life (h)
Hydrocortisone	30	Up to 12 h
Cortisone	37,5	
Short-acting substances		
Prednisone	7,5	
Prednisolone	7,5	
Cloprednol	7,5	Up to 36 h
Fluocortolon	7,5	
Methylprednisolone	6	
Intermediate-acting substances		
Triamcinolone	6	Up to 48 h
Long-acting substances		
Dexamethasone	1,125	Up to 72 h
Betamethasone	1,125	

Th.:   A) Substitution therapy:

Administration of cortisone at doses corresponding to the physiological requirements for the natural glucocorticoids cortisol and cortisone.

Daily doses in adrenocortical insufficiency: 37,5 mg of cortisone acetate or 30 mg of hydrocortisone, with 2/3 of the dose given in the morning and 1/3 in the afternoon; in all stress situations the dose must be increased (danger of an acute adrenal cortex insufficiency = Addison crisis; details see adrenal cortex insufficiency).

B) Pharmacologic therapy:

Corticosteroid administration in unphysiologically high doses. Some rules to remember with such therapy:

1. Corticosteroids only act symptomatically, not causally!
2. The longer the duration of use and the higher the dose, the greater the risk of side effects
3. The initial dose is dependent on the activity and severity of the disease to be treated: Acute and severe diseases require high to very high doses (extreme case: high-dose i.v. bolus therapy = pulse therapy with circa 1 g of prednisolone/d for several days, e.g. in rejection crises). In chronic disorders small initial doses are sufficient. After clinical improvement gradual dose reduction is performed until maintenance dose = smallest dose, which still shows a pharmacologic effect: For continous therapy the upper dose limit is considered to be 7,5 mg of prednisolone dose equivalence per day (Cushing threshold dose).This dose should not be exceeded in a long-term therapy, because otherwise increasing side effects have to be expected. But also in a long-term low-dose corticosteroid therapy with 5 mg of prednisolone dose equivalence/d side effects (e.g. osteoporosis) cannot be avoided.
4. Therapy recommendations for long-term therapy of more than 2 weeks:
   ▶ The following has to be considered to avoid the development of a secondary adrenocortical insufficiency in a long-term therapy:
   - To use only those preparations for the systemic therapy, that have a relatively weak inhibitory effect on the adrenal feedback mechanism: e.g. prednisone, prednisolone
   - Circadian therapy: Administration of the total daily dose at one time in the morning before 8 o´clock.
   An alternating therapy is even more beneficial: The total dose is only given every 2. day in the morning before 8 o´clock; this regimen however does not always have the desired clinical effect in severe conditions.
   - No use of combination preparations (corticosteroid + second substance), because they do not take into account the circadian rhythm of cortisol secretion and prevent individual dose adjustments.
   - No i.m. administration, no corticosteroid depot preparations → long-lasting interference with the adrenal feedback mechanism, furthermore poss. trophic tissue damage at the injection site.
   ▶ In long-term therapy osteoporosis prophylaxis
   ▶ In risk patients of gastric/duodenal ulcers (ulcus history, alcohol abuse, simultaneous NSAID therapy) ulcus prophylaxis (see there)
5. Abrupt withdrawal of the glucocorticoids after a therapy duration of > 2 weeks can result in an adrenocortical insufficiency. Therefore gradual withdrawal of therapy over several days.

The dose must be reduced basically very slowly in a therapy > 4 weeks or the adrenal feedback circle must be examined before discontinuation of the therapy. (24 h cortisol profile, ACTH test). A more rapid discontinuation is only possible in case of a normal functioning feedback mechanism.

<u>Unwanted effects:</u>
1. <u>Side effects of a high-dose long-term therapy:</u>
   - Exogenous <u>Cushing´s syndrome</u> with doses above the Cushing threshold dose: Redistribution of the fatty tissue: Fat loss in the extremities, increased fat deposition in the trunk, neck, face
   - <u>Hypernatraemia, water retention, oedemas, weight gain, hypertension</u>
   - Hypokalaemic alkalosis
   - Manifestation of a <u>diabetes mellitus</u>, deterioration of a diabetic metabolic state
   - <u>Increased risk of infection, wound healing defect</u>
   - <u>Steroid acne</u>, skin atrophy (up to ´parchment-like` appearance of the skin), hirsutism
   - <u>Risk of ulcers</u> under steroids alone slightly raised, in combination with nonsteroidal antiphlogistics it increases 15fold. However, ulcer complications such as penetration or perforation are often masked by steroids.
   - Protein catabolic effect with atrophy of the muscles
   - <u>Steroid myopathy</u> (relatively rare): Acute form charcterized by muscle weakness after start of a high-dose steroid therapy; chronic form characterized by muscle weakness of the proximal muscles of the extremities (shoulder and pelvic girdle)
   - <u>Osteoporosis</u>, aseptic bone necroses. In case of longer corticosteroid therapy the risk of fractur already increases from doses of 2,5 mg/d on → prophylaxis with calcium + vitamin D.
   - Growth inhibition in children, menstrual disorders in women
   - <u>Psychological disturbances</u> (dysphoria and psychoses)
   - <u>Eyes:</u> Posterior subcapsular cataracts and glaucoma (→ monitor intraocular pressure)
   - Increased capillary fragility and increased thrombosis tendency

2. <u>Side effects after discontinuation</u> of a high-dose long-term therapy:
   - Acute adrenocortical insufficiency
   - Latent adrenocortical insufficiency, which only becomes manifest in stress situations (traumas, infections, surgical procedures)
   - Exacerbation of the underlying disorder
   - Corticoid withdrawal syndrome with fever, arthralgias, myalgias, fatigue

<u>CI:</u>   <u>for a pharmacologic therapy:</u>
   - Gastroduodenal ulcer
   - Osteoporosis, corticosteroid induced myopathy
   - Psychoses
   - Various infectious diseases
   - 8 weeks before to 2 weeks after active immunization
   - Glaucoma
   - Recurrent thromboses/emboli

Some of the contraindications are not absolute but relative (always weigh out risk/benefit relation). No contraindications exist in life-threatening conditions (e.g. anaphylactic shock, status asthmaticus), particularly because the corticoid therapy is only short-term. In long-term therapy an individual risk/benefit analysis must be made. For post-tuberculosis conditions isoniazid prophylaxis should be given in case of long-term glucocorticoid therapy. A planned long-term therapy > 6 months should be accompanied from the beginning with calcium and vitamin D administration for osteoporosis prophylaxis, since the main loss of bone substance occurs in the first months.

[E24.9]

Classification and aetiology:

    I. Exogenous (iatrogenic) Cushing´s syndrome (frequent):
        Caused by long-term therapy with glucocorticosteroids or ACTH

    II. Endogenous Cushing´s syndrome (rare):

        Caused by increased secretion of cortisol or ACTH

        1. ACTH-dependent form with secondary adrenocortical hyperplasia:
            1.1 Central Cushing´s syndrome (= Cushing´s disease): 70 % of the endogenous Cushing´s syndromes; predominantly in middle-aged women. In 80 % of cases it is a microadenoma of the anterior pituitary, which not always is neuroradiologically detectable. In the other cases a primary hypothalamic hyperfunction is discussed. In some patients autoantibodies against anterior pituitary cells can be detected.
            1.2 Ectopic (paraneoplastic) ACTH secretion:
                Secretion of ACTH in tumours, most commonly small cell bronchial carcinomas and carcinoids
            1.3 Less common ectopic CRH secretion
            1.4 Alcohol induced Cushing´s syndrome (reversible after alcohol avoidance)
        2. ACTH independent primary form: Adrenal Cushing´s syndrome
            2.1 Cortisol producing adrenal cortex tumours (in adults predominantly adenomas, in children often carcinomas)
            2.2 In rare cases micronodular dysplasia
            2.3 In rare cases macronodular hyperplasia

Cl.:
1. Lipid metabolism: Redistribution of depot fat: ´Moon face`, ´buffalo hump´, truncal obesity; supraclavicular fossas covered by fat cushions; hypercholesterinaemia
2. Protein metabolism: Osteoporosis with poss. bone pains, myopathy with muscle wasting, adynamia; "strong appearance" + adynamia lead to misjudgement of being a malingerer. Simple test: Let the patient stand up from a squatting position without the help of the arms (not possible in marked myopathy).
3. Carbohydrate metabolism: Diabetogenic metabolic state
4. Haematopoietic system: Leuko-, thrombo- and erythrocytes ↑
   Eosinophiles and lymphocytes ↓
5. Hypertension (85 %)
6. Skin: Poor wound healing, tendency to acne, furunculosis, ulcers, occurrence of striae rubrae, atrophy of the skin (thin skin on back of hand – in case of obesity thick skin)
   (DD: Bright striae in genuine adipositas)
7. Virilism, hirsutism, menstrual disorders in women
8. Growth inhibition in children
9. Psychological, poss. psychotic changes
10. Hypokalaemia (5 %) due to overproduction of mineralocorticoids is relatively rare and if occurring it is suspicious of an adrenal cortex tumour or ectopic ACTH production.

In primary hypercortisolism due to an adenoma of the adrenal cortex usually only the glucocorticosteroids are increased.

In the secondary hypercortisolism with increased ACTH secretion and bilateral adrenocortical hyperplasia – as well even more marked in adrenal carcinomas – also the androgens (and less aldosterone) are in addition increased, so that androgen related features occur (virilism, hirsutism, menstrual disorders etc).

DD:
- Adiposity (normal dexamethasone suppression test, see below)
- Adrenal incidentalomas = coincidentally discovered tumours of the adrenal gland (most frequently endocrine inactive adenomas of the adrenal cortex)
- Slightly elevated cortisol levels in patients with depressive syndrome (cortisol determination in the 24 h urine)
- Elevated cortisol levels during the taking of contraceptives: Increase of the cortisol binding globulin (CBG) with an increase of the total hormone level while maintaining a normal level of free hormones.
- Administration of corticosteroids without the doctor´s knowledge (→ chromatographic detection of synthetic steroids)

Di.:   **a) of hypercortisolism:**
    • Clinical picture
    • Low dose dexamethasone suppression test: After administration of 1 mg of dexamethasone around midnight only an insufficient suppression of the serum cortisol occurs, which is determined at 8 o´clock of the following day (> 80 mmol/l). Occasionally though the test also is pathologic in patients with endogenous depression, under stress, after administration of hormonal contraceptives (oestrogens) or antiepileptics, in alcohol abuse and adiposity.
    • Free cortisol in the 24 h urine ↑, cease of the circadian rhythm of the serum cortisol in the daily profile

b) Aetiologic classification of the hypercortisolism:
1. CRH test: ACTH determination before and after administration of CRH (see chart)
2. High-dose dexamethasone suppression test:
   In the central Cushing´s syndrome, a suppression of the serum cortisol by at least 50 % is achieved after administration of 8 mg of dexamethasone at midnight for 2 days. In case of adrenal tumours or in the ectopic Cushing´s syndrome this suppression does not take place. This test does not permit a differentiation between pituitary microadenoma and hypothalamic hyperfunction, since also the adenomas of the anterior pituitary are to some degree subject to a feedback mechanism.

Diagnostic	Hypothalamic hyperfunction and adenoma of the anterior pituitary = central Cushing´s syndrome	Ectopic (paraneopl.) Cushing´s s.	Adrenal cortex tumour = adrenal Cushing´s syndrome
ACTH in plasma	n - ↑	↑ - ⇑	↓
ACTH increase after CRH administration	Yes	No	No
Cortisol decrease after high doses of dexamethasone	Yes	No	No
Localizing diagnostic	of the sella: CT, MRI (microadenomas not always detectable)	Tumor search !	of the adrenal cortex: sonography, CT, MRI, scintigraphy, angiography

Poss. additional diagnostic:
- In hypothalamic hyperfunction and adenomas of the anterior pituitary an ACTH concentration gradient or a difference of each sides is found after CRH stimulation if blood is taken at different levels - from the venae jugulares internae or the petrosal sinus (sinus petrosus catheter), however not in paraneoplastic Cushing´s syndrome.
- In paraneoplastic ACTH syndrome the so-called lipotropin (LPH) is found in some cases, a metabolite of the ACTH synthesis which can be used as a tumour marker.

Th.:
- Hormonal active adrenal cortex tumours: Adrenalectomy (surgical or endoscopic)
  Peri- and postoperatively a glucocorticoid substitution is temporarily (up until 2 y.) required, until the atrophic contralateral adrenal gland has again recovered.

- Hypothalamic hypophyseal Cushing´s syndrome:
  - Therapy of 1. choice: Transnasal/transsphenoidal surgical removal of the adenoma
  - Proton irradiation of the pituitary gland if surgery fails or in case of contraindications against surgery.
    Postoperative test whether therapy has been successful: Normalizing of the ACTH level

  Note: A bilateral adrenalectomy only removes the effector organ, not the cause.
  Disadvantages: 1) Life-long substitution of glucocorticoids necessary
                 2) Development of invasively growing ACTH producing pituitary tumours and brown skin pigmentation (Nelson syndrome, Nelson tumour) in circa 20 % of cases

- Inoperable adrenal cortex carcinoma and paraneoplastic ectopic ACTH secretion:
  Adrenostatic substances (blockage of the cortisol synthesis):
  - Ketoconazole (Nizoral®) + octreotide (Sandostatin®)
  - o-p-DDD (Mitotane®)
  - Aminoglutethimide (Orimeten®)
  - Metopirone

INCIDENTALOMAS OF THE ADRENAL GLAND

Def.: Coincidentally discovered tumour of the adrenal gland during an imaging diagnostic (U/scan, MRI, CT).

Ep.: 1 – 2 % in the age ≥ 65 years

Aet.:
1. Hormonally active adenomas and hyperplasias (circa 60 %)
2. Hormonally inactive tumours: Phaeochromocytoma, Cushing adenomas, Conn adenomas (circa 30 %)
3. Adrenenal cortex carcinomas (circa 5 %): Rule of thumb: Tumors > 6 cm ⌀ are very suspicious of being a carcinoma, tumours < 3 cm are probably benign
4. Metastases of bronchial, breast, renal carcinomas, carcinomas of the gastrointestinal tract, malignant melanomas (inspection of skin !)
5. Other rare causes:
   Cysts, pseudocysts, haematomas, myelolipomas, haemangiomas, tuberculomas, neuronal tumours etc

Di.:
- Imaging diagnostic: (Endo-)sonography, MRI, CT
- Hormone analysis: Exclusion of a phaeochromocytoma (catecholamines in 24 h urine), a Cushing´s syndrome (low-dose dexamethasone suppression test), a Conn´s syndrome (aldosterone/renin quotient) and of an increased androgen production (DHEAS)

Th.:
- Hormonally active tumours: see there
- Hormonally inactive tumours:
  < 3 cm ⌀ : Monitoring of the course
  > 6 cm ⌀ : Surgical or endoscopic removal
  3 - 6 cm: No standardized strategy. In case of the slightest suspicion of a carcinoma as well surgery.

ADRENAL CORTEX CARCINOMA

Internet: *www.nebennierenkarzinom.de; www.firm-act.org*

Syn.: Adrenocortical carcinoma

Ep.: Incidence: rare: 0,1/100.000/y.; m : f = 1 : 1,5

Cl.: Dependent on endocrine activity:
1. Hormonally active tumours (80 %): Signs of the hormone excess (sometimes several)
   - Glucocorticoids (Cushing´s syndrome or subclinical)
   - Sex steroids (f: virilizing, m: gynaecomasty)
   - Rarity: Aldosterone excess with hypokalaemia and hypertension
2. Hormonally inactive tumours (20 %): Signs of the abdominal space occupation (pains, nausea, constipation)

DD: Benign incidentalomas of the adrenal gland, phaeochromocytoma, adrenal metastases (esp. of bronchial Ca and mamma Ca etc)

Di.:
▶ Endocrine diagnostic: Exclusion of
   - Glucocorticoid excess (1 mg dexamethasone suppression test, urine free cortisol in 24 h urine collection)
   - Sex steroid excess (Serum: Androstenedione, 17-OH-progesterone, DHEAS, testosterone, 17β-oestradiol)
   - Mineralocorticoid excess (Aldosterone/renin quotient)
▶ Imaging diagnostic: Sonography, CT, MRI
   Rule of thumb: Tumors > 6 cm ⌀ are very suspicious of being a carcinoma, tumours < 3 cm are probably benign.
   Optional: Fluorodeoxyglucose PET

Th.: Surgical: R0 resection only curative approach (also after recurrence)
Adjuvant therapy: In high-risk patients (tumour > 8 cm, high mitosis rate, invasion of the vessels) also after R0 resection: Mitotane and/or irradiation of tumour site.
In case of distant metastases or after incomplete resection: Mitotane combined with streptozotozin or etoposide, doxorubicin, cisplatin (EDP)

Prg.: Depending on stage. The 5 year survival is in stage I - II (localized disease) circa 60 % and in stage IV (distant metastases) circa 15 % (median survival circa 12 months).

HYPOCORTISOLISM = ADRENOCORTICAL INSUFFICIENCY [E27.4]

Aetiology and classification:

1. Primary form:
    ACTH increased:
    - Addison´s disease [E27.1]: Autoimmune adrenalitis (80 %): Destruction of the adrenal cortex due to autoimmune processes with detection of autoantibodies against the adrenal cortex, which often aredirected against the 17α-hydroxylase (= key enzyme of the steroid synthesis).
      Some of these patients suffer from a polyendocrine autoimmune syndrome (see below)
    - Carcinoma metastases (especially of bronchial carcinomas, malignant melanomas, renal cell carcinomas)
    - Infectious diseases: Tuberculosis, CMV infection in AIDS patients
    - Aplasia or hypoplasia of the adrenal cortex, enzyme defects; therapy with substances, which inhibit the cortisol synthesis (e.g. aminoglutethimide).

    Causes of an acute adrenocortical insufficiency:
    - Waterhouse-Friderichsen syndrome = haemorrhagic infarction of the adrenal glands due to meningococcal septicaemia
    - Haemorrhages (coumarines, newborns)
    - Surgical removal of the adrenal glands
    - Failure of dose adjustment in patients with adrenocortical insufficiency and infections, other stress factors, unconsciousness, vomiting and diarrhoea etc

2. Secondary forms:
    ACTH decreased:
    - Insufficiency of anterior pituitary or hypothalamus.
    - Long-term therapy with corticosteroids! (Never withdraw corticosteroids abruptly in long-term therapy! → Danger of Addison crisis!)

    In primary adrenocortical insufficiency usually the production of all corticosteroids ceases; whereas the aldosterone production is only minimally affected in the secondary forms due to ACTH deficiency, so that in this case the electrolyte disturbances are in the background. Furthermore in hypophyseal insufficiency often the other glandotropic hormones are decreased, so that complex endocrine deficiency syndromes develop. Unlike in Addison´s disease pituitary insufficiency is characterized by a pale and unpigmented skin. This is due to a deficiency of POMC peptides (= peptides derived from proopiomelanocortin), which have MSH activity (MSH = melanocyte stimulating hormone).

Cl.: Addison´s disease:
Clinical symptoms usually only occur when 90 % of the adrenal cortex are destroyed. According to duration and severity of the adrenocortical hypofunction the symptoms can range from absence of any symptomatology under normal living conditions, through adynamia to unexpected Addisonian crisis occurring upon stress[E27.2]: 4 stages of the disease:
1. Latent adrenocortical insufficiency
2. Manifest adrenocortical insufficiency
3. Endocrine crisis
4. Endocrine coma

4 Cardinal symptoms of the manifest adrenocortical insufficiency (which are present > 90 % of cases):
1. Weakness and rapid fatigue
2. Pigmentation of the skin and mucous membranes, poss. vitiligo
3. Weight loss and dehydratation
4. Low arterial blood pressure
Further:
• Abdominal symptoms (nausea, vomiting, abdominal pain, diarrhoea, constipation)
• Loss of axillary and pubic hair in females (androgen deficiency) amongst others

Patients with unrecognized latent adrenocortical insufficiency are particularly at risk: Unusual stress can cause acute decompensation at any time:

Addisonian crisis
In addition to those symptoms mentioned before:
• Exsiccation, blood pressure drop, shock, oliguria
• Pseudoperitonitis
• Poss. diarrhoea and vomiting
• Hypoglycaemia, metabolic acidosis
• Initially subnormal temperatures, later fever due to dehydration
• Delirium, coma

Lab.: Serum Na^+ ↓ / K^+ ↑ ($Na^+/K^+ < 30$)
Poss. hypercalcaemia (30 %), lymphocytosis, eosinophilia
basal (morning) serum cortisol ↓

DD: • Adynamia and weight loss of other origin
• Abdominal disorders
• Hypoglycaemia, hyponatraemia/hyperkalaemia of other origin
• In Addisonian crisis additionally shock and acute abdomen of other origin
• In infants AGS (adrenogenital syndrome)

Di.: 1. ACTH test (Synacthen® test):
Serum cortisol determination before and 60 minutes after administration of 0,25 mg of ACTH (Synacthen®). In Addison´s disease the basal level is decreased or in the lower normal range and does not increase after ACTH (an increase of at least 7 µg/dl is considered to be normal). This also applies for a longer existing secondary adrenocortical insufficiency, in which the absence of the ACTH stimulation has led to an adrenal cortex atrophy.
2. Plasma ACTH:
In primary adrenocortical insufficiency (Addison´s disease) the basal plasma ACTH is significantly increased. In the secondary adrenocortical insufficiency the basal plasma ACTH is decreased or in the lower normal range and shows no or an insufficient increase in the CRH test.
3. Diagnostic tests for clarification of the aetiology:
- Search for adrenal cortex autoantibodies (up to 80 % positive)
- Imaging diagnostic of the adrenal gland:
Sonography, plain abdominal X-ray (calcification of the adrenal glands in Tb), CT, poss. angiography (carcinoma metastases?)

Th.: Substitution of the glucocorticoids and in Addison´s disease additionally of the mineralocorticoids:
1. Glucocorticosteroid: Cortisone acetate 15 - 20 mg in the morning, 10 - 15 mg in the afternoon
or hydrocortisone = cortisol 10 - 15 mg in the morning, 5 - 10 mg at noon and in the evening
2. Mineralocorticoid: Fludrocortisone (Astonin H®): Dose to be chosen that the plasma renin activity normalizes (0,05 - 0,2 mg/d).
3. Additional administration of DHEA = dehydroepiandrosterone: As an androgenic steroid it can be useful in women, who complain about loss of libido.

Monitoring of therapy regarding the exact dose of fludrocortisone: Physical well-being, normalization of blood pressure when lying and standing (Schellong test), sodium, potassium and plasma renin.
In all conditions of stress (infections, surgery etc) the doses of the glucocorticosteroid must be increased (depending on each stress situation 2 – 5fold of the normal daily dose)!
As with all hormone deficieny disorders requiring substitution patient education and emergency card are obligatory! (Prednisolone suppositories in case of vomiting in the hand luggage, in case of vomiting and diarrhoea to go to hospital for parenteral steroid substitution.)

Th.: of the Addison crisis: Immediate therapy after collection of a blood sample for determination of cortisol and ACTH!
1. 0,9 % NaCl and 5 % glucose infusion – total dose of 0,9 % NaCl and glucose infusion dependent on the degree of the hypotonic dehydration (CVP, serum sodium) and on blood sugar levels. No K+ containing solutions! Slow correction of the hyponatraemia (danger of the central pontine myelinolysis). Poss. correction of a metabolic acidosis.
2. Hydrocortisone: 100 mg i.v., afterwards 200 mg/24 h (in glucose 5 %) or equivalent dose of synthetic glucocorticosteroid.

Appendix

Polyendocrine autoimmune syndromes

Syn.: Polyglandular autoimmune syndromes (PAS), autoimmune polyglandular syndromes (APS)

Def.: In PAS (or APS) an autoimmune process of unknown cause leads to insufficiency of various endocrine organs.

• PAS (or APS) type 1 = juvenile form (Blizzard syndrome or APECED syndrome = autoimmune polyendocrinopathy candidiasis ectodermal dystrophy syndrome); very rare, autosomal recessive disorder; manifestation in childhood; mutation in the autoimmune regulator gene (AIRE)
- Addison´s disease
- Hypoparathyroidism
- Mucocutaneous candidiasis
- Lymphocyte functional defects

• PAS (or APS) type 2 = adult form (Carpenter syndrome): Manifestation in 3. decade of life, association with HLA-B8/DR-3
- Addison´s disease
- Diabetes mellitus type 1
- Autoimmune Hashimoto thyroiditis or Basedow´s disease

Note: Schmidt syndrome [E31.0] = Addison´s disease + autoimmune Hashimoto thyreoiditis
Often in addition also other organ-specific autoimmune disorders are present.

<u>Th.:</u> Substitution of the absent hormones, in APS 1 immunosuppressive therapy

ANDROGENITAL SYNDROME (AGS)   [E25.9]

<u>Def.:</u>   Autosomal recessive hereditary defect of the cortisol synthesis in the adrenal cortex.

<u>Ep.:</u>   Prevalence of the classic form of the AGS circa 0,1 ‰, frequency of heterozygosis 2 %.

Classification:
a) 21-hydroxylase deficiency (90 % of cases) and 3ß-dehydrogenase deficiency (rare):
2 clinical variants:
▶ "Simple-virilizing" form = uncomplicated AGS (rarer)
Cardinal symptom: Only virilization
▶ "Salt wasting" form = AGS with salt loss syndrome (more common)
Cardinal symptoms: Virilization + salt loss syndrome
b) 11 -hydroxylase deficiency (5 % of cases):
Cardinal symptoms: Virilization + <u>hypertension</u> (salt retaining form of the AGS due to excess in 11-desoxycorticosterone)
c) Rare: Adrenal cortex tumour

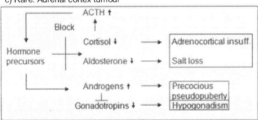

Clinical picture of the 21-hydroxylase deficiency:
• Virilization:
The increased androgen production in the AGS leads to an isosexual defect in boys and to an intersexual defect in girls.
<u>m.:</u> The hypogonadism contrasts the hyperdevelopment of the secondary sexual charcteristics: <u>Precocious pseudopuberty.</u>
<u>w.:</u> Hypertrophy of the clitoris with internal female genitalia (uterus, ovaries) = <u>female pseudohermaphroditism, virilization, primary amenorrhea, absent breast development.</u>
The patients are <u>tall in childhood and short in adulthood</u> (premature closure of the epiphysial plates).
• Salt loss syndrome in the neonatal period (50 % of cases)
- Defects in electrolyte metabolism (Na^+ ↓ / K^+ ↑)
- Vomiting, diarrhoea, exsiccation → incorrect diagnosis: Pyloric stenosis

Types of the disease:
▶ Classic form with manifestation in infancy
▶ "Late onset" forms with manifestation of the symptoms in puberty
▶ "Cryptic" forms: Enzyme deficiency with typical hormone profile, but without significant symptomatology

<u>DD:</u>   - Polycystic ovary syndrome (PCOS) = Stein-Leventhal syndrome: Hyperandrogenaemic ovarian insufficiency with increased LH/FSH quotient (> 2): Hirsutism, oligo-/amenorrhea, infertility, acne, alopecia. In circa 45 % a metabolic syndrome exists amongst others at the same time → *www.pco-syndrom.de*
- Androgen-producing ovarian tumours
- Androgen-producing adrenal cortex tumours (extremely rare) = acquired AGS

<u>Di.:</u>   Clinial picture + laboratory
● Cortisol ↓, ACTH ↑
● Detection of an overproduction of hormone precursors:
C-21-H deficiency: 17α-hydroxyprogesterone ↑
C-11-H deficiency: 11-desoxycortisol ↑

Late onset AGS and cryptic forms can in most cases only be recognized by the ACTH stimulation test: Increase of 17α-hydroxyprogesterone only after administration of ACTH.

- Screening for heterozygous carriers and genetic counselling:
  - The classic AGS is caused by the mutation CYP21.
  - HLA typification: All affected individuals of a family are identical regarding their HLA genotype
  - ACTH test: Heterozygous genetic carriers show normal basal values for 17α-hydroxyprogesterone and at the same time an excessive increase of its hormone precursors.

- In case of a repeated pregancy, prenatal AGS diagnostics:
  - Determination of 17α hydroxyprogesterone in the amniotic fluid
  - HLA typification of cultivated amniotic or chorionic cells
  - Analysis of the 21-hydroxylase gene from chorionic villae

Th.: Lifelong substitution therapy with glucocorticosteroids (patient card): Some part of the daily dose should be taken in the evening in order to suppress the morning ACTH peak (and thus to suppress the androgen production of the adrenal gland); in aldosterone deficiency additional administration of mineralocorticoids, in female patients additional treatment of the virilization with antiandrogens. In order to optimize the glucocorticoid substitution, 17α-hydroxyprogesterone is monitored in the serum or saliva or alternatively its metabolite pregnanetriol in the 24 h urine. A poss. necessary mineralocorticoid substitution is monitored via the plasma renin level. In all situations of stress the glucocorticoid dose is to be increased for a short period!

HIRSUTISM

Distinctions:
- Hypertrichosis: [L68.9] Androgen independent increase of the hair on the whole body without site of predilection
- Hirsutism [L68.0]: Abnormal increase of the androgen dependent male-type hair (chin, upper lip, breast, inner side of upper thigh and pubic area) in females without symptoms of virilization. A hirsutism which develops rapidly is suspicious of an androgen-producing tumour (ovary, adrenal cortex), if exogenous androgen supply is excluded.
- Virilism [E25.9] (virilization): Hirsutism + masculinization of the voice, larynx, body proportions; hypertrophy of the clitoris and amenorrhea due to overproduction of androgens, alopecia

Aet.: of the hirsutism

1. Idiopathic (90 %), genetic disposition/descent (Mediterranean countries → family history): Testosterone level and fAI (free androgen index) normal

2. Secondary:
   - Ovarian:
     a) Androgen-producing ovarian tumours
     b) Polycystic ovary syndrome (PCOS) = Stein-Leventhal syndrome → Rotterdam diagnostic criterias: 1. hyperandrogenaemia, 2. anovulation, 3. polycystic ovaries
     Sy.: Oligo-Amenorrhea, infertility, hirsutism; in 45 % a metabolic syndrome is present.
     Di.: Testosterone and fAI, androstenedione increased, dehydroepiandrosterone (DHEAS) normal, androgens increase after administration of HCG (ovarian origin !).
     Gynaecological examination, U/scan, CT
   - Adrenal: Androgen-producing adrenal cortex tumours (extremely rare), Cushing´s syndrome, adrenogenital syndrome (AGS), adiposity and type 2 diabetes
     In case of increased plasma testosterone a high dehydroepiandrosterone fraction indicates that the adrenal cortex is the cause of the androgen overproduction (androstenedione normal). In hyperplasia of the adrenal cortex there is in that case an ACTH dependent stimulation of the androgens (ACTH test). This finding is also typical for the adrenogenital syndrome (AGS). In adenoma and carcinoma of the adrenal cortex on the other hand there are found excessively high levels of androgens without ACTH dependency. In adiposity and type 2 diabetes mellitus (diabetic hirsutism) in most cases a bilateral adrenal cortex hyperplasia with moderately increased androgen production is present, while the symptoms frequently appear however only after the menopause.
   - Other endocrine causes: e.g. acromegaly, hypothyroidism
   - Drug related:
     - Testosterone and anabolics
     - Gestagens (progesterone derivatives)
     - Glucocorticosteroids and ACTH
     - Nonsteroidal drugs: Phenytoin, minoxidil, diazoxide, spironolactone, acetazolamide, ciclosporin, penicillamine etc

Di.: - Family/drug history – clinical picture
      - Consultation by internist/gynaecologist

Lab.: Dehydroepiandrosterone (adrenal: ↑ / ovarianl: normal) - testosterone and SHBG (sexualhormone binding globulin) - androstenedione

The free androgen index (fAI), which correlates with the free testosterone, is calculated from total testosterone and SHBG.

> *Note:* Free androgen index (fAI): 100 x total testosterone (nmol/l) : SHBG (nmol/l)
> Normal range: Women who are spontaneously ovulating 0 - 8,5; Men 14,8 - 94,8
> Hirsutism: 1,7 - 20,6

Th.: • Secondary hirsutism:

Causal therapy, e.g. omitting of causative drugs, surgical removal of androgen-producing tumours, treatment of an existing Cushing´s disease or of an AGS.
Adrenal form of the hirsutism: Inhibition of the early morning adrenal cortex stimulation via dexamethasone: 0,25 - 0,50 mg before going to bed.

• Ovarian form of the hirsutism: Poss. antiandrogenic hormonal contraceptive (e.g. Dianette 35®)

• Idiopathic hirsutism:
- Cosmetic: Epilation, hair bleaching agent, depilatory cream, shave
- Drug therapy: Ovulation inhibitor with antiandrogenic effect; spironolactone (100 mg/d – beware of hyperkalaemia); in severe hirsutism poss. additionally antiandrogens (e.g. cyproterone acetate)
- Photoderm therapy

GYNAECOMASTIA [N62]

Def.: Uni- or bilateral enlargement of the male mammary glands.

Pat.: Active form with epithelial hyperplasia
Fibrotic form with hypocellular fibrotic stroma (not reversible)

Ep.: Symptom free physiological gynaecomastia in 1/3 of all men; increasing with age.

Aet.: 1. Physiologic: Neonatal period, puberty, old age

2. Pathologic:
Shift of the oestrogen-testosterone ratio towards oestrogen.
   ▶ Oestrogen excess:
   Oestrogen therapy, oestrogen- or HCG-producing tumours of the testicles and of the adrenal glands, paraneoplastic syndrome (in small cell bronchial carcinoma), increased oestrogen conversion from androstenedione and testosterone in liver cirrhosis
   ▶ Androgen deficiency:
   Anorchia, castration, hypogonadism, Klinefelter syndrome (e.g. XXY), hyperthyroidism, seldom androgen receptor defects (androgen resistance → testicular feminization), prolactinoma
   ▶ Drug induced:
   Oestrogens, antiandrogens, spironolactone, cimetidine, ranitidine, omeprazole, finasteride, in rare cases also digitalis, beta-blockers, calcium antagonists, methotrexate etc
   ▶ Cannabis consumption (phytooestrogens)

3. Idiopathic (50 %)

DD: - Male breast cancer
- Gynaecomastia caused by recovery from malnourishment or pseudogynaecomastia due to increased adipose tissue depot: Weight gain in (previously undernourished) men

Di.: - History (drugs, weight gain, liver cirrhosis, hypogonadism)
- Examination of the breasts and testicles (palpation, U/scan of the testicles, mammography)
- Laboratory screening including liver/thyroid function
- Hormone analysis: Oestradiol, testosterone, LH end FSH, β-HCG, prolactin

	LH	Testosterone
Primary hypogonadism	↑	↓
Secondary hypogonadism or oestrogen-producing tumour	↓	↓
Androgen resistance or LH-secreting tumour	↑	↑

- Screening for tumours of the adrenal gland and lungs, exclusion of a prolactinoma
- Specialized examinations (e.g. chromosome analysis if Klinefelter syndrome is suspected)

Th.: - Provided that causes are recognized: Causal therapy, e.g. omitting of causative drugs, surgical removal of tumours, androgen substitution only in hypogonadism with testosterone deficiency
- A physiological and in most cases also an idiopathic gynaecomastia do not require treatment. Indication for surgery (subcutaneous mastectomy): Severe psychological/cosmetic problems as well as suspicion of carcinoma.
- In case of painful gynaecomastia poss. short-term attempt with the antioestrogen tamoxifen (2 x 10 mg/d for 6 weeks)

HYPOTHALAMUS and PITUITARY GLAND PITUITARY TUMORS   [D44.3]

Ep.: Circa 10 % of all brain tumours; incidence: 3 - 4/100.000/year; coincidentally discovered pituitary adenomas (= incidentalomas) are found in autopsies and in MRI examinations in circa 10 %.

Classification:
- Endocrine inactive pituitary tumours (40 %):
  In a broader sense, tumours close to the sella (e.g. craniopharyngiomas, which radiologically can show calcifications) belong to the group of endocine inactive pituitary tumours.
- Endocrine active pituitary tumours (60 %):
  - Prolactin-producing pituitary tumours = prolactinoma (40 %)
  - Growth hormone(= GH)-producing pituitary tumour with acromegaly (15 %)
  - ACTH-producing pituitary tumour = central Cushing´s syndrome (5 %)
  - TSH- and gonadotropin (LH/FSH)-producing pituitary tumours are extremely rare

Hi.: The classic light-micoscopic classification in chromophobic (endocrine inactive pituitary tumours, prolactinoma), eosinophile (acromegaly) and basophile adenomas (Cushing´s disease) has been replaced by the direct immunohistochemical detection of the corresponding hormones.

DD: Syndrome of the "empty sella" (empty sella syndrome): [E23.6] CSF-filled sella due to a malformation of the diaphragma sellae (causing an incomplete separation of the sella from the CSF space   this gradually leads to a displacement of the pituitary by CSF); found in circa 10 % of all autopsies. In most cases they are coincidentally (radiologic) findings; sometimes a hyperprolactinaemia occurs, only very rarely it comes to an anterior pituitary hypofunction. Occasionally secondary after pituitary surgery or necrosis of a pituitary adenoma.

Di.: MRI / CT

Endocrine inactive pituitary tumours   [D44.3]

Aet.: Endocrine inactive chromophobic adenomas, craniopharyngioma, (epi-)dermoid cysts, teratomas, metastases

Cl.: - Signs of the hypopituitarism (see below)
- Central diabetes insipidus
- Visual disturbances: Suprasellar growth can lead to pressure onto the optic chiasm with temporal upper quadrant hemianopia, bitemporal hemianopia, poss. scotoma, optic nerve atrophy or amaurosis.
- Poss. headaches

Di.: - MRI, CT
- Ophthalmologic examination (visual field !)
- Endocrinologic diagnostic in order to exclude a hypopituitarism and a diabetes insipidus (see below)

Th.: - To wait and monitor in case of a small coincidentally discovered pituitary tumour (incidentaloma) without visual disturbances.
- Transsphenoidal pituitary surgery or transfrontal craniotomy if tumour extension larger
- Irradiation therapy (in recurrence or CI for surgery; craniopharyngioma are radioresistant)
- Substitution therapy in hypopituitarism (see there)

Proclactinoma [D35.2]

Def.: Prolactin secreting adenoma of the anterior pituitary:
Microprolactinoma: Serum prolactin < 200 ng/ml, tumour size < 1 cm $\varnothing$
Macroprolactinoma: Serum prolactin > 200 ng/ml, tumour size > 1 cm $\varnothing$

Occ.: Most common endocrine active pituitary tumour (40 %). Approx. 20 % of the secondary amenorrheas are caused by a hyperprolactinaemia! f : m = 5 : 1; peak occurrence of disease: predominantly 3. and 4. decade of life.

Cl.: • Women: - <u>Secondary amenorrhea</u>, anovulation with infertility and poss. osteoporosis
                 - Poss. galactorrhea
                 - Loss of libido

    • Men: Loss of libido and virility, poss. gynaecomastia (indirectly via the hypogonadism)

    In pituitary tumours poss. signs of space occupancy (headaches, visual field defects) and hypopituitarism (see below)

DD: • Hyperprolactinaemia

    A) <u>Physiological: e. g.</u>
        - Pregnancy = oestrogen effect (10 – 20fold increase compared to normal)
        - Manipulation of the nipples/breasts, breast feeding
        - Stress

    B) <u>Pathologic causes: e. g.</u>
        - Prolactinoma
        - Para-/suprasellar tumours with impairment of the production or of the transport of dopamine = prolactin inhibiting factor (PIF)
        - "Empty sella syndrome" (= CSF- filled sella)
        - Severe primary hypothyroidism
        - Chronic renal insufficiency

    C) <u>Pharmacologic causes: e.g.</u>
        - Oestrogens
        - Neuroleptics, tricyclic antidepressants, opiates
        - Reserpine and α-methyldopa
        - Dopamine antagonists (e.g. metoclopramide)
        - Cimetidine, antihistamines etc

    • Other causes of a secondary amenorrhea

    • In case of galactorrhea exclusion of a breast carcinoma!

Di.: - <u>Multiple determination of the basal prolactin: Values > 200 ng/ml are almost conclusive, 25 - 200 ng/ml require further clarification</u>
    - Prolactin after TRH administration (in prolactinoma usually no increase)
    - Drug history to exclude a drug induced hyperprolactinaemia, a hypothyroidism and a renal insufficiency
    - Ophthalmologic diagnostics
    - Localizing diagnostic (<u>MRI</u>, CT)
    - If prolactinoma is detected, testing of the other pituitary partial functions

Th.: Multidisciplinary collaboration between gynaecologist, endocrinologist, radiologist, ophthalmo-gist, neurosurgeon!
The therapy is primarily based on <u>dopamine agonist</u> drugs: <u>Bromocriptine (Pariodel®), quinagolide (Norprolac®), cabergoline (Dostinex®)</u>. With this therapy in more than 95 % of the patients a normalization of the serum prolactin and retrogression of the tumour size occurs. As well in visual impairments initially a short-term trial of drug treatment is indicated, during which visual field defects often regress.

In pregnancy and microprolactinoma dopamine agonists usually are to be discontinued and tight prolactin controls are to be performed, since in pregnancy a sudden increase in tumour size can develop (oestrogen effect).
Transsphenoidal/transfrontal pituitary surgery is only indicated, if there is no response to dopamine agonists (especially in the case of impaired vision).

ACROMEGALY [E22.0]

Syn: Hyperpituitarism [E22.9]

Ep.: 0,3/100.000/year (rare disorder), patients in the 4. and 5. decade of life are predominantly affected

Aet.: <u>Somatotrophic adenoma of the anterior pituitary</u> with overproduction of growth hormone = (GH) = somatotrophic hormone (STH)

PPh: GH is secreted most during the sleep (esp. in puberty). The blood concentrations during the day are low. Hunger (hypoglycaemia), physical strain and stress stimulate release of GH; ingestion (hyperglycaemia) decreases the GH concentration.
The secretion of GH is stimulated by the <u>GH-releasing hormone (GHRH))</u> and inhibited by <u>somatostatin</u>.
The effect of the GH is caused mainly indirectly by IGF-1 (insulin like growth factor 1) = <u>somatomedin C</u>, which is synthesized in the liver. IGF-1 causes an inhibition of the GH-secretion (<u>negative feedback</u>).

Cl.: On average 5 – 10 years pass until the diagnosis is made. If a hyperpituitarism becomes manifest before the end of the linear growth, gigantism results (body height more than 2 m); in adulthood the GH excess manifests only in <u>acro- and visceromegaly</u>. The onset of the disease is insidious!

1. Cardinal symptom:
   - Change of the physiognomy with coarsening of the facial features, thickened and wrinkled facial skin (cutis gyrata): Comparison with former photos!
   - Enlargement of hands, feet and skull (shoes, gloves and hats no longer fit)
   - Enlargement of the tongue and widely spaced teeth (altered voice), poss. OSAS
   - Enlargement of the visceral organs (splanchnomegaly)

2. Possible symptoms:
   - Head aches, hypertension (up to 30 % of cases)
   - Visual impairment, visual field defects (bitemporal hemianopia) → ophtalmologic diagnostics
   - Poss. carpal tunnel syndrome (compression of median nerve with predominantly nocturnal pain + paraesthesias of the first 3 fingers + thenar atrophy), poss. joint pain
   - Hyperhidrosis, hypertrichosis
   - Poss. pathologic glucose tolerance (65 % of cases), diabetes mellitus (15 %)
   - Secondary amenorrhea
   X-ray: - Enlarged paranasal sinuses
           - Hands/feet: Cortical thickening of the bones
           - Heart enlargement seen in chest X-ray

DD: Athletic type (norm variant)

Di.: ▶ Hormone analysis:
   - Serum GH ↑; due to the pulsatile secretion more readings must be taken in the daily profile.
   - Absent suppression of the GH concentration after glucose load (oral glucose tolerance test). A serum GH < 1 ng/ml usually excludes a diagnosis of acromegaly.
   - Corrected value of IGF-1 according to age ↑
   - Verification of the remaining pituitary partial functions, in order to exclude an insufficiency.

   ▶ **Localizing diagnostic:** Detection of a pituitary adenoma: MRI, CT

Th.: 1. Surgical: Transsphenoidal adenomectomy

2. Irradiation therapy: Conventional X-ray therapy or stereotactic radiosurgery (e.g. "gamma-knife")

3. Drug based inhibition of the GH secretion or GH effect:
   Ind.: In insufficient efficacy or CI of 1 + 2
   - Dopamine agonists: e.g. bromocriptine (Pariodel®) are only successful in 20 % of the cases.
   - Somatostatin analogues: e.g. octreotide (Sandostatin Lar®) or lanreotide (Somatuline®) in form of depot injection: 1 x/month s.c.
     Disadvantage: Subcutaneous application, high costs
     Ind.: Preoperative preparation, bridging the time until radiation therapy starts becoming effective; poss. in older patients with CI against radiosurgery
     SE: Occasionally local reactions at injection site, gastrointestinal SE, blood glucose increase in diabetes mellitus etc
   - GH receptor antagonists: Pegvisomant (Somavert®) normalize the increased IGF-1 level and lead to clinical improvement, they do not however reduce the size of the anterior pituitary adenoma.
     SE: Occasionally AB production, gastrointestinal SE, influenca-like symptoms, increase of transaminases etc
     Administration: Daily s.c. injections

Prg.: If untreated, the life expectancy is reduced by circa 10 years, esp. because of cardio- and cerebrovascular complications, increased occurrence of breast and colon carcinomas (→ preventive examinations). Normalizing of the growth hormone IGF-1 system improves the prognosis.

HYPOPITUITARISM [E23.0]

Def.: - Panhypopituitarism = complete failure of functions of the anterior pituitary with full clinical picture
- Incomplete hypopituitarism = failure of single partial functions of the anterior pituitary (most common form)

Aet.: 1. Hypophysial space occupying lesions (most common):
   - Pituitary adenomas (endocrine active or inactive); usually the macroadenomas (≥ 1 cm ∅) are the ones, which are space occupying.
   - Craniopharyngiomas (developing from cell rests of the Rathke pocket) often already become symptomatic in childhood.
   - Further: Meningiomas and other tumours, metastases

   Beware: In case of suprasellar growth visual field defects (initially temporal upper quadrant anopia, later bitemporal hemianopia), optic nerve atrophy

2. Traumatic and vascular causes (second most common)
   - Traumas of the pituitary (accidents, surgical procedures), resulting from radiation

- Sheehan syndrome = ischaemic pituitary necrosis resulting from larger peripartal blood losses (in rare cases also other states of shock, e.g. shock due to burns, can lead to a anterior pituitary necrosis).
Early symptoms: Agalactia, secondary amenorrhea, no regrowth of poss. shaved pubic hair
A hypopituitarism following a Sheehan syndrome can manifest sometimes only after years and in those cases remains unrecognized for a long period or just is diagnosed during the crisis.
3) Inflammatory-infiltrative causes (rarer)
- In the context of systemic granulomatous disorders (e.g. Wegener granulomatosis, sarcoidosis, Langerhan´s Histiocytosis, TB etc)
- Pituitary involvement in haemochromatosis
- Autoimmune hypophysitis (with lymphocytic infiltration, typical occurring in the second half of pregnancy)
4) Rare hereditary forms, e.g. due to mutations of the PROP1 gene, idiopathic hypogonadotro-phic hypogonadism (IHH) poss. in combination with absence of olfactoric neurones (= Kallmann syndrome)

CI.: 1. Poss. symptoms of the space occupancy due to tumours (headaches, visual disturbances, see above)
2. Hormone deficiency symptoms in case of failure of any single or all 5 hormonal axes of the anterior pituitary
A) Chronic hypopituitarism:
80 % of the anterior pituitary must be destroyed, before clinical symptoms present due to deficiency of the peripheral hormones.
Should a hypopituitarism develop as a result of a anterior pituitary adenoma, then the hormonal partial functions often fail in the typical order: GH - gonadotropin - TSH - ACTH
- GH deficiency during the growth period: Pituitary dwarfism (intelligence + body proportions normal).
- Syndrome of the GH deficiency in the adult: Abdominal fat storage; muscle mass ↓; adynamia; LDL ↑/HDL ↓, risk of arteriosclerosis ↑, risk of osteoporosis ↑
- Secondary hypogonadism (gonadotropins LH and FSH ↓): Secondary amenorrhea, loss of libido and virility, disappearance of the terminal hair (axilla and pubic hair), poss. depression, osteoporosis
- Secondary hypothyroidism (TSH ↓): Cold intolerance, bradycardia, fatigue etc
- Secondary adrenocortical insufficiency (MSH and ACTH deficiency): Adynamia, weight loss, waxen paleness due to depigmentation, arterial hypotension, hypoglycaemia etc
The glucocorticoid deficiency causes an uncontrolled ADH secretion (see SIADH) with hyponatraemia.
- Prolactin deficiency in breast feeding women leads to agalactia.

Note: The patient´s face appears without expression; an absence of the lateral eyebrows is a typical sign. In advanced cases poss. weight loss (25 % of cases).

Remember: "Six `A´s" due to deficiency of:
1. Gonadotropins: Axilla/eyebrow hair loss, amenorrhea, agalactia
2. TSH: Apathy
3. ACTH: Adynamia
4. MSH: Alabaster coloured paleness

B) Acute hypopituitarism and hypophysial coma:
Deficiency of GH, LH, FSH or MSH never leads to an acute crisis-like situation. Under stress however it can come due to an ACTH or TSH deficiency to an acute pituitary coma with drowsy-stuporous clinical picture. Triggering factors could be infections, traumas, operations, vomiting and diarrhoea, esp. If in such cases substitution therapy is insufficient.

Sy.: • Hypotension, bradycardia
• Hypothermia, hypoglycaemia
• Hypoventilation with hypercapnia
• Waxen paleness, absence of terminal hair

DD: - Polyendocrine autoimmune syndrome (see there)
- Severe general diseases (liver/renal insufficiency) with endocrine defetcs
- In pituitary coma: Myxoedematous coma (note: no hypoglycaemia) and Addison crisis (brown skin pigmentation)
- Anorexia nervosa (see there)
- Other causes of a pituitary microsomia: Hypothyroidism, Turner syndrome (karyotype 45,XO), coeliac disease etc

Di.: 1. History and clinical picture

2. Diagnostic of the endocrine function
Besides the decreased basal level of the pituitary hormones there is an insufficient stimulation after application of the releasing hormones.
- Thyrotropic function: Low basal TSH and thyroxine with absent or insufficient increase after administration of TRH.
- Corticotropic function: Low basal ACTH/cortisol with insufficient increase after CRH administration or insulin hypoglycaemia test
- Somatotropic function: IGF-1 decreased and insufficient increase of the GH after administration of GRH (GRH test) or in the insulin hypoglycaemia test.
- Gonadotropic function: LH and FSH basal and after LHRH administration decreased, reduced basal testosterone or oestradiol

- Lactotropic function: Prolactin determination basal and after TRH administration. In panhypopituitarism also the prolactin is decreased, in hypothalamic processes however rather increased due to the failure of dopamine (= prolactin inhibiting factor = PIF).

3. Localizing diagnostic of the pituitary (exclusion of a tumour): MRI, CT

Th.: A) Causal therapy: e.g. treatment of a pituitary tumour

B) Substitution of the deficient hormones:
Patient education + issue of medical emergency card!
- Gonadotropic function:
Men: 250 mg of testosterone (e.g. testoviron depot®) every 3 - 4 weeks i.m. or 1.000 mg testosterone undecanoate (Nebido®) every 3 months i.m. or in form of a patch (e.g. Androderm®) or gel (50 – 75 mg/d): Androtop®, Testogel®
Women: Oestrogen-gestagen combination under gynaecological supervision
- Thyrotropic function: L-thyroxine (see chapter on the thyroid)
- Corticotropic function: Cortisone acetate 25 - 37,5 mg/d or hydrocortisone 20 - 30 mg/d; with 2/3 of the dose given in the morning, 1/3 in the afternoon
- Somatotropic function: In children with microsomia, but also in adults with marked STH deficiency substitution with genetically engineered GH.

The dose of the substitution therapy is based on the clinical findings and the levels of the replaced peripheral hormones (thyroxine, cortisol etc.)
A problem of the replacement therapy is the adaption to unusual stress situations, e.g. infections, operations etc., while the dose of the replaced glucocorticosteroids is increased multiple times; if oral intake is no longer possible: Parenteral substitution!
In the pituitary coma the rapid administration of hydrocortisone (100 mg as a bolus and 100 mg/24 h) i.v. and fluid replacement are especially important and poss. therapy of a hypoglycaemia. Only 12 - 24 h later also replacement of levothyroxine.

DIABETES INSIPIDUS [E23.2]

Def.:: Reduced ability of the kidneys to produce concentrated urine in case of withdrawal from water due to ADH deficiency (central D.I.) or failure of the kidneys to respond to ADH (renal D.I.).

Aet.: A) Central diabetes insipidus (most common form):
1. Idiopathic (circa 1/3 of cases), some of these cases are dominantly inherited, in some cases autoantibodies against vasopressin-producing cells are found.
2. Secondary (circa 2/3 of cases):
- Tumors of the pituitary gland or of its vicinity, metastases
- Traumas, neurosurgical operations
- Encephalitis, meningitis etc

B) Nephrogenic (renal) diabetes insipidus (NDI) (rare disorder)
1. Congenital form, 2 genetic variants:
- X-chromosomal recessive NDI: Mutated gene (at Xq28) for the vasopressin type 2 receptor
- Autosomal recessive NDI. Deficient water transport channel "aquaporin 2" of the renal collecting tubule
2. Acquired renal disorder with tubular damage, hypokalaemia, hypercalcaemia, drug induced (lithium carbonate)

Pg.: Central diabetes insipidus is caused by a deficiency in antidiuretic hormone of the poterior pituitary (ADH = arginine-vasopressin). Thus the ADH dependent urine concentration in the distal renal tubules and collecting tubules is not possible and it leads to an increased excretion of a diluted urine (polyuria) with at the same time an inability to concentrate urine (asthenuria). Due to an osmoregulative effect, compulsive polydipsia is induced.
Nephrogenic D.I. is caused by a failrue of the distal tubule to respond to ADH (deficiency of the ADH receptors).

Cl.: Typical triad:
- Polyuria (5 - 25 l/24 h)
- Compulsive thirst with polydipsia (frequent drinking)
- Asthenuria (inability to concentrate the urine)

Beware: Extended periods without drinking results in hypertonic dehydration. In infants < 2 y. diarrhoea instead of polyuria is observed! An absent nocturia virtually excludes a diabetes insipidus !

DD: 1. Psychogenic polydipsia
2. Diabetes mellitus (osmotic diuresis)
3. Misuse of diuretics
4. Hypercalcemic crisis

Di.: 1. Determination of the urine osmolarity in a water deprivation test and after exogenous ADH administration:

- Water deprivation test: In healthy individuals there is an increase of the urine osmolarity due to osmoregulation via ADH sectretion. In D.I. the urine osmolarity remains < 300 mOsmol/l , while the plasma osmolarity increases to > 295 mOsmol/l. In this case a test dose of ADH or desmopressin is given (contraindicated in coronary heart disease because of vasospastic effect!). After this treatment the urine osmolarity increases in the central D.I. (unlike however in the nephrogenic D.I.).
- If fluid intake during water deprivation test is suspected, poss. challenge with hypertonic sodium chloride solution (test according to Hickey-Hare): Physiologic and pathophysiologic processes analogue to the water deprivation test.

2. ADH determination (rarely necessary in the water deprivation test)
3. Localizing diagnostics in order to exclude a tumour in the area of the pituitary/hypothalamus: MRI, CT

		Central D.I.	Renal D.I.	Psychog. Polydipsia
Water deprivation test	Urine osmolar.	Remains low	Remains low	↗
	Plasma osmolar.	↗	↗	Normal
	Serum ADH	Remains low	↗	↗
Test dose desmopressin		Urine osmolar. ↗	Without effect	Without additional effect

Th.: 1. of the central D.I.:
- Causal: Treatment of the underlying disease in the symptomatic forms
- Symptomatic: Desmopressin (Minirin®), a vasopressin analogue for intranasal or oral administration

2. of the renal D.I.:
- Causal therapy !
- Symptomatic: Trial with thiazide diuretics or nonsteroidal antiphlogistics

SYNDROME OF INAPPROPRIATE ANTIDIURETIC HORMONE (SIADH)   [E22.2]

Def.: Pathologically increased ADH secretion with water retention and dilution hyponatraemia.

Aet.: - Paraneoplastic (esp. small cell bronchial carcinoma - 80 % of cases)
- Uncontrolled pituitary ADH secretion: Pulmonal processes (e.g. pneumonias), central nervous system related disorders (meningitis, apoplexia etc), drug induced (thiazide diuretics, vasopressin, oxytocin, serotonin re-uptake inhibitors = SSRI, tricyclic antidepressants, carbamazepine, vincristine, cisplatin etc)
- Idiopathic (without recognizable cause)

Cl.: Some cases are asymptomatic (coincidental laboratory finding). The following symptoms can occur:
- Loss of appetite
- Nausea, vomiting, headaches, muscle cramps
- Irritability, personality changes
- Poss. water intoxication with neurologic symptoms (stupor, convulsions)
- No oedemas, since water retention only amounts to approx. 3 - 4 ℓ.

Lab.: • Hyponatraemia (often < 110 mmol/l), hypoosmolality of the serum
- Despite a hypotonic extracellular fluid a concentrated (hypertonic) urine is excreted (> 300 mOsm/kg H2O).
- Normal function of the adrenal cortex and of the kidneys
- Plasma ADH normal to ↑ (in other forms of the hyponatraemia ↓)

DD: - Other causes of a hyponatraemia (see there)
- Hypothyroidism, Addison´s disease
- In case of water intoxication DD of neurologic/psychiatric disorders

Di.: History – clinical picture/laboratory

Th.: A) Causal:
e.g. in inflammatory diseases, spontaneous improvement during therapy of the underlying illness. Poss. discontinuation of triggering drugs etc
B) Symptomatic:
- No therapy in asymptomatic mild hyponatraemia
- Fluid restriction (500 - 800 ml/d)
- Vasopressin receptor antagonists: Vaptans, e.g. tolvaptan, conivaptan etc. are under clinical trials and can lead to an aquaresis with increase of the serum sodium.
- Only in life-threatening water intoxications (serum Na$^+$ < 100 mmol/l), careful infusion of hypertonic NaCl solution (serum sodium may increase by 10 mmol/l maximum in 24 h) and administration of furosemide to increase the diuresis (In a too rapid increase of the serum sodium danger of the pontine myelinolysis!)

Disorders of the arteries:
• Peripheral vascular disease
• Thrombangiitis obliterans (Winiwarter-Buerger disease)
• Cerebrovascular disease
• Atherosclerotic occlusive disease of the visceral arteries
• Vasculitis
• Diabetic angiopathies (refer to chapter „ diabetes mellitus")
• Abdominal aortic aneurism and dissecting aortic aneurysm
• Raynaud's-syndrome

Peripheral atherosclerotic occlusive disease of the limbs (AOD) [I73.9]

Def.: Stenosing and occluding changes of the aorta and the arteries of the limbs; > 90% in the lower limbs; > 95 % are caused by atherosclerosis.

Ep.: Prevalence: approx. 3% of the population >60 years have got symptoms (usually smokers = single largest risk factor), higher incidence among the older population; m : f = 4 : 1

Aet.: • Usually chronic obliterating arteriosclerosis (> 95 %)
main risk factors: 1. Nicotine abuse and 2. diabetes mellitus
also: arterial hypertension, hyperlipidaemia (for further risk factors: refer to chapter „coronary heart disease")
• Rare causes (<5 %): Thrombangiitis obliterans = Winiwarter-Buerger disease(refer to the according chapter), Takayasu-arteritis (vasculitis)

PPh: The remaining blood supply (or haemodynamic compensation) depends on:
• Length of the occlusion, severity of stenosis, number of affected vessels
• Collateral circulation
• Blood supply requirement of the affected tissue
In healthy people the blood supply of the limbs can be increased 20fold via dilation of the precapillary resistance (arterioles). The difference between blood supply at rest and the maximal possible blood supply is called the reserve.

Loc: Mainly (> 90 %) lower limbs:

A) One segment only:

type (frequency)	localisation	absent pulses	ischaemic pain
aortoiliac type = pelvic type: 35 %	aorta / iliac artery	from groin downwards	buttocks, thigh
upper leg: 50 %	femoral artery/ popliteal artery	from popliteal artery downwards	calf
peripheral type: 15 %	arteries of lower leg and feet	pedal pulses	sole of foot

B) Several segments:

Specials:
Diabetes mellitus: arteries of the lower leg (50 %) and deep femoral artery
Thrombangiitis obliterans: arteries of the lower leg (90 %), also arteries of the forearm

Sym.: Stenosis > 90%: there is no palpable pulse distal to the stenosis. If there is sufficient collateral circulation and/or the patients can't exercise/walk a lot due to other underlying diseases (cardiac/ pulmonal failure; neurological or orthopaedic diseases), then they might be asymptomatic (grade I). The main presenting symptom is ischaemic pain triggered by exercise (St. II), distal to the stenosis, causing the patient to stop walking after a certain distance (intermittent claudication; claudere = to limp). The feet are pale and cold, there is thinning of the skin (St. III) and poor wound healing.

Note: The pain in intermittent claudication is ischaemic muscular pain, triggered by exercise; it resolves as soon as the patient stops walking/ or rests!

Grade III: distal localised pain at rest, particularly at night, and when the leg is being lifted. Grade IV: necrosis/gangrene; first appearing on the toes/pressure points.

Complication: ischaemic neuropathy (painful paraesthesiae, atrophic paralysis etc.)
Obstruction of the aortic bifurcation (Leriche-syndrome) can cause sciatic-like pain and erectile dysfunction.

Grades of the peripheral arterial disease (Fontaine-Ratschow):
I.   asymptomatic (75 %)
II. exercise related pain = intermittent claudication
     a) pain free walking distance > 200 m
     b) pain free walking distance < 200 m
III. ischaemic muscular pain at rest   III and IV: severe ischaemia
IV. necrosis/gangrene/ulcer

DD: • of intermittent claudication (grade II):
        - Orthopaedic diseases (vertebral root related diseases, tilted pelvis, difference in length of legs, flat feet, arthritis
          of knees and hips): Sometimes this is misdiagnosed as intermittent claudication! In these cases the patient feels
          relief when shifting the weight and taking on a different posture, avoiding painful movements
        - Neurological diseases (spinal stenosis, peripheral sensorineural nerve damage): no obvious relation to
          exercise/effort)
        - Pronounced venous obstruction (venous claudication): improving when the leg is being raised
        - Necrosis and ulcers from St. II (= complicated St. II) due to trauma must be distinguished from St. IV (severe
          ischaemia).
      • of the non-exercise related pain (at rest) (St. III): e.g.
        - vertebral root syndrome
        - gout of the MTP joint of the big toe
        - diabetic polyneuropathy

Di.: • Examination: skin colour, skin temperature, skin thinning, dry gangrene or wet (infected) gangrene

      • Pulse: loss of pulse if reduction of the lumen ≥ 90 %

      • Auscultation: systolic bruit if outflow lumen is > 60 - 70 %

      • Systolic Doppler blood pressure at rest of both arms and lower legs (most important technical investigation)
        Usually the systolic pressure of the posterior tibial artery (Pankle) is approximately 10 mmHg higher than the
        pressure of the upper arm → Ankle-Brachialis-Index (ABI) normal > 0,9 - 1,2. Minor arterial obstruction has an
        index of 0,9 - 0,75. When the patient is free of any pain, we call it asymptomatic AOD (St. I). Moderate obstruction
        causes an index of < 0,75 - 0,5. Index < 0.5 or post-occlusive systolic pressures < 50 - 70 mmHg can be found in
        severe ischaemia with a risk of necrosis and amputation (St. III and IV)! blood pressure difference between arms:
        The higher reading of the two should be used. Doppler readings shouldn't be performed unless the patient has
        been supine for at least 15 minutes, otherwise they might be too low. Mönckeberg's media-sclerosis (90% of type
        2-diabetic patients): Often the readings are too high due to impaired compressibility of the arteries. In these cases
        a pulsatility index should be calculated (Gosling) from the doppler results (severe ischaemia if index < 1,2) or the
        Toe-Brachial-Index (TBI), which is usually > 0,6. The toe pressure is obtained via photo-plethysmography or laser-
        Doppler.

      • CW-doppler pressure measuring after exercise (e.g. standing on one's toes 20 times): In healthy individuals there
        is a drop in the ankle pressure by max. 35 %, which returns to the basic pressure after 1 minute. There is a bigger
        drop in pressure and delayed recovery in patients with peripheral arterial disease.

      • Standard exercise tests via metronome or treadmill (diagnostics for St. II):
        To measure the pain free exercise tolerance.

      • Measure percutaneous pO_2:
        normal readings for pO_2 (supine): 64 ± 10 mmHg (when in an upright position: 30 mmHg higher)
        readings < 30 mmHg reflect severe pathology (severe ischaemia). In advanced disease these readings cannot be
        improved by inhaling 100% oxygen.
        Screening test using pulse oximetry: O_2-saturation of the big toe is pathological, when it is 2 % lower than the
        saturation on the index finger.

      • Directional Doppler US with typical curve:
        Distal to an arterial stenosis there is a loss of the early diastolic backflow and a reduced amplitude of the systolic
        pre-flow; occ. there is a holodiastolic outflow. Within the stenosis the flow is much increased (> 180 cm/sec.).

      • Radiography to localise a stenosis:
        Coloured Duplex US and 3 D-MRI-angiography
        arteriography (DAS) only if indicated and when the team is prepared for interventional therapy

      • Diagnostics to exclude CHD and AOD of the cerebral arteries. The most common cause of death of AOD is CHD!

Th.: A) Causal:
        Eliminate any risk factors = basic therapy for all patients: stop smoking; aim for very good control of diabetes
        mellitus, hypertension, hyperlipidaemia (LDL-Cholesterol < 100 mg/dl is desirable → use CSE-inhibitors!)

      B) Symptomatic
        Treat AOD according to its stage:
        - Ergotherapy (walking training): St. II
        - Drug therapy
        - Revascularisation

- Treat any wounds and infections
- Last resort: amputation

1. Conservative:
   - **Ergotherapy:** Stimulation of blood flow <u>to generate new collaterals by daily programmed walks of 1 - 1½ hours!</u> To pause when the pain sets in (walking sport-/coronary sport-/AOD-group).
     Ind: St. II (only in the presence of sufficient haemodynamic compensation)
     CI: poorly compensated St. II, St. III and IV
   - Drug treatment:
     - <u>Thrombocyte aggregation inhibitors</u> are indicated from grade II, but they also make sense at grade I: Acetylsalicylic Acid 75 - 300 mg/day; Acetylsalicylic Acid-intolerance: Clopidogrel 75 mg/day (SE + CI refer to "thrombocytes aggregation inhibitors")

       Note: thrombocyte aggregation inhibitors reduce the vascular total mortality by 20 % (Antiplatelet Trialist's Collaboration-Study).

     - <u>Anticoagulants</u> are only indicated in special situations: relapse prophylaxis of arterial (cardiac) embolism, arterial occlusion with a mainly thrombotic component etc.
     - <u>Prostanoids i.v.:</u> (no evidence of efficiency)
       Ind: St. III and IV, when revascularisation is not possible or not successful.
       ◦ Alprostadil = Prostaglandin E1 = PGE 1 (Prostavasin®)
       ◦ Iloprost, a Prostacyclin derivate (Ilomedin®)
       <u>Mode of action:</u> vasodilatation → improved circulation via collaterals, inhibition of the thrombocytes aggregation, promotes positive metabolism in the ischaemic area
       CI: cardiac failure, cardiac arrhythmia, CHD, hepatic disease, pregnancy, breast feeding etc.
       SE: blood pressure drop, tachycardia, occ. triggers angina pectoris, deterioration of cardiac failure, central nervous SE etc.
     - <u>Isovolaemic haemodilution:</u> (no studies to prove efficiency)
       target: reduce Hct to 35 - 40 %
       Ind: only in polycythaemia, polycythaemia rubra vera
       CI.: anaemia, dehydration
       principle: 500 ml venesection + give <u>simultaneously</u> 500 ml volume substitution
     - <u>Vasoactive substances</u>, e.g. <u>Naftidrofuryl</u> is supposed to improve microcirculation by a complex mechanism; there is no supporting evidence
       Ind.: St. II, when walking is not possible, and the systolic ankle pressure is > 60 mmHg. CI: iliac artery occlusion, established cardiac failure.
       SE + CI (e.g. cardiac failure) have to be considered.
     - <u>Cilostazol:</u> a phosphodiesterase (PDE) 3-inhibitor, has thrombocyte aggregation inhibiting and vasodilating properties; no longterm studies.
       SE: headache, diarrhoea, occ. tachycardia etc.
       IA: Erythromycin, Ketoconazole, Diazepam, Cimetidine will increase plasma levels.
       CI: cardiac failure etc.
   - <u>Treat any cardiac failure</u> (to improve the pumping function) and <u>treat any pulmonary disease</u> (improves the arterial O_2-saturation)
   - <u>Omit any medication, which might worsen the peripheral circulation</u> (e.g. Betablocker, Dihydroergotamine).
   - <u>Local treatment:</u> thorough podiatry (moisturisation of dry skin, careful podiatry, comfortable shoes), prophylaxis and consequent therapy of any injuries; St. III and IV: legs must be in a low position, heels must not touch the mattress or stool, padding with cotton wool, care with bedding to avoid pressure sores.
     <u>No</u> measurements to increase circulation, no heat treatments (increased need for O_2, risk of burns).
     local treatment of necrosis, ischaemic ulcers: wound cleaning, remove any necrosis, daily change of dressing
     <u>local infection requires systemic antibiotic therapy</u> (take swab for antibiotic sensitivity); local antibiotics are not indicated.

2. Catheter procedures for revascularisation:
   - <u>Percutaneous transluminal angioplasty (PTA) and stent-therapy is the standard method</u>
     Ind: St. II and above; short, little calcified stenosis and occlusion < 10cm length. For bilateral stenosis of the iliac artery PTA with „kissing balloons" + stents. Subsequent therapy with Clopidogrel
   - <u>Further catheterising methods:</u> rotation-, laser-, ultrasound-angioplasty etc.
     Ind: long stenosis
   - <u>Combination of local lysis sometimes with aspiration-thrombectomy + subsequent PTA</u>
     Ind: arteriosclerotic stenosis + apposition thrombus; acute thrombotic occlusion, sometimes post PTA
     <u>time limits for thrombolysis:</u> upper leg/upper arm: 2 months; lower leg/lower arm: 1 month; foot/hand: few days

3. Operative therapies:
- • Operative revascularisation
  - thrombendarterectomy (TEA) = disobliteration = scraping (incl. vascular intima); e.g. using a ring stripper
    Ind: stenosis of the iliac artery or femoral artery (e.g. disobliteration of a narrowed femoral artery)
  - Bypass-operations:
    - • Vessel replacement using an autologous saphenous vein to bypass a stenosis in the upper leg and lower leg
    - • Vessel replacement using artificial material (Teflon = Polytetrafluoroethylen = PTFE);
    Ind: High infrarenal aorta obstruction with involvement of the iliac arteries. The operative mortality of an aorto-bifemoral Y-bypass is 1 %.

  Ind. for operation: St. III and IV
  Co.: After operative revascularisation:
  - Generally: venous thrombosis, pulmonary embolism
  - Locally: postoperative haemorrhage, displacement of the prosthesis, infection, relapse (up to 50 % within 5 years post PTA or operation)
- • Amputation: last resort in St. IV, when inflow and outflow are not sufficient for revascularisation.
  Note: always consult a vascular specialist before any amputation!

4. Experimental therapy:
  - Intra-arterial application of fibroblast-growth factor (rFGF-2) or vascular growth factors (VEGF)
  - Therapy with bone marrow stem cells and endothelial progenitor cells (EPC)

Prg.: Depends on:
- Severity (stage)
- Presence or ongoing risk factors:
  Without smoking cessation the prognosis is poor. Poorly controlled diabetes and smoking increase the risk of relapse and amputation!
- Any further manifestations of generalised arteriosclerosis (coronary heart disease, cerebro-vascular disease) and other diseases (e.g. cardiac failure in CHD, COPD in smokers).

Note: patients in St. II also suffer from coronary artery stenosis in ca. 50 %; in St. III and IV 90 % suffer from coronary heart disease! (CHD - diagnostics!) and in 50 % arteriosclerotic changes of the extracranial cerebral arteries. The majority will die from myocardial infarction (ca. 60 %) and stroke (ca. 10 %). Ca. 10 years reduction of life expectancy.

Thrombangiitis obliterans [I73.1]

Syn: Winiwarter-Buerger syndrome, Buerger's disease

Def.: Not-atherosclerotic, smoking associated immune mediated endarteritis, causing secondary thrombosis of the vessel lumen.

Ep.: • Patients with TAO make up ca. 2% of patients with peripheral arterial disease in West Europe, in East Europe 4 %, in the mediterranean and Israel 6 %, and in Japan 16 %.
• Men > women; almost all patients are heavy smokers! Onset of disease before the age of 40

Aet.: Unknown; 3 factors seem to play a role: smoking, genetics (HLA-A9 and B5) and immunopathogenesis (circulating immunocomplexes in the blood).

Sym.: • Pain, occ. plantar claudication (misdiagnose: orthopaedic diseases)
• Cyanosis, cold distal phalanges
• Phlebitis migrans (or phlebitis saltans)
• Necrosis, ulcers, gangrene at the finger tips and sometimes toes
  DD: embolism (exclude any cardiac source of embolism)

DD: Exclude any peripheral arterial embolism (transoesophageal echocardiography)

Di.: Smoking - symptoms - doppler, MR-angiography (multiple occlusion of the hand-/foot arteries with „cork screw-collaterals")

Th.: • Stop smoking! (most important part of management!) – make use of smoking cessation courses and Nicotine replacement therapy!
• Prostaglandin E1: Alprostadil (Prostavasin®) and Iloprost (Ilomedin®)
• Acetylsalicylic acid (100 mg/day)
• Occ. sympathectomy (no evidence)

Prg.: The amputation rate is up to 30 %; reduced life expectancy due to smoking complication. Most patients are not successful in stopping smoking!

CEREBROVASCULAR DISEASE [I67.2] AND STROKE

Def.:
- Extracranial cerebral arteries: Arteries between aortic arch and the base of the skull (supraaortal branches of the aortic arch): Truncus brachiocephalicus, subclavian artery - vertebral artery, common and internal carotid artery. Most frequent location of impaired circulation: internal carotid artery (ICA): 50 % (particularly: carotid bifurcation)
- Intracranial cerebral arteries: Circle of Willis and its branches; most frequently: middle cerebral artery (ACM): 25 %
- Stroke is a primary clinically defined, polyaetiological syndrome; there is a sudden onset of focal-neurological deficit, with a vascular cause. Clinical criteria cannot differentiate between an ischaemic cerebral infarction (ca. 80 %) and a haemorrhagic stroke (ca. 20 %):

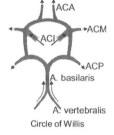
Circle of Willis

  - Spontaneous intracerebral haemorrhage [I61.9] (15% of all strokes), is mostly seen in hypertension (massive haemorrhage), but may also be due to fibrinolysis or anticoagulation therapy: Rapid progressive loss of consciousness.
  - Subarachnoidal haemorrhage (SAH) (5 % of all strokes; incidence 15/100.000/year) in 80 % caused by a ruptured aneurysm, usually located at the cerebral base; excruciating headache, neck stiffness, often drowsiness; blood stained cerebrospinal fluid, CT, MRI, angiography

  Specific stroke therapy is only possible after radiographic diagnostics, and it must not be initiated before hospital admission.

Ep.: Incidence of strokes in industrial countries (highest in East Europe and Germany):
- Total incidence: 50 - 60/100.000/year
- 55 - 64 years: 300/100.000/year
- 65 - 74 years: 800/100.000/year

Lifelong prevalence is ca. 15 %, with the numbers increasing rapidly after the age of 60 (m > f).
Stroke is the second most common cause of death in Germany (3rd in the USA) - after CHD/MI - and contribute to the most common causes for invalidity in the elderly.

Aetiology of ischaemic cerebral infarction (Encephalomalacia):

1. Arteriosclerosis and arterial thrombosis (ca 70 %)
   hypertension (particularly systolic) is risk factor No 1: a rise in the systolic blood pressure of 10 mmHg will increase the risk of suffering a stroke by ca. 30 % (Physicians' Health Study). People with arterial hypertension are 4 times more likely to suffer a stroke than those with normal blood pressure. 75% of all stroke patients have arterial hypertension. Further risk factors: family history (first degree relatives) of stroke before the age of 66, old age, CHD and its risk factors (e.g. diabetes mellitus, smoking etc.), heavy alcohol intake (moderate alcohol intake seems to be protective), oestrogen based contraceptives, migraine which started before the menopause

2. Arterial embolism (25 %) → 2 sources of embolism:
   2.1 cardiac: left atrium in atrial fibrillation, mitral- and aortic valve disease, myocardial infarction, myocardial aneurysm, bacterial endocarditis, cardiac catheter manipulations. Chronic atrial fibrillation without thrombo-embolism prophylaxis carries a 6 %/year risk of cerebral embolism.
   2.2. Arterio-arterial: Ulcerated plaques or carotid stenosis

3. Other causes (5 %), particularly in younger patients: Paradoxical embolism in ASD, persistent foramen ovale (PFO) or extracardial right-left-shunt; dissection of the extracranial cerebral arteries, vasculitis, Antiphospholipid-syndrome, M. Fabri, Cocaine- or Amphetamine abuse, Moyamoya-angiopathy (headache, TIA, stroke; angio: typical „cloudy" collateral arteries); chiropractic manoeuvres of the cervical spine with injuries to the vertebral artery

Pg.: Micro- and macroangiopathy:
1. Cerebral microangiopathy of the small arteries:
   a) Small lacunar infarctions
   b) Subcortical arteriosclerotic encephalopathy (SAE) = Binswanger disease; which may lead to dementia.
   main risk factors: arterial hypertension and diabetes mellitus
   neurological disorders/deficits are often only mild and settle spontaneously.
2. Macroangiopathy
   a) Macroangiopathy of the intracranial cerebral arteries (10 %)
      main locations: carotid siphon and middle cerebral artery (ACM)
   b) Macroangiopathy of the extracranial cerebral arteries (90 %)
      only stenosis/occlusions of the internal carotid artery (ICA) are relevant for the pathogenesis. Infarctions due to disorders of the ICA, are usually caused by arterio-arterial embolism = thrombotic material being flushed from the ICA.

Stenosis with a reduced lumen < 75 % are asymptomatic in most cases.

Unilateral advanced stenosis/occlusion of the ICA will cause a drop of the perfusion pressure. Only in case of a severely impaired function of the circle of Willis will this drop be so critical, that it will lead to an infarction; the trigger can be a drop in blood pressure (e.g. cardiac failure, dehydration, antihypertensive therapy).

The extent of the neurological deficit depends on:
- Localisation of the stenosis/occlusion
- Presence of any compensating anastomosis
- Blood pressure and blood viscosity: sudden cerebral ischaemia will cause a collapse of the cerebral autoregulation and hence the cerebral perfusion. This results in vasoparalysis; cerebral circulation then depends on arterial blood pressure and blood viscosity. A sudden drop in blood pressure or a high haematocrit can lead to underperfusion of the poststenotic tissue.
- Extent of the ischaemic brain area

PPh.:
- Territorial infarction: caused by thromboembolic occlusion of the large cerebral arteries; they are either located in the basal ganglia or cortical/subcortical (wedge-shaped).
- Extraterritorial infarctions: caused by extracranial stenosis/occlusions, 2 types:
  a) Bordering territory infarctions can be found between 2 terminal circulation territories of the anterior/middle/posterior cerebral artery.
  b) Terminal territory infarctions will be seen in non-collateralised medullary arteries in the periventricular/sub-cortical medulla
- Lacunar infarction in cerebral microangiopathy (hypertension, diabetes mellitus)

Sym.: Staging terminology of "TIA, PRIND and infarction" cannot be done with radiographic diagnostics, since cerebral lesions are found in all of them. Nevertheless they are commonly used.

4 stages of extracranial cerebral artery disease:
St. I: Asymptomatic stenosis
St. II: TIA [G45.9] = transitory ischaemic attack: short reversible neurological dysfunction, which usually settles spontaneously after 10 minutes, max. within 24 hours: e.g. impaired vision, arm- and/or leg weakness, slurred speech. Diffusion weighted (DWI) MRI reveals minor cerebral lesions in ca. 50 % (DWI-lesions). Ca. 20 % of all strokes were preceded by a herald TIA

Note: stroke risk after TIA: 12 % within one month, up to 20 % within one year and ca. 40 % within 5 years. A TIA must be investigated immediately!

St. III: (P)RIND: (Prolonged) reversible ischaemic neurological deficit: complete regression of the neurological deficit takes > 24 hours.
St. IV: Complete cerebral infarction: partial or no regression of the neurological deficits

Symptoms depend on the localisation:

Main symptoms of a stroke: reduced consciousness, (hemi-)paralysis, speech- and sensory disorders. Also: vegetative symptoms, circulatory- and respiratory disorder.

A) occlusion of extracranial cerebral arteries:
  1. Carotid type (occlusion of the internal carotid artery (ICA) often in the bifurcation; 50 %):
     Unilateral ICA-occlusions can be asymptomatic, when there are plenty of collaterals. Unilateral Amaurosis fugax is typical for ICA-stenosis! Cerebral infarctions cause contralateral sensori-motor hemi-paralysis with reduced reflexes; later there will be spasticity with increased reflexes and positive signs of the pyramid tract (Babinski). Large infarctions also cause speech disorder and reduced consciousness; occ. the head or eyes are turned to the side of the infarction.
  2) Vertebro-basilar-type (15 %):
     Vertigo, falls, nystagmus, vomiting, visual disorder, paralysis etc.
     occlusion of the posterior inferior cerebellar artery = PICA (originates from the vertebral artery) →
     Wallenberg-syndrome: Ipsilateral paralysis of the soft palate, pharynx and vocal cords; Trigeminal dysfunction, nystagmus; Horner-syndrome; limb-ataxia, dysmetria (inability to judge any distance); contralateral dissociated impaired sensibility for temperature and pain (neck and trunk).
     Subclavian-steal –syndrome [G45.8]:
     Subclavian artery occlusion proximal to the vertebral artery branch will lead in the beginning to an asymptomatic reversal of the flow in the vertebral artery (subclavian-steal-phenomenon). Any additional obstruction to the cerebral circulation may cause dizziness and impaired vision when the arm is being used. Subclavian occlusion causes a difference in blood pressure between the arms > 20 mmHg.

B) Occlusion of the intracranial cerebral arteries:
  most frequently seen in the middle cerebral artery (MCA) territory: 25 %. Symptoms are similar to those caused by ICA-occlusions (apart from amaurosis fugax). Occlusions of the anterior cerebral artery (ACA) (contralateral hemi-paralysis affecting the leg) or of the posterior cerebral artery - PCA (hemi-anopsia) are rare
  acute occlusion of the basilar artery causes progressive reduction of consciousness, disturbed oculomotorius nerve function and papillary paralysis, occ. visual disturbance, hemi-paralysis, dysarthria, dizziness, ataxia etc.

DD: - Hypertensive encephalopathy
- Sinus vein thrombosis (swelling of both eyelids! Look for any facial focus of infection!)
- Subdural haematoma (headache, slowly increasing loss of consciousness, occ. history of minor injury → Echo, CT, MRI, angiography)
- Intracranial lesions (tumour, abscess → CT, MRI)
- Epileptic fit with transient (Todd') paralysis (wetting, tongue bite)
- Migraine attack with aura
- Head injury
- Meningoencephalitis (fever, meningism, drowsiness, cerebrospinal fluid diagnostics)
- Disseminated encephalitis
- Hypoglycaemic shock, diabetic coma (test blood sugar!)
- Intoxication (take history from witnesses, family etc.)
- Neurosyphilis (positive TPHA-test) ; cerebral vasculitis caused by tick born Borreliosis

Di.: 1. History + symptoms: Neurological examination: grade of consciousness, pupils, eye movements, cranial nerves, limb mobility, neck stiffness etc., fundoscopy (papillary oedema ?)

Initially the following three functions always have to be checked:
- Facial paralysis: ask the patient to smile and show teeth
- Arm paralysis: ask the patient to lie down, and to lift each arm separately (up to 45°)
- Speech: ask the patient to repeat a sentence, ask him to name an object

This quick test has a sensitivity of 80%, and a specifity of 90% for the diagnosis of stroke (positive in the presence of at least one pathological sign); it matches the following in-patient investigations very well.

When the quick-test is negative, at least four further functions should be checked:
- Paralysis of the external eye muscles: ask the patient to look to the right and left
- Reduced vision: test the right and left visual field (separately for each eye)
- Paralysis of the leg: ask the patient to lie down and to lift one leg at a time (up to 30°)
- Hemi-paraesthesia: test sensitivity to the touch for either side separate (face, arm, leg)
also:
- Check pulse: irregular? → suspect atrial fibrillation; auscultation, particularly of the carotid artery (2/3 of carotid stenosis cause an audible murmur)
- Blood pressure of both arms (aortic arch syndrome: difference > 20 mmHg)

2. Radiographic cranial investigation: immediate CT is compulsory (Angio-CT / Spiral-CT): to differentiate between ischaemic infarction, haemorrhage and tumour; to show the localisation and extent of any infarction.

Haemorrhage: hyperdense areas;

Infarction: hypodense areas

(CT can be normal in the first 12 hours; MRI enables an early diagnosis within the first 30 mins.).

Note: A rapid cranial-CT is necessary to determine the aetiology of a stroke, and to start the right therapy!

MRI (and PET): to see if tissue is alive, has suffered an infarction or suffering from acute poor perfusion ("Penumbra", which may be rescued by lysis).

3. Radiographic diagnostics of the extra- and intracranial vessels: US scan of the carotid artery, transcranial doppler US scan of the intracranial vessels with contrast medium; MR-angiography; DSA is only indicated when invasive procedures are planned

4. Cardiac diagnostics: ECG (atrial fibrillation?), transoesophageal echocardiography (TEE) → to exclude any cardiac source of embolism. To find any cardiac right-left-shunt causing paradoxical embolism in ASD or PFO: TEE (transoesophageal Echo) and TCD (transcranial US); use both investigations with contrast medium.

5. Exclude any rare cause in younger patients

6. Dysphagia-screening: Up to 50% of all stroke patients initially present with dysphagia. This may result in aspiration pneumonia, and hence an increased risk of death. Dysphagia-screening in the acute phase is mandatory. This includes inspection of the mouth/pharynx (look for any salivary pools, test tongue- and pharyngeal motoric function, gag- and cough reflex). When these tests are normal, and conscience is not impaired, one may proceed and let the patient drink small sips of water.

Th.: Emergency call-Tel. 112/999: admit the patient into a "stroke unit" with immediate CT- and neurological diagnostics. During transport the patient should be accompanied by qualified medical staff; the upper body should be slightly raised (30°); lay the patient on their side if vomiting; pad the paralysed limb; consult a neurologist. Don't give any Heparin, Acetylsalicylic Acid or i.m.-injections before you have the CT-result. Don't lose any time; i.v.-lysis only makes sense within the first 3 hours ("Time is brain")!

A) General measurements:
- Protect respiration/circulation
   Note: Don't perform any venesection on the paralysed arm, because of the increased thrombosis risk.
- Monitor respiration, circulation, water-/electrolyte balance, blood sugar, blood gas analysis. Give O_2, when O_2-saturation < 95% (pulse oximetry); respiratory dysfunction: intubation and controlled ventilation, avoid hypoxia and/or hypercapnia

- Dysphagia or reduced consciousness associated with increased risk of aspiration: tube feeding or parenteral nutrition. Persistent dysphagia > 2 weeks: PEG-tube (percutaneous endoscopic gastrostomy).
- Monitor bowel- and bladder function (catheter)

- Thromboembolism prophylaxis when the patient is bed ridden (low dose heparin, elastic stockings, physiotherapy)
- After evidence of an ischaemic stroke: early start thrombocytes aggregation inhibitors (e.g. 100 - 300 mg Acetylsalicylic Acid/day) will reduce mortality.
- Treat any other diseases: cardiac failure, arrhythmia
- Ensure normoglycaemia, because increased blood sugar levels can raise intracranial pressure!
- Ensure (high)normal or slightly increased blood pressure in the acute phase:
  Hypertension in the acute phase is frequently reactive; don't start any antihypertensive therapy within the first 24 hours (regular BP-monitoring!). Indication to reduce blood pressure carefully: only when the blood pressure is very high (> 220/110 mmHg) or in a hypertensive emergency, when the patient's life is at risk due to hypertensive encephalopathy, angina pectoris or pulmonary oedema. Always lower the blood pressure carefully, not more than ca. 20% of the initial reading!
- Reduce any fever (tepid sponging of legs, Paracetamol etc.)
- Pathologically increased haematocrit occ. requires isovolaemic haemodilution (for details refer to chapter "AOD").
- Treat any increased intracranial pressure:
  · Conservative: raise the upper body (30°) and keep head/neck straight; osmotherapy (e.g. Mannitol 50g i.v. every 6 hours) and occ. intubation and ventilation, but no longterm hyperventilation (deterioration of the cerebral perfusion)
  · Neurosurgery:
    - Decompression craniotomy after large media infarction; brain stem decompression after large posterior infarction
    - Temporary ventricle drainage after cerebellum infarction followed by occlusion hydrocephalus
- Pressure sore prophylaxis (pad the paralysed limbs, regular turning of the patient, antidecubitus-mattress), Prophylaxis of contractures and foot drop by adequate positioning
- Early physiotherapy and occ. speech therapy, respiratory exercises

B) Revascularisation therapy:
  B1. Intravenous thrombolysis therapy:
      Acute cerebral infarction, thrombosis of the basilar artery: i.v.-thrombolysis, consider any CI and SE (consult a neurological centre). Occlusion of the basilar artery is usually fatal; thrombolysis is the only chance for the patient!

      Note: i.v.-thrombolysis only has a chance of being successful for a short period of time (the first 3 hours after the onset of stroke). First choice: rt-PA (Alteplase, e.g. Actilysis®) 0,9 mg/kg i.v. (max. total dose 90 mg); give 10% as a "stat-dose", 90% as an infusion over 90 min.
      precondition:
      1. CT result to exclude any intracranial haemorrhage
      2. Start within 3 hours since the onset of symptoms
      3. Absence of any early signs of a large media-infarction (> 1/3 of the area receiving blood supply from the middle cerebral artery); no major reduced consciousness
      4. Exclude any CI: Coumarin/Warfarin therapy, severe head injury < 4 weeks; major operation < 4 weeks; severe gastrointestinal haemorrhage < 4 weeks, tumour, pregnancy, age > 80 years etc.

  B2. Intra-arterial lysis:
      There is a grade-A recommendation for the 6 hours-gap (only in special clinics)

C) Anticoagulation with Heparin after cerebral embolism, and overlapping relapse prophylaxis with Coumarins (see further down) - Ind.: cerebral embolism (consider any SE + CI)

D) Rehabilitation:
  D1. Early rehabilitation in a regular hospital:
      physiotherapy, respiratory physiotherapy, prevent contractures, train sitting/standing etc.
  D2. Rehabilitation clinic

Prg.: Prognosis depends on the extent of brain damage. Up to 50% of all stroke patients die before arriving at a hospital (similar pre-hospital death rate to myocardial infarction). Hospital mortality in stroke units is up to 5 %. 1/3 of surviving patients make a full recovery, 1/3 become invalid, 1/3 are restricted in their daily routine. That means, that 2/3 remain permanently disabled. Ca. 30% become depressive. 30% die within one year after the stroke.
Subsequent mortality risk is 9% per year. 25% will suffer a second stroke within 5 years. Up to 30% will die from CHD/myocardial infarction!

Pro.: 1. Primary- and secondary prevention: All risk factors for arteriosclerosis have to be treated carefully. Smoking increases the stroke risk by a factor of 2 - 3. Smoking cessation will reduce the risk accordingly. Diligent blood pressure stabilisation will reduce the risk by 40%, particularly the risk of intracerebral haemorrhage. A diet high in vegetables and fruits will reduce the risk by 35% (m) and 25% (f) (Hiroshima/Nagasaki Life Span Study).
Aim for LDL-Cholesterol < 100 mg/dl → use CSE-inhibitors! Good diabetes control is important. Exclude sleep apnoea-syndrome as a cause of stroke!

2. Thrombocyte aggregation inhibitors for secondary prevention (post TIA, PRIND or non-embolic cerebral - infarction); they reduce the risk by 30%.
- Acetylsalicylic acid: standard drug; Dose: 100 - 300 mg/day (beware SE + CI).
- Adverse reactions to Acetylsalicylic acid: give Clopidogrel instead; Dose: 75 mg/day (refer to index for details)
- High risk of recurrence: ESPS-2-study recommends a combination of Acetylsalicylic Acid and Dipyridamol. Dose: e.g. Aggrenox® (1 retard capsule = 25 mg Acetylsalicylic Acid + 200 mg Dipyridamol): 2 x 1 retard capsule/day) – (beware SE + CI!)

3. Primary prevention in chronic atrial fibrillation using anticoagulants (target INR 2 - 3) : this reduces the risk by ca. 70%; anticoagulants used in cardiac valve prosthesis (refer to chapter "thromboembolism prophylaxis")

4. Paradoxical embolism: interventional occlusion (with catheter) of the ASD or PFO

5. Elimination of a severe asymptomatic ICA-stenosis:

   Ind.: • Symptomatic patients with at least 50% stenosis (haemodynamically significant)

   • Asymptomatic patients with at least 80% stenosis

   mortality of carotid artery surgery is < 3%. mortality of PTA/stent implantation is < 1% in good centres. Intra-operative stroke risk (up to ca. 3 %) can be reduced by cerebral protection systems, and also the patient being awake during the operation.
   Life expectance should be > 5 years.

   Methods:
   1. Carotid-TEA = thrombendarterectomy and dilation (gold standard)
   2. Carotid-PTA and stent implantation in the absence of any contraindications (e.g. new thrombotic plaques)
   Note: The 3-year-stroke risk for an asymptomatic carotid stenosis of ≥ 80 % is almost 10% (for stenosis 70% - 79% ca. 2%). Carotid TEA reduces the risk of an asymptomatic (> 70%) stenosis by ca. 6 %.

Visceral Arterial Disease
[K55.1]

PPh: The superior mesenteric artery provides blood supply for the bowel from the duodeno-jejunal flexure to the left colonic flexure. In slowly developing arteriosclerotic stenosis of the superior mesenteric artery circulation can be secured via collateral vessels (1. Via the pancreatic-duodenal arcade of the coeliac trunk or 2. Riolan-anastomosis via the middle colonic artery and left colonic artery of the inferior mesenteric artery).
Most stenoses will remain asymptomatic. However, acute occlusion of the superior mesenteric artery will usually cause intestinal infarction (occlusive mesenteric ischaemia = OMI). Advanced stenosis of the inferior mesenteric artery can cause ischaemic colitis.

Aet.: 1. In most cases arteriosclerosis of the mesenteric arteries and acute arterial thrombosis (elderly)
2. Occ. acute arterial embolism (e.g. in atrial fibrillation, endocarditis)
3. Rarely aortic aneurysm, -dissection
4. Very rarely aortitis, e.g. in Takayasu-arteritis, Polyarteritis nodosa
5. Extremely rare: compression of the coeliac trunk by the medial arcuate ligament

Sym.: 4 stages:
St. I: Asymptomatic (arteriography or duplex US; coincidental finding)
St. II: **Abdominal angina** = intermittent, postprandial abdominal pain caused by ischaemia
St. III: Permanent abdominal pain + malabsorption syndrome, occ. associated with **ischaemic colitis**
St. IV: **Acute mesenteric arterial occlusion and mesenteric infarction**; there are 3 stages:
   1. Initial severe colicky abdominal pain, nausea
   2. Almost no symptoms for several hours
   3. Paralytic ileus, peritonitis and acute abdomen, diffuse tenderness, rebound, shock, occ. bloody stool

DD: - Mesenteric vein thrombosis accounts for 10% of cases of mesenteric infarction (diagnostic and therapeutic procedures are the same as for AOD of visceral vessels)
- Ischaemia in the territory of the mesenteric arteries without arterial occlusion: Non occlusive mesenteric ischaemia (NOMI") caused by a drop in cardiovascular circulation with vasoconstriction of the mesenteric blood vessels (cardiac failure, myocardial infarction, circulation collapse, after cardiac operations and the use of extracorporal circulation). Digoxin can cause mesenteric vasospasm, despite peripheral vasodilatation elsewhere.

Di.: • History:
- Often progressive, recurrent postprandial abdominal pain
- Old age
- Cardiac diseases (CHD, cardiac failure, total arrhythmia)
- Hypertension, Diabetes mellitus, hypercholesterolaemia etc. risk factors (refer to CHD)
- Treatment with Digoxin or Ergotamine (intestinal vasospasm)
- Circulation collapse (blood pressure drop)
- Postoperative: after panproctocolectomy or aortic aneurysm-OP
• Auscultation: occ. bruits in the upper abdomen
• AXR: air bubbles, isolated dilated small bowel loops, thickened intestinal wall
• CXR, ECG
• US-abdomen (free fluid, immobile bowel loops)
• Doppler, Angio-/3 D-MRI, i.a.-DSA: in suspected ischaemic colitis colonoscopy should be performed with great care (risk of perforation!): mucosal oedema, ulcers with purple discolouration
• Laboratory: in mesenteric infarction: lactate, CK, LDH ↑

Note: Suspected mesenteric infarction is an acute emergency, just like myocardial infarction and stroke, and requires rapid diagnostics and therapy. Don't do any unnecessary and time-consuming diagnostics! Ischaemia tolerance of the bowel is max. 6 hours

Th.: Clinical suspicion of acute mesenteric arterial occlusion with intestinal infarction requires angiography and explorative laparotomy.
According to the intraoperative findings, the treatment is either embolectomy, disobliteration or bypass-operation; bowel necrosis requires bowel resection. Minor ischaemic colitis can be dealt with using conservative therapy (symptomatic).
For widespread bowel resection with subsequent short bowel syndrome, there is the possibility of bowel transplantation in specialised centres, and for selected cases (young patient) (1-year survival ca. 75 %).

Prg.: Operations in stage II have a good prognosis, operation mortality is fairly small (5 %). mortality of mesenteric infarction is 30 %, after 12 hours of ischaemia, and > 85 % after 24 hours! Patient's age, accompanying diseases, the length of the ischaemic bowel and operation time are important factors for the prognosis!

Pro.: Reduce the arteriosclerosis risks, give thrombocyte aggregation inhibitors (e.g. Acetylsalicylic Acid)

ABDOMINAL AORTIC ANEURYSM (AAA) [I71.4]

Ep.: Ca. 1% of the population > 50 years
Up to 10% of all male patients suffering from hypertension > 65 years;
Peak: 6. - 7. life decade, increasing incidence in industrialised countries

Aet.: Usually arteriosclerotic cause + hypertension

Loc: > 95% infrarenal, in 30% the pelvic arteries are involved

Di.: US scan: external diameter of the abdominal aorta ≥ 3 cm; occ. supporting diagnostics:
Angio-/3D-MRI and Angio-CT in spiral technique
∅ > 5 cm → risk of rupture is 10%/ year
∅ < 5 cm → risk of rupture is 3%/ year) → perform regular US monitoring
Indication for operation: so far 5,0 cm ∅ had been considered to be the critical limit; the ADAM-study recommends operation from 5,5 cm ∅ for men (for women from 4,5 cm ∅)

Th.: 1. Operation (Y-prosthesis aortic-bi-iliacal) = gold standard. The mortality of an elective operation is ca. 4%. An important complication (same as after stent-therapy) is postoperative spinal ischaemia with neurological deficits.
2. Endovascular therapy (stent-based Dacron prosthesis); mortality up to 3%; late complication: Endoleaks (up to 10%), thrombosis, embolism, stent displacement, further progression of the aneurysm; secondary interventions up to 30%
3. Cardiovascular risk factors must be treated/eliminated.

Prg.: Patients with AAA have a 2 fold cardiovascular mortality (→ exclude CHD!)

Pro.: Rigorous hypertension treatment!
There are current studies as to if an additional drug treatment can stop the AAA increasing (e.g. Doxycyclin).

THORACIC AORTIC ANEURYSM (TAA) [I71.2]

Def.: Dilated aorta > 3,5 cm ∅

Loc: Ascending aorta - aortic arch - descending aorta

Ep.: Rare disease; ca. 6/100.000/years; it only makes up to 3% of all aortic aneurysms, mainly found in older patients > 60 years

Aet.: 1. Rarely congenital (e.g. Marfan-syndrome, Ehlers-Danlos-syndrome)
    2. Usually acquired:
       - Aorta dissection and arteriosclerosis
       - Rarely cystic media necrosis (Gsell-Erdheim), Takayasu-arteritis, Syphilis (Mesaortitis)

Sym.: Usually silent (coincidental finding in Echo, CT, MRI)

Co.: 1. Aortic valve insufficiency seen in aneurysms of the ascending aorta
    2. Acute dissection with ischaemia of the distal organs
    3. Rupture (usually fatal)
    The risk of rupture and dissection depends on the diameter and morphology:
    ca. 3%/ year for ∅ 5,0 - 5,9 cm
    ca. 7%/ year for ∅ ≥ 6,0 cm.

Di.: Echo / TEE, cardiac-MRI

Th.: Aim for normal blood pressure!
    1. Operation:
       Indication for operation: aorta diameter between 5 - 6 cm
       Technique: Dacron prosthesis: (aneurysm of the ascending aorta should receive aortic valve prosthesis)
       Hospital mortality after elective operation:
       - Replacement of the ascending aorta or the arch: up to 5 %
       - Replacement of the descending aorta: up to 15 %
       Descending aorta replacement: risk of postoperative paraplegia is up to 5 %.
    2. Endovascular aorta revascularisation (EVAR) using a stent:
       Ind: Aneurysm of the descending aorta; hospital mortality is up to 10%, risk of paraplegia is less than after an operation; however there is the risk of late stent complications (refer to AAA)

DISSECTING AORTIC ANEURYSM [I71.0]

Def.: Acute life threatening disease of the thoracic aorta. Unlike a true aneurysm, in which all layers of the vascular wall are dilated, the dissecting aneurysm shows a tear in the intima causing intramural haemorrhage into the media, and hence a second false aortic lumen; this will spread further distal and/or proximal.
    ▶ 2 types of localisation (Stanford-classification):
       - Proximal type A (60 %): aortic arch including the ascending aorta
       - Distal type B (40 %): distal of the aortic arch = descending aorta
    ▶ classification of the European Task Force (pathogenetic aspects into 5 classes):
       1. Classic AOD
       2. Intramural haematoma
       3. Localised, localised AOD with an eccentric pouch
       4. Plaque rupture with AOD and/or aortic rupture
       5. Traumatic AOD (after whiplash injury), iatrogenic AOD (after catheter investigations)

Ep.: Incidence 3/100.000/year, mainly > 50 years (exception: Marfan-syndrome: peak ca. 30 years).

Aet.: Risk factors for aortic dissection:
    - Arterial hypertension (70 %)
    - Marfan-syndrome: Autosomal-dominant inherited defect of the Fibrillin-1-gene with floppy connective tissue. The classic skeletal anomalies like arachnodactylia, abnormally long limbs, over-extensive joints are absent in 30 % of the patients.
    - Post aortic valve replacement, correction of an aortic isthmus stenosis
    - Aortitis of various origin (Takayasu-syndrome etc.)

Sym.: Excruciating, sometimes migrating thoracic pain, of sharp and tearing character; in type A mainly retrosternal, in type B mainly in the back, with radiation into the abdomen. Type A: occ. pulse- and blood pressure difference between both arms. In complicating aortic insufficiency: diastolic heart murmur.

Co.: Type A: pericardial effusion, aortic valve regurgitation, occlusion of the coronary arteries (→ myocardial infarction), stroke
    Type B: haemothorax, mediastinal or abdominal haemorrhage, occlusion of the renal and/or mesenteric arteries (→ renal failure, mesenteric infarction).

DD: • Myocardial infarction = DD and occ. complication! (Troponin I or T, CKMB, ECG)
• Pulmonary embolism (pO$_2$, Echo, Spiral-CT)

Di.: Symptoms + radiographic diagnostics:
TEE, CT with contrast medium, MRI

Th.: 1. Reduce blood pressure to 100 - 110 mmHg systolic (e.g. with Betablocker i.v.); analgesics
2. Operation (artificial prosthesis – always for type A; in type B it is only required when complications are likely) – hospital mortality 5 - 30 %
3. Aortic stent implantation in selected cases for type B-dissection (there are no longterm studies)

Prg.: Only 50% survive the first 48 hours if untreated, 80% die within 2 weeks after aortic rupture.

RAYNAUD'S SYNDROME (RS) [I73.0]

Def.: Primary RS (> 50 %) = cold or emotionally triggered attacks of painful vasospasms with ischaemia of the fingers, lasting up to max. 30 min.
Secondary RS (< 50 %) = same symptoms, asymmetric, usually with organic changes of the digital arteries; there are various underlying diseases:
- Collagenosis (most frequently in scleroderma and Sharp-syndrome)
- Vasculitis (Winiwarter-Buerger)
- Damage due to vibration, carpal tunnel syndrome, Sudeck' dystrophy
- Peripheral arterial embolism
- Peripheral arterial disease (AOD)
- Medication: e.g. β-Blocker, Ergotamine, decongesting nose drops, Bleomycin, Cisplatin etc.; drugs: Nicotine, Amphetamine, Cocaine
- Haematological/oncological diseases: e.g. polycythaemia, thrombocytosis, cold agglutinin syndrome, cryoglobulinaemia, paraproteinaemia (multiple myeloma, Waldenström's disease)

Ep.: Ca. 3 % of the total population suffer from primary RS
f : m = up to 5 : 1; age of manifestation of primary RS: 20 - 40 years

Cl.: The ischaemic attacks come in 3 phases:
1. Paleness caused by vasospasm of the finger arteries (exception: thumb !)
2. Acrocyanosis caused by paralysis of the venules
3. Redness caused by reactive vasodilatation
There are not always the 3 phases of the classical „Tricolour-phenomenon", particularly in the organic fixed stenosis there is no reactive hyperaemia.

Criteria for the diagnosis of primary RS:
- Fingers are affected symmetrically
- No necrosis
- Triggered by cold or emotional stress
- Symptoms have been present for> 2 years without proof of any underlying disease

DD: Embolism (ischaemia lasts > 30 min.)
AOD (DD and cause of a secondary RS)

Di.: - Ask the patient to raise their hand; then compress the wrist; ask the patient to make a fist 20 times; sometimes this can trigger some fingers to turn pale; after releasing the wrist, the re-perfusion is delayed in some places (test both sides).
- Allen-test: to prove isolated occlusion of the radial or ulnar artery. Only compress the radial or ulnar artery. Ask the patient to make a fist. When the hand turns white, occlusion of the uncompressed artery is likely.
- Cold exposure test: ask the patient to place his hands in ice-cold water for 3 minutes; this may trigger attacks of vasospasm.
- Capillary microscopy: capillary diameter is raised; this investigation is of particular importance for the diagnosis of secondary RS. Scleroderma: giant capillaries, areas without any capillaries, haemorrhages. 12% of patients suffering from scleroderma have this histology result before any presenting with any clinical sign of systemic scleroderma. Similar results are found in other types of collagenosis. Lupus erythematosus: bundled capillaries and convoluted capillaries.
- MR-angiography: reveals vasospasms. Stenosis that persists despite injection of α-blockers, are likely to be due to organic vascular changes.
- Laboratory: to exclude secondary RS:
  - Unspecific signs of inflammation: ESR, CRP
  - Blood count + thrombocytes, protein- and immune electrophoresis
  - Cold agglutinins, cryoglobulines
  - ANA and anti-DNS-antibodies in SLE

- Anti-SCL70 in scleroderma
- Anti-U1RNP in Sharp-syndrome

Diagnosing primary RS:
Fingers of both hands are affected symmetrically; pain never lasts longer than the actual attack, no atrophic changes of the skin, capillary microscopy and laboratory normal (ESR, ANA), secondary RS/AOD has been excluded

Th.: • General treatment:
- Sports, hot and cold baths
- Protect hands from cold /stop smoking
- Avoid stress
• Treatment for primary RS:
- Calcium antagonists: e.g. Amlodipine 5 - 20 mg/day
- Alpha1-receptor blockers: e.g. Prazosin 1 - 5 mg/day
- Nitrates: creams or spray
- Angiotensin II-receptor blockers: e.g. Losartan 25 - 100 mg/day
• Treatment for secondary RS:
- Treat any underlying disease
- Symptomatic: PGE1, if indicated also Bosentan (ET1-receptor antagonist)

VENOUS DISEASES

- Varicose veins
- Superficial thrombophlebitis
- Deep vein thrombosis (phlebothrombosis) (DVT)
- Pulmonary embolism
- Chronic venous insufficiency (CVI)

VARICOSE VEINS (VV) [I83.9]

Syn: Varix = knot

Def.: WHO: VV = diverticle-shaped or cylindrically dilated, superficial (epifascial) veins; the VV can be local or extend over a longer distance; they are often distorted by tortuosity.
1. Primary VV (95 %) = idiopathic without any obvious cause
2. Secondary VV (5 %) = acquired; they are usually collaterals due to obstruction of the deep veins (after venous thrombosis).

CEAP-classification of chronic varicose veins (varicosis, CVI):
(too complicated for practical purpose; more of scientific interest)

clinical (C):
C0   no visible or palpable vein disease
C1   teleangiectasies or reticular VV
C2   varicose veins
C3   oedema
C4   skin changes, due to chronic venous insufficiency (pigmentation, white atrophy, eczema, lipodermatosclerosis)
C5   skin changes as in C4, scarred ulceration
C6   skin changes as in C4, ulceration
This is accompanied by the letters
A   asymptomatic
S   symptomatic

etiology (U):
EC   congenital
EP   primary (unknown cause)
ES   secondary: - post thrombotic
                    - posttraumatic
                    - other causes

anatomy (A):

As   superficial veins
- 1   teleangiectasies, reticular VV
- 2   long saphenous vein above knee
- 3   long saphenous vein below knee
- 4   short saphenous vein
- 5   does not belong to the saphena territory

Ad   deep veins

Ap   communicating veins

pathology (P):

PR   reflux

PO   obstruction

PR,O   reflux and obstruction

Ep.: Ca. 20% of all adults; increasing prevalence with advanced age; f:m = 3:1; manifestation usually in the 3rd decade of life

Aet.: - Genetic (family history - 50 %):
- Age
- Hormonal factors in women (e.g. pregnancy)
- Standing or sitting

There are different opinions regarding obesity as a risk factor.

PPh: Primary VV: incompetence of the valves of the epifascial veins, so the blood will start flowing backwards from deep to superficial.

This results in abnormal venous circulation in the leg: at the proximal point of insufficiency (in the groin) blood does not flow from the long saphenous vein into the deep vein, but flows in a retrograde manner from the common femoral vein into the long saphenous vein down to the distal point of insufficiency; from there it flows via branch VV and the communicating veins back into the deep veins.

Anatomy: 3 venous systems of the leg:
1. Superficial: the long saphenous vein starts at the medial aspect of ankle and ends at the saphenofemoral junction just below the groin; short saphenous vein and its branches (lower leg).
2. Deep veins carry 90% of the venous flow; this works mainly due to muscle contractions and joint movements. Venous valves prevent retrograde flow.
3. Communicating veins: connection between the superficial and deep system. Venous valves secure the physiological flow from superficial to deep veins. 3 important groups:
   - Dodd-group: inner side of the mid thigh
   - Boyd-group: inner side of the lower leg, just below the knee
   - Cockett-group: 3 communicating veins on the medial lower leg (lower third), ca. 7, 14 and 18 cm above the sole of the foot.

Types of VV:

- VV of the main stem and branches (most common), mainly affecting the long saphenous vein (medial thigh and lower leg) and short saphenous vein (posterior lower leg) and their branches. Ca. 15% of the population are affected.
- Reticular VV = web-like, localised, superficial venous dilatation with a diameter of 2-4 mm, mainly found in the popliteal fossa and the lateral upper and lower leg.; quite common; cosmetic nuisance.
- Thread veins = a web of tiny intradermal VV with a diameter < 1 mm, particularly seen in the posterior upper leg; quite common; can be a cosmetic nuisance
- Varicocele = dilated plexus pampiniformis
- Vulval VV and suprapubic VV: seen in pregnancy

Sym.: - Heavy and tense legs (improved by lying down and exercising)
- Ankle oedema in the evening
- Occ. pruritus and sensation of pressure in the area of the insufficient communicating veins
- Nocturnal foot- and calf cramps

Note: venous diseases show a typical increase of symptoms in the evening, after standing or sitting for a long time or in warm weather; unlike AOD longer walks will improve the symptoms

Stages of stem varicosis of the long saphenous vein (Hach, 1996) are determined by the distal extent. During the Valsalva-pressure test (= reflux stops at the next distal intact valve and branch VV) the distal point of insufficiency determines the stage:

I.   Valvular incompetence at the saphenofemoral junction only

II.   VV with reflux above the knee

III.   VV below the knee

IV.   VV below the ankle

- After some years the <u>recirculating blood volume</u> of the main VV (long saphenous vein) increases the volume in the popliteal and femoral vein, and cause dilation and valve insufficiency. This is called secondary <u>popliteal- and femoral vein insufficiency.</u>
- In <u>incomplete stem varicosis of the long saphenous vein</u> the saphenofemoral junction is sufficient, however distal insufficient communicating veins or side veins cause reflux from deep veins into the superficial long saphenous vein.
- <u>Stem varicosis of the short saphenous vein</u> is rare. Large VV at the back of the calf are possible.
- <u>Varicosis of the communicating veins</u> is often combined with other forms of primary and secondary VV.

Clinical staging (Marshall):

stage. I: no symptoms, cosmetically significant
stage. II: sensation of congestion, nocturnal cramps, paraesthesiae
stage. III: oedema, thickened skin, pigmentation, healed ulceration
stage. IV: venous ulceration

Co.: Thrombophlebitis, deep vein thrombosis (occ. with pulmonary embolism), chronic venous insufficiency, venous ulceration

Di.: • <u>History, inspection, palpable</u> gaps where the insufficient communicating veins pass through the fascia; occ. protruding communicating veins ("blow-out"-phenomenon)
• <u>Functional tests</u> have been replaced by Duplex scan; e.g. <u>Trendelenburg-test</u> to show any insufficient venous valves and the <u>Perthes-test</u> to check for deep vein function.
• <u>Duplex-US scan</u> → to answer 2 questions:
1. Intact deep veins?
  - Vein compressible? (compression US scan)
  - Does the flow vary with respiration? (S [= spontaneous] sounds)
  - Increased flow after distal and proximal compression (A [= augmented] sounds)
2. Are the vein valves of the main vein competent? (= flow stops during Valsalva-pressure test) or insufficient (=reflux during Valsalva-pressure test) → for staging: find the <u>distal point of insufficiency</u> (the distal point where the reflux murmur ends).
• Occ. ascending pressure phlebography

Th.: A) <u>Conservative:</u>
- <u>Compression hosiery/stockings</u> (compression class II is usually sufficient, equals an ankle pressure of ca. 30 mmHg); tailored stockings are only necessary for extremely large or thin legs.
- <u>Avoid sitting and standing, prefer walking and lying.</u>

Note: There is no really efficient medication to treat VV!

<u>Surgery:</u>
<u>Ind:</u> patient symptomatic
<u>Condition:</u> proven intact deep veins (coloured Duplex, phlebography)
<u>CI:</u> occlusion of the deep veins etc.

<u>Methods:</u>
• <u>Saphenofemoral ligation</u> = tie all branches at the saphenofemoral junction in the groin, to prevent relapse
• <u>Vein stripping</u>, e.g. using a Babcock-tube or Vollmar-ring stripper
• <u>Tie all insufficient communicating veins</u>
• Separate removal of any further VV (which must have been marked <u>pre</u>-operatively with the patient standing up)
<u>Current clinical trials:</u> endovenous radiofrequency- and laser-therapy

C) <u>Sclerotherapy or laser-therapy:</u> thread veins, reticular VV and small branches can be removed as an outpatient treatment (cosmetic indication).

Prg.: Relapse rate after sclerotherapy > 50% after 5 years
Competently performed operations have a small relapse rate (< 5%).
Operative/postoperative mortality is 0.02 %.

| Chronic Venous Insufficiency | (CVI) [I87.2]

Syn: Chronic venous congestion syndrome, calf pump failure syndrome, postphlebitic limb

Def.: Venous hypertension (patient standing up) associated with veins- and cutaneous changes

Aet.: - Post-thrombotic syndrome
- Primary or secondary valve insufficiency of the deep veins
- Venous angiodysplasia (congenital defects/absent vein valves)

PPh: - Venous valve insufficiency with raised static venous pressure
- Retrograde blood flow, recirculation
- Pathological collateral circulation
- Venous areas cut out from circulation
- Peripheral venous insufficiently supported by the pumping system (muscle- and joint pump)
- Disorder of the microcirculation and lymphatic circulation

Loc: Deep venous insufficiency of the upper leg determines the pathogenesis. Clinical symptoms are found mainly in the lower leg- and foot.

Sym.: **3 Stages** (Widmer)

Stage I:
• Reversible oedema
• Corona phlebectatica (dark blue skin discolouration of the lateral and medial foot)
• Perimalleolaer venous changes

Stage II:
• Irreversible oedema
• Skin purpura and haemosiderosis of the lower leg (red-brown pigmentation)
• Dermatosclerosis, lipodermatosclerosis (occ. with inflammatory redness)
• White atrophy blanche (depigmented, atrophic skin, usually bilateral above the ankles)
• Eczema: with pruritus and tendency to allergic reactions
• Cyanosis

Stage III:
Acute or healed ulcerations (venous ulcer)
Preferred location: on top of insufficient communicating veins of the Cockett´ group above the medial malleolus

Co.: - Patients are prone to cellulitis
- Arthrogenic congestion syndrome (secondary reduced mobility of the ankle, with subsequent further reduction of the ankle vein pump)

DD: - Leg oedema of different origin
- Arterial leg ulcer in AOD

Di.: • History and symptoms
• Duplex- and coloured Duplex US scan:
Shows flow and passage of the deep veins. Reveals reflux in the case of incompetent vein valves.
• Indirect investigations have been replaced by Duplex scan.
• Ascending phlebography before an operation.

Th.: A) Causal: treat all varicose veins

B) Symptomatic:
• General measurements:

Note: **sitting** with bent knees and **standing** reduces the venous flow.
Lying (ideally the legs should be higher than the upper body) and **walking** (calf-muscles and ankle joint act as a pump) improve venous flow.

Heat causes unwanted venous dilatation → no sauna, no direct sunbathing. Cold showers promote venous constriction (good).
Foot exercises to avoid any secondary stiffening of the ankle joint.

• Compression therapy:
– Compression taping with 8 -10 cm wide tape
– Zinc tape dressing for several days
– 2-way elastic supportive stockings:
class II (~ 30 mmHg ankle pressure): CVI stage II (Widmer) - suitable for most patients
class III (~ 40 mmHg ankle pressure): CVI stage III (Widmer)
The compliance of compression therapy is max. 50 %.
CI of compression therapy: AOD with ankle pressure < 80 mmHg, decompensated cardiac failure, septic phlebitis, Phlegmasia coerulia dolens

• Therapy of venous ulcer:
– Remove any necrotic tissue and clean the ulcer, e.g. with water and sugar or enzymatic wound cleaning or Hydrocolloid dressings. Use topical treatment with care, because of the risk of allergic eczema!
– Afterwards: compression taping using foam rubber tape, which also should compress the neighbouring vein (cover the edge of any ulcer with zinc paste). Without compression there is hardly any chance of healing!
– When the ulcer doesn't heal, refer the patient to a dermatologist or a plastic surgeon

Prg.: The prognosis is good when therapy is consistent.
(general measurements, treatment of VV, compression taping)

THROMBOPHLEBITIS [I80.0]

Def.: Inflamed superficial (epifascial) veins with thrombotic occlusion of the affected veins.

Aet.: - Legs (90 %): usually seen in patients with underlying VV of the long and short saphenous vein and their side veins; caused by (micro-)trauma.
- Arms: mostly caused by infected intravenous catheters or injection/Infusion of hyperosmolar or intima irritating infusions.
- Thrombophlebitis saltans sive migrans
  Def.: Recurrent thrombophlebitis with changing localisation (sometimes arms, rarely visceral) in normal veins (i.e. not VV).
  Prev.:- Early stage of Thrombangiitis obliterans (Winiwarter-Buerger)
          - Sometimes associated with malignant tumours (e.g. Pancreas Carcinoma)
- Mondor disease: idiopathic thrombophlebitis of the lateral thoracic veins, a tender visible and palpable string (self limiting).

Sym.: Signs of inflammation: erythema, paleness, pain, tumour (painful palpable, firm [thrombosed] string of a vein); unlike DVT there is no swollen leg, since 90% of the blood flow is via the deep vein system.

Co.: – In up to 20% thrombophlebitis can spread along insufficient communicating veins or into the long saphenous vein via the saphenofemoral junction into the deep veins.
– Rarely bacterial infection + abscess, sepsis

DD: Deep vein thrombosis (venous thrombosis)

Di.: Symptoms, Duplex-US scan to exclude that the thrombus of the long saphenous vein invades (via the saphenofemoral junction) the superficial femoral vein.

Th.: • In the community: mobilise the patient. (No bed rest! Immobilisation promotes growth of the thrombus into the deep vein system!)
• Eliminate any causal factors (e.g. venflons, catheter!)
• New thrombophlebitis (< 7 days): incision, remove the thrombotic material, compression dressing + mobilisation.
• Old thrombophlebitis: (> 7 days): only compression dressing + mobilisation
• Indications for Heparin therapy: Thrombophlebitis of the long saphenous vein (risk of ascending thrombosis) and bedridden patients.
• For severe pain: Anti-inflammatories, e.g. Diclofenac
• Infected venous cannula of the arm: antiseptic dressings
• Fever: use antibiotic against Staphylococci.

Pro.: Treat any underlying VV, take care when performing injections/infusions; only leave any cannulas in situ as long as absolutely necessary.

DEEP VEIN THROMBOSIS (DVT) [I80.2]

Syn: Phlebothrombosis

Definition of thrombosis: intravasal, focal coagulation of blood

Definition of DVT: thrombosis of the deep leg vein with the associated risk of
                        – pulmonary embolism
                        – post-thrombotic syndrome (chronic-venous insufficiency)

Ep.: The risk of DVT depends on age and risk factors; average risk at < 60 years 1:10.0000/year; at > 60 years up to 1:100/year. DVT and pulmonary embolism are a major cause of morbidity and mortality during hospital stay, but they also occur after day surgery operations.

Loc: Primary localisation (4 levels): iliac vein 10%, femoral vein 50%, popliteal vein 20%, vein of the lower leg 20%.
2/3 of all DVTs appear in the left leg (obstructed flow where the left iliac vein and right iliac artery cross each other, causing a septum-like fold in the venous lumen in 20% of all adults = May-Thurner-syndrome). Up to 20% of untreated lower leg vein thrombosis progress into the upper leg vein, and ca. 20% of all femoral vein thrombosis cause an ascending iliac vein thrombosis.
> 90% of all emboli originate in the territory of the inferior vena cava - 30% in the iliac vein and 60% in the femoral vein. Up to 50% of all patients who had a proximal DVT will develop (mostly asymptomatic) pulmonary emboli (PE).

Pg.: Virchow' triad:
1. endothelium changes: inflammation, trauma
2. changes in blood flow:
   – turbulence (VV)
   – reduced flow (local stop, cardiac failure)

3. altered blood composition and imbalance between coagulation and fibrinolysis → refer to "causes of thrombophilia" - etiology)

PPh.: Types:

▶ Platelet thrombus:
Adhesion and aggregation of thrombocytes to an <u>endothelium defect</u> (initial trigger of thrombosis)
features: • Firmly attached to the wall of the vessel.
• Does not obstruct the entire lumen
• Thrombus low in erythrocytes (<u>white</u>) with an uneven surface.

▶ Coagulation thrombus:
significant pathogenic factor: reduced blood flow
features: • No firm attachment (risk of embolism!).
• Obstructs the entire lumen.
• Red thrombus with smooth surface.

▶ Mixed thrombus:
white head and red tail

Aet.: 1. Medical predisposing factors for a DVT (risk factors):

• History of <u>DVT or PE (up to 30 fold)</u>
• Polycythaemia vera
• Hyperviscosity-syndrome
• Forced diuresis with dehydration
• Obesity (BMI > 30)
• Immobilisation (up to 20fold)
• Stroke
• Myocardial infarction, cardiovascular circulation shock, cardiac failure
• severe VV

• Acquired Protein C-deficiency:
liver cirrhosis,DIC, Coumarin/Warfarin therapy
• Acquired Protein S-deficiency: liver cirrhosis,
Pregnancy, oestrogen therapy
• Acquired AT-deficiency: liver cirrhosis,
nephrotic syndrome, exsudative enteropathy, DIC
• Heparin induced thrombocytopenia type II
• Neuroleptics (e.g.
Phenothiazines or Butyrophenones)

• Malignancies, particularly abdominal (e.g. pancreas- and prostate-carcinoma). → in cases of idiopathic thrombosis always consider the possibility of a tumour and search for it!
• Treatment with oestrogens, combined oral contraceptive pill (this risk is enhanced by smoking! → smoking is a relative CI for the use of ovulation inhibitors.)
• Pregnancy and up to 6 weeks postnatal

2. Postoperative risk of thrombosis (without Heparin-prophylaxis):

General Operation risk	DVT	PE	Death
Low patients < 40 years, short operation (<30 min.), arthroscopy, plaster of Paris	2 %	0,2 %	0,02 %
Moderate general surgical/ urological/ gynaecological operations (> 30 min.)	10 - 40 %	1 - 4 %	0,4 - 1 %
High polytrauma, pelvic-/knee-/hip-operation	40 - 80 %	4 - 10 %	1 - 5 %

Prophylactic Heparin can prevent 3 out of 4 DVTs (risk reduction by 75%).

3. Compression of the popliteal vein during long periods of sitting in cars, buses, trains, planes can cause flight thrombosis in patients at risk: "Economy class syndrome".

4. Hereditary causes of thrombophilia (up to 60% !) → signs to suspect a hereditary cause:
Younger patients DVT; recurrent DVT, atypical localisation, positive family history

type of defect	pathomechanism	prevalence	risk of thromboembolism
APC-resistance / Factor V-Leiden-Mutation = Factor V-G1691A-Mutation	reduced inactivation of factor Va by activated protein C (APC)	Ca. 30% of all thrombosis patients	Up to 8 fold ↑ heterozygote Up to 80 fold ↑ homozygote
Prothrombin (FII)-G20210A-mutation	increased plasma-Prothrombin-level	Ca. 10% of all thrombosis patients	3 fold ↑ in heterozygote 6 fold ↑ in homozygote
Protein C-deficiency various mutations	inhibitor deficiency (reduced inactivation of factor Va and VIIIa)	Ca. 5 % all thrombosis patients	8 fold ↑

Protein S-deficiency	reduced Protein-C-activity due to reduced co-factor activity	Ca. 2 % all thrombosis patients	8 fold ↑
Antithrombin- (AT) deficiency (Syn: = AT III-deficiency)	type I: AT-level reduced by ca. 50 % type II: AT-level normal, but reduced AT-activity	< 1 % of all thrombosis patients	5 fold ↑ in heterozygotes 50 fold ↑ in homozygotes

- APC-resistance: most frequent cause of thrombophilia; prevalence 7 %
- Persistent increased F. VIII will increase the thrombosis risk (5fold).
- Antiphospholipid-syndrome: refer to according chapter

All those clotting disorders are inherited autosomal dominant. Heterozygotia is more common; there is ca. 50% reduced concentration of the affected protein. Apart from the (rare) homozygote APC-resistance and AT-deficiency, all other disorders only carry a moderately increased thrombosis risk; however this can cause thrombosis in the presence of additional risk factors (1-3)!

Sym.: – Sensation of heaviness and tension, "sore muscles" (calf, popliteal, groin); better in a horizontal position.
- Swelling (difference in circumference of the leg!) cyanotic shiny skin, „Pratt-Warn-veins" = collateral veins on the shin
- Warm limb
- Tenderness over the deep veins
- Calf compression pain: manual (Meyer' sign) or blood pressure cuff (Lowenberg-May' sign)
- Calf pain on dorsal flexion of the foot (Homans' sign)
- Plantar pain triggered by pressure on the medial foot sole (Payr' sign)
- Sometimes fever, ESR ↑, leucocytosis, tachycardia

Note: Clinical signs are in 50%not reliable! Signs of arrested circulation (leg) are only present in large proximal thrombosis (upper leg); the typical triad: swelling, pain, cyanosis is only present in 10%. The absence of clinical signs does not exclude a DVT (particularly in bedridden patients): Only 1/3 of all cases of pulmonary embolism are preceded by a clinical DVT!

Co.: 1. Pulmonary embolism: Up to 50% of all patients with DVT have at the time of diagnosis a (mainly asymptomatic) pulmonary embolism (scintigraphically detectable)! Pelvic vein thrombosis carries the highest risk of PE.
2. Post-thrombotic syndrome (PTS) (symptoms: refer to "chronic-venous insufficiency CVI") is present in ca. 40% after a thrombosis (of these 25% will develop leg ulcers): Rare in lower leg vein thrombosis; usually seen after an extended thrombosis (50% and more).
3. Thrombosis relapse

DD: – Post-thrombotic syndrome with chronic-venous insufficiency
- Lymphatic oedema (toes are swollen)
- Muscle tear and posttraumatic swelling (history, skin not warm, not cyanotic)
- Sciatic-syndrome (history, pain radiation, Lasègue sign, neurological Status)
- Ruptured Baker-cyst (pouch in the synovia in the popliteal fossa → US)
- Acute arterial occlusion (no pulse, pale and cold skin, no oedema)

Di.: • History (risk factors?) + symptoms
Determine the clinical probability of DVT

Parameter	scores
Active malignancy	1
Paralysis, recent joint immobilisation (e.g. plaster of Paris)	1
Bed rest (> 3 days) or major operations	1
Swelling of the whole leg	1
Difference of the calf circumference > 3 cm	1
Oedema (worse in the symptomatic leg)	1
Visible superficial non-varicous collateral veins	1
Alternative diagnosis is more likely than DVT	-2
Score interpretation	
Low possibility	< 1
Medium possibility	1 – 2
High possibility	≥ 3

- D-Dimer: seen after a recent DVT, but also postoperative, in malignancies and DIC
A positive D-Dimer-test is suspicious for thrombosis, but it is no proof (there are other possible causes). A negative D-Dimer-test (< 500 µg/l) excludes DVT in the presence of few clinical signs. A negative D-Dimer-test in the presence of significant clinical signs requires further radiographic diagnostic tests.

- Radiographic diagnostics:
  - (coloured Duplex-)US scan: ideal investigation; in case of thrombosis there is a typical absent/ reduced compressibility of the vein lumen (compression US). When blood flow is normal, Duplex scan of the femoral and popliteal veins show a variable flow profile parallel to respiration. Coloured Duplex-US can visualize that part of the thrombus, which is surrounded by blood. In complete vein obstruction the Doppler-US scan cannot detect any blood flow.
  - MR- and CT-phlebography: Good tests, but more cumbersome than US
  - Ascending phlebography: only indicated when coloured Duplex-US scan is equivocal
- Find the causes of DVT (thrombophilia in younger patients, occ. search for malignancy in older patients)
- Thrombophilia diagnostics:

  a) Who should be tested? DVT in the following cases:
     - Young patients
     - Multiple relapses
     - Significant family history
     - Atypical thrombosis (localisation, extent, absence of a cause)
     - Suspected Antiphospholipid syndrome (thrombocytopenia, miscarriages, autoimmune-phenomenon, prolonged aPTT or reduced INR).
  b) What to test?
     - TP (thromboplastin time) and aPTT
     - Plasma mixing test when aPTT is prolonged or INR is reduced
     - APC-resistance. In case of abnormal result: genetic factor V-Leiden-test
     - Protein C- and Protein S activity
     - AT, factor VIII-activity
     - Homocysteine (fasting and sometimes after Methionine-intake)
     - Diagnostics of Antiphospholipid-syndrome
     - Prothrombin Mutation 20210
  c) When to test?
     Not earlier than 3 months after all signs of DVT have ceased. Ovulation inhibitors must be stopped before the test, and the patient must not be pregnant.
     Coumarins/Warfarin should be stopped 2 weeks before the test (if this is clinically acceptable). When thrombophilia has been proven, then family screening is recommended. Anti-Specialist-clinics are the ideal place for these tests.

Th.:   Therapy targets:
1. Prevent pulmonary embolism
2. Avoid any expansion of the thrombosis
3. Recanalisation of the thrombosed blood vessel; retain the vein valves, and prevent post-thrombotic syndrome.

A) General measurements:
   compression: start with elastic tape, later use supportive stocking (whole length, compression class II) → improves the venous/lymphatic flow, improves wound adherence of the thrombus (CI: pAOD and Phlegmasia coerulea dolens).
   Mobilisation after DVT does not depend on its localisation (distal or proximal) and thrombus morphology ("mobile", "fixed", "occluding"); no strict bed rest is necessary, unless it is to reduce symptoms, when the leg is very painful or swollen; or to perform therapeutic procedures. When anticoagulation, compression and mobilisation are acceptable, there is no increased risk nor severity of PE.
   Local warmth application is obsolete! Prevent constipation (to avoid pressing!).

B) Anticoagulant therapy with Heparin in therapeutic doses reduces the risk of pulmonary embolism by 60%.
   advantage: 4 x less intracerebral haemorrhage rate (0,2%) than lysis (0,8%)
   disadvantage: re-opening rate of the thrombosed vein is much rarer and only incomplete compared to lysis therapy → 2 x likely to develop post-thrombotic syndrome.
   The incidence of PE after anticoagulation and lysis are almost the same → hence: anticoagulants are the first choice in acute therapy of thrombosis. SE + CI: refer to chapter "thrombosis prophylaxis"!

   Procedure: 2 alternatives:
   - Unfractionated Heparin (UFH) i.v.
     Dose for UFH: Initial 70 IU/kg i.v. stat, followed by 30.000 - 35.000 IU/24 hours for 5 days according to the aPTT (target should be ca. 1.5 - 3,0fold of the normal aPTT; results vary between different laboratories). On day 1 or 2 start Coumarin/Warfarin. Stop Heparin, when INR is > 2.0 on 2 subsequent days.
   - Low molecular weight Heparin (LMWH):
     advantages compared to UFH: equally effective; s.c.-use; laboratory monitoring is not necessary, less SE
     CI: renal failure (accumulation with the risk of haemorrhage) → use UFH in this case
     Use of low molecular weight Heparin and their successors in DVT and PE

generic name	brand name	dose/day	licence	
			DVT	PE
low molecular weight Heparins				
Enoxaparin	Clexane®	2 x 10 mg/10 kg	yes	yes
Tinzaparin	Innohep®	1 x 175 IU/kg	yes	yes
Nadroparin	Fraxiparin®	2 x 0,1ml/10 kg	yes	no
Certoparin	Mono-Embolex®	2 x 8.000 IU	yes	no
Factor Xa-inhibitors				
Fondaparinux	Arixtra®	5-10 mg	yes	yes
Thrombin inhibitors				
Desirudin	Revasc®	2x15 mg	yes (only for DVT after joint replacement)	no
Heparin induced thrombocytopenia HIT				
Danaparoid	Orgaran®	Refer to manufacturer recommendations	yes	yes
Hirudin	Refludan®		yes	yes

Note: Thromboembolism prophylaxis requires lower doses of LMWH (refer to chapter "thrombosis prophylaxis").

PTT is unsuitable for monitoring therapy with LMWH. Monitoring of anti-factor Xa-level (3-4 hours post injection) is indicated for underweight (< 50 kg) or overweight patients (> 100 kg); also in patients with reduced renal function and those who develop therapy induced haemorrhages or those who have recurrent thrombosis. The target level of anti-factor Xa is between 0.6 – 1.0 IU/ml (drug administration 2x/day) and 1.0 – 2.0 IU/ml (administration 1x/day).

Any oral anticoagulants can be started on the 1. or 2. day. Stop Heparin, when INR has been >2 for 2 days.

C) Recanalisation therapy:
- Therapy with fibrinolysis activators (Fibrinolytics, „lysis-therapy")

 Ind: - Phlegmasia coerulea dolens
  - new proximal DVT (age < 10 days) with massive swelling
  - pulmonary embolism St. III and IV
  - new myocardial infarction
  - new stroke (up to 3 hours) (strict conditions)

Contraindications to thrombolysis in patients with pulmonary infarction
(Guidelines ESC 2002):
- Absolute contraindications
  • Acute internal haemorrhage or spontaneous intracerebral haemorrhage
- Relative contraindications
  • Major operations, organ biopsies or puncture of a non compressible vessel within the last 10 days
  • Ischaemic stroke within the last 2 months
  • Gastrointestinal haemorrhage within the last 10 days
  • Major trauma within the last 15 days
  • Neurosurgical or ophthalmological operation within the last month
  • Poorly controlled hypertension (syst. BP > 180 mmHg, diast. BP > 110 mmHg)
  • Cardiopulmonary resuscitation
  • Thrombocytes < 100.000/mm^3
  • Prothrombin time < 50 %
  • Pregnancy
  • Bacterial endocarditis
  • Diabetic haemorrhagic retinopathy

Fibrinolytics:
 • Streptokinase (SK) = a protein derived from β-haemolysing Streptococci, which forms a activator-complex together with Plasminogen, which in turn activates Plasminogen into Plasmin.
 • tPA = tissue-type plasminogen activator = Alteplase
 • Further fibrinolytics: rPA = Reteplase, nPA = Lanoteplase, TNK-tPA = Tenekteplase

Dose: Standard SK lysis: initially 250.000 U Streptokinase i.v. over 30 min., followed by 100.000 U/hours over ca. 3 days. Maximum period of treatment is 1 week, because otherwise anti-streptokinase antibodies will develop. A change to standard Urokinase lysis is possible.
Ultrahigh Streptokinase short lysis: 9 million U over 6 hours. This can be repeated after one day, when the first treatment has not been successful.

Beware: No short lysis for iliac vein thrombosis, because there is a higher rate of PE.

Dosage of other fibrinolytics: refer to manufacturer's advice.

A short lysis is followed by Heparin- and overlapping Coumarin/Warfarin therapy. Standard lysis: Heparin therapy starts during lysis therapy once the initially increased PTT is normal again.

SE: Haemorrhage: The incidence of intracerebral haemorrhage during lysis-therapy is 0.8% (during anticoagulation 0.2%), mortality due to haemorrhage: 0.5-1%; in case of haemorrhage the antidote are antifibrinolytics (Aprotinin or Tranexamic acid)

Beware: Allergic reactions can be seen in SK lysis after previous Streptococci infection or previous SK lysis within the last 6 months.

- Regional hyperthermic fibrinolytic-perfusion via cardio-respiratory machine. Indicated when systemic fibrinolysis is contraindicated (can only be performed in a few specialised centres).

- Thrombectomy via Fogarty-catheter + temporary arterio-venous shunt to prevent a thrombosis relapse. The intraoperative risk of PE is low due to overpressure ventilation.
Ind.: Phlegmasia coerulea dolens, V. cava-thrombosis, descending iliac vein thrombosis (when lysis is contraindicated). mortality is ca. 3%.

Thromboembolism prophylaxis with oral anticoagulants: for period of treatment and details refer to the according chapter

SPECIAL TYPES OF TROMBOSIS :

PAGET-VON-SCHROETTER-SYNDROME [I82.8]

Syn: Deep arm vein thrombosis (DAVT)

Def.:   Thrombosis of the axillary or subclavian vein

Aet.:   1. Thoracic-outlet-syndrome (TOS) = compression vein, artery and nerve in the shoulder causing paraesthesiae, pain, weakness of arm and hand; arterial involvement can cause finger necrosis.
        causes:
        – Cervical rib
        – Scalenus-anterior-syndrome = narrowed scalenus muscle triangle
        – Costo-clavicular syndrome = abduction causes narrowing between 1. rib and clavicle
        – Hyperabduction syndrome = minor pectoral muscle tendon causes compression
        – Post clavicle fracture callus, or exostosis of the 1st rib
        Di.: Adson-maneuvre (scalenus-test): In thoracic-outlet-syndrome the radial pulse vanishes during abduction and elevation of the arm, and simultaneous reclining and contralateral turning of the head. Further tests: Wright's hyperabduction test, Falconer's costo-clavicular test
        2. Longterm central vein catheter (history!), hyperosmolar infusion or intima-irritating infusion, pacemaker-carriers
        3. „Thrombosis par effort" = triggered by repeated exercise, e.g. chopping wood, bodybuilding or carrying a rucksack, violin players, rifleman

DD: Mediastinal axillary and clavicular tumours,

Sym.:   Triad: Pain (lower-/upper arm, shoulder), swelling + cyanosis ; occ. visible collateral veins

Di.:   Symptoms - doppler, venography

Th.:   Rest the arm in an elevated position, immobilise the arm, Heparin i.v. for a few days, overlapping start of Coumarin/Warfarin therapy (for 6 months, occ. longer). Fibrinolysis is only indicated in advanced thrombosis.
        Thoracic-outlet-syndrome: occ. transaxillary resection of the first rib, and severing of the muscle insertion of the anterior, posterior and subclavian scalenus muscles.

Phlegmasia coerulea dolens [I80.2]

Def.:   Acute occlusion of all veins of a limb with secondary compression of the arterial circulation (caused by rapidly developing oedema).

Sym.:   limb is extremely swollen, painful, cyanotic, cold, pulse is not palpable.

Co.:   – Hypovolaemic shock and DIC
        – Gangrene
        – Acute renal failure

Di.: Symptoms + doppler scan, phlebography is contraindicated

Th.: Volume replacement, prophylaxis to prevent pulmonary embolism, rapid surgical intervention: thrombectomy, fasciotomy; occ. thrombolysis

ANTIPHOSPOLIPID-SYNDROME (APS)   [D68.8]

Def.: Diagnosis criteria (consensus-conference 2005)
Clinically:
1. One or multiple thrombosis
2. Complications in pregnancy:
    a) Early miscarriage after 10 weeks
    b) At least one premature birth before 34 weeks due to (pre-)eclampsia or placental insufficiency
    c) At least 3 miscarriages before 10 weeks without chromosomal, anatomical, hormonal cause
Serology:
1. IgG- or IgM-anti-Cardiolipin-antibodies = ACA (> 40 IU)
2. IgG- or IgM-β2-Glycoprotein 1-antibodies = β2-GPI-antibodies positive
3. Positive lupus anticoagulant-test

APS is diagnosed in the presence of 1 clinical + 1 serological criteria (must be 2 x positive within 3 months!).

Ep.: Ca. 2-5% of the total population are APLA-positive; they often have only low antibody levels of equivocal relevance. Only some of those people will ever develop any symptoms; f:m = 2:1

Aet.: 1. Primary APS without underlying disease,
2. Secondary APS: with underlying disease (malignancies, AIDS, SLE etc.)
"Catastrophic APS": > 3 organ systems are affected

Pg.: Hypercoagulability due to antibodies against Phospholipids (coagulation factors, thrombocyte receptors etc.)

Sym.: - Thromboembolism (30%), myocardial infarction (up to 20% of infarction patients under 45 years of age), cardiomyopathy, cardiac valve thickening (with occ. increased risk of embolism)
- Proteinuria, renal hypertension
- Thrombocytopenia (usually < 50.000/µl), haemolysis (antibody related or microangiopathic haemolytic anaemia as in TTP/HUS), paradoxical haemorrhage (< 1%)
- Strokes (ca. 30% of all stroke patients < 50 years), loss of vision, hearing loss, convulsions, migraine
- Raynaud's, Livedo reticularis, skin ulcers and -necrosis
- Early miscarriage, (pre-) eclampsia

DD: Other causes of thrombophilia

Lab.: Various results: raised aPTT (due to in-vitro-interaction with Phospholipids), thrombocytopenia, occ. haemolysis

Di.: Refer to "diagnosis criteria"

Th.: 1. Thrombosis: oral longterm-anticoagulation, aim for INR 2.3 - 3.0
2. Stroke: Acetylsalicylic Acid, occ. oral anticoagulation
3. Clinically significant thrombocytopenia: there are graded therapies as in ITP: steroids, Dapson, Azathioprine, Cyclophosphamide
4. Recurrent miscarriage: preferred therapy: low-dose Acetylsalicylic Acid + low dose-Heparin
5. Catastrophic APS: trial of plasmapheresis + Cyclophosphamid-stat-dose
6. Asymptomatic patients: longterm Acetylsalicylic Acid; always give thrombosis prophylaxis where there is increased thromboembolism risk (refer to chapter "DVT").

THROMBOEMBOLISM PROPHYLAXIS

I. General treatment:
- Early postoperative mobilisation and active physiotherapy
- Bed rest is only indicated when there is an obvious benefit
- Antithrombotic stockings
- Intermittent external pneumatic compression
- Treat any varicose veins of the legs and any cardiac failure, be careful when treating oedema with diuretics
- Stop any thrombosis provoking medication (e.g. oestrogens)
- Treat any thrombocytosis or any polycythaemia
- Younger patients or recurrent thrombosis: thrombophilia must be excluded (refer to "DVT")

II. Medication:

1. Anticoagulants:

► **Heparin:**

MoA.: Prophylactic treatment with Heparin in patients at increased risk of thrombosis reduces risk of DVT by 75%, and for PE by 50%. Heparin is a mucopolysaccharide, which enhances the effect of antithrombin (AT). Heparin-AT-complex inhibits thrombin and factor Xa. Hence, in AT-deficiency the antithrombotic effect of Heparin is reduced! Heparin does not cross the placenta, therefore it can be used in pregnancy.

Protamine is the counteracting drug : 1 ml Protamine 1000 Roche® neutralises 1.000 IU Heparin - SE of Protamine: drop in blood pressure, occ. anaphylactic reaction; after overdose: occ. risk of haemorrhage (since too high doses of Protamine inhibits fibrin polymerisation).

	unfractionated Heparin (UFH)	low molecular weight Heparin (LMWH)
mean molecular weight (Dalton)	5.000 - 30.000	3.000 – 6.000
mode of action	final phase of coagulation = inhibits Thrombin	pre-phase of coagulation = inhibits factor Xa
anti-IIa/Anti-Xa	1 : 1	up to 1: 8
half time after s.c.-injection	1 – 2 hours	4 hours accumulation in renal failure!

SE of Heparin:
- Haemorrhage: risk of haemorrhage depends on the dose (high risk after overdose → close monitoring). Risk of LMWH < UFH. Intracerebral haemorrhage after full Heparin dosage in 0.2%
- Reversible increase in transaminases in up to 60% of cases, osteoporosis in longterm therapy, rarely hair loss etc.
- Heparin induced thrombocytopenia (HIT): 2 forms:
  Type I: Non immunological early HIT in the first 2 days of treatment with UFH seen in ca. 25%. Drop in thrombocyte count < 30% of the baseline count. There is spontaneous remission when Heparin treatment is discontinued. Continued Heparin treatment is possible.
  cause: pro-aggregation effect of Heparin, due to inhibition of the Adenylatcyclase.
  Type II: Immunological (antibody) HIT:
  incidence: UFH up to 3 %; LMWH ca. 0.1 % (30 x rarer than UFH !) → prefer LMWH = prophylaxis for HIT II
  Manifestation in non sensitised patients is usually between day 5 and 14.
  symptoms: „White clot syndrome" associated with life threatening thrombosis in ca. 50%! The ratio of venous : arterial thrombosis is ca. 5 : 1, most commonly PE.
  Note: Thromboembolic complication during Heparin treatment always consider HIT II! Disastrous misdiagnosis: Heparin dose is too low → increase Heparin dose → fatal white clot syndrome!

DD of HIT II:

Diagnosis	Diagnostic hints
Pseudo-thrombocytopenia	Normal thrombocyte count in citrate blood sample, aggregate in the blood film
Non immunological HIT I	1 - 2 days after therapy with UFH. Rarely the thrombocyte count is < 100.000/µl or drops > 30 % (excludes the diagnosis)
Thromboembolic complications during therapy with Heparin	Thrombocyte count antibody detection
DIC	History, symptoms, thrombocytes, antibody detection, fibrin monomers, INR, PTT
Autoimmune thrombocytopenia of different origin	- Idiopathic (ITP) - Secondary: malignant lymphomas, HIV-infection - SLE and Antiphospholipid-syndrome
GP-IIb/IIIa-inhibitor-induced Thrombocytopenia	Starts within 12 hours after the application of GP-IIb/IIIa-inhibitors, thrombocytes < 20.000/µl, haemorrhage
Post-transfusion-purpura (PTP)	7 – 14 days after transfusion of pre-immunised patients (> 95 % women), thrombocytes < 20.000/µl, haemorrhage

Di.:   - Thrombocyte count < 50 % of the original count, usually < 100.000/µl
       - HIT-antibodies:
         - Functional test: e.g. HIPA = Heparin induced platelet activation test +
         - Antigen test: binds antibodies of the patient's serum to PF4-Heparin

Th.: Stop Heparin at once! No heparin containing medication/ointments/catheter rinse! Diagnosis is confirmed by the detection of <u>heparin induced platelet-antibodies</u> (HIPA-Test and Heparin-PF4-ELISA). Change to a different antithrombotic drug: Danaparoid (Orgaran®) or Hirudine: Lepirudin (Refludan®) or Desirudin (Revasc®) or Argatroban (Argatra®). Severe thrombosis occasionally requires fibrinolysis. Avoid thrombocyte transfusion (white clot syndrome)!
Provide an allergy alert certificate or bracelet!

IA: Increased blood sugar levels during Heparin therapy

Ind: - <u>In full dose:</u> e.g. DVT, pulmonary embolism (PE) St. I + II
- <u>In low dose</u> for thrombosis prophylaxis

CI: Full dose: similar to the CI for fibrinolysis; known allergy to Heparin and HIT II; patients who suffered HIT II due to unfractioned Heparin must not be treated with LMWH either.

Dos: • UFH:
a) <u>thrombosis prophylaxis:</u> 3 x 5.000 IU s.c./day or 2 x 7.500 IU s.c./day
b) <u>therapy of deep vein thrombosis</u> (DVT): refer to "DVT"
• LMWH/anti-Xa-dose:

Thromboembolism prophylaxis using low molecular weight Heparin:

Substance	brand name	dose/day
Dalteparin	Fragmin®	refer to manufacturer
Enoxaparin	Clexane®	refer to manufacturer
Tinzaparin	Innohep®	1 x 3.500 IU
Nadroparin	Fraxiparin®	1 x 2.850 IU
Certoparin	Mono-Embolex®	1 x 3.000 IU
Reviparin	Clivarin®	1 x 13,5 mg

PTT is not suitable for monitoring therapy. If in doubt, anti-Xa-levels should be tested, this applies particularly to patients with renal failure. This is no necessary when the renal function is normal.

▶ **Fondaparinux** (Arixtra®) is a Heparin analogue (Pentasaccharid) with F. Xa-inhibition. Due to its long half life (17 hours) it is administered only 1 x/day s.c. Protamine is not effective as an antagonist. Coagulation or thrombocyte count monitoring is not necessary.

Ind: 1) Thromboembolism prophylaxis in major orthopaedic operations of the legs (e.g. THR)
2) Treatment of DVT and pulmonary embolism (PE)

Dos: Thrombosis prophylaxis: 2,5 mg s.c. (1x/day), start 6 hours postoperatively
treatment of DVT and PE: depends on body weight (refer to manufacturer)

SE: Occ. anaemia, thrombocytopenia, hepatic disorder, oedema etc.

CI: advanced renal failure etc.

▶ **Coumarins/Warfarin:** <u>Phenprocoumon = PPC</u> with T50 of 3-7 days (Marcumar®, Falithrom®), Warfarin with T50 of 1.5-2 days (Coumadin®)

MoA.: Coumarins are <u>vitamin K-antagonists.</u> Vitamin K is a co-factor in the γ-carboxylation of Glutaminacid rests in the N-terminal ending of the <u>factor of the Prothrombin complex</u> (= Factors II, VII, IX, X) and of the proteins C and S. In vitamin K-deficiency the liver produces dysfunctional preliminary stages of the clotting factors (PIVKA = Prothrombin induced in vitamin K-absence), which lack the γ-carboxylation of the Glutamyl side chains. PPC is absorbed at a rate of almost 100%. Due to the long half life of PPC, a change in dose will not alter the INR for another 3 - 4 days.
When PPC is stopped, the coagulation time will return to normal after 7 - 14 days. No studies have shown that slow weaning of Coumarin therapy is any better than sudden disconnection.
<u>PPC overdose or the need to improve coagulation, requires a step by step procedure:</u>

A) Without haemorrhagic complication:
• omit the medication for 1 - 2 days
• 1-10 mg Phytomenadion (= Vitamin K1) orally is possible: normalises the coagulation after 1 - 2 days
B) With haemorrhage / emergency operations:
PPSB infusion: normalises the coagulation immediately, + give vitamin K1

SE: • Haemorrhage
• Hair loss
• Rarely hepatitis
• Rarely adverse reactions, gastrointestinal symptoms
• <u>Coumarin induced skin necrosis:</u> Protein C has a shorter half life than the factors of the Prothrombin complex. Therefore in protein C-deficiency there can be temporary hypercoagulability during the initial

phase of Coumarin treatment, associated with increased risk of thrombosis and Coumarin-necrosis.

Th.: Stop the Coumarin effect by giving Vitamin K1, give Heparin and corticosteroids

Pro.: During the initial treatment with Coumarin, give also Heparin until the target INR has been reached.

IA: Causes increased efficiency of Coumarins/Warfarin (INR ↑):
- Competes for the bond with protein (non-steroidal anti-inflammatories)
- Reduced intestinal vitamin K-production (antibiotics) and -resorption (exchange resins)
- Thrombocyte aggregation inhibitors, Heparin, fibrinolytics
- Several other medications (e.g. Clofibrate, local anaesthetics, Allopurinol, Cimetidine, Amiodarone etc.) →
always be aware of IA, tight INR-monitoring must follow any change in medication!

Causes of reduced efficiency of Coumarins (INR ↓):
- Hepatic enzyme induction (Barbiturates, Antiepileptics, Rifampicin etc.)
- Other medication (e.g. Digoxin, diuretics, corticosteroids)
- Food rich in vitamin K (e.g. spinach, cabbage)

Ind: Thromboembolism prophylaxis; Coumarins normalize the risk of relapse after DVT; they are also effective in cases of APC-resistance, AT-, Protein C- and Protein S-deficiency.

CI: • Diseases associated with increased haemorrhage risk (e.g. haemorrhagic diathesis, liver parenchyma diseases, renal failure, severe thrombocytopenia)
• Diseases with an increased risk of bleeding (e.g. gastro-duodenal ulcers, poorly controlled hypertension, stroke, trauma after CNS surgery, retinopathy with risk of haemorrhage, cerebral artery aneurysms, active endocarditis lenta)
• Cavernous pulmonary tuberculosis
• Epilepsy
• Chronic alcoholism
• Nephrolithiasis
• Poor patient compliance
• Pregnancy, breast feeding

Dos: The test for therapy monitoring is Thromboplastin time (Quick). Since there are various Thromboplastins, and they cannot be compared, an internationally comparable standard has been launched, called "International Normalized Ratio".

INR = PTT (patient) : PTT (control)

Quick and INR are inversely proportional: rising Quick means drop of INR and vice versa. Therapeutic INR in standard-anticoagulation is between 2,0 - 3,0. With this INR haemorrhage is less frequent than in more intense anticoagulation with higher INR.

When Quick/INR is normal (pre-treatment readings must be available), then adults can receive e.g. the following dose: Marcumar®, Falithrom® (1tbl = 3mg): 2 tbl/day for 3 days, further doses according to the Thromboplastin time (standardised INR). Patients don't need to avoid food rich in vitamin K (cabbage, broccoli, spinach), but they should avoid excessive intake and eat them regularly in moderate doses. They shouldn't take any medication, which might increase the risk of haemorrhage (e.g. thrombocyte aggregation inhibitors). Suitable patients receiving unlimited anticoagulation can be trained to perform home INR tests (CoaguCheck®) and therefore improve their monitoring. The risk of haemorrhage and thromboembolic complications will be diminished and the survival rate will increase (ESCAT-study).

Recommended oral anticoagulation (OAC) for cardiac valves (American College of Chest Physicians)	INR
A) Mechanic valves	
a) Bicuspid (St. Jude) and tilted valves (Medtronic)	
In aorta position	2,0 - 3,0 (2,5 - 3,5 in AF)
In mitral position	2,5 - 3,5 (+ Acetylsalicylic Acid 100 mg in AF VHF)
b) „Caged ball"- valves (Starr-Edwards)	2,5 - 3,5 + Acetylsalicylic Acid (100 mg)
c) Mechanic valves + embolism	2,5 - 3,5 + Acetylsalicylic Acid (100 mg)
B) Bio-prosthesis (for 3 months)	INR 2,0 - 3,0

Recommended treatment with OAC for chronic atrial fibrillation = AF : INR 2.0 - 3,0

Recommended treatment with OAC for thromboembolism (DVT, PE): refer to "pulmonary embolism"

Score to estimate the risk of haemorrhage during OAC:

Marker	Scores
Age > 65 years	1
Post gastrointestinal haemorrhage	1
Post stroke	1
Comorbidity: - Diabetes mellitus *or* - Post myocardial infarction *or* - Renal failure *or* - Anaemia	1
risk of haemorrhage	**significant haemorrhages in 4 years**
0 scores	3 %
1 - 2 scores	ca. 10 %
3 - 4 scores	ca. 50 %

2. thrombocyte aggregation inhibitors

• Acetylsalicylic acid

MoA.: Irreversible Cyclooxygenase (COX)-1-inhibitor. The antithrombotic effect is due to synthesis inhibition of the aggregation promoting Thromboxan A2.

SE: Erosive gastritis, ulcers, gastric bleeding (due to inhibition of the production of mucosa protective prostaglandins and due to inhibition of thrombocyte aggregation); pseudoallergic asthma is seen in predisposed patients (the same applies for all COX-inhibitors); increased risk of haemorrhage (if at all possible, stop one week before elective operations).

IA: Reduced antithrombotic effect of Acetylsalicylic Acid is possible, when the patient also takes any other COX-inhibitors (e.g. Ibuprofen); anticoagulants increase the risk of haemorrhage

Sym.: Haemorrhagic diathesis, gastroduodenal ulcer, Acetylsalicylic Acid-intolerance; relative CI: Asthma, renal damage, third trimester of pregnancy

Ind: Secondary prevention of arterial thrombosis in patients with atherosclerosis

Dos: 75 - 325 mg/day

• Clopidogrel (Plavix®)

MoA. :Irreversible inhibitor of thrombocyte ADP-receptors (P2Y12). The antithrombotic effect is due to the inhibition of aggregation stimulating ADP. Prevents thrombotic complications in patients with atherosclerosis (similar to Acetylsalicylic Acid).

SE: gastrointestinal symptoms, headache, dizziness, increased risk of haemorrhage (if at all possible, stop week before an elective operation), very rarely thrombotic-thrombocytopenic purpura

Ind: 1. alternative prophylaxis to prevent arterial thrombosis (in patients who had adverse reactions to Acetylsalicylic Acid)
2. treatment of unstable angina pectoris in combination with Acetylsalicylic Acid
3. temporary use after PTCA-/stent insertion

Dos: 75mg/day (after 1 x 300mg stat)

• Glycoprotein IIb/IIIa-receptor inhibitors (GPIIb/IIIa-antagonists)

Two groups:
1. antibody fragment: Abciximab (ReoPro®)
2. low molecular GPIIb/IIIa-antagonists: Eptifibatid (Integrilin®), Tirofiban (Aggrastat®)

MoA.:prevents fibrinogen to be bound to the activated GPIIb/IIIa-receptor: intense thrombocyte aggregation inhibition.

Ind: Temporary use in special indications: PTCA-/stent insertion, unstable angina pectoris/acute coronary syndrome

SE: (e.g. haemorrhage); also beware CI

EMBOLISM

Def.: The following material can be carried in the blood flow and cause an embolisation: thrombotic material (thromboembolism), Cholesterol from atheromatous plaques, fat and air (post-traumatic), septic material due to bacterial endocarditis, tumour tissue (a common mechanism of metastasation), amniotic fluid during childbirth, foreign body (e.g. pieces of a torn catheter), talcum powder (i.v.-drugs).

Note: Venous thrombosis: the highest embolic risk is within the first 8 days = the thrombus has not been fixed yet by granulation tissue.

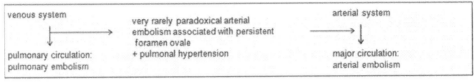

venous system	very rarely paradoxical arterial embolism associated with persistent foramen ovale + pulmonal hypertension	arterial system
pulmonary circulation: pulmonary embolism		major circulation: arterial embolism

Aet.: 1. Venous embolism: Refer to "DVT"

2. Arterial embolism: see below

localisation arterial embolism: preferred locations: vascular bifurcations and physiological narrowing

results: ▶ Venous embolism: pulmonary embolism

    ▶ Arterial embolism: depending on the localisation of the embolism:
- Brain (cerebral embolism)
- Limbs (acute arterial occlusion of a limb)
- Kidneys (renal infarction with occ. loin pain + haematuria)
- Spleen (splenic infarction with left loin pain, occ. perisplenitic rub)
- Intestines (mesenteric infarction with acute abdomen, bloody diarrhoea, known visceral angina)

Th.: Revascularisation therapy: e.g. embolectomy, aspiration embolectomy, local lysis

Relapse prophylaxis: • anticoagulants

        • thrombocyte aggregation inhibitors only for arterial thrombosis (exception: cardiac thrombosis → use anticoagulants)

        • eliminate any risk factors (refer to "etiology")

CHOLESTEROL EMBOLISM

Aet.: Angiography, vascular surgery; in 50% spontaneously caused by rupture of aortic, arteriosclerotic plaques.

Inc.: Rare; elderly patients with arteriosclerosis, m > f;

Sym.: • Skin: blue-toe-syndrome, Livedo reticularis
- GI-tract: abdominal pain, ischaemia
- Eyes: Hollenhorst-plaques (retinal Cholesterol-embolism)
- Kidney: raised urea/creatinine
- Brain: TIA, embolism

Lab.: Occ. leucocytosis, eosinophilia, hypo-complementaemia, occ. slow (days/hours, up to weeks post angiography) deterioration of renal function

Di.: History, symptoms; Certain diagnosis is only possible after biopsy (often at post-mortem).

Th.: Symptomatic

Prg.: Depends on localization

ACUTE ARTERIAL OCCLUSION   ARMS [I74.2] LEGS [I74.3]

Vo.: Most frequent angiological emergency

Aet.: 1. Embolism (70 %) - in 90% the heart is the origin (infarction, atrial fibrillation, mitral valve disease, endocarditis, valve replacement with artificial prosthesis, aneurysm). In 10% the embolism comes from arteriosclerotic plaques of the abdominal aorta or iliac artery, or from arterial aneurysms (arterio-arterial embolism).

2. Arterial thrombosis due to AOD (20 %)

3. Other causes: e.g. external arterial compression; trauma, arterial puncture, arterial prosthesis; arteritis; Heparin induced thrombocytopenia type II; use of oestrogens (ovulation inhibitors) etc.

Sym.: • Incomplete ischaemia syndrome without sensomotoric deficits
- Complete ischaemia syndrome: 6 x "P" (Pratt)
  1. Pain (sudden, excruciating)
  2. Pallor (+ cold skin)
  3. Paraesthesiae
  4. Pulselessness
  5. Paralysis
  6. Prostration (shock)

- Embolism: sudden onset + known cardiac disease
- Thrombosis: slow onset + known arterial disease
- Localisation: ischaemic pain and pulselessness distal to the stenosis
  · Aortic bifurcation (10 %)
  · Femoral bifurcation (45 %)
  · Popliteal artery (15 %)
  · Lower leg-/foot arteries (20 %)
  · Arm arteries (10 %)

Co.: • Shock, ischaemic necrosis
  • Tourniquet-syndrome [T81.1]:
    Reperfusion after complete ischaemia for longer than 6 - 12 hours can cause rhabdomyolysis and metabolic acidosis, hyperkalaemia, myoglobinuria, acute renal failure (life threatening complication)

DD: Phlegmasia coerulea dolens (also absent pulse!)

Di.: • History (cardiac diseases, e.g. arrhythmia due to atrial fibrillation → suspected embolism; intermittent claudication → suspected arterial thrombosis in underlying AOD)
  • Symptoms; check the presence/absence of the pulse (→ localisation), systolic blood pressure reading at the ankle
  • Coloured Duplex US scan,
  • Occ. intra-arterial DSA (arteriography): If history + symptoms clearly indicate an embolism, then embolectomy is performed immediately without arteriography.
    In the presence of thrombosis or in unclear cases, an arteriogram is performed first.

Th.: A) Immediate actions:
  • Inform the surgical team, keep patient nil by mouth
  • Keep the leg in a low position (increased perfusion pressure) and padding (no cold/no warmth, no pressure!)
  • Analgesics i.v.
  • Shock prophylaxis (volume substitution)
  • 10.000 IU Heparin i. v. (to prevent apposition thrombus)

  B) Revascularisation:
  • Embolectomy inclusive of the attached thrombus with a Fogarty-balloon catheter, if possible within the first 6 hours; but even later embolectomy (up to max. 10 hours post occlusion) is sometimes successful.
  • Local fibrinolysis: alternative therapy in occlusion of the lower arm/lower leg, particularly in incomplete ischaemia syndrome; sometimes in combination with aspiration-thrombembolectomy

Pro.: Search for the source of the embolism and remove it, eliminate any risk factors of arteriosclerosis; anticoagulation for recurrent embolism; in case of arterial disease: revascularisation and thrombocyte aggregation inhibitors.

PULMONARY EMBOLISM (PE) [I26.9]

Def.: Occlusion of a pulmonary artery due to a travelling embolism been caught in the artery (= detached thrombus). In > 90% the embolism originates from the catchment area of the V. cava inferior (DVT of a leg- or pelvic vein). Embolism from the catchment area superior vena cava (central venous catheter) and the right heart (pacemaker cable) are rare.

There are 2 conditions to be fulfilled to cause a pulmonary embolism:
1. Presence of a deep vein thrombosis (DVT)
2. Embolisation of the thrombus to the lung

Ep.: Incidence: 60 - 70/100.000/year; fulminate PE: 1/100.000/year

10% of all post mortems reveal a PE. 1 - 2% of all inpatients suffer a PE. Up to 50% of all patients with proximal DVT have a scintigraphically proven (mainly asymptomatic) PE! Postoperative mortality caused by PE is 0,2 - 0,5% despite prophylaxis; hence PE is one of the major causes of morbidity and mortality during a hospital stay. PE happens in the community as well (mainly in patients who underwent day surgery or stayed in hospital for a short time). PE is the most common cause of maternal mortality during pregnancy (in industrialized countries). PE is common, but it also frequently misdiagnosed. In many cases they go clinically unrecognised! Just 1/4 of fatal PEs had been diagnosed whilst the patient was alive. Particularly small signal-embolisms with transient symptoms are missed, although they often herald larger embolisms!

Loc: The right pulmonal artery is affected in most cases (typical localisation: right lower lobe).

Aet.: DVT

Note: Absent signs of DVT does not exclude an embolism.
Bed ridden patients often don't have any clinical sign o DVT. Only 25% of all DVTs show clinical symptoms (!) before the onset of pulmonary embolism!

Trigger factors:
Getting up in the morning, straining (e.g. defecation !), sudden physical exercise

PPh: Thromboembolism causes obstruction of the main pulmonary artery or its branches → sudden increase of the pulmonary resistance (afterload) → drop in cardiac output and hypotension; the area of functional underperfusion is increased (ventilation without perfusion). Hypoxaemia is either the result of a massive PE or previous small PEs. Reflex mechanisms and mediators, released from thrombocytes (Thromboxan, Serotonin etc.) cause additional spasm of the pulmonary vessels with further increase of the afterload.

3 phases of pulmonary embolism:

---
1. Obstruction of the pulmonary artery (afterload ↑) → right ventricular pressure increases
   (acute Cor pulmonale)
2. Underperfusion → arterial hypoxaemia and myocardial ischaemia
3. Acute cardiac failure (cardiac output ↓) → circulation shock (BP ↓, pulse ↑)

Right ventricular pressure rises and myocardial ischaemia may cause right heart decompensation.
Pulmonary infarction (death of pulmonary tissue) after PE is just seen in 10%.
The compensatory blood supply between bronchial- and pulmonary arteries prevents pulmonary infarction even in larger embolism with occlusion of the medium pulmonary arteries.
Embolisms of smaller segmental arteries distal of the anastomosis with the bronchial circulation can cause wedge-shaped, subpleural haemorrhagic pulmonary infarction, particularly in the presence of cardiac failure.
Pulmonary atelectasis can develop within von 24 hours due to surfactant factor reduction.

Sym.: Acute onset of symptoms (single symptoms have a low specifity):
- Dyspnoea/tachypnoea and tachycardia (90%)
- Thoracic pain (70%), occ. infradiaphragmatic pain projection!
- Fear, chest tightness (60%)
- Cough, auscultation: crackles (50%), haemoptysis (10%)
- Severe sweating (30%)
- Syncope, shock (15%)

Note: The majority of fatal embolism presents in phases. Typical for recurrent pulmonary embolism: dizziness, short syncopes, unexplained fever and tachycardia! Important: suspect PE and then arrange the necessary diagnostics!

Co.:
- Pleurisy with respiratory related thoracic pain, pleural effusion
- Pulmonary infarction with haemoptysis (blood stained sputum)
- Infarction related pneumonia, abscess
- Right cardiac failure
- Relapsing embolism (without anticoagulation in ca. 30% !)
- Chronic "Cor pulmonale" seen in recurrent PE

Lab.:
- D-Dimer (D-Dimer-antigen) can be found in recent DVT and in pulmonary embolism due to spontaneous fibrinolysis.
  In the following cases the D-Dimer can also be positive, without DVT/PE: trauma, operations (< 4 weeks), aortic dissection, anticoagulation or fibrinolytic treatment, DIC, disseminated malignancy, sepsis, pneumonia, cellulitis, pregnancy etc.
  A negative D-Dimer-test most likely excludes a pulmonary embolism!

- Troponin T/I and BNP as prognostic parameters:
  Negative Troponin-test and normal BNP are a sign of moderate PE, and they usually exclude severe PE, particularly in the presence of normal echocardiography.

Blood gases (BGA): The diagnostic importance is limited, because there are frequent false negative results. Underlying cardiopulmonary diseases will make the diagnosis (PE?) much more difficult. pO_2 and pCO_2 ↓; normal pO_2 (> 80 mmHg): severe pulmonary embolism (St. III and IV) is unlikely. In severe pulmonary embolism (from St. III onwards), it is very difficult to improve hypoxia with O_2.

ECG: Only in 25% are there typical changes. Pre-ECG and tight monitoring are very important! The changes are often just temporary.
- Sinus tachycardia (90 %)
- SIQIII-type (Mc Ginn-White-syndrome) or SISIISIII-type due to dilation of the right ventricle with clockwise rotation of the heart around its longitudinal axis (compare with pre-ECG: 10%)
- Incomplete right bundle branch block: 10 %

- ST-elevation with terminal negative T-wave in lead III (DD: posterior infarction)
- Negative T-wave in V1,2 (3)   } 50%

- Pulmonal P-wave = dextroatrial P-wave (P ≥ 0,25 mV in II) : 10 %
- Arrhythmia, particularly extrasystoles, occ. atrial fibrillation

Echo with coloured Duplex:

1. Excludes any other diseases (left ventricular pumping dysfunction), aortic dissection, pericardiac effusion, mitral valve tear etc.)
2. Echo is not very sensitive/specific for the diagnosis of PE. When the obstruction of the pulmonary flow is > 30 %, there can be the following signs :
   Right ventricular (RV) dysfunction (of prognostic importance) is ell visible in transthoracic echocardiography (TEE).
   Calculation of the systolic pulmonal artery pressure is possible by using the maximal speed of blood regurgitation through an insufficient tricuspid valve (Beware: Not in the case of left ventricular failure and mitral insufficiency.
   - Occ. direct thrombus detection in the right heart or in the pulmonal artery (transoesophageal echocardiography = TEE)
   - indirect sign for acute increase of pressure of the right ventricle in haemodynamic significant PE:
     ◦ Dilatation and in most cases also hypokinesia of the right ventricle
     ◦ Diastolic shift of the ventricular septum towards the left ventricle („D-form" instead of the round form of the left ventricle)
   - Thoracic US scan: sometimes there is a triangular low echogenic lesion on the pulmonary surface with central reflex (pulmonary infarction)

Radiography: CXR can provide non-specific signs in just ca. 40%. Congested pulmonal artery, acute enlarged heart, unilateral raised diaphragm, „vascular gaps" of large pulmonary artery branches; Westermark' sign: temporary local translucency; unilateral small pleural effusion, in pulmonary infarction there is marked (rarely triangular) peripheral shadowing (Hampton's hump), atelectasis.

Detection of embolism:

- CT-angiography or MR-angiography: shows the pulmonal artery all the way up to the subsegmental arteries: preferred method
- Pulmonary perfusion scintigraphy: microembolisms in pulmonary capillaries can be shown using $99m$Technetium-marked Albumin macroaggregates (this will not disturb pulmonary circulation and -function): comparing this with a more recent CXR and (if in doubt) with a ventilation scintigram can help to diagnosis any perfusion defects as the result of embolism! Is not available everywhere, it doesn't actually show the embolism. In COPD, pneumonia, atelectasis etc. there will be more equivocal investigations; but: normal scintigraphy practically excludes PE (highly negative predictive - as with the D-Dimer).
- Pulmonal artery angiography and DAS are rarely indicated (filling gaps, abruptly ending vessels), Indicated only when other tests have been equivocal and if it has therapeutic consequences.

Find the DVT which caused the PE:

Compressions- and coloured Duplex scan (gold standard); in 90% die leg- or iliac veins are the source of embolism

Staging of pulmonary embolism:

	stage I	stage II	stage III	stage IV
Symptoms	haemodynamically stable without RV-dysfunction	haemodynamically stable with RV-dysfunction	shock BPsyst. <100mmHg Pulse > 100/min.	needs resuscitation
PA-mean pressure (mmHg)	normal < 20	usually normal	25 - 30	> 30
Pa O$_2$ (mmHg)	> 75	occ. ↓	< 70	< 60
vascular obliteration	peripheral branches	segmental arteries	PA-branch or several lobular arteries	PA-branch and several lobular arteries (PA-trunk)
Mortality	Low	< 25 %	> 25 %	> 50 %

DD: Depends on the symptoms:
- Acute shortness of breath: pulmonary oedema, asthma attack, spontaneous pneumothorax, psychogenic hyperventilation etc.
- Acute thoracic pain: myocardial infarction/angina pectoris, pericarditis, pleurisy, dissecting aortic aneurysm (note: in a large PE Troponin can be positive!)
- Acute upper abdominal pain: biliary colic, perforated ulcer, pancreatitis, posterior infarction etc.
- Collapse/shock: DD of unexplained shock
- Haemoptysis: haemorrhage from nose/pharynx, oesophagus, stomach, bronchi system/ lungs
- Any hospital acquired pulmonal infiltration will require to find the diagnosis: pulmonary embolism/pulmonary infarction or pneumonia !

	pulmonary embolism	myocardial infarction
History	longterm immobilisation (e.g. postoperative, thrombosis, heart disease)	angina pectoris known CHD
Onset	Suddenly	often slowly
Pain	pleuritic pain (increased by inspiration)	pain with radiation (shoulder, arm, neck, upper abdomen); not increased by breathing
Dyspnoea	suddenly, severe	Minor
Laboratory	Troponin I/T positive and BNP ↑ in a large PE	CK-MB ↑ and Troponin I/T positive
ECG	sometimes similar to the one seen in posterior infarction	ECG: typical ECG-changes (refer to "myocardial infarction")
Doppler scan	right ventricular dysfunction in large PE	hypo- or non-kinetic infarction areas are mostly located left ventricular

Di.:
- History (predisposing factors!)
- Symptoms

  Estimation of clinical probability for PE (Wells):

Clinically	Score
Symptoms of a recent DVT	3,0
PE is more likely than any other diagnosis	3,0
HR > 100/min	1,5
Operation or immobilisation of the leg within the last 4 weeks	1,5
History of DVT or PE	1,5
Haemoptysis	1,0
Tumour (present treatment/ life expectancy < 6 months/ palliative)	1,0
probability for PE	
Low	< 2,0
Moderate	2,0 - 6,0
High	> 6,0

- D-Dimer and prognostic laboratory results (Troponin, BNP)
- Proof of right ventricular dysfunction (Echo)
- Proof of embolism (occ. TEE, Angio-CT, Angio-/3 D-MRI)
- Proof of venous thrombosis (source of embolism) (US)

Th.: Two targets:
1. Prevent embolism relapse

Memo: 70% of fatal pulmonary embolism come in phases!

2. Reduce mortality with suitable treatment

Emergency treatment of the acute pulmonary embolism:
- Semi-sitting positioning of the patient + careful transport to the hospital (treat the patient like a raw egg, to avoid further embolisms!)
- Sedation (occ. 5 mg Diazepam slowly i.v. – SE: respiratory depression), painkillers
- O_2-nasal tube (6 l/min) and pulse-oximetry; respiratory failure: intubation and ventilation
- Central venous access (monitor CVD and pulmonal artery pressure); no i.m.-injections!
- 5.000 - 10.000 IU Heparin i.v. stat
- Occ. shock treatment: Dobutamin (4 - 8 µg/kg/min), occ. Noradrenalin
- Cardiovascular arrest due to fulminate pulmonary embolism: cardio-respiratory resuscitation for a longer period (→ fragmentation of the embolism) + thrombolysis

B. Specific actions:

a) Conservative:
- Heparin: in the absence of any contraindications, in St. I and II (haemodynamic stabile patients) Heparin is the first choice therapy to prevent further embolisms and to reduce mortality.

  spontaneous fibrinolytic activity of the lungs will reopen the congested blood vessels within days - weeks.

  2 alternatives:
  □ unfractionated Heparin or
  □ low molecular Heparin: Tinzaparin or Enoxaparin
  dose: refer to "DVT"
  period: 7 - 10 days, overlapping start of treatment with Coumarins/Warfarin

  period for treatment with Coumarin/Warfarin after thromboembolism
  (German society for angiology, 2005):
  first thromboembolism
  Transient risk factor (DVT proximal and distal; PE): 3 months

> Idiopathic cause or thrombophilia: 6 - 12 months
> Combined thrombophilia or Antiphospholipid-antibody-syndrome: 12 months
> Recurrent thromboembolism or active cancer: no time limit

- Fibrinolysis (thrombolysis):
  Target: To dissolve the embolism (recanalisation) + to dissolve any thrombus which could be the cause (eliminate the source of relapses).
  Ind: Massive pulmonary embolism in St. III and IV (haemodynamic unstable/shock/ cardiovascular arrest, RV-dysfunction in the echocardiogram, Troponin positive, BNP ↑)
  stipulation: no contraindications (refer to "DVT")
  procedure: stop Heparin when using Streptokinase or Urokinase for lysis;
               continue Heparin when tPA is used
  Dose: e.g. Alteplase 10 mg stat (2 min.) and 90 mg over a period of 2hours i.v., followed by Heparin, and overlapping start of treatment with Coumarin/Warfarin.

b) Catheter methods:
   Ultrapidall-Thrombolysis, mechanical fragmentation of the embolism via right cardiac catheter, local fibrinolysis
   Ind: Massive pulmonary embolism (St. III and IV)

c) Surgery:
   Pulmonal embolectomy without (Trendelenburg) or with the support of an extracorporal cardio-respiratory machine. This way the pulmonary arteries can be cleared right into the subsegments.
   Ind.: When all conservative treatments have failed after the first hour, you may consider operative embolectomy (after angiography): mortality: Ca. 25 %!

Prg.: The prognosis of pulmonary embolism depends on:
1. Severity (refer to table)
2. Age and underlying diseases
3. How quickly diagnosed and start of therapy
4. Complications and occ. relapse
When the acute event has been successfully managed, treatment of the underlying disease and prophylaxis of further embolisms is indicated, because the relapse rate is at least 30%!

Pro.: • Primary prophylaxis = thrombosis prophylaxis
          • Secondary prophylaxis after pulmonary embolism and/or DVT:
            - Anticoagulation using Coumarins/Warfarin
            - AT- deficiency: substitute AT in those at risk.

DISEASES OF THE LYMPHATIC VESSELS

Lymphangitis   [I89.1]

Def.: Inflammation of lymphatic vessels due to tissue infection spreading into the lymphatic vessels or due to pathogens being flushed into the lymphatic vessels. Lymphangitis of the limbs is usually caused by infected wounds. Once the infection has been cured, the affected lymphatic vessels will close up.

Ep.: Frequent disease

Aet.: Outside tropical/subtropical areas usually Streptococci, also Staphylococci and other pathogens. In tropical/subtropical areas also Filariasis (Wuchereria bancrofti and Brugia malayi) etc.

Sym.: A red stripe running from an infected injury towards the tender, swollen regional lymph nodes; occ. fever.

Co.: Sepsis, rarely lymph node abscess; occ. lymphatic oedema as a late complication

Di.: Symptoms (red stripes on the skin: stets look peripherally for the infection focus and the associated lymph node).

Th.: Streptococci-/Staphylococci infection: give Penicillins effective against Staphylococci (e.g. Oxacillin, Di- or Flucloxacillin); immobilisation of the affected limb, disinfecting dressings, clear the infection!
Filariasis: Diethylcarbamazin

Erysipela [A46]

Def.: Erysipela is an inflammation of the skin and subcutaneous tissue caused by β-haemolytic A-Streptococci (sometimes S. aureus) which spread via the main lymphatic vessels (surface lymphangitis).
Predisposing factors: entries through the skin (e.g. small wounds, ulcers or tinea pedis), lymphatic oedema, chronic venous insufficiency, morbid obesity, dermatological diseases

Sym.: - Erythematous, warm, oedematous, tender marked cutaneous area, with or without vesicles; central clearing (pale); occ. pruritus; occ. regional lymphangitis/-adenitis
localisation: Am most common lower leg; occ. arms, face etc.
- Occ. there are tiny injuries (entries) (e.g. interdigital athlete's foot)
- General symptoms: fever, malaise, raised ESR, CRP

Co.: Bullous ulcer, gangrenous ulcer, migrating ulcer
Secondary diseases due to Streptococci-allergy: Acute post-streptococcal-glomerulonephritis;
relapses are common
lymphatic oedema is a late complication

> **Note:** Erysipelias can be the cause of subsequent lymphatic oedema. The other way round, erysipela is the most common complication of lymphatic oedema.

DD: Cellulitis (undefined edges)
Necrotising fasciitis (e.g. β-haemolysing A Streptococci): severe soft tissue infection with necrosis (MRI), rapidly progressive, life threatening! Complication: toxic shock syndrome.

Di.: Symptoms, laboratory (leucocytes ↑, ESR and CRP ↑, Anti-DNAseB = ADB-titre ↑ in 90%), search for the infection entry

Th.: - Penicillin; Penicillin allergy: Erythromycin (2 weeks). Suspected S. aureus-infection: Staphylococci-Penicillin or Cephalosporin.
- Rest the limb and apply local cold, symptomatic treatment of pain and fever
- Treat any entry wounds

Pro.: Treat any predisposing factors

Lymphatic Oedema [I89.0]

Def.: Swelling of subcutaneous tissue associated with congestion of the lymphatic fluid due to reduced transport capacity of the lymphatic vessels (obstruction, destruction, hypoplasia)

Aet.: - Primary lymphatic oedema (10% are hereditary: Nonne-Milroy-syndrome, Meige-syndrome): disorder of the lymphatic vessels
Ca. 85% women, peak of manifestation: 17 years; develops from distal to proximal
- Secondary lymphatic oedema (majority): develops from proximal to distal
Caused by tumour, operation, trauma, inflammation, infections, venous congestion, radiation

> **Note:** Any new onset of lymphatic oedema after the age of 18 is suspicious of a malignant tumour!

Cl.: 4 stages:
  0   Latent stage: reduced transport capacity of the lymphatic vessels without swelling
  I.   Soft swelling (there is a dent on pressure) without secondary tissue conversion = reversible
  II. Fibrosis of the skin. Pressure cannot produce a dent, the oedema can be removed with diuretics, and fibrosis can be partially reversible with intensive therapy.
  III. Lymphostatic elephantiasis = irreversible: very fibrotic thick skin.

DD: Refer to chapter "oedema"

Di.: • Symptoms:
    - Lymphatic oedema of the leg (unlike in venous oedema): the toes are affected as well (square toes)
    - Deep horizontally engraved lines of the toes
    - The dorsal surface of the toes is often rough (like a wart) (Papillomatosis cutis)
    - Stemmer' sign positive = no skin fold can be lifted above the toes
  • Radiographic diagnostics:
    - Indirect lymphography with water soluble contrast medium
    - Isotope lymphography (functional lymphatic scintigraphy)

Th.: A) Conservative
    Target: remission of the disease into the latent stage is only possible during St. I.
    - The limb must be raised
    - Complex physical decompression therapy → 3 phases:

298

phase I: decongestion - phase II: optimising - phase III: conservation
1. Skin care
2. Manual lymphatic drainage
3. Compression therapy
4. Decongesting exercise therapy
- Tailored compression stockings, once the oedema is in complete remission

CI for KPE: acute inflammation, cardiac decompensation, malignant lymphoma

CI for compression therapy: AOD with ankle pressure below 80 mmHg

B) Surgery when conservative therapy has failed:
Resections - drainage - autologous lymphatic vessel transplantation

Tumours of the lymphatic vessels

Primary: Benign lymphangioma
              Very rarely malignant lymphangiosarcoma
Secondary: Lymphangiosis carcinomatosa found in various cancers

Internet-Infos:
German: *www.rki.de* (Robert-Koch-Institut)
                www.pei.de (Paul-Ehrlich-Institut)
                www.dtg.mwn.de (Deutsche Gesellschaft für Tropenmedizin)
                www.crm.de (Zentrum für Reisemedizin)
                www.auswaertiges-amt.de (Auswärtiges Amt, Berlin)
                www.dghm.org (Deutsche Gesellschaft für Hygiene und Mikrobiologie)

International: *www.idsociety.org* (Infectious Disease Society of America)
                www.cdc.gov (Centres of Disease Control)
                www.who.int (WHO)

EXANTHEMATOUS INFECTIOUS DISEASES

5 exanthematous infectious diseases, occurring predominantly in children: Scarlet Fever, Rubella, Fifth Disease, Measles, Chicken Pox.

SCARLET FEVER [A38]

PPh.: Streptococcus pyogenes = β-haemolysing Streptococci of Lancefield group A, 80 serotypes

Infections caused by Streptococci A (Streptococcus pyogenes):
- Asymptomatic carriers: During winter months up to 20 % of the population
- Non-invasive infections: various mucosae (tonsillitis, pharyngitis, otitis media, sinusitis), skin (impetigo, streptococci-pyodermia)
- puerperal sepsis
- invasive infections: e.g. (postoperative) wound infections, necrotising fasciitis, Fournier's gangrene of scrotum-/perineum
- After entering the blood circulation: sepsis, meningitis, toxic shock syndrome

Ep.: peak age 3 – 10; endemics in nurseries etc; seasonal peak: October - March.

Inf.: Usually airborne; rarely from pus, contaminated milk, contaminated objects etc.

Inc: 2 - 4 days

Con.: Ceases 24 hrs after starting antibiotic therapy.

Pg.: Streptococci infections primarily cause a local infection, and hence they don't leave an antibacterial, but only an antitoxic immunity against the erythrogen toxin. In order to cause scarlet fever, the pathogenic S.pyogenes strain has to carry 1 or more bacteriophages in its genome, coding the erythrogen toxins (type A and C). If the patient is not immune to the relevant erythrogen toxin, he will develop scarlet fever (if he is immune, he will just suffer from tonsillitis). The existence of various erythrogen toxins explains how patients may develop scarlet fever for a second time (1 - 4 %).

Sym.: • Sudden onset with a sore throat, cough, vomiting, high fever, tachycardia, headache and abdominal pain
- Pharyngitis, tonsillitis with enanthema (including the uvula), submandibular lymphadenopathy
- coated tongue in the beginning, from day 4: strawberry tongue
- day 2 or 3: Rash: very fine, beginning in the area of the axilla, groins, then ascending towards the neck; intensive flushed cheeks with perioral pallor; after 2 - 4 weeks peeling of the hands (palms) and feet (soles).
- Rumpel-Leede test positive (petechiae on the lower arm post inflation of a blood pressure cuff above the diastolic pressure for 5 minutes)

Co.: • Toxic Disease: Vomiting, diarrhoea, cardiovascular collapse (myocarditis), convulsions, drowsiness
- Ulcerative tonsillitis, purulent sinusitis, otitis media
- Septic disease, cerebral sinus thrombosis, meningitis
- Post-Streptococcal diseases: rheumatic fever, rheumatic carditis, Chorea minor, acute glomerulonephritis

DD: Rubella/Measles: See 'Rubella'
Staphylococci Scarlet Fever, toxic shock syndrome, Kawasaki-syndrome (a rare vasculitissyndrome) etc.

Di.: symptomatic (pharyngitis, rash, enanthema)

Lab.: • Leucocytes with basophile blotches and toxic granulations, eosinophilia, raised ESR
- Antibody detection: ≥ 4 times raised titre of antistreptolysin 0 (ASO = ASL)

> ***Memo:*** ASL-titre ↑ (mainly throat infections)
> ADB-titre ↑ (mainly skin infections)

- microbiology culture: nose/throat swab: β-haemolysing Streptococci A (test twice)

<u>Th.:</u> Gold standard is <u>oral-Penicillin (Phenoxymethylpenicillin or Propicillin) for 10 days.</u>
<u>adult dose: 3 x 1 Mio. IU/d.</u> Penicillin allergy: Erythromycin.
If the therapy fails (betalactamase producing bacteria or very rarely Penicillin resistant Streptococci): change to Cephalosporins.
<u>2 weeks post onset of symptoms check for any haematuria!</u>
The intrainfectious microscopic haematuria (= interstitial nephritis) is harmless, the postinfectious microscopic haematuria indicates the very serious acute glomerulonephritis!

<u>Prq.:</u> An early and long enough lasting (10 days) antibiotic therapy is the best prophylaxis against rheumatic carditis. It is not always possible though, to avoid a glomerulonephritis.

<u>Pro.:</u> Prophylaxis with Penicillin for any people who are in close contact to the patient – return to nursery etc. 48 hrs post onset of antibiotic therapy and in the absence of any symptoms.

RUBELLA   [B06.9] Congenital infection is notifiable

<u>Syn:</u> Rubella, Rubeola

<u>PPh.:</u> Rubivirus = RNA-Virus from the family of Togaviridae

<u>Ep.:</u> Contagious < 50 %, peak amongst school children, 80 - 90 % of all adults over 20 years of age are immune.

<u>Inf.:</u> Airborne

<u>Inc:</u> 2 - 3 weeks

<u>Con.:</u> 1 week before until up to 1 week post onset of the rash

<u>Sym.:</u> <u>Rubella embryopathy (Gregg) = congenital rubella syndrome (CRS):</u>
Highest risk during the first trimester of pregnancy: 1. - 6. week: 55 %; 7. - 12. week: 25 %; 13. - 17. week: 15 -10 %; > 17. week usually no fetal damage.
Inner organs damages: <u>Eyes:</u> 70 % (retinopathy, cataract, rarely glaucoma); <u>ear:</u> 60 % (deafness); <u>heart:</u> 50 % (open Ductus Botalli, septum defects, pulmonal stenosis); cerebral damage and mental retardation: 45 %; growth retardation, reduced birth weight: 75 %

<u>Postnatal Rubella infection:</u>
Children: 50 % asymptomatic
> slow onset (unlike scarlet fever!), oft unapparent course without fever or rash
> <u>Maculopapulous exanthema:</u> medium sized spots (between measles (gross) and scarlet fever (pinpoint-sized), usually <u>not</u> confluent, starts behind the ears, lasts: ca. 3 days
> <u>lymphadenopathy</u> (particularly retroauricular): 50% splenomegaly.

<u>Rubella-re-infections:</u>
Occasionally a long time after a first infection or after a vaccination; they are usually asymptomatic.

<u>Co.:</u> ▶ Rubella-encephalitis (frequency 1 : 6000)
▶ Rubella-purpura (caused by transient thrombocytopenia (good prognosis)
▶ Rubella-arthritis (adults, good prognosis)

<u>DD:</u>

	Scarlet fever	Measles	Rubella
Begin	high fever sore throat (tonsillitis)	high fever bad cough occ. sore throat	mild fever; slightly unwell
Exanthema	fine rash spreading from caudal to cranial (perioral pallor)	coarse confluent rash, spreading from cranial to caudal (begin retroauricular)	mild non-confluent rash of neck/chest
Extras	strawberry tongue	Koplik´s spots	large nuchal lymphadenopathy

<u>Di.:</u> • Leucopenia, lymphadenopathy, plasma cells
• pathogen detection (no routine diagnostics): Virus detection, PCR
• AB-detection
- <u>new infection:</u> Sero-conversion or ≥ 4 times titre raise of IgG-AB in 2 specimens, IgM-AB-raise; IgM-AB may persist up to > 1 year (low titre).
- Rubella-re-infection: titre-raise of IgG-AB (IgM usually negative)
• <u>antenatal rubella-diagnostics</u> (from 11. week): detection of virus-RNA (e.g. by PCR) from amniotic fluid or biopsy-tissue from chorionic villi; also from week 22: IgM-AB-detection and PCR from fetal blood.
• rubella <u>diagnostics of the newborn:</u> Usually IgM-AB-detection + additional virus-detection

Th.: Rubella immunoglobuline in complicated cases

Pro.: Active: 2 x vaccination with live vaccine: e.g. combined Measles-Mumps-Rubella-(MMR-) vaccination (at 12 - 15 months and pre-school booster). Side effects: local- and generalised reactions, sometimes arthralgia etc. Ind: All non-immune girls should be vaccinated before onset of menarche. Protection is probably lifelong. All women of childbearing age should have a serology test. Sufficient antibody-titre 1 : ≥ 32 (haemagglutinin-inhibition-test). In case of absent immunity: Only vaccinate, if the woman is not pregnant. She must use contraception for 3 months post vaccination. However-there are no known cases of rubella-embryopathy caused by vaccination. Then check immunity (serology). Vaccinate all seronegative staff working in obstetrics, paediatrics, nurseries.

Passive: Post-exposition-prophylaxis (PEP) with hyper-immunoglobulin
Ind: pregnant women with negative or unknown protection, who have been in contact with patients. Give hyper-immunoglobulin between exposure and viraemia, in order to avoid embryopathy! It is futile, if given after the onset of the rash!

PARVOVIRUS - 19 INFECTION   [B08.3]

Syn: Erythema infectiosum, Fifths Disease

PPh.: Parvovirus B 19 – target cells: erythropoetic cells of the bone marrow, which will be partially destroyed; temporary anaemia; receptor: blood group P-antigen

Ep.: age peak: school children, seasonal peak: spring; occ. epidemics in nurseries etc.; up to 50 % of all 15 year old, and up to 80 % of all 50 year old have antibodies against Parvovirus B 19.

Inf.: airborne and parenteral (e.g. haemophilia patients!)

Inc: 6 - 18 days

Cont: Ca. 50 % of contacts in a household

Sym.: Asymptomatic in 30% of children. Adults: More severe symptoms, 20 % sustain virus persistency.
- Erythema infectiosum: circle- or garland shaped maculopapulous rash, starting as a purple discolouration on the cheeks ("slapped cheek disease"), possibly butterfly-shape, periodical blanching and recurrence of the rash, for about 10 days (very rarely longer).
- Possibly petechial rash of hands and feet
- Possibly arthritis
- Lab.: There is always a temporary anaemia and reduced reticulocytes count, possibly also temporary thrombocytopenia, granulocytopenia (DD: SLE)

Co.: 1. Aplastic crisis in patients with chronic haemolytic anaemia
2. Infections during pregnancy:
   1st trimester often miscarriage
   2nd trimester aplastic anaemia, fetal hydrops and fetal death
   3rd trimester often just temporary aplastic phases without fetal damage
3. Immune-deficient patients: Chronic anaemia, arthritis, thrombocytopenia, granulocytopenia or pancytopenia, pure red cell aplasia, myocarditis/DCM, hepatitis

Di.: • Diagnostics of the acute infection:
Symptoms + detection of IgM-antibodies and virus-DNA (PCR)
Post-infectious: DNA- and IgM-antibody-serology negative; IgG-antibody positive.

• Diagnostics of the antenatal infection:
Ultrasound follow-up and surveillance of the pregnancy. If a fetal infection is being suspected: Check amniotic fluid, ascites and fetal blood for IgM-antibodies, virus-DNA and fetal Hb.

Th.: • symptomatic therapy
• aplastic crisis: blood-cell transfusions; give 7S-immunoglobulins (also for patients post organ- or bone marrow-transplantation)
• antenatal infection at risk of a fetal hydrops: intrauterine exchange-transfusion.

Prg.: usually benign; complications in risk groups (as above)

Pro.: keep away people suffering from immune deficiency, chronic anaemia and pregnant women!

MEASLES (MORBILLI)   [B05.9] Notifiable disease

PPh.: Paramyxovirus (RNA-virus) ; only 1 serotype, but 8 classes (A - H) with 23 genotypes (important for epidemiological investigations)

Ep.: Worldwide > 1 million children die from measles, particularly in the poor countries (Africa). airborne infection; highly contagious, very virulent, high chance of immunity → typical disease in children (infants up to 8 months are protected by their mother's antibodies); lifelong immunity.

Inf.: Infectiousness begins 5 days before the onset of rash and lasts until 4 days post onset

Inc: 8 - 10 days until onset of prodromi, 14 days until onset of rash

Sym: 1. prodromi: rhinitis, conjunctivitis, fever, barking cough, light phobia, swollen face, enanthema of the palate + Koplik's' spot (whitish spots on the buccal mucosa opposite the molars)

2. rash (caused by immuno-complex- phenomenon): coarse, confluent, brownish-pink, starts behind the ears, cranio-caudal spread, later fine peeling of the skin; another temperature spike during the onset of the rash
• cervical lymphadenopathy; occ. mild abdominal discomfort

Lab.: Reduced leucocytes, lymphocytes, eosinophiles; transient thrombocytopenia.
A previously positive Heaf test may become temporarily negative.

Co.: ca. 15 % in western countries. Usually bacterial secondary infections due to the temporary (ca. 6 weeks) immune deficiency:
1. most frequently otitis media ca. 10 %
2. pneumonia : 3 varieties:
    - viral pneumonia (Hecht's giant cells)
    - bacterial secondary infection (→ antibiotics)
    - giant cell pneumonia in patients with immune deficiency (poor prognosis)
3. occ. laryngotracheitis & croup (risk of suffocation)
4. encephalitis: 3 varieties:
    - acute postinfectious encephalitis: onset ca. 4 - 7 days post onset of rash (1 : 1.000 patients aged > 1 year) - fatality up to 20%; permanent damage up to 20% (e.g. impaired intelligence/concentration, epilepsy etc.); temporarily abnormal EEG in 50 %, permanent abnormal EEG in 3%.
    - encephalitis in the immunosuppressed after 5 weeks to 6 months
    - subacute sclerosing panencephalitis (SSPE) after ca. 6 - 8 years = "slow virus-infection" with virus mutants containing an 'M-antigen'; no antibodies can be produced against this antigen (though they develop antibodies against other virus proteins);
    frequency is unknown (1 : 10.000 - 1 : 100.000?); demyelinising, lethal .
5. very rarely malignant course: fading of the rash and cardiovascular collapse due to immune deficiency

DD: weak rash in partially immunosuppressed patients)
DD Rubella

Di.: symptoms + serology: detection of IgM-AB or virus-RNA in a new infection (or rise of titre in second specimen)
encephalitis: virus detection (PCR) in cerebrospinal fluid.
SSPE: high antibody-titre in cerebrospinal fluid (usually no virus detection)

Th.: symptomatic, antibiotics for bacterial complications
Return to nursery not earlier than 5 days post onset of rash.
Non-vaccinated persons cohabiting with the patient have to be excluded from nursery for 14 days.

Pro.: • vaccination (live vaccine):
    Ind: 1. children over 12 months of age (e.g. Measles/Mumps/Rubella - MMR) and booster after 4 weeks. (**Note:** In the UK there is no booster after 4 weeks, but instead children have a pre-school booster; different vaccination programmes may apply to other countries). All non-immunised children should be vaccinated until the age of 17.→ WHO intends to erase measles.
    2. seronegative staff looking after people are at increased risk of infection(paediatrics, oncology, caring for immunosuppressed people)
    3. post exposure vaccination within 3 days of exposition
    SE.: 5 – 14 days post vaccination mild measles-like infection in up to 5% of cases. Occasionally febrile convulsions or allergic reactions.
• passive immunisation with human immunoglobulin in the immunosuppressed: within 3 days of exposition; if given after 4.-7. days→ reduced severity of disease.

HERPES VIRUSES

The family of herpes viridae contains ca. 100 different viruses, of these there are 7 human herpes viruses, i.e. human beings are the natural host:

- Varicella-Zoster-Virus (VZV)
- Herpes Simplex-Virus (HSV), type 1 and 2
- Epstein-Barr-Virus (EBV)
- Cytomegaly virus = CMV (= HHV-5)
- Human Herpes virus 6 (HHV-6) = Exanthema subitum (Sixth disease)
- Human Herpes virus 7 (HHV-7) not causing any disease
- Kaposi-Sarcoma-Herpes virus (KSHV = HHV-8): not pathogen in healthy individuals, in the immunosuppressed it may cause the Kaposi-Sarcoma .

All herpes viruses share the important capacity of a lifelong persistence in the host's target cells. This may result in a reactivation of the virus from latency, if the host suffers a temporary immune deficiency.

VARIZELLA-ZOSTER-VIRUS-INFECTIONS

PPh.: Varicella-Zoster-Virus (VZV)

Ep.: Chickenpox in childhood (highly contagious - 90%). > 95% of all adults have antibodies. Up to 25% of all adults suffer a local reactivation later in life (usually at an older age) Herpes zoster (shingles).

Inf.:: Chickenpox is an airborne disease). It is contagious from one day before the onset of the rash until all scabs have disappeared. Zoster vesicles are not as contagious.

Inc. of varicella: 8 - 28 days, peak at age 2- 6 years, 90% of all infections before the age of 20.

Sym.: usually uncomplicated in otherwise healthy individuals.

Severe disease/complications (usually seen in high-risk-patients):
- congenital or acquired T-cell defect
- pregnant women
- eczema
- congenital varicella-syndrome
- neonatal varicella (chickenpox contracted just before birth).

1. Varicella (Chickenpox) [B01.9]:
primary infection causing a widespread vesicular rash and fever. The appearance of the rash is polymorph, because it comes in phases (roseolae - papules - vesicles –scabs); if not being scratched, the lesions heal without scarring, scalp and oral mucosa are also affected. The densest rash is to be found on the trunk, there is only mild illness in children, but often severe pruritus.

2. Zoster (Shingles) [B02.9]:
After the primary infection, the VZV remains dormant within the spinal ganglia; if the individual goes through a temporary cellular immunodeficiency, it may be reactivated. Zoster affects mostly elderly people or those suffering from some sort of immunodeficiency (cancer, leukaemia, Hodgkin, AIDS, patients on cytostatic or immunosuppressant therapy); other triggers are: sun exposure, stress and trauma. The most frequent location is the thorax (50%).
Most infections are limited to one or a few dermatomes (usually T3 – L3) unilateral. Elderly: In 20% the cranial nerves are affected. There may be associated fever. The affected areas are extremely painful before, during and after the development of vesicles (which sometimes also can be haemorrhagic- and then causing scarring).

Co.: Varicella complications:
- Varicella embryopathy (maternal infection up to 20. week) - risk 1%
- neonatal varicella (maternal infection 5 days before to 2 days after delivery: Haemorrhagic rash, lethality up to 30%
- bacterial secondary infection of the rash
- otitis media
- meningeal irritation (good prognosis)
- acute cerebellar ataxia (risk 1 : 4.000), good prognosis
- encephalitis (risk 1 : 40.000), poor prognosis
- pneumonia (viral and sometimes secondary bacterial; lethal in up to 30%)
- thrombocytopenia
- immunodeficient patients: malignant course affecting internal organs

Zoster complications:
- post herpetic neuralgia (up to 70 % in the elderly): pain > 4 weeks (steady and burning or paroxysmal and stabbing)
- zoster ophthalmicus (risk of corneal damage)
- rarely intraocular complications (e.g. in HIV positive patients) –may cause blindness

- zoster oticus (may include VIIth nerve palsy) (Ramsay-Hunt-Syndrome)
- meningoencephalitis, rarely granulomatous angiitis causing hemiplegia; very rarely myelitis
- in the immunosuppressed patient there may be disseminated zoster with associated visceral dissemination (e.g. pneumonia, hepatitis)

DD: - Varicella: Infections caused by <u>Orthopox viruses (animal pox viruses)</u>, which in the immunosuppressed resemble smallpox!
- Zoster before the rash: various types of neuralgia
- Eczema herpeticatum caused by HSV in patients suffering from eczema
- Strophulus infantum (unknown origin)

Di.: • usually clinical criteria alone
• virus detection: DNA (PCR), antigen detection (virus isolation is expensive)
• <u>AB detection:</u> serological diagnosis can be made by demonstrating IgM-AB or quadrupling of IgG in 2 specimens (only makes sense in primary infections).

Th.:• uncomplicated cases: only symptomatic treatment:
   • <u>Varicella:</u> Sedatives and antihistamines (no scratching = no scars → cut fingernails!)
   • <u>Zoster:</u>
   - <u>Antiviral therapy</u> can reduce severity, shorten disease and limit complications, if given early-preferably within 2 - 3 days of onset of symptoms
   Ind.:   • any Zoster in patients > 50 years
        • cervical or cranial Zoster :any age
        • severe and haemorrhagic Zoster of the trunk or limbs, more than one dermatome affected, mucosa affected
        • Zoster in the immunodeficient patient
        • Zoster in patients suffering from eczema
        • Zoster in children on longterm steroids or salicylates
   Relative indication:
   Simple Zoster of the trunk or extremities (patients < 50 years)

Virostatic	Daily dose for adults (course of 7 days)
Aciclovir i.v.	3 x 10 mg/kg body weight for immunocompromised patients or severe disease
Aciclovir (oral)	5 x 800 mg
Brivudin (oral)	Immunocompetent patients: 1 x 125 mg
Famciclovir (oral)	Immunocompetent adults: 3 x 250 mg Zoster ophthalmicus: 3 x 500 mg Immunocompromised patients >25 y: 3 x 500 mg
Valaciclovir (oral)	Immunocompetent patients: 3 x 1.000 mg

- give immune globulin and Interferon-Beta for severe cases.

Beware patients on high-dose corticosteroids! Screen for disease that might affect the immune system!

• <u>Post herpetic neuralgia:</u> therapy with Carbamazepine or Desipramine or Amitriptyline (start on low dose), and analgesics

Prg.: • Varicella and Zoster have a good prognosis in immunocompetent patients. The long lasting post herpetic neuralgias can be a therapeutic challenge.
• risk patients : the course can be severe and lethal. The prognosis can be improved by an early antiviral therapy.

Pro.: • <u>prophylaxis:</u> Children suffering from Varicella are not allowed to attend school or nursery for one week after the onset of the rash.

• <u>Active immunization:</u>
   - <u>attenuated live vaccine:</u> OKA-strain of VZV
   Ind:1. Standard vaccination for all children. (Not in the UK)
        2. Seronegative persons at increased risk and persons in close contact to a patient: e.g. women of childbearing age, tumour patients, eczema, patients who will undergo immunosuppressive therapy or organ transplantation. Leukaemia patients (only those in remission for >1 year, lymphocytes ≥ 1.200/µl). Their cytostatic maintenance therapy must be stopped one week before and one week after the vaccination. Seronegative medical staff + nursery staff.
   SE: Local- and generalized reaction, occ. mild 'Vaccination Varicella'.
   CI: Immunocompromised individuals, those on immunosuppressive therapy (vaccinate during therapy-pause), pregnancy, Framycetin-allergy

Dose: children < 13 years: 1 dose, after that 2 x 1 dose (6 weeks apart)
  - Higher concentrated vaccine is indicated for those >60 years of age
  • Passive immunization with immunoglobulin for exposed seronegative individuals (e.g. newborns whose mothers suffered perinatal Varicella, pregnant women who have been in contact with a patient, immunocompromised individuals who have been infected).
  Hyperimmunoglobuline has to be given within 96 hours of exposition.

HERPES SIMPLEX INFECTIONS   [B00.9]

Internet-Infos: *www.herpesinfo.com*

PPh.: Herpes simplex-virus, type 1 and 2 (HSV-1 and HSV-2), is a DNA herpes virus

Ep.: HSV 1: infection often begins during childhood, > 95 % of the adult population have been infected.
HSV 2: infection starts after puberty, 10 - 30% of the adult population have been infected.

Inf.: HSV 1: Oral (airborne)
HSV 2: Sexual and perinatal

Inc.: Primary infection with HSV1: 2-12 days

KL.: **A) Primary infection:**
  - asymptomatic (> 90 %)
  - symptomatic (< 10 %)

  HSV 1: herpetic gingivostomatitis (Stomatitis aphthosa)
    Usually young children 1 - 4 years: Fever, painful ulcerating blisters of mouth and throat; localized lymphadenopathy.

  HSV 2: • newborn infants:
    - congenital HSV-2-infection:
      herpetic sepsis with fever and disseminated vesicles, jaundice, hepatosplenomegaly, intradermal haemorrhage, encephalitis; untreated cases are always lethal.
    - infection during delivery:
      very severe disease; in 30% lethal
  • genital herpes of adolescents and adults:
    Genital herpes infections: usually caused by HSV-2, but increasingly by HSV-1 (USA: 30%, Norway: 70%); f : m > 2 : 1.
    risk factors:
    ▪ risk proportional to increasing number of sexual partners
    ▪ i.v.-drugs abuse, HIV-infection
    50% are asymptomatic, 30% causing typical symptoms, 20% causing equivocal symptoms (e.g. dysuria).
    - women: vulvovaginitis herpetica (pain or burning sensation), dysuria, fever, localized lymphadenopathy
    - men: Herpes progenitalis: vesicles on glans penis; (beware anal HSV-2-infections in homosexual men)

B) Endogen reactivation:
  After the primary infection the virus persists inside the local neural ganglia (trigeminal ganglion Gasseri in HSV-1, lumbosacral ganglia in HSV-2) (latent infection). 1/3 of all human beings suffer from recurrent oral HSV-lesions.
  Trigger factors for a reactivation:
  Infections, fever, UV light (solar herpes), injuries, hormonal changes (e.g. period, pregnancy), stress, psychological problems, immunodeficiency etc.

  - asymptomatic virus excretion
  - symptomatic virus excretion:
    HSV-1: labial herpes: Perioral vesicles, turning into scabs and healing without scars.
    HSV-2: genital herpes: perigenital, perianal vesicles and ulcerations, sometimes associated with malaise and fever. HSV-2 is more prone to relapse than HSV-1.

Co.: • herpetic keratoconjunctivitis (may cause corneal damage)
  • anogenital herpes can lead to neurological complications (urinary retention etc.)
  • eczema herpeticatum is a severe infection in infants suffering from eczema
  • benign meningitis
  • herpetic encephalitis = most frequent viral encephalitis; most affected areas limbic system and the temporal lobe - usually HSV-1. An early diagnosis (MRI and PCR-test of cerebrospinal fluid) and early start of therapy (blind therapy for suspected infection) are essential for the outcome (lethality > 80 %)

- there is a possibility, that idiopathic VII-nerve-palsy is caused by HSV-1.
- immunocompromised patients (AIDS, those on immunosuppressants) may suffer severe disseminated HSV infection or HSV-pneumonia
- possible HSV-infections in HIV positive patients: necrotizing skin- and mucosa-lesions, keratoconjunctivitis, retinitis, uveitis, meningoencephalitis

DD: - gingivostomatitis herpetica may be caused by Coxsackie A-virus
- genital herpes: other sexually transmitted diseases= STD (TPHA-test, GO-diagnostics, HIV-test)
- genital herpes associated with dysuria may be misdiagnosed as cystitis/urethritis.
- corneal lesions may be caused by Adenovirus (Epidemic Keratoconjunctivitis)

Di.: Mainly done clinically + occ. Virus detection from vesicles or ulcers:
- detection of HSV-antigen or virus-DNA
- virus isolation (cell culture is more reliable than electron microscopy)
- HSV-AB-detection is an inadequate test, since so many people are carriers. It only makes sense for a primary infection (IgM-AB + seroconversion)

Th.: A) treat any underlying immune deficiency

B) antiviral chemotherapy
- systemic: gold standard is Aciclovir: acts on replicating HSV, but not on dormant HSV-infection. Famciclovir und Valaciclovir are available as oral medication.
  Dose: - Aciclovir: 5 x 200 mg/day for 10 days (herpetic encephalitis: 3 x 10 mg/kg body weight i.v. for 2 weeks)
  - Famciclovir : 3 x 250 mg/day for 5 days (oral)
  - Valaciclovir : 2 x 500 mg/day for 10 days (oral)
  Ind: severe disease, herpetic encephalitis, active genital herpes, gingivostomatitis, keratitis etc.
- local: Aciclovir-cream (also available as eye ointment for herpetic keratitis)
- Pregnant women with genital herpes must be delivered via Caesarean section before the rupture of membranes
- newborn infants with suspected HSV-infection need empirical treatment

Prg.: - good prognosis in localized infection
- often poor prognosis and high lethality rate in disseminated infection and in encephalitis.

Pro.: - prophylaxis with Aciclovir or Famciclovir for the immunodeficient; no vaccination available yet.

EPSTEIN-BARR-VIRUS (EBV)-INFECTION) [B27.0]

PPh.: Epstein-Barr-Virus (EBV) type 1 and 2, is a DNA-virus of the family of Herpes viridae. Target cells: naso- and oropharyngeal epithelium and B-lymphocytes, which carry the CD-21-antigen (EBV-receptors). Most infected B-lymphocytes will be rapidly destroyed by the immune system. However-some surviving B-lymphocytes may result in a lifelong virus persistency.

Ep.: In Western Europe > 95 % will have been infected by the age of 30 → peak in adolescents. In Central Africa most children will have been infected by the age of 3.

Inf.: via saliva (e.g. in nurseries, and by kissing - "Kissing disease")

Inc: 10 - 14 - 50 days

Sym: Young children- usually asymptomatic; later on in life- typical symptoms:

Infectious Mononucleosis = Glandular fever:
Triad: - febrile tonsillitis/pharyngitis
- lymphadenopathy
- typical blood film with 'virus cells'

Various types of courses:
- glandular type: widespread lymphadenopathy (50 %), frequently splenomegaly (risk of splenic rupture!) and tonsillitis.
- exanthemous type (3 %), petechial enanthema of the hard palate
  Amoxicillin and Ampicillin are contraindicated in glandular fever, because they cause a rash!
- hepatic type: Hepatitis (sometimes with jaundice)- 5 % (good prognosis)

Co.: - mild granulocytopenia, thrombocytopenia
- rarely infection-associated haemophagocytic syndrome (IHS)- pancytopenia and haemorrhage.
- rarely splenic rupture
- meningoencephalitis, myocarditis
- TINU-syndrome (tubulo-interstitial nephritis + uveitis) of children and adolescents

- chronic mononucleosis:

very rarely; ongoing virus replication:

fever, weight loss, lymphadenopathy, hepatosplenomegaly, sometimes thrombo-/granulocyto-penia, haemolytic anaemia, T_4/T_8-ratio↓ etc.

- EBV-infection in the immunodeficient patient:
  · uncontrolled proliferation of EBV-infected, immortalized B-lymphocytes result inside polyclonal lymphoproliferative diseases of the B-lymphocytes.
  · X-chromosomal recessive lymphoproliferative syndrome (XLP, Duncan's disease, Purtilo-syndrome): The immune system is unable to defeat an EBV-infection → subsequent (usually lethal) self destruction of the immune system.
- EBV-associated malignancies:
  · post transplantation lymphoproliferative disease (PTLD): Patient who had undergone a kidney-, heart- or liver transplantation have a 1-5% risk to suffer from PTLD due to the fact, that the immune suppressive therapy prevents them from developing an appropriate EBV-immunity → they will have an impaired balance between EBV-infected B-cells and cytotoxic T-cells, keeping them under control.
  According to the chronological sequence we distinguish between:
  1. Early PTLD within the first 12 months post transplantation: EBV-positive
  2. Late PTLD 5 - 10 years post transplantation, mostly EBV-negative.
  · EBV-associated B-cell-lymphomas in patients with AIDS
  · EBV is always to be found in tumour cells of the nasopharyngeal carcinoma (this carcinoma is endemic in the south of China and Alaska) and in the Burkitt-Lymphoma, which is endemic in Equatorial Africa. EBV is probably a co-factor.
- oral hairy leucoplakia is a benign epithelial proliferation in AIDS-patients, caused by EBV.

DD: - streptococcal tonsillitis
- acute HIV-infection
- Vincent's disease
- diphtheria
- cytomegaly-infection
- agranulocytosis
- acute leukaemia (monotonous blood count; Glandular Fever: mixed blood film, erythrocytes and thrombocytes are normal)

Di.: • symptoms + laboratory tests: blood count: leucocytosis (40-90% mononuclear cells and activated T-lymphocytes
• serological antibody detection:
  1. new EBV-infection (infectious Mononucleosis):
    Anti-VCA (viral capsid-antigen) (IgG + IgM)
    Note: Heterophil IgM-antibodies agglutinating sheep erythrocytes (Paul-Bunnell-reaction, Monospot) can be detected in 80 % in adults and in 50% in children, but they are not specific for EBV infection.
  2. previous EBV-infection:
    - Anti-VCA (IgG)
    - Anti-EBNA-1-IgG (EBV nuclear antigen)

Th.: symptomatic

Prg.: good prognosis in healthy individuals; the course can be severe in immunosuppressed patients.

| CYTOMEGALOVIRUS INFECTION | [B25.9]

PPh.: The Cytomegalovirus (CMV) is a DNA-virus, also called HHV-5; it is thought to be carcinogen; following primary infection the virus persists for life as a latent infection and can be reactivated during a period of a weakened immune system.

Ep.: In 3rd World countries > 90% of the population are infested with CMV, in Europe ca. 50%. In high risk groups (AIDS, prostitutes, homosexuals) ca. 90 % are found to be antibody positive.
Congenital CMV-infection is the most common congenital virus infection. It is mostly caused by a primary infection during pregnancy, which occurs in ca. 3% of all seronegative women. The highest risk for the unborn child if the mother acquires the infection around the conception and in the first two trimesters. This leads to a severe congenital CMV-infection in ca. 10%, and in ca. 10% of all asymptomatic children it will cause delayed damage (usually deafness). Ca. 10% of seropositive women have a reactivation of their CMV-infection during pregnancy; in these cases there is less risk of fetal infection, and < 1% of the newborns show mild symptoms.
Postnatal infection can usually be found in the following high risk groups: Immunodeficiency caused by malignancies (e.g. leukaemia, Hodgkin, Non-Hodgkin-lymphoma), acquired (AIDS) and congenital immunodeficiency, immunosuppression after transplantations. Seronegative recipients of a graft of a positive donor are at the highest risk to suffer a severe course of infection.

Inf.: intrauterine (congenital)
postnatal via close contacts with body fluids (blood transfusion, organ transplantation, sexual).

Inc: not known for sure (4 - 12 weeks?)

PPh.: Interstitial lymphoplasmacellular inflammation with giant cells + viral inoculation bodies

Sy.: The disease can affect various organs:
- newborns (congenital infection): premature birth, hydrocephalus, cerebral calcification, chorioretinitis, microcephaly, visceral symptoms with jaundice, hepatosplenomegaly, anaemia and thrombocytopenia. In ca. 25 % nerve deafness, mental retardation, minor neurological deficiencies.
- adults (postnatal infection):
  a) in > 90 % symptom free, occ. Symptoms similar to mononucleosis with lymphadenopathy or mild hepatitis; occ. lethargy lasting weeks to several months.
  b) immunosuppressed patients: severe course of infection:
  - fever, mononucleosis-like disease
  - myalgia, arthralgia
  - leucopenia, thrombocytopenia
  - retinitis (most frequent CMV-manifestation in AIDS): Cotton-wool-exudates and haemorrhage
  - encephalitis
  - interstitial pneumonia with a very high lethality (50%) (second frequent CMV-manifestation in AIDS, most frequent cause for pneumonia post bone marrow transplantation)
  - oesophagitis, gastritis
  - colitis with ulcerations (frequent in AIDS)
  - hepatitis
  - delayed haematopoietic restitution (with pancytopenia) post bone marrow transplantation

Lab.: Frequently leucopenia with relative lymphocytosis and atypical lymphocytes; occ. thrombocytopenia

DD: hepatitis and pneumonia of a different origin, mononucleosis, HIV-infection

Di.: ▶ symptoms, fundoscopy
▶ antenatal diagnostics of the fetus after primary infection during pregnancy:
- fetal sonography
- if indicated: detection of IgM-antibodies and CMV-DNA in the amniotic fluid or cord blood
▶ diagnostics of the newborn:
IgM-antibody and virus detection in urine + pharyngeal secretion
▶ diagnostics of postnatal infection:
  - virus-, pp65-antigen- and CMV-DNA-detection: from urine, blood, broncho-alveolar lavage, biopsy tissue: diagnosing an active CMV-infection in immunosuppressed patients!
  - antibody detection:
    - primary infection: CMV-IgG and - IgM positive
    - persistent infection: CMV-IgG is a useful marker to determine CMV carriage.
    - reactivation: rise of IgG titre and possibly recurrent detection of IgM antibody. Serology is of limited use in the immunosuppressed patient, since there may be no raise of antibodies.

Hi.: from biopsy specimens: virus inoculations in infected giant cells

Th.: • in the immunocompetent host, no specific antiviral treatment is required.
• pregnant seronegative women who have been exposed to CMV may be given CMV-immunoglobulin
• immunosuppressed patients: Ganciclovir (nephro- und myelotoxic) and CMV-immunoglobulin. AIDS-patients with CMV-retinitis may be given Valganciclovir.
Second choice: Cidofovir (nephrotoxic), Foscarnet and Fomivirsen (ultima ratio for CMV-retinitis for intravitreal application)
AIDS-patients should receive Ganciclovir to prevent relapse.

Pro.: 1. Prophylaxis of CMV-infection in the immunosuppressed CMV-seronegative recipient of transplants and transfusions:
• take transfusions and transplants from CMV-seronegative donors
Erythrocyte concentrate cleared of leukocytes (filter) is supposed to reduce the infection risk.
• give CMV-immunoglobulin
• supervision of organ recipients with PCR or with an antigen-test. Give Ganciclovir early if infection is suspected!
3. protect seronegative pregnant women from infection

[A09]

PPh.: 1. Bacteria and their toxins
- ▶ Salmonella (frequent); in 5 - 10 % pathogen of traveller's diarrhoea
- ▶ Escherichia coli (EC): important types :
  1. Enterotoxin producing EC (ETEC): ca. 40 % pathogen of traveller's diarrhoea!
  2. Enteropathogen EC (EPEC): diarrhoea in infants
  3. Enteroinvasive EC (EIEC): Dysentery-like diarrhoea (cramps and runny or bloody stools)
  4. Enterohaemorrhagic EC (EHEC)
  5. Enteroaggregative EC (EAEC = EaggEC): Enteritis in infants and toddlers
- ▶ Campylobacter jejuni; in 5 - 10 % pathogen of traveller's diarrhoea
- ▶ Yersinia enterocolitica (rarely Y. pseudotuberculosis):
  Colicky abdominal pain (DD: appendicitis), can be associated with arthralgia and erythema nodosum
- ▶ Clostridium difficile: pathogen of the Clostridium-difficile-associated diarrhoea (CDAD), leading to pseudomembranous enterocolitis (PMC) in 20%.

  Note: All antibiotics can potentially cause diarrhoea; 20% are caused by toxin producing Clostridium difficile.
- ▶ Staphylococcus aureus, Bacillus cereus and Clostridium perfringens: These are toxin producing bacteriae causing food poisoning. After a very short incubation period (few hours) there will be vomiting, diarrhoea and dehydration.
- ▶ Shigella: 5 - 10 % pathogen of traveller's diarrhoea
- ▶ Vibrio cholerae

2. Viruses:
- ▶ Noroviruses (previous name: Norwalk-like viruses): They make up to 50% of the non-bacterial gastroenteritis in adults
- ▶ Rotaviruses: More than 70% of infectious diarrhoea in children. In developing countries it is a cause for the high infant mortality
- ▶ SRSV (small round structured viruses), Astroviruses etc.

3. Protozoa:
- Giardia lamblia
- Entamoeba histolytica (Amebiasis)

  Note: Ongoing diarrhoea post return from tropical/subtropical countries should always be investigated for G. lamblia and E. histolytica!
- Cryptosporidiae (particularly in the immunosuppressed patient), Cyclospora cayetaneus, Isospora belli

4. Funghi (Candida, Aspergillus)

Ep.: Travellers going to tropical/subtropical countries with a low hygienic standard face a 30 - 50% chance to catch infectious traveller's diarrhoea; in >30 % no pathogen can be detected.

PPh: 2 types:
1. Secretory diarrhoea – impaired intestinal ion-transport: e.g. from activation of the membrane Adenylatcyclase by enterotoxins (e.g. Vibrio cholerae) or viruses
2. Exsudative inflammatory diarrhoea -mucosal lesions caused by Shigella, Salmonella etc.

Sy.: I. **Dysenteric diarrhoea**
Abdominal pain/diarrhoea mixed with blood/mucus/pus
1. Amebiasis - like (Entamoeba histolytica)
Gradual onset of symptoms; relapsing course with symptom-free periods.
2. Type "bacterial dysentery" (Shigella, EHEC, EIEC)
Acute or peracute onset of symptoms

II. **Non-dysenteric diarrhoea**
Acute onset of moderate symptoms; occ. excretion of mucus and undigested food.

1. enterotoxic type
Acute onset, may be accompanied by vomiting
pathogen: ETEC, Salmonella, Enteroviruses, food poisoning bacteria, Vibrio cholerae

2. impaired absorption type
Large volumes of frothy faeces, sometimes mixed with fat and undigested food
Pathogen: Giardia lamblia.

Co.: Dehydration, reduced electrolytes, orthostatic dysregulation, occ. collapse, thrombosis, embolism, septicaemia, acute renal failure, arthritis etc.

course: • acute diarrhoea: settles spontaneously after 2-10 days.

• chronic diarrhoea: lasts 10 - 20 days
- consider Entamoeba and Lamblia after a tropical holiday!
- AIDS-patients suffering from diarrhoea: there is a large variety of possible pathogens; the most common are Cryptosporidium, Microsporidiae, CMV, Mycobacteria (MAI). Quite often several bacteria contribute to the infection (DD: 1. Side effects from medication, 2. Idiopathic diarrhoea)

DD: • non-infectious causes of diarrhoea, particularly in chronic diarrhoea (see chapter ‚diarrhoea')
• tropical feverish traveller's diarrhoea : always exclude Malaria!

Di.: 1. (travel-) history + symptoms: e.g.
• diarrhoea & vomiting after eating: food poisoning caused by bacterial toxins (Staphylococci toxins have the shortest incubation of 2 - 3hours)
• watery diarrhoea + tropical holiday: consider Cholera!
• bloody diarrhoea + fever: e.g. Shigella, Amebiasis etc.
• feverish diarrhoea after a course of a broad spectrum antibiotic: consider antibiotic induced CDAD or PMC caused by Clostridium difficile
→ Di.: colonoscopy and check stool for Clostridium difficile and its toxins A and B

2. pathogen detection, e.g.
• bacteria: stool cultures (exception: V. cholerae via microscopy). Microsporidiae can be seen in an electron microscope.
• Entamoeba and Lamblia: microscopic detection of Lamblia or of its cysts and of Entamoeba in fresh, warm stools or fresh duodenal liquid; perhaps duodenal biopsy; antigen detection in stool
• viruses: Rota- and Noroviruses in stool (RNA- or antigen detection). A serological antibody test is just of retrospective importance (rise of titre after 2 - 3 weeks in immune competent patients).

Th.: A. treatment e.g.
• antibiotics are not medically indicated for moderate traveller's diarrhoea; but they can shorten the course of the disease, which may be relevant on holiday.
absolute indication: bloody diarrhoea, severe disease, fever, diarrhoea in infants and the elderly. If possible use a specific antibiotic as per stool culture; in very acute and severe cases use a broad spectrum antibiotic.
1. Choice: Ciprofloxacin: active against Shigella, Salmonella and E. coli (SE + CI: see index).
• alternative: Trimethoprim (TMP) or TMP/Sulfonamide = Cotrimoxazole
SE: allergic reactions caused by the Sulfonamide (→ hence pure TMP is sufficient !), gastrointestinal symptoms (nausea, vomiting etc.), haematotoxic effects caused by impaired folic acid balance, hyperkalaemia, rarely agranulocytosis
CI: renal failure, hepatic failure, allergy against Sulfonamides, pregnancy etc.
Uncomplicated traveller's diarrhoea often just requires antibiotic therapy for 1-2 days.
• Metronidazole for Amebiasis or Lambliasis
• If antibiotic induced CDAD or. PMC (Clostridium difficile) is suspected → stop the responsible antibiotic and start Metronidazole (alternative: Vancomycin), isolate the patient and follow strict hygiene regulations!

B. symptomatic
• fluid- and electrolyte substitution
This is the most important and possibly life saving intervention in acute diarrhoea! Infants and toddlers dehydrate rapidly!
According to the severity of disease rehydration is either oral or parenteral. The WHO recommends the following for oral rehydration: NaCl 3,5 g - NaHCO3 2,5 g - KCl 1,5 g - Glucose 20 g – add 1000ml of water. Readily mixed drug: e.g. Elotrans®-powder
• Anti motility drugs (e.g. Loperamide) delay the excretion of the infectious pathogen and hence they should only be given for a very short period on journeys.
• Spasmolytics for abdominal pain, e.g. N-Butylscopolamin (Buscopan®).

Prg.: Prognosis depends on the patient's immune status, the pathogen, how early therapy is being started and complications. Very young children, malnourished, immunosuppressed and elderly people are most at risk.

Pro.: traveller's diarrhoea:

• Most important are drinking water-, food, and personal hygiene!
- Quite safe:
Boiled or disinfected water (use also for tooth brushing!); bottled and sealed drinks, freshly cooked and well-done food.

- Avoid:
  Unboiled water; ice cubes, ice cream; cold buffet, raw or not well-done food (meat, fish, seafood); sauces, salads, mayonnaise, ready peeled fruits, melons (they may have been injected with water to appear firmer).

  Note for travellers into tropical countries: "Cook it, peel it or leave it!"

- Immunisation against:
  - Typhoid (oral live vaccine or parenteral inactivated vaccine)
  - Cholera (oral live or inactivated vaccine) not recommended for the average tourist.

  Note: Prophylaxis with antibiotics is not recommended, since it would enhance the selection of antibiotic resistance-plasmids in the bowel flora.

EHEC infections [A04.3] | Suspected disease, proven disease and death of HUS are notifiable |

Syn: STEC - diseases (Shigatoxin producing EC)

PPh.:   Entero-haemorrhagic Escherichia coli (= EHEC); they produce Shigatoxin; various serotypes: In USA + UK mainly O157:H7, in Germany mostly O103 + other serotypes ($O_2$6 etc.)

Ep.:   Global distribution, mainly affecting young children; incidence in mid Europe ca. 1/100.000/year.

Inf.:   1. Contaminated faeces
        2. Eating EHEC-containing food (cattle, sheep, goats): non-pasteurised milk, raw (minced) beef; children and the elderly are particularly at risk.
        Contagiousness: as long as pathogens can be detected in stool samples (3 samples)

Inc: 1 - 8 days

Sym.:   Various courses:
        1. Asymptomatic (frequently in adults)
        2. Watery diarrhoea (children: 80 %)
           Bloody-watery diarrhoea (children: 20%, elderly: frequently)

Co.:   • Intestinal complications: Haemorrhagic colitis (DD. Ulcerative colitis)
        • Extraintestinal complications (postinfectious syndrome):
          Seen in 10% of children <6years of age and frequently in the elderly
          - **complete haemolytic-uraemic syndrome (HUS = Gasser-Syndrome)** → Triad: Haemolytic anaemia, thrombocytopenia, acute renal failure (this is the most frequent cause of acute renal failure in children)
          - **incomplete HUS:** only 2 of the above symptoms
          - **thrombotic-thrombocytopenic purpura (TTP) = Moschcowitz-Syndrome:** HUS + cerebral symptoms (e.g. convulsions) → see chapter ‚thrombocytopenia'
          - Possible late symptoms: renal failure (can be chronic-dialysis, arterial hypertension, proteinuria, nervous system damage)

Di.:   symptoms + laboratory (pathogen detection in stool, antibody detection, Shigatoxin detection)

Th.:   • no antibiotics, no motility reducing drugs!
        • symptomatic/supportive therapy (fluid and electrolyte substitution and correction; dialysis etc.)
        Patients with von-Willebrand's disease will need FFP (fresh frozen plasma) when plasmapheresis cleavage-factor (vWF) is too low or if they have antibodies against vWF

Prg. of HUS: 50% of all children will need dialysis ; fatality up to 5% ; after 10-15 years about 40% of patients will be suffering hypertension and chronic renal failure.

Pro.:   -don't eat raw meat or drink raw milk
        -isolate affected patients/obey hygiene regulations
        -don't return the patients to any institution unless they had 3 negative stool samples

SALMONELLA/TYPHOID | Notification: see 'infectious diarrhoea' | [A02.9]

Def.:   pathogen of the Typhoid-Paratyphoid enteritis- (TPE-)group

PPh.:   Salmonella enterica: Gram negative, non-sporing mobile bacilli with numerous long peritrichate flagella; there are 3 different antigens:
        O-Ag (surface antigen) - H-Ag (flagella antigen) - Vi-Ag (virulence antigen)
        These antigens determine the serotype (there are > 2000) (Kauffmann-White classification).

        1. Typhoid fever: Salmonella enterica, Serovar typhi = S. typhi
        2. Paratyphoid: Salmonella enterica, Serovar paratyphi B (A and C only in tropical countries)

3. Enteritis: Salmonella enterica, Serovar enteritidis (the most frequent; mostly lysotype 4/6) and Serovar typhimurium (lysotype DT104 is resistant against Tetracycline, Chloramphenicol, Sulfonamide, Betalactam-antibiotics).

- for Typhoid- and Parathypoid bacteria humans are the only host. They penetrate from the bowel lumen into the blood circulation and cause a cyclic general malaise with septicaemia.
- enteritis: humans and animal are main hosts ; the bacteria remain inside the bowel (localised infection of the small bowel causing diarrhoea and vomiting after a short incubation period); therefore there is no subsequent antibody immune response.

4 variations of courses:
a) cyclic general malaise
b) septic general malaise
c) gastroenteritis
d) asymptomatic carrier

TYPHOID FEVER [A01.0]　　 Notification of suspected disease, proven disease, death

Syn: (Typhoid = fog, mist → delirium of a feverish disease)

Def.: cyclic, systemic infectious disease, caused by Salmonella enterica Serovar typhi =

PPh.: Salmonella typhi is part of serogroup D1 , (= serogroup 9); it is equipped with the O-antigens O9 and O12, the H-Antigen Hd and in most cases also the Vi-antigen.

Ep.: worldwide > 30 million cases/year; the highest rates are to be found in India, Nepal, Indonesia. Cases in West Europe usually have been imported from tropical/subtropical countries (50 - 100 cases/year in Germany). Main host is human.

Inf.: S. typhi is only transmitted from human to human:
- direct infection: sources of typhoid transmission are excreting chronic or convalescent carriers and the acutely infected; transmission via hand (anus to mouth)
- indirect infection through contaminated water or food
Contagiousness: As long as there are detectable pathogens in the stool (3 samples)

Inc: 3 - 60 days, mean: 10 days
The higher the infective dose, the shorter the incubation period.

Sym.: • gradual onset of symptoms (unlike fulminate onset of Leptospirosis) with slowly rising temperature (always consider typhoid for every fever of unknown origin for more than 4 days). The fever shows little diurnal variation without any shivers (unlike most other bacterial septic diseases).
- splenomegaly, rose spots on the abdominal wall (= septic foci)
- headache, cough (may be wrongly diagnosed as URTI – typhoid fever can't be lowered by ASS unlike a fever in a URTI)
- coated tongue (grey/yellow with red margin ("typhoid tongue"), initial enanthema of the anterior palate
- drowsiness
- **bradycardia (despite the fever)**
- a frequent early symptom is **constipation**; diarrhoea often doesn't start before the 2nd week (necrotic inflammation of the Peyer' patches in the small bowel)
- leucopenia (unlike most other bacterial infections; Paratyphoid causes leucocytosis) with toxic granulation, ESR is often normal, absolute eosinopenia, transaminases can be slightly raised

Infection incubation period 1-3 weeks	St. incrementi	St. fastigii	St. decrementi	
	fever with antibiotics		without antibiotics	
Primary bacteremia	1. Secondary bacteremia	2. pea mush stool	3. cleaning of the bowel ulcers	4. week
pathogen in the blood	+	+	(+)	—
pathogen in the stool and urine	—	(+)	+	+
detectable antibody (Widal-Agglut.-Reakt.)	1:100		1:400-800	

Co.: meningitis, intestinal haemorrhage, perforation of the intestinal ulcers with subsequent peritonitis, myocarditis, haemodynamic/renal failure, thrombosis, typhoid relapse, Salmonella septicaemia (in AIDS-patients), metastatic abscesses in bones, joints, reactive arthritis, delayed reconvalescence, hair loss

<u>Salmonella-excreting chronic carriers</u>
<u>Def.:</u> Those patients who still excrete S.typhi in their stools after 10 weeks post onset of the disease
<u>Inc.:</u> ca. 4 % of all infected/diseased patients
<u>2 types:</u> a) excreting via bile (2/3)
        b) excreting through small bowel (1/3)
Chronic carriers have an increased risk of developing a gallbladder carcinoma.

DD:
- fever of a different origin:
  - after foreign journeys: <u>Malaria</u> and other tropical diseases
  - influenza, pneumonia, <u>bacterial endocarditis, miliar tuberculosis</u> etc.

- <u>Paratyphoid</u> caused by S. parathyphi A, B, C: typhoid like disease (d-Tartrate-positive variety S. Java only causes gastroenteritis). Just like typhoid, most cases are imported).

- <u>other enteritides</u>

- ulcerative colitis

Di.: ask for foreign travel, symptoms (fever, blood count)
Pathogen detection – serology: demonstration of antibodies (tires from 1 : 2.000 or 4fold rise in titre)

Note: Send <u>blood sample</u> in suspected cases of typhoid (stool sample should be sent later on); if Salmonella-enteritis is suspected: send <u>stool sample.</u>

Th.: <u>multidrug resistant strains of S. typhi</u> are an increasing problem (in Asia this can be 50 - 80 % of all cases). 1st Choice antibiotic is <u>Ciprofloxacin</u>; alternatives: 3rd generation Cephalosporins e.g. Cefotaxim); treat for at least 2 weeks.
After discharge from hospital the patients will be followed up by the health authorities until they have provided three negative stool samples.

<u>Therapy of chronic carriers:</u>
Try to eradicate by giving e.g. Ciprofloxacin for more than 1 month. Those patients who excrete through the small bowel, should also be given Lactulose.

Prg.: Without treatment there is a fatality of up to 20%, with treatment <1 %, depending on age, nutrition and immune system.

Pro.: <u>food- and water hygiene, wash and disinfect hands frequently</u> (in a hospital you can become infected from your own hands!), find the source of infection/disinfection.
Staff handling food need to be screened by an occupational health department.
In Germany chronic Salmonella-carriers are under the supervision of the health authority; they must obey personal hygienic regulations. They are not allowed to be employed for food handling jobs.

<u>Active vaccination:</u>
<u>Ind:</u> 1. For travellers to endemic areas
      2. For epidemics and deployments
- Oral live vaccine (not used in the UK)with the strain Ty21a: Typhoral L®, Vivotif®: 1 capsule on day 1,3,5 one hour before meals (don't take any antibiotics or antimalarials during and 3 days after this period). Vaccination should be completed 10 days before travelling.
Provides immunity for 1 year in 60%
<u>SE:</u>   gastrointestinal symptoms, very rarely allergic reactions
<u>CI:</u>   acute infections, pregnancy, immune deficiency
- <u>parenteral inactivated vaccine</u> with Vi-polysaccharide: Typhim Vi®, Typherix®: 1 dose i.m. or s.c. provides immunity for 3 years in 60%
<u>SE:</u> occ. local- and generalised reactions, very rarely allergic reactions
<u>CI:</u> allergy towards the vaccine, current infection, pregnancy
Note: There are combined vaccines against Hepatitis A and Typhus available (e.g. Hepatyrix®, Viatim®)

SALMONELLA-GASTROENTERITIS   Notifiable   [A02.0]

PPh.: Salmonella type Serovar enteritidis: There are > 2.400 serovares, but only 25 are of practical importance: Most commonly S. enteritidis, lysotype 4 (LT 4), then S. typhimurium DT 104. Serovar DT 104 is developing and spreading multiresistant lysotypes (part. in the USA). Salmonella can survive for many months and they are not destroyed by freezing (thaw from frozen poultry often contains Salmonella!)

Ep.: This is the second most frequent (after Campylobacter) notifiable infectious diarrhoea caused by food poisoning. Incidence in Germany is 65/100.000/year.
Peaks are in summer and young children.

Inf.: • The organism usually comes from animal sources (raw eggs/egg products, raw/not well-done meat: poultry, mussels, mince etc.)
• Rarely from temporary carriers

Sym.: 5 - 72 h (depending on the infectious dose) after intake of infected food the endotoxines will cause violent diarrhoea and vomiting, abdominal pain, fever and headache.

Co.: Dehydration, cardiovascular collapse; systemic disease with Salmonella-septicaemia in the immunosuppressed (AIDS etc.) affecting the endocardium, pleura, meninges, bones, joints; reactive arthritis; Salmonella-carriers are very rare and temporary (1 : 1.000).

DD: food poisoning caused by enterotoxin producing bacilli (s. Staphylococci enteritis)

Di.: pathogen detection from food leftovers (soups, salads), stool, vomit; take a blood culture if the course is feverish. The path of infection may be discovered via lysotyping.

Th.: Correct fluid and electrolyte balance, oral electrolyte-/Glucose solution (e.g. Diorrhalyte, Elotrans®); young children dehydrate quickly, hence they should be treated parenteral; keep nil by mouth.
Antibiotics are not indicated in mild disease, since they don't shorten the course of infection, but they can prolong the excretion of Salmonella; excretion > 6 months is very rare.
Indication for Antibiotics: Severe disease, infants/young children, elderly and immunosuppressed patients → 1. choice: Ciprofloxacin; alternatives: Cotrimoxazole or Ampicillin i.v.

Pro.: foodstuff hygiene, personal hygiene, eat freshly prepared food as soon as possible, heat food (poultry, eggs, egg products) appropriately (> 10 minutes > 70°C).
Strictly keep any food that might contain Salmonella apart from other food. Separate clean and dirty working areas in the kitchen! Don't interrupt the cold chain and mind the expiry dates of food!
Staff been diagnosed as a carrier, are not allowed to handle food.

CAMPYLOBACTER-ENTEROCOLITIS   [A04.5]   notifiable

Syn: Campylobacter-enteritis

PPh.: Campylobacter jejuni (ca. 90%) and C. coli (ca. 10%), thermopile pathogen, that doesn't reproduce at temperatures > 30°C. It is found in a wide variety of warm-blooded animals, infections are acquired via contaminated food or contact with animals (poultry or unpasteurized milk).

Ep.: It is the most common bacterial infection of the gut in industrialized countries. Incidence in Germany ca. 75/100.000/year. Peak in summer.

Inc: 2 - 5 days

Sym.: After a short prodromal period of fever and headache, there is explosive watery, and sometimes bloody diarrhoea with severe abdominal pain lasting up to 1 week. A brief relapse occurs in 10%.

Co.: Reactive arthritis (< 1 %); very rarely Guillain-Barré syndrome

DD: • infectious of different origin
• ulcerative colitis

Di.: pathogen isolation, antigen detection in stool (can be found for ca. 2 weeks)

Th.: oral fluid- and electrolyte substitution. No antibiotics, unless the course is particularly severe, and for patients at high risk: Macrolide-antibiotics.

Pro.: avoid unpasteurized milk and not properly cooked poultry; watch hygiene in the kitchen (as for Salmonella)

FOOD POISONING CAUSED BY ENTEROTOXIN PRODUCING BACTERIA

Notifiable: see infectious diarrhoea

[A05.9]

Def.: This is not an infection, but food poisoning caused by enterotoxin producing bacteria (mainly Staphylococcus aureus, rarely Bacillus cereus or Clostridium perfringens in food that has gone off (e.g. milk-/egg products, meat, potato salad etc).
The toxins of S. aureus (A - I) are heat resistant and won't be destroyed even after heating up to 100 °C for 30 minutes!

Ep.: Quite common disease with a high number of unreported cases, usually 2 or more people are affected, if they had a common meal within the last 1 - 16 hours. The source of infection with S. aureus are usually human (ca. 30% carry S. aureus inside their nose).

Inc: 1 - 6 hours: enterotoxin producing S. aureus or B. cereus
8 - 16 hours: enterotoxin producing Cl. perfringens or B. cereus

Sym.: After a short incubation period of only a few hours, there is acute onset of nausea, vomiting, diarrhoea, possibly abdominal pain; usually no fever

Co.: electrolyte- and fluid deficiency, impaired orthostatic regulation, collapse

DD: • infectious diarrhoea caused by Salmonella, Norovirus etc.
• combination of gastrointestinal + neurological symptoms (part. Double vision, dysphagia) ; consider Botulismus intoxication!
• poisoning with toadstools or heavy metal

Di.: symptoms + history: acute gastroenteritis, affecting two or more people, if they had the same meal within the last 16 hours.
Possibly detection of enterotoxins in food leftovers.

Th.: symptomatic: fluid- and electrolyte replacement

Prg.: The disease usually just lasts 1 - 2 days.

Pro.: foodstuff hygiene (see Salmonella-gastroenteritis); consume freshly prepared food immediately.

NOROVIRUS INFECTION   Notification: See infectious diarrhoea

PPh.: Noroviruses (obsolete: Norwalk-like-viruses) are RNA-viruses divided into 5 genogroups; 3 of the are human pathogen (GG I, GG II und GG IV); further subdivision > 20 genotypes

Ep.: worldwide distribution, typical infection in winter and spring, often epidemics in institutions like hospitals, residential homes, cruising ships. This is the most common cause of a non-bacterial acute gastroenteritis in adults.

Inf.: airborne, faecal-oral

Inc: 10 - 50 hours

Sym.: Triad: Acute watery diarrhoea, nausea, projectile vomiting; often sever malaise, sometimes joint- and muscle ache, abdominal pain, rarely fever

Lab.: Leucocytosis

Co.: Dehydration (children, elderly); prolonged/chronic course in the immunosuppressed

DD.: gastroenteritis caused by food poisoning, Salmonella-gastroenteritis etc.

Di.: • symptoms and epidemiological situation
• detection of the pathogen in stool (RNA-, antigen detection)

Th.: symptomatic: fluid- and electrolyte substitution

Prg.: good prognosis (self limiting disease of 1-3 days); immunosuppressed patients may experience a chronic course. Fatality is low (< 0,1%); these are usually young children and the elderly.

APPENDIX:

BOTULISM   [A05.1] NOTIFIABLE DISEASE

Def.: food poisoning caused by 7 different neurotoxins (usually type A, B or E) produced by Clostridium botulinum

Ep.: This is a very rare disease now in western industrialized countries thanks to commonplace food hygiene (Germany < 10 reported cases/year)

Path.: Clostridium botulinum: Anaerobe, gas producing sporing bacteria. The heart resistant spores are ubiquitously distributed. They are able to germinate under anaerobic conditions and produce the toxins. For example on contaminated food, like smoked fish and sausage (botulus = sausage), particularly in jars and tins (food botulism), rarely in the bowel of infants (infant botulism) and in wounds (wound botulism). The toxins can be inactivated by cooking at 100°C for 15 minutes. Cans with a raised lid are highly suspicious! Otherwise the contaminated food looks normal!

PPh: Botulinum toxin irreversibly inhibits the release of acetylcholine of the peripheral cholinergic nerve endings until new nerve endings have grown.

Inc: • food botulism: usually 12 - 36 hours
• wound botulism: ca. 10 days

Sym.: The disease usually starts with gastrointestinal disturbance (nausea, vomiting); this is followed by peripheral weakness and paralysis: Paralysis descends from cranial nerves (dilated pupils), diplopia, ptosis, dysarthry, dysphagia. Within hours or days the paralysis travel caudal leading to respiratory failure, dry mouth, constipation, paralytic ileus, urinary hesitancy. Sensation and consciousness are preserved, there is no fever.

DD: Myasthenia gravis (Edrophonium Chloride-test etc.), diphtheria, poliomyelitis, Atropine poisoning, stroke etc.

Di.: Typical neurological symptoms (sometimes several people!) after intake of (homemade) canned or smoked food. Detection of Botulinum-toxin in food leftovers, vomit, gastric contents, stool or serum. Mice are inoculated extraperitoneally with and without mixing of antitoxin and observed for signs of botulism. This test takes 1-2 days, and it can be false negative.
Wound botulism: grow the pathogen from wound culture.

Th.: causal:
Toxin elimination from the gastrointestinal tract (empty stomach and intestines). Even in a suspected case of food botulism equine antitoxin should be given, intending to bind the free circulating toxin (give a test conjunctival dose to exclude allergy).
Wound botulism requires surgical intervention and Penicillin.
Symptomatic: e.g. ventilation for those with respiratory failure

Prg.: Food botulism: < 10% deaths if treated, 70% when not treated.

Pro.: keep vacuum wrapped food in a cool place (< 8°C no reproduction). Meat that is going to be preserved, should be heated twice in order to inactivate the spores.
Infants under 1 year of age shouldn't be fed honey, since this can be a source for infant botulism.
Watch the expiry date, discard any suspiciously looking cans; cooking for 15 minutes at 100°C will destroy the toxin.
Note: Botulinum-toxin type A is the strongest bacterial toxin. It is being used medically to treat muscle spasms.

SHIGELLOSIS   [A03.9] Notification: see infectious diarrhoea

PPh.: 4 Serogroups:
- Shigella dysenteriae (13 Serovares) (the tropics/subtropics) producing endotoxin (→ large bowel ulcers) and exotoxin (→ cardiovascular impairment, lethality up to 60% !)
- Sh.boydii (18 Serovares) (India, North Africa)
- Sh.flexneri (8 Serovares) (less frequently, not as dangerous, eastern countries and USA)
- Sh.sonnei (1 Serovar) (fairly harmless, Western Europe)

Ep.: This enteric infection is a disease of wartime and famine (reduced immunity), it is an epidemic due to lack of hygiene. Humans are the main host. In Germany there are ca. 1.000 reported cases/year, ca. 75% have been imported, in most cases from Egypt. Homosexual men are more at risk.

Inf.: Faecal-oral, particularly from contaminated water and food (particularly dairy)

Inc: 1–4 days; as long as there are bacteria found in stool, the patient is contagious (1-4 weeks)

Sym.: Mild course: watery diarrhoea; severe course: bloody, frothy, purulent diarrhoea, abdominal pain, painful bowel motions, fever

Co.: intestinal haemorrhage and perforations, loss of electrolytes and fluid, reactive arthritis; HUS (see there) caused by Shiga-toxin

Di.: send rectal swab (dry or special medium) (Shigella will survive a dry transport, they will die within hours in wet stool – for Cholera it is the other way round!).

Th.: Improve immunity, correct water and electrolyte balance, Chinolones or Ampicillin i.v. Shigella sometimes can be multiresistant (R-multiplasmides), then the antibiotic might have to be changed according to sensitivity. Don't give Loperamide.

Pro.: Hygiene: Accurate handling of drinking water and food, wash and disinfect hands; take measures for disinfection), dispose of faeces properly

AMOEBIASIS [A06.0]   Notifiable: see infectious diarrhoea

PPh.: Entamoeba histolytica: 2 very different groups:
- E. dispar (90 %) and E. moshkovskii: Harmless; not pathogen
- E. histolytica sensu stricto (10 %): pathogen of amoebic colitis and amoebic abscesses

There is no microscopic difference between these two species, they can only be distinguished by PCR.
There are 2 phases in the life of Entamoeba:
- cystic form: Cysts can remain viable in the environment for several months. They are resistant to gastric acid. Infected, asymptomatic patients excrete the cysts in their stool.
- adult form: cysts will be transformed into trophozoites inside the colon (= minuta forms). Trophozoites, which have digested erythrocytes (Phagocytosis), are called magna form or erythrocytophagous. Trophozoites are only excreted in stool, if the patient suffers from dysentery (not infectious).

Ep.: Common parasitosis in tropical/subtropical countries: 50 million cases/year; imported traveller's infection in industrialised countries.

Inf.: Faeco-oral via ingestion of cysts directly or indirectly through contaminated food or water. Inside the colon, the four-nucleated cysts transform into trophozoites containing one nucleus with the ability of replication. Mature cysts are excreted with stool, and they may remain infectious for weeks in the environment.
Another source of infection: infected kitchen staff and food traders.

Inc: Amoebic colitis: 1 - 4 weeks
Hepatic Amoebiasis: months - years

PPh.: Amoebic colitis: Ulcerations of the colonic mucosa (flask shaped in cross section); sometimes granulomatous inflammation, with a tumour like appearance (Amoeboma).
Hepatic abscesses: Singular or multiple, mainly in the right lobe

Sym.: Most infections are asymptomatic. Only 20% lead to invasive amoebiasis.

  1) Intestinal form:
  ▶ Acute amoebic dysentery: stools vary from semi formed to watery containing mucus and blood; there is abdominal pain and tenesmus; 30% suffer from fever.
  ▶ Chronic: relapsing colitis

  2) Extraintestinal Form:
  In > 95% the liver is affected: Hepatic abscesses, pain in the right hypochondrium and right lower thorax, mild fever. The onset of a hepatic abscess after the infection is insidious and can extend over several months or years. Only about 30% give any convincing history of dysentery! Consider this in patients with a history of foreign travel presenting with abdominal pain!

  Note: Every not self limiting diarrhoea after a tropical journey should be investigated for Amoebiasis (as well as Giardia lamblia!). The absence of a previous dysentery does not exclude a hepatic abscess!

Co.: Dysentery: Violent course leading to toxic megacolon, colon perforation and peritonitis
Hepatic abscess: Rupture and emptying into the abdominal cavity, pleura or (rarely) pericardium.

Lab.: Unspecific parameters: ESR, CRP, leucocytes ↑
Hepatic abscess: possibly transaminases and cholestasis enzymes ↑

DD: Dysentery: Shigellosis and other infectious dysenteries; ulcerative colitis etc.
Hepatic abscess: Bacterial abscess, Echinococcus cyst, congenital hepatic cyst

Di.: 1) Intestinal form: Microscopic pathogen detection in blood stained, mucous bits of fresh stool or endoscopic samples. Magnaforms (erythrophagous forms) will prove the amoebic dysentery. PCR (showing E. histolytica-DNA) enable to classify the species: E. dispar is non-pathogen, only E. histolytica sensu stricto is the causing pathogen!

  2) Extraintestinal form: Ultrasound, CT or MRI of the liver + serology: Combination of at least 2 different antibody tests (ELISA, IFAT, IHA)

Th.: • Intestinal form: Imidazole derivates (e.g. Metronidazole) for 10 days. In some cases Amoeba can remain inside the bowel lumen after a therapy with Imidazole derivates. In this case a contact amoebicide should be given (e.g. Paromomycin, which is more effective than Diloxanide). Check successful eradication by follow up stool tests.
• Extraintestinal form: Same as intestinal form. More than 90% of those being treated with Imidazole derivates show an immediate improvement within 72 hours. Afterwards a contact amoebicide. Only in the case of an impending perforation should drainage of the abscess be considered (CT).
• Asymptomatic carriers (those with cysts or minuta forms in their stool) should only be treated in case of infection with E. histolytica sensu stricto (Contact amoebicide for 10 days).

Pro.: water- and food hygiene

PPh.: 1. Vibrio cholerae Serovar 01 (2 biovares (= biotypes):
  - classical biotype
  - El Tor (2 serotypes)

  2. Vibrio cholerae 0:139 Bengal

Ep.: Worldwide 6 million cases/year; > 100.000 deaths/year. Breeding ground: Ganges delta, historic 7 pandemics: 1883 epidemic outside Alexandria: Robert Koch discovered the pathogen; 1892 epidemic in Hamburg (8.600 deaths). The current El-Tor-pandemic started in 1961 in Celebes, reached Africa in 1970, and Peru/South America in 1990. Serovar O139 "Bengal" started in Bangladesh and India (1992); it is geographically limited and the numbers are declining. Poor hygienic conditions are responsible for the spread of Cholera. Man is the only host.
Tourists are at minimal risk to catch Cholera in endemic areas, since it mainly affects malnourished people and those suffering from other diseases (disease of the poor).

Inf.: V. Cholerae resides in brackish surface water - transmission:
  1. contaminated water, sea food and foodstuff
  2. human to human (faeco-oral) by Cholera patients or asymptomatic carriers

Pg.: Enterotoxin activates Adenylatcyclase → increases cyclical AMP → hypersecretion and hypermotility of the small bowel

  Note: Virulence genes for toxin production is transmitted by viruses (bacteriophages) – similar to Diphtheria.

Inc: Several hours to 5 days
Infectiousness: As long as pathogens can be found in stool.

Sym: 1. Many infected people are asymptomatic carriers. Only ca. 15% follow a symptomatic course.
  2. Mild forms (Cholerine): In 90% El Tor-infection is mild and can't be distinguished from other forms of diarrhoea.
  3. Severe form: 20 - 30 watery diarrhoea and vomiting → dehydration → anuria; reduced body temperature to 20°C!, leg cramps, aphonia
  4. Most severe form = Enterotoxin poisoning; leading to death within a few hours

Di.: In suspected cases, contact a bacteriologist, because even suspected cases have to be reported to the WHO.
Take a rectal- or stool swab, and send it to the microbiology lab in 1 % Pepton solution (diagnose will be complete within 6 hours). If you don't have any Pepton solution, the rectal swab needs to be in the lab within one hour (needs to be kept cold during transport)!! (Vibrio dies off very quickly in a dry environment).

Th.: Isolate the patient and start therapy!
  1. Substitution of water and electrolytes is the most important measure: WHO-solution (see infectious diarrhoea)
     Even oral substitution will improve the outcome!
  2. Also antibiotics: Chinolones or Macrolide antibiotics

Prg.: mean fatality 1-5% (fatality without therapy up to 40%), fatality increased in patients with malnutrition and when therapy has been started rather late.

Pro.: Food, water and personal hygiene; Provision of clean water; if this is not possible-use a drinking water filter, dispose of sewage safely.

  Active immunisation: There are two vaccines available, providing limited protection for short periods. They are both not licensed in Germany!
  Ind: Obligation only if the country of immigration requests a vaccination. No indication for the average tourist.
  CI: acute and chronic disease, infectious diseases, allergy to vaccine, infants < 6 months; no live vaccine for
    pregnant women and the immunosuppressed patient.
  - oral live vaccine (e.g. Orochol Berna®):
    SE: mild diarrhoea
    Interaction: Up to 1 week after vaccination there will be pathogens in stool → don't take any antibiotics/malaria
      drugs
    Dose: 1 oral dose
  - oral inactivated vaccine (e.g. Cholerix®, Dukoral®)
    SE: indigestion
    Dos: 2 x 1 oral doses 2 weeks apart
  Note: None of these will protect from infection with Cholera Type O 139 (Bengal)!
  Protection: 6 days to 6 months post vaccination

YERSINIOSIS   [A28.8] Notification: see infectious diarrhoea

PPh.: Yersinia enterocolitica: the most frequent serotypes in Europe are 0:3 and 0:9 -, less frequently 0:5,27. Yersinia pseudotuberculosis is very uncommon in West Europe, but it can be found more frequently in East Europe + Russia.

Ep.: Worldwide distribution; incidence in Germany ca. 7/100.000/year. Represents ca. 1% of all cases of diarrhoea (peak in January). Many animals are hosts, source of infection are contact with animals and contaminated food of animal origin (dairy, raw pork). Y. Still is able to replicate at a temperature of +4°C in the fridge (meat, sausage). Y. enterocolitica can be transmitted via blood transfusion.

Inc: 10 days

Sym: • gastroenteritis (young children)
  • pseudoappendicitis form (older children, adolescents): acute mesenteric lymphadenitis and terminal ileitis (DD: appendicitis)
  • enterocolitic form:
    Diarrhoea for 1-2 weeks, often associated with colic like lower abdominal pain, sometimes chronic diarrhoea (DD: Crohn's disease)

Co.: • reactive arthritis and/or erythema nodosum, mostly in HLA-B27 positive patients
  • rarely sepsis (usually in immunodeficient patients)

Di.: pathogen detection: stool, mesenteric lymph nodes (postoperative), bowel biopsy, blood (in septic cases), Yersinia-DNA can also be detected.
serology: check 2x titre antibodies against Y. enterocolitica 0:3 and 0:9, and Y. pseudotuberculosis.

Note: Antigens of the serotype 0:9 cross-react with those of Brucella. If antibodies against Serotype 0:9 is found in the absence of pathogen in stool, a Brucellosis has to be excluded.

Th.: Oral fluid and electrolyte substitution
no antibiotics, unless for patients at high risk or a complicated course of disease → therapy: Fluorchinolones, maybe combined with Cephalosporins of the 3rd generation

CRYPTOSPORIDIOSIS   [A07.2] Notifiable disease

PPh.: Cryptosporidium parvum is an obligate intracellular protozoon. They produce oocysts containing sporozoites.

Ep.: In industrialized countries Cryptosporidium makes up only ca. 2% of infectious diarrhoea of immunocompetent people; HIV-patients are much more affected. Zoo-keepers, veterinary surgeons and people travelling to countries with a poor hygienic standard are more at risk. Swimming pools can be another source of infection.

Inf.: Oral infection (contaminated food and water); infected calves and other farm animals and pets; transmission from human to human is possible

Inc: 1-12 days

Sym: Watery diarrhoea, abdominal pain, mild fever

course: -immunocompetent patients: asymptomatic infection or self limiting disease over 1–2 weeks and lifelong immunity
  - infants or immunodeficient patients: severe and long disease, no immunity

Co.: fluid and electrolyte deficiency, malabsorption syndrome, infection of the bile ducts (raised γ-GT and alkaline phosphatase); bronchopulmonal infection

DD: • infectious dysentery of different origin
  • consider microsporidiosis, mycobacteria (MAI), CMV etc. In AIDS-patients.

Di.: Microscopic detection of oocysts in stool, antigen detection (3 stool samples)

Th.: there is no known efficient therapy; there are clinical trials taking place with Nitrazoxamid; symptomatic therapy: electrolyte-/fluid substitution; HIV-infection requests optimal antiviral therapy

Pro.: boil drinking water if in an epidemic area (chlorine is not effective. Wash hands thoroughly (the usual disinfectants for hands don't work).
AIDS-patients receiving a prophylactic drug against atypical mycobacteriosis (Rifabutin + Clarithromycin), are less at risk from contracting cryptosporidiosis, than those who are not.

There is a number of endemic intestinal parasites in central Europe.
Guest workers, refugees, asylum seekers from endemic areas, and increasing tourism into countries outside Europe are the cause for an increased risk of infection for the individual, and also for small epidemics even in industrialised countries. For this reason immigrants have to be screened. Tourists travelling to tropical or subtropical countries are advised to follow strict hygienic rules, particularly uncooked vegetables, fruits and water must be avoided.
Patients presenting with abdominal pain always have to be asked for foreign travel, and intestinal parasites have to be excluded.

Disease /Pathogen	Mode of infection	Symptoms	Diagnosis⁾	Therapy
Ascariasis: Ascaris lumbricoides = roundworm	orally, eggs, contaminated food, auto-infection is possible	flu-like symptoms, abdominal pain, occ. ileus, urticaria, eosinophilia, cholestasis	stool: eggs, adult worms sputum: larves X-ray: pulmonal shadows parasite inside bowel lumen	Mebendazole Pyrantele
Trichuriasis: Trichuris trichiura = whip worm	orally, eggs, contaminated food, vegetables	abdominal pain eosinophilia	stool: eggs (look like lemons), worms	Mebendazole
Oxyuriasis (Enterobiasis): Enterobius vermicularis = pinworm, threadworm	orally and via inhalation, eggs: indirect faecal infection, auto-infection	anal itching, occ. vulvo-vaginitis, rarely appendicitis	rectal exam + adhesive tape egg sampling (eggs not found in stool)	Mebendazole Pyrantele Repeat therapy after 3 and 6 weeks - Treat the whole family!
Ancylostomiasis: Ancylostoma duodenale, Necator americanus = hookworm	percutaneous invasion of larvae through contact with moist soil	pulmonary oedema, dermatitis, abdominal pain, iron-deficiency anaemia (intestinal haemorrhage)	stool: eggs (microscopy) culture of larvae	Mebendazole Tiabendazole
Strongyloidiasis: Strongyloides stercoralis	percutaneous invasion of larves, auto-infection	dermatitis, bronchitis, enterocolitis, urticaria, eosinophilia	duodenal fluid and stool: larvae in culture, skin tests,	Mebendazole Tiabendazole
Cestodes (Tapeworms): a) Taenia saginata (beef tapeworm) b) Taenia solium (pork tapeworm) c) Diphyllobothrium latum (fish tapeworm)	orally, cysts in raw meat: a) beef b) pork (auto-infection leads to cysticercosis) c) fish	often no symptoms, no eosinophilia, cysticercosis (cysts of Taenia solium in muscles, brain, eyes), fish tapeworm: vitamin B12-anaemia	-stool: proglottids and eggs -differentiation between species- count uterine lateral branches: T. solium: 7 – 10 T. saginata ≥ 12	Niclosamide Praziquantel Mebendazole

321

INTESTINAL PARASITES IN CENTRAL EUROPE II

Disease / Pathogen	Mode of Infection	Symptoms	Diagnosis	Therapy
Cystic hydatid disease: Echinococcus granulosus (dog tapeworm) Alveolar hydatid disease: Echinococcus multilocularis (fox tapeworm) (notifiable diseases !)	a)direct or indirect from dogs and wolves (faeces) b) direct or indirect from foxes (raw berries and mushrooms from the forest) and cats	Cysts in liver (70 %) upper abdominal pain, poss. jaundice. Cysts in lung (20 %) cough etc. Occ. Hydatid cysts in the brain and other organs Also: occ. urticaria, rarely eosinophilia Fox tapeworm infiltrates the organs like a carcinoma.	ultrasound, CT, MRI; antibody detection; - alveolar disease: antigen detection (Em2). PNM-classification of alveolar disease: liver (P), neighbouring organs (N), metastases (M)	Therapy only in specialised clinics: cystic disease.: cystectomy or Cannulation - Aspiration - Injection - Reaspiration (Injection of 95% alcohol) only after exclusion of a cystobiliar fistula + chemotherapy (Albendazole) Alveolar disease: curative resection (1/4 of all cases) + long term treatment with Albendazole
Trichin(ell)ose: Trichinella spiralis	Orally: raw meat (mince) from pigs, bears, seals etc.	Muscle pain from 10th day of infection, occ. fever, periorbital oedema, eosinophilia, CK ↑ Co.: myocarditis, meningoencephalitis	Antibody detection 3 - 4 weeks post infection, muscle biopsy (pathogen), PCR	Mebendazole Albendazole Prophylaxis: meat inspection, heat > 65 °C, freeze (10 days for -23°C)
Amoebiasis: Entamoeba histolytica	orally, cysts, contaminated food (flies!), occ. water	intestinal form: diarrhoea (with blood and mucous), tenesm extraintestinal form: hepatic abscess, fever + leucocytosis	pathogen detection (fresh stool, bowel biopsy) amoebic abscess: ultrasound, CT, antibody detection	Imidazole derivates: Metronidazole or Nimorazole or Tinidazole
Lambliasis (Giardiasis) Giardia lamblia	orally, cysts, faecal infection, food, water Prophylaxis: boil drinking water	often no symptoms; diarrhoea (occ. frothy), abdominal pain, flatulence, occ. malabsorption	duodenoscopy + biopsy, duodenal liquid: adult form (trophozoit), stool: cysts and Giardia-antigen	Imidazole derivates

Drugs:
Albendazole; Mebendazole; Praziquantel; Tiabendazole; Metronidazole; Niclosamide; Pyrantel; Tinidazole

INFLUENZA [J11.1]

Internet-Infos: *www.grippe-info.de*
www.dgk.de/agi
www.grippe-online.de

PPh.: Myxovirus influenzae is a RNA-Virus. Types A, B and C are antigenetical distinct (nucleoprotein- (NP) and matrix- (M) antigen). These are inside the virus. Influenza A-virus is further divided into subtypes, according to two glycoproteins in the viral envelope (haemagglutinin (H) and neuraminidase (N).

- The rod-shaped haemagglutinin (H) enables adhesion to the host cell.
- The fungus-shaped neuraminidase (N) enables the release of viruses from infected cells and distribution in the respiratory   tract.
There are 16 known H-subtypes and 9 N-subtypes. However, so far only 6 haemagglutinin-types (H1, H2, H3, H5, H7, H9) and 3 neuraminidase-subtypes (N1, N2, N7) have been found in human epidemics. It appears, that historically there was also N8. New subtypes can develop at any time, e.g. the H5N1-virus of the avian flu. Subtypes and variations are determined by: Type, first place where it was found, a current number, year and an antigen formula (antigens haemagglutinin (= H) and neuraminidase (= N)), e.g. Influenza A/Singapore/6/86 (H1N1).

Ep.: Influenza A and B are ubiquitously distributed and they occur at different times on the two hemispheres (South: May - October / North: November - April). Immunoincompetent patients and the elderly frequently die. Immunity is subtype- or variant specific, hence repeated infections over a lifetime are possible.

▶ Influenza A is the most common cause of epidemics and pandemics. In influenza virus type A minor changes of the antigen have been observed. These are due to single mutations – exchange of single amino acids in the haemagglutinin and/or in the neuraminidase (antigen drift). This leads to new variants of the virus subtype, with subsequent epidemics 2-3 years apart. A new subtype of the virus (antigen shift) can develop, if major genetic parts between viruses are exchanged (reassortment). The genome of influenza A-viruses consists of 8 RNA-strands (segmented), enhancing reassortment, when a cell is being infected with two different influenza A-viruses. Reassortment between human and animal influenza A-viruses (e.g. birds) causes pandemics every 10-40 years with millions of deaths worldwide, since the new virus hits an unprepared and unprotected population. 4 pandemics in the 20th century:
1) 1918/19: Spanish H1N1-pandemic
2) 1957/58: H2N2-Asia-pandemia
3) 1968/69: H3N2-Hongkong-pandemia
4) since 1977 H1N1-spreading

▶ Avian influenza (avian flu) consists of ca. 15 different avian influenza viruses of group A. The current H5N1-influenza a zoonic pandemic, highly infectious only for birds. Reassortment of the H5N1-virus with a human influenza virus could cause a human pandemic.
There are only sporadic transmissions from birds to humans. Human to human transmission of H5N1-viruses has not been reported.

▶ Influenza B mainly affects children and adolescents; the course is less severe. Antigen shift have not been observed, but antigen drifts have been. Influenza B does not affect animals.

▶ Influenza C doesn't play any active role (sporadic cases only).

PPh.: Before the virus can enter the host cell, an enzymatic breakup of the haemagglutinin in the viral envelope is required. Certain bacteria, particularly Staphylococci and Streptococci, produce proteases able to brake up the haemagglutinin. This means, that an existent bacterial infection of the respiratory tract can enhance the risk of acquiring an influenza . infection.

Inf.: airborne (high concentration of viruses in the nasopharyngeal secretion)

Inc. 1-3 days

Sym.: In 80 % the infection is asymptomatic, or there may be minor flu-like symptoms.

• sudden onset with shivers, fever, severe malaise, laryngo-tracheo-bronchitis with dry cough, rhinitis, occ. epistaxis, pharyngitis, conjunctivitis, photophobia, tiredness, headache, joint- and muscle ache, swollen face
• occ. Gastrointestinal symptoms
• fever with one peak lasting 2-3 days; a second feverish peak usually indicates a secondary bacterial infection.
• sputum: scarce, thick, occ. blood stained

Lab.: uncomplicated course: CRP, ESR, leucocytes usually normal; iron ↓
bacterial secondary infection: CRP, ESR ↑, leucocytosis; iron normal

Co.: Particularly at risk: children, elderly with underlying disease and immunosuppressed patients !

• 3 pneumonia types: primary-haemorrhagic influenza pneumonia (often fatal), interstitial influenza pneumonia and

secondary-bacterial influenza pneumonia (most common type); bacteria: Staphylococcus aureus, Pneumococci, Haemophilus influenzae etc.
- sinusitis, otitis media (children), pseudo croup in young children, exacerbation of asthma
- purpura Schönlein-Henoch after Influenza A
- myopericarditis (can lead to sudden death), meningoencephalitis, orthostatic hypotension.
- very rarely peracute lethal course in young adults

prolonged reconvalescence is not uncommon (weakness, tiredness and hypotension can sometimes last for weeks)

DD:
- acute respiratory infections = 'common cold' = upper respiratory tract infection (URTI) are mostly caused by Rhinoviruses (~ 40%), Adenoviruses, Parainfluenza-Viruses, Respiratory Syncytial-Viruses (RSV), Corona viruses.

  Note: Influenza is associated with a sudden onset, high fever (> 38,5°C) and muscular/ joint pain. Infections similar to influenza with sudden onset are called influenza like illness. Common colds are associated with only mild fever or normal temperature and a mild course of the disease.
- pneumonia of different origin (see ‚pneumonia')
- pertussis

Di.: sudden onset, fever (>38,5°C), cough, headache, muscle ache during an influenza epidemic (80 % probability)
- pathogen detection:
  - Influenza A/B-test (e.g. detection of Influenza A- and B-RNP): high specifity, sensitivity ca. 75 %
  - virus detection from nasal or throat swab
  - antigen detection or nucleic acid (PCR)
- antibody detection (e.g. HAH-test): at least 4 fold titre rise 1-2 weeks apart

Suspected avian influenza:
A) clinical features (acute onset, fever > 38°C, cough, occasionally dyspnoea, in 50% gastrointestinal symptoms
B) positive exposition history (recent travel to an area of avian influenza + been in contact with diseased or dead birds/poultry or their excretions)
C) Influenza A quick test, pathogen detection

What to do: isolate patients, protect medical staff (protective dressing, FFP2-mask, goggles) + phone the health authorities; antiviral therapy

Th.: • antiviral therapy:
Neuraminidase inhibitors:
Effective against Influenza A- and B, if started within 24 - 48 hours.
- Zanamivir (Relenza®); Dose: 2 x 10 mg/d – powder inhalation for 5 days; SE: rarely bronchospasm
- Oseltamivir (Tamiflu®); Dose 2 x 75 mg/d for 5 days (doses needs to be reduced in renal failure)
  SE: gastrointestinal symptoms (e.g. nausea), rarely mental confusion – resistance has been observed.
  Ind: therapy with neuraminidase inhibitors is recommended if
  a) there is a known regional epidemic,
  b) there are typical symptoms with fever,
  c) other severe infections have been excluded
  d) treatment can be commenced within 48 hours.

- symptomatically:
  - fluid substitution (dehydration caused by fever!), possibly reduction of temperature(Paracetamol, tepid wrapping)
  - suspected bacterial secondary infection: antibiotics (e.g. Macrolides)
  - children must not take Acetylsalicylic acid (beware Reye-syndrome: acute encephalopathy and hepatic failure, is usually fatal)
  - severe course and impaired immune system: maybe immunoglobulines.
  - bedridden patients: thrombosis prophylaxis.

Prg.: Influenza causes ca. 1 million deaths worldwide/year, in pandemics it can be several millions: The 1918/19 pandemic caused > 20 million deaths (more than the 1. World war). Certain groups of patients (as above) are particularly at risk. Avian influenza has 50% mortality.

Pro.: annual active vaccination with a inactivated vaccine. Manufacturers of vaccines follow the most recent recommendations of the WHO, to produce an antigen combination according to the current epidemic strains. Protection for people < 65 years: 70%; less for older people. People > 60 years: reduced mortality by 50 %! Cardiovascular mortality (myocardial infarction, stroke) seems to be reduced as well.
  SE: Occ. mild malaise, sometimes pain at the injection site, rarely allergy to egg protein; very rarely thrombocytopenia or vasculitis, Guillain-Barré-syndrome (1 in 1 million) (life threatening acute, demyelinising polyneuropathy caused by autoimmune pathogenesis)
  → Th.: high dose immunoglobulines and plasmaphoresis)
  Ind: 1) everybody > 60 years
  2) patients suffering from cardiopulmonary diseases or impaired immune system
  3) exposed people
  4) people in direct contact with poultry and birds (no protection against avian flu, but it protects from double infection)
  5) during and epidemic: EVERYBODY

Adults are being vaccinated before onset of the cold seasons. Annual vaccination considering the current subtypes.

Note: HIV-tests can be false positive up to 3 weeks after the vaccination.

- preparation/plans for a pandemic of the WHO and certain countries can be looked up on the internet
- avian influenza: don't come into contact with suspicious (alive or dead) birds/poultry + measures to prevent epidemics

Whooping cough – Pertussis   [A37.0]

PPh.: Bordetella pertussis; airborne infection

Ep.: infants and young children, who have not been vaccinated are particularly at risk. The incidence in the adult population is ca. 150 - 500/100.000/year.

Inc: 7 - 20 days

Sym.: 3 stages: I.   St. catarrhalis (1 - 2 weeks)
                       II. St. convulsivum (staccato cough, occ. ending in vomiting; 4 - 6 weeks)
                       III. St. decrementi (6 - 10 weeks)
consider pertussis in adults, if they present with a non-settling dry cough after an URTI (bronchitis) (DD: hyper reactive bronchial tract).

Co.: subconjunctival haemorrhage, pneumonia, otitis media, convulsions; rarely death caused by apnoa in infants

Di.: history/clinical features + laboratory: lymphocytosis; pathogen detection from nasopharyngeal swab; antibody detection from the 2nd stage (IgG, IgA, titre rise)

Th.: Macrolide-antibiotic + symptomatic therapy

Infectiousness: End of incubation up to 5 days after start of antibiotic therapy (= patient needs to be isolated)

Pro.: Anon-cellular vaccine:
Ind:
1. Infants from the age of 3 months (UK: 2 months) receive a primary course of 3 doses, with a booster 3 years after this course
2. Adults without immunity (women before they become pregnant, those in close contact 4 weeks before a child is born; staff in nurseries etc.). Adults receive one dose of combined vaccine (Tdap =Tetanus+ Adsorbed Diphtheria or TdapIPV = Tetanus + Adsorbed Diphtheria + Inactivated Poliomyelitis Vaccine).

Coxsackie Virus Infection   [B34.1]

PPh.: Coxsackie A and B are RNA- (Picorna) viruses from the Enteroviridae group.
It was first isolated in Coxsackie/USA, humans are the reservoir. 2 main groups: A (serotypes 1-22 and 24; serotype 23 was reclassified as ECHO virus type 9) and B (serotypes 1 - 6).

Ep.: Worldwide distribution, occ. small epidemics (e.g. Bornholm).
Many carriers, many asymptomatic courses of infection

Inf.: Faeco-oral transmission

Inc: 2 - 6 days

Sym.: Mostly asymptomatic

Coxsackie A:
- Herpangina, acute onset with myalgia, headache and fever, then vesicles on the soft palate, uvula and tonsils with erythema. Rarely bronchitis or otitis.
- summer flu : fever, headache, myalgia, pharyngitis, laryngitis, often haemorrhagic conjunctivitis (DD: adenovirus infections)
- rhinitis
- lymphocytic meningitis, meningoencephalitis and poliomyelitis-like symptoms with paralysis (these are very rare complications)
- hand-foot-mouth-disease (type A 16), particularly in children with rash on the palms, soles of the feet and intraoral mucosa; occ. vesicles on mouth, lips and genital mucosa
- gastrointestinal symptoms

Coxsackie B:

- Bornholm-disease (pleurodynia, pleurisy): acute onset with sharp thoracic and epigastric pain, fever; pain is enhanced by breathing and cough; headache.
- myocarditis, perimyocarditis are severe complications
- lymphocytic meningitis, meningoencephalitis
- summer flu

DD: • summer flu can be caused by ECHO viruses etc.
        • pleurisy: consider other causes for thoracic pain (angina pectoris)

Di.:   diagnosis from swabs from pharyngeal vesicles, stool or liquor is possible; routine diagnostics is not indicated
        Antibody titre (positive 1 ≥ 64). PCR only indicated new infection and remains positive for only ca. 4 - 5 months.

Th.:   symptomatic, bed rest; treat any (peri-) myocarditis

PAROTITIS EPIDEMICA (MUMPS)   [B26.9]

PPh.: Paramyxovirus parotitidis, RNA-virus

Ep.:   Worldwide distribution, highly contagious, predominantly in the winter months; infections mainly at the age 4 - 15 years; afterwards 90 % of the population have a lifelong immunity.

Inf.:   airborne infection

Inc: 14 - 25 days

Sym.: In ca. 35 % the infection is subclinical or asymptomatic.
        - prodromi: subfebrile temperatures, tiredness, headache, sore throat, ear pain
        - painful swollen parotid (= parotitis) (75% bilateral), protruding ear lobe; painful chewing
        - other salivary glands can be affected as well (siladenitis)

Lab.: amylase ↑ → to exclude pancreatitis: elastase 1 and lipase normal, normal abdomen

Co.: • pancreatitis
        • orchitis (25 % of men), can lead to infertility, oophoritis (5 % of women), mastitis
        • Frequently the CNS is affected: usually meningitis (10%) with a good prognosis, rarely (1 ‰) meningoencephalitis with a poor prognosis. 50 % of the meningitis cases don't present with a parotitis at all!
        • complications of other organs are rare (e.g. thyroiditis, labyrinthitis etc.).
        • sensorineural deafness (1 : 10.000) → perform an audiogram after parotitis!

DD: • ductus parotideus-calculus, purulent parotitis, parotid tumour, dental infection (US, ENT referral)
        • Sjögren-syndrome (symptoms, SS-A-/SS-B-antibodies)
        • Mumps-orchitis: consider other causes of testicular pain (e.g. testicular torsion → doppler-US, admit to hospital)

Di.:   ▶ clinical features
        ▶ serology: ≥ 4 fold titre rise of the IgG-antibodies in 2 sample; IgM-antibodies ↑ proves a new infection !
        ▶ pathogen detection (no routine diagnostics): virus isolation, PCR

Th.:   - symptomatic, apply warm packs onto the parotid gland, soft food, oral hygiene
        - give immunoglobuline to patients at high risk or those with complications
        - pancreatitis: keep nil by mouth (see pancreatitis)
        - orchitis: rise testicles, antiphlogistics and corticosteroids
        - patients are not allowed to attend any public institutions for at least 9 days after onset of the parotid swelling

Pro.: active: vaccination with an attenuated live vaccine, e.g. combined vaccine with measles and rubella: MMR: 1st vaccine at the age of 12-15 months; 2nd dose at the age of 15 - 23 months. All seronegative staff in nurseries etc. Should be vaccinated.

SE: local reactions, malaise, occ. mild mumps-like illness for 1- 4 weeks.
        passive: mumps-immunoglobulin, e.g. for newborns of mother with active infection

DIPHTHERIA   [A36.9]   Notifiable: Suspected disease, disease, death!

PPh.: Corynebacterium diphtheriae (long gram positive rod with polar bodies). Humans are the reservoir.
        Pathogenesis: Diphtheria toxin with subfraction A (= active, toxic component) and B (binds the toxin to the cell receptors). Not all strains of diphtheria produce toxins. Bacteriophages transmit the ability to produce a toxin. The diphtheria toxin damages the myocardium, nerves, liver and kidneys.

Ep.: There have been epidemics after a long period of ca. 30 years, it has been very rare in Germany after 1955. The GUS-epidemic in the 90's took a toll of ca. 10.000 deaths; contagiousness index 10 - 20 %, healthy carriers in epidemic times up to 7%! (source of infection); airborne infection.

Inf.: airborne infection, direct contact in the case of cutaneous diphtheria
contagiousness: as long as there are pathogens in nose-/ throat swabs (3 samples).

Inc: 2 - 7 days

Sym.: • mostly facial diphtheria: tonsillitis with whitish plaques, spreading into the nasopharynx, contact bleeding (pseudomembranes). laryngeal diphtheria (risk of suffocating)
• sweet smell of fermenting apples
• blood stained rhinitis (nasal diphtheria of infants)
• rarely wound diphtheria

Different forms of courses:
1. local infection: throat/tonsils, nose (nasal diphtheria with blood stained rhinitis), eyes, larynx (→ Croup with inspiratory stridor); infants: umbilicus, wounds (wound diphtheria)
2. generalised intoxication: onset 4-5 days after the local infection; high fever, vomiting, croup: barking cough

Co.: • oedematous swelling of neck (bull neck) leading to airway obstruction
• myocarditis ! (often leading to an heart block): early myocarditis 8 - 10 days and late myocarditis 4 - 8 weeks after beginning of the illness (sometimes late death due to cardiac failure)
• polyneuropathy: paralysis of the motoric cranial nerves, palatal paralysis (!), dysphagia, paralysis of accommodation, oral numbness
• rarely renal damage and acute renal failure

DD: - infectious mononucleosis (full blood count, Monospot)
- Vincent's disease (disproportion: well patients and – usually unilateral ulcerative tonsillitis, foul breath; throat swab: Treponema vincentii + Fusobacterium)
- pseudo croup of young children = subglottical laryngitis (mainly Parainfluenza-infection)
- Streptococci tonsillitis (swab → haemolytic group A Streptococci → therapy: Penicillin V
  3 x 500mg for 10 days; in case of Penicillin allergy: Macrolides)
- sexually transmitted diseases: Gonococci-pharyngitis, herpes-pharyngitis, primary syphilis
- agranulocytosis (blood film)

Di.: history/symptoms + pathogen detection (culture): throat-/nose swab before the start of therapy (take swab of the tissue below the plaques!). Detection of diphtheria toxin or nucleic acids (PCR) of the diphtheria toxin-gene

Th.: isolate any diseased patient and those with suspected disease!
  ▶ suspected infection:
    Start diagnostics (throat-/nose swab); start therapy immediately: give antitoxin, which neutralises the circulating toxin (not the one already adherent to cells).
    • Heterologous equine diphtheria antitoxin: Intracutanous- or conjunctiva test beforehand, in order to avoid an allergic or anaphylactic reaction
      Dose: 500 - 2.000 IU/kg i.m.
    • Penicillin (in case of allergy: Erythromycin)
  ▶ asymptomatic contact persons must receive a postexposure prophylaxis (PEP):
    take throat-/nasal swab, then give prophylactic antibiotics disregarding any previous vaccinations. In the case of non-immunity, active immunisation should be provided; booster after 5 years.

Prg.: depends on resistance, onset of therapy and complications; the Russian-epidemic caused < 5 % deaths.

Pro.: active immunisation with Aluminium-Formalin-Toxoid
Ind: vaccination for everybody, particularly for travellers into countries with a high risk!

active vaccination is usually a combined vaccination
- infants and toddlers get the inactivated vaccine D (30 IU toxoid) as a part of the combined vaccine
- from the age of 6 they get the vaccine d (only 2 IU toxoid), in form of Td-vaccination against tetanus and diphtheria (if a tetanus booster is needed).

After 3 primary doses given in infancy, there are booster doses every 10 years. Titre control after vaccination is not necessary. Antitoxin titres ≥ 0,1 IU/ml serum will protect from diphtheria, titres > 1,0 IU/ml will provide longterm protection.
  SE: occ. localised- and general reactions; rarely allergic reactions; very rarely side effects affecting the peripheral nervous system
  CI: refer to vaccination table

| LEPTOSPIROSIS | [A27.9]

PPh.: Leptospira interrogans: 23 serogroups, > 200 serovares: e.g.
L. canicola, L. grippotyphosa, L. pomona, L. icterohaemorrhagicae. Immunity after an infection will only be developed against this very serotype.

Ep.: Worldwide zoonosis; natural reservoir: rats, mice and other rodents; certain serotypes are transmitted by dogs or pigs, who spread the pathogens in their urine via wet soil (e.g. flooded areas!) and water. Germany: ca. 40 cases/year.

Inf.: Humans catch the infection from urine of infected animals via cutaneous and mucosal lesions, conjunctivae and via contaminated aerosols. Anglers are at risk, people enjoying water sports and certain professions (e.g. sewage-, canal- and field workers)(→ report as an occupational disease)

Inc: 2 - 20 days

Pg.: Leptospirosis is an anthropozoonosis, causing sepsis and subsequent infestation of liver, kidneys and CNS. Nephritis causes the excretion of infectious urine.

Sym.: The disease varies between mild + short (few days) and severe + long (~ 3 weeks); often the course is biphasic. The severe illness is called Weil's disease.
1. early stage (bacteraemia):
   - abrupt onset of very high fever
   - conjunctivitis, rash
   - calf pain, headache (particularly retrobulbar)
2. affected organs:
   - hepatitis (often jaundice); unlike virus hepatitis, the patient is quite unwell at the onset of jaundice!
   - nephritis
   - meningitis/encephalitis, myocarditis, iridocyclitis etc.

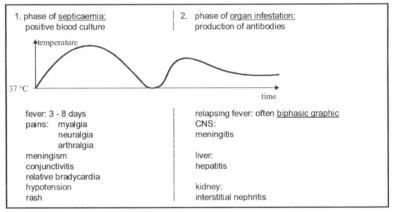

Co.: renal failure, hepatic failure
thrombocytopenia and haemorrhagic diathesis

DD: • mild illness: virus influenza, malaria etc.
• severe illness: Hantavirus-infection, renal diseases, hepatitis, meningitis of different origin, septicaemia, typhoid etc.

Di.: 1. professional-/recreational history
2. pathogen detection from blood and cerebrospinal fluid (only during the 1st week) and urine (from 2nd week)
3. antibody detection from 2nd week (4 fold tire rise within 2 weeks; detection of IgM-antibodies)

Th.: Penicillin G or Cephalosporins of the 3rd generation; high dose for 10 days; antibiotics are only effective when given early - treat for suspected illness

Prg.: severe illness can lead to death in > 20 %.

Pro.: exposition prophylaxis, inform the public about infectious diseases; active vaccination for pets and farm animals

BRUCELLOSIS   [A23.9] Notifiable disease

PPh.: Gram negative, immobile rods; intracellular replication
1. Brucella melitensis (most frequently), 3 biovares: Malta fever; source of infection: sheep and goat milk and dairy.
2. Brucella abortus, 7 biovares: Bang's disease – source of infection : cow's milk and dairy, occupational exposition (cattle)
3. Brucella suis, 5 biovares (porcine brucellosis) and Brucella canis (canine brucellosis) are very rare.

Ep.: Ubiquitous anthropozoonosis, domestic animals are the natural reservoir. Humans are the final host of the infectious chain (no source of infection). In Germany all farmed animals are free from brucellosis. Certain occupational groups are particularly affected by brucellosis: farmers, shepherds, butchers, milkers, veterinary surgeons (notifiable occupational disease). Tourism is the cause for imported disease (in Germany ca. 30 cases/year, mainly from Turkey):
Br. melitensis: farmed sheep/goats: Mediterranean, Spain, Portugal, Central- and South America, Africa
Br. abortus: cattle arms in mild + tropical climate
Br. suis: North America

Inf.: 1. Infection via contact (farmers and lab staff are at risk) (notifiable occupational disease). Invasion through mucosa, (microscopic) skin lesions!
2. oral infection through unheated/unpasteurised milk (dairy) of infected animals (Brucella can survive up to 6 months in sheep-/goats cheese).
3. rarely transmission through breast milk (infected women).

Inc: 5 – 60 days

PPh.: Epitheloid cells in non-caseating granulomas in lymphnodes, spleen, liver (reticulo-endothelial system) and in vascular walls.

Sym.: 90 % of all infections are subclinical (diagnosis only by antibody detection). There are acute or chronic courses of illness (very variable disease)
1. no prodromal stage Br. melitensis: uncharacteristic general symptoms
2. general stage (bacteraemia):
   • fever (irregular temperature , rarely undulant fever), relatively slow heart rate, heavy sweating
   • hepatosplenomegaly, lymphadenopathy, gastrointestinal symptoms, occ. pleuritis/pericarditis, headache, muscular- and joint pain, epistaxis and bleeding gums, occ. rash
3. infected organs: granulomas in spleen, liver, bones (Brucella-arthritis) etc.

Co.: endocarditis, osteomyelitis, sacroiliitis, spondylitis, meningoencephalitis, splenic abscess, orchitis etc.

DD: typhoid, sepsis, malaria, other acute febrile illness, hepatosplenomegaly of different origin, malignant lymphomas etc.

course: 1. inactive brucellosis (primary latent course)
         2. active brucellosis: - acute (<3 months)
                              - subacute (3 - 12 months)
                              - chronic (> 12 months)

Di.: 1. Occupational/ travel history+ clinical features
2. pathogen detection (culture, Brucella-DNA-detection) from blood and other body fluids, bone marrow- or lymphnode biopsy
3. antibody detection (titre > 1:80 or titre rise in second sample), PCR, ELISA; distinguish between IgG and IgM-antibodies (acute illness)

> **Note:** False-positive reactions can be found in Yersinia enterocolitica- and Cholera infections, and after Cholera vaccination due to cross-over antigens. - Serological diagnostic may fail, when antibodies are incomplete (perform Coombs test- positive!).

4. histology of organ biopsy

Th.: Doxycyclin + Rifampicin (6 - 12 weeks, longer course for neurobrucellosis or endocarditis)

Prg.: The pathogen settles inside the RES, hence there is often no complete cure, but just a balance between micro- and macroorganism (chronic brucellosis), leading to intermittent flare ups depending on the patient's overall health (sometimes for several years) → chronic illness up to 20 years! Deaths are rare.

Pre: Eliminate diseased animals, active vaccination of healthy animals, hygiene at work + protective clothing for those at occupational risk; don't eat raw meat or unheated/unpasteurised milk from endemic areas.

TOXOPLASMOSIS [B58.9] Notify congenital disease!

PPh.: Toxoplasma gondii is an intracellular growing protozoon;
two hosts for two different stages of development; intermediate hosts: mice, pigs, sheep, cattle and humans infected with cysts in the muscles and other organs. Final host: cats, excreting infectious oocysts with their faeces.

Ep.: After primary infection and development of host's immunity, idle bradyzoites may persist lifelong! In middle Europe and USA up to 50 % of people aged 40 got antibodies against Toxoplasma gondii; in the UK this number is only ca. 20 %, in France up to 90 %.
Only first infections during pregnancy may lead to prenatal fetal infection in 50 %. The incidence of prenatal infections is 0,1 - 2 ‰ worldwide.

Inf.: transmission to humans:
- raw meat of infected animals containing cysts (e.g. mince – up to 25 % of pork is infested with cysts!)
- contact with oocysts containing faeces of cats; unwashed lettuce/vegetables
- transplacentar infection:
There is only a risk for the unborn child, if the mother catches an infection during pregnancy. The more advanced the pregnancy is, the higher the risk of a transplacentar transmission; on the other hand there is a reciprocal risk for the fetus the more advanced the pregnancy becomes (fetal abnormality decreases):

Fetal risk of infection (transmission) and subsequent damage:
- 1. trimester 15 %, leading to miscarriage or sometimes severe damage of the infant
- 2. trimester 30 %, leading to moderate or severe damage of the infant
- 3. trimester 60 %, leading to minor or delayed damage of the infant

Inc: several days - weeks

Sym.: A) postnatal toxoplasmosis:
    1. immunocompetent individuals: lifelong persistency in the form of bradyzoites, mainly in the CNS.
       - mostly chronic latent toxoplasmosis without symptoms
       -symptomatic toxoplasmosis (1%): infected lymphnodes with lymphadenopathy (often nuchal, cervical), headache, myalgia, fever; occ. ocular symptoms (uveitis), hepatitis
    2. immunocompromised patients and AIDS: reactivation of the latent infection, severe illness with cerebral toxoplasmosis (particularly basal ganglia) and occ. septic spread (heart, liver, spleen). Bradyzoites will differentiate into replicative tachyzoites causing fulminate and rapidly fatal CNS involvement; interstitial pneumonia, myocarditis

    Note: toxoplasmosis is the most common CNS infection in AIDS patients

B) congenital toxoplasmosis:
early fetal infection is fairly rare: miscarriage or severe illness:
- hepatosplenomegaly, jaundice, myocarditis, interstitial pneumonia, miscarriage, stillbirth
- encephalitis with triad: hydrocephalus, chorioretinitis, intracerebral foci

Late fetal infection is more frequent: mild illness. Ca. 80% of the infected infants (asymptomatic at birth) will develop delayed damage up to 20 years later (strabismus, retinochorioiditis, deafness, psychomotoric retardation, epilepsy).

DD: lymphadenopathy of different origin!

Di.: ▶ diagnosis of congenital and postnatal toxoplasmosis:

- serological antibody detection:
  - IgG:
  Sabin-Feldman-test, indirect immune fluoresce test, direct agglutination: positive 2 weeks after infection, 6-8 weeks later very high titre (> 1 : 1.000), later on descending titre, which usually remain lifelong at a low level (down to 1 : 64). Recent infection can only be revealed by a significant titre rise.
  - IgM: recent infection and congenital toxoplasmosis

  Note: Immunocompromised patients (e.g. AIDS) and isolated chorioretinitis - absent IgM and a significant IgG-titre rise. Here it is necessary to isolate the pathogen and a positive therapy test. For infants: compare maternal/infant antibody profile (Immunoblot).

- pathogen detection (e.g. cerebrospinal fluid or blood): for congenital toxoplasmosis and Toxoplasma encephalitis Detection of Toxoplasma-DNA (real time PCR) in blood proves a new infection. Quantitative detection is useful for the treatment follow up.

- lymphnode histology: lymphatic toxoplasmosis Piringer-Kuchinka' lymphadenitis with epitheloid cell clusters.
  Note: histology is not specific for toxoplasmosis, it can be found in mononucleosis and brucellosis.

- suspected cerebral toxoplasmosis: CT, MRI: circular lesions

► Toxoplasma-antibody-screening for pregnant women(not done in the UK):
seronegative women should be screened every 2 months during their pregnancy (to detect any infection during pregnancy) and the woman should be briefed about prophylaxis. Antibody detection before pregnancy means protection for the unborn child. Only experienced laboratories should perform the tests:

Assessment of antibody concentration				
IgG low	IgG low	IgG high	IgG high	IgG low
IgM low	IgM low	IgM low	IgM high	IgM high
↓	↓	↓	↓	↓
inactive infection (latent infection) control		subsiding infection after 2 - 3 weeks	active infection	acute infection

► antenatal diagnostics:
 - fetal ultrasound
 - detection of Toxoplasmosis-DNA in amniotic fluid or fetal blood; IgM only positive in 20 %

Th.: no treatment for chronic carriers; lymphatic toxoplasmosis in the immunocompetent patient is usually self limiting without any intervention

indication for antibiotic therapy:
• toxoplasmosis with clinical symptoms (fever, myalgia etc.)
• primary infection during pregnancy
• immunosuppressed and AIDS-patients with acute toxoplasmosis
• congenital toxoplasmosis
1. choice: Pyrimethamine + Calciumfolinate to prevent myelotoxic SE + Sulfadiazine
treat immunocompetent patients for 4 weeks
therapy during pregnancy: up to 16th week: Spiramycine, then a combination of Pyrimethamine (+ Calciumfolinate) + Sulfadiazine in form of an interval therapy (4 weeks therapy, 4 weeks pause). This can reduce the risk of congenital toxoplasmosis by 90%. Sulfonamide allergy: give Spiramycine.

Pre: • screening of all pregnant women for toxoplasmosis
• seronegative pregnant women, immunosuppressed and AIDS-patients:
  Don't eat raw or bloody meat! Avoid contact with cats! Wash fruits and vegetables thoroughly. After working in the garden or kitchen: wash your hands with soap.
• AIDS-patients:
  - CD4-cell count < 100 - 200/μl requires primary prophylaxis: e.g. Cotrimoxazole (also protects the patient from catching Pneumocystis carinii-pneumonia)
  - give secondary prophylaxis after a completed toxoplasmosis infection ,by using the drugs for treatment of the disease

---

LISTERIOSIS   [A32.9]   Notify congenital infection!

PPh.: 7 Listeria-species, but only Listeria monocytogenes plays a significant human pathogen role; 3 of the 15 serovares play a role for humans (4b, 1/2a, 1/2b). Ca. 10% of humans and animals are healthy intestinal carriers of L. monocytogenes. Listeria are able to replicate at a temperature of + 4°C (fridge)!

Ep.: Infective disease is rare. Immunocompetent humans usually don't get ill, since the widespread existence of pathogen Listeria many humans are immune. Individuals at risk are more likely to become infected: those with a weak immune system (leukaemia, AIDS, patients taking immunosuppressants), newborns, elderly and pregnant women (opportunistic pathogen)

Inf.: Consumption of contaminated food (animal and plant origin) (e.g. unpasteurised milk(dairy), cheese, raw salads, raw sausages, smoked fish, vacuum sealed salmon etc.); sporadic cases or minor epidemics.
Neonatal infection is transmitted transplacentar, or during childbirth or postnatal contact.

Inc: 3 - 90 days

Sym: 1. postnatal listeriosis is a flu-like illness: fever, myalgia; occ. diarrhoea + vomiting
  Co.: Sepsis, purulent meningitis, rarely encephalitis (→ MRI); occ. local abscesses
2. neonatal listeriosis:
  a) early infection: premature birth or stillbirth; onset of symptoms in the first week of life: sepsis, respiratory distress syndrome, skin lesions (granulomatosis infantiseptica)
  b) late infection: onset of symptoms in the second week of life, often with meningitis, granulomatosis infantiseptica

Di.: symptoms, pathogen detection (culture, PCR) from blood, cerebrospinal fluid, pus, vaginal discharge, lochia, stool, meconium or from autopsy samples

Th.: Amoxicillin + Aminoglycosides for at least 2 weeks
alternatives: Cotrimoxazole, Macrolides

Prg.: 30% of septic illness are fatal.

Pre: 1. hygiene when handling and producing food. Cooking, frying, sterilizing will destroy the pathogen.
2. hygiene in the kitchen, wash hands before handling food; separate areas for the preparation of meat and raw vegetables
3. individuals at risk should not eat unpasteurised cheese, and they should remove the peel around cheese, avoid soft cheeses; heat ready meals before consumption

TICK BORNE DISEASES

1. Borreliosis (Lyme-Borreliosis)

2. Viral Meningoencephalitis

3. Human granulocytary Ehrlichiosis (HGE)

Notification by name of infections with Borreliosis and viral meningoencephalitis!

Human Granulocytary Ehrlichiosis [A79.8]

PPh.: Human granulocytary Ehrlichia, is a bacterium, which can be transmitted by ticks, it multiplies obligatory intracellular in vacuoles of granulocytes.

Ep.: endemic areas: USA, sporadically in Europe; forestry workers and foresters are particularly at risk

Inc: 10 - 30 days

Sym.: > 5 % asymptomatic; variable:
• fever and flu like symptoms, headache, arthralgia, myalgia
• sometimes abdominal pain, nausea
• sometimes dry cough
• sometimes rash

Co.: HGE may be immunosuppressive, occ. pneumonia in patients with existing immune deficiency

Lab.: CRP and ESR ↑, sometimes leuco-/thrombocytopenia and transaminases ↑,rarely LDH and CK ↑

DD: fever of unknown origin

Di.: • pathogen detection:
- HGE-DNA in blood (PCR)
- detection of inoculated bodies (Morulae/in Granulocytes)
- antibody detection (IgM and titre rise)
• blood film: intracytoplasmatic inoculation bodies in leucocytes (Murulae) in 20 %
Note: Often there is at the same time a false positive Borrelia-serology due to cross-over reaction.

Th.: e.g. Doxycyclin (2 x 100 mg/d for 2 weeks)

Prg.: in 80 % mild illness; immunosuppressed and old people: severe illness, which can be fatal if not being treated

Pro.: avoid tick bites

	TICK-BORRELIOSIS (LYME-BORRELIOSIS) [A69.2]	VIRAL MENINGO-ENZEPHALITIS [A84.1]
pathogen	Borrelia burgdorferi, 4 species in Europe: B. sensu stricto, B. garinii, B. afelii, B. spielmanii	Virus (notifiable disease!)
max. altitude of risk for infection	1.000 m	800 m
Carrier	ticks (Ixodes ricinus)	
percentage of infected ticks	5 - 35 % in endemic areas After a tick bite, there is a ca. 10 % risk of infection, and a 1% risk of illness	up to 5 % in endemic areas (more in „hot spots") natural reservoir: mice
epidemiology Viral : Lyme-B. ~ 1 : 100 to 300	Ubiquitous in Central-, East-, North Europe, North America, Australia Incidence: 50 - 100/100.000/year	Endemic areas: e.g. Russia, Baltic countries, East Europe, Bavaria, Baden-Württemberg, Carinthia, Balkan
seasons	erythema migrans: March – November; peak July - August; late illness: all year round	March – November; peak July – September
Incubation	1. stage:1 - 6 weeks 2. stage: months 3. stage: years	5 - 28 days
clinical features **Note:** Not every stage is mandatory. Any stage can appear without the previous one for the first time! St. 1 + 2 = early stage St. 3 = late stage	<u>1. st.:</u> Erythema migrans (antibodies only in 50 %!) <u>2. st.:</u> Lymphocytic meningoradiculitis Bannwarth, sometimes facial nerve palsy, meningoencephalitis, myelitis, rarely cerebral vasculitis; myocarditis (occ. heart block), arthritis (mainly knees and ankles); rarely lymphocytoma, e.g. on ear lobes) <u>3. st.:</u> Acrodermatitis chronica atro- phicans (caused by B. afzelii), polyneuropathy, encephalomyelitis	• asymptomatic (70 - 90 %) • symptomatic illness, double peak fever graphic 1. symptoms(10 - ascending fever with flu like 30 %). afebrile interval for several days 2. ascending fever with meningitis (10 %) or meningoencephalitis, rarely me- ningomyelitis 10 % permanent damage
DD	polyneuropathy, meningitis, MS, myocarditis, arthritis of different origin	meningitis/encephalitis, myelitis of different origin
diagnosis	history (tick bite, travel to area where ticks are common) clinical features + serology (IgM ↑), pathogen detections (culture, PCR); Borreliosis: cross over reaction with Treponema pallidum (TPHA-Test !) etc. Borrelia-DNA-detection in urine, synovia, skin biopsy; detection of Borrelia-specific T-lymphocytes (Elispot); neuroborreliosis: symptoms, cerebrospinal fluid; lymphocytic pleocytosis, protein ↑, Borrelia antibodies positive; pathogen detection in cerebrospinal fluid is not very reliable	
therapy	1. stage: Doxycycline (2 x 100 mg/d) or Amoxicillin for: 2 weeks 2. stage: preferably Ceftriaxon i.v. for: 3 – 4 weeks	only symptomatic
Fatality		1% of patients with meningoencephalitis
prophylaxis	• protection from tick bites: appropriates clothes, avoid undergrowth and high grass, search body after a walk in these areas. Post expositional prophylaxis of borreliosis is possible with Doxycycline (1 x 200 mg) • after tick bite: remove tick without squeezing it; don't use any oil or glue; disinfect bite!	
Immunisation	no vaccine available (check success of therapy in the urine – antigen detection)	active immunisation with inactivated vaccine are advised for individuals at risk (3 doses), booster after 3 years. Beware SE + CI!

BACTERIAL MENINGITIS [G03.9]

| Suspected disease, proven disease + death of Meningococci meningitis is notifiable ! |

Syn: purulent meningitis

Def.: inflammation of the meninges, caused by a bacterial infection

PPh.: wide spectrum, depending on the environment and patient's age:

A) community acquired:
- infants < 1 month: E. coli, group B-Streptococci, Listeria
- toddlers: Haemophilus influenzae (if not vaccinated), Meningococci (> 50 %), Pneumococci (Streptococcus pneumoniae) etc.
- adults: Pneumococci (~ 50%), Meningococci*) (~ 30 %), Neisseria meningitidis, Listeria (particularly the elderly, those with a weak immune system) etc.

*) Neisseria meningitidis (12 serogroups), in Germany predominantly serogroup B (65 %) and C (30 %). Other countries have a higher relative incidence of serogroup C (up to > 50 %). Type ET15 of the serogroup C can cause severe septic illness. Up to 10 % of all healthy people are carriers of mostly non-pathogen Meningococci in nose and throat).

B) nosocomial (in hospitals) acquired:
Enterobacteriaceae, Pseudomonas aeruginosa, Staphylococci

C) immunosuppressed/immunodeficient patients:
also: Listeria monocytogenes, Cryptococcus neoformans etc.

Ep.: Incidence varies from country to country 1 - 10/100.000/year. (more frequent in children). Meningococci-meningitis is a worldwide disease(particularly meningitis belt in Central Africa, Saudi-Arabia (during the Hadj -pilgrimage), Asia, South America), often epidemics caused by serogroup A. In Europe mostly serogroups B and C. Up to 80 % of the meningococcal diseases will affect people < 20 years.

Inf.:
- airborne: Meningococci-meningitis, sometimes epidemic
- haematogenic: e.g. Pneumococci pneumonia
- per continuitatem: e.g. otitis, sinusitis
- direct infection: e.g. after an open head injury

Inc: Meningococci-meningitis: 2 - 10 days, usually 2 - 4 days

Sym.: headache, neck ache, irritability (towards light and painful triggers), fever (note: not obligatory), nausea, vomiting, confusion, convulsions impaired consciousness. Meningococci meningitis presents with a sudden onset of illness, the most severe malaise, petechial lesions (containing Meningococci), often on the legs. Older patients and alcoholics are often oligosymptomatic.
Signs of meningism (can be absent in the elderly, infants and those in a coma !):
- neck stiffness on passive forward flexion
- passive lifting of the stretched leg causes : -the patient to flex his knee (Kernig-sign)
  -or causes pain in the leg, gluteal muscle or sacrum (Lasègue-sign)
- passive neck flexion causes a refectory knee flexion (Brudzinski-sign)

Co.: cerebral oedema, hydrocephalus, cerebral abscess, septic sinus vein thrombosis, deafness, cranial nerve palsy fulminate meningococci septicaemia (up to 10 % of all invasive meningococci infections)
= Waterhouse-Friderichsen-Syndrome leading to multi organ failure and adrenal necrosis (DIC with cutanous-/mucosal purpura and -haemorrhage, cardiovascular failure); up to 85 % fatality

Lab.:
- general inflammatory markers: (leucocytosis, CRP and ESR ↑)
- cerebrospinal fluid (CSF): puncture after an increased ICP has been excluded (fundoscopy, occ. CT), indispensable if meningitis is suspected (risk of trapped brainstem post lumbar puncture). After assessing the appearance of the fluid, 2 samples have to be sent to the laboratory. One is for clinical chemistry (glucose, protein, lactate) and for microscopy (cell count, -differentiation, gram- and Methylen-Blue- slides), the other one is used for pathogen detection, culture, antigen detection, PCR). A third sample is kept in the fridge for any further investigations.

	Bacterial M.	Viral M.	Tuberculous M.
CSF appearance	cloudy	clear	with clots
cell count/µl	several thousand	several hundreds	several hundreds
cell type	granulocytes	lymphocytes	lymphocytes, monocytes
glucose	↓ (< 30 mg/dl)	normal	↓ (< 30 mg/dl)
protein	↑ (> 120 mg/dl)	normal	↑ (> 120 mg/dl)
lactate	> 3,5 mmol/l	< 3,5 mmol/l	> 3,5 mmol/l

Bacterial meningitis: unlike viral meningitis lactate is raised, glucose is reduced (< 40 % of serum glucose), CSF is cloudy.

DD:
- viral meningoencephalitis (Coxsackie-, Echo-, Mumps-viruses, spring meningoencephalitis, HIV, measles, CMV, VZV, HSV). Treatment for HSV- or VZV-encephalitis (encephalitic symptoms): give i.v. Aciclovir ASAP (suspected disease is sufficient to launch antiviral therapy).
- tuberculous meningitis (see chapter)
- brain tumour, stroke, migraine

Di.:
- history/clinical features (4 leading symptoms: headache, fever, stiff neck (meningism), lethargy- often there are only 2 or 3 of them!), petechial rash in meningococci-meningitis
- CT/MRI (indispensable to exclude increased ICP before lumbar puncture)
- investigation of CSF
- pathogen detection from CSF and blood (culture, antigen-/virus-DNA-detection) - serological antibody detection (titre rise, IgM)
- search for foci (pneumonia?, otitis?, sinusitis? → consult ENT!, head injury?, throat swab)

A quick diagnose of herpetic meningoencephalitis is most important for treatment (temporal lobe-syndrome with Wernicke-aphasia, confusion, temporal lobe epilepsy; MRI, CSF-PCR); the same applies to Meningococci meningitis. In both cases the prognosis depends on an early diagnosis and treatment. Treat on the slightest suspicion, if necessary add to antibiotic therapy (until the pathogen has been found).

Th.: A) initial blind antibiotic therapy for adults:
   Start immediately after bloods for culture and CSF have been taken!
   - ("community acquired"):
   Cephalosporin of the 3. generation (e.g. Cefotaxim or Ceftriaxon) + Ampicillin (areas with a high percentage of Penicillin-resistant Pneumococci – e.g. France Spain, Hungary etc.: start with a combination of Ceftriaxon+ Rifampicin or Ceftriaxon + Vancomycin). Treat for 10 days.
   - nosocomial acquired (e.g. neurosurgery or open head injury, shunt-infection):
   Vancomycin + Meropenem (or Vancomycin + Ceftazidim)

   B) symptomatic therapy:
   In suspected Pneumococci meningitis fatality + permanent damage can be reduced by giving Dexamethasone: give 20 minutes before or parallel to the antibiotic therapy (4 x 10 mg/day for 4 days).
   Treat an increased ICP (see chapter 'stroke'); regulate fluid and electrolytes; give thrombosis prophylaxis etc.

   Patients with suspected Meningococci meningitis must be isolated. Comply to hygienic regulations!

Prg.: Death rate of Meningococci meningitis is 10 %, higher in high risk patients, in Waterhouse-Friederichsen-syndrome up to 95%; death rate of Pneumococci meningitis is ca. 25%, in Listeria meningitis up to 50%. Risk factors for a severe disease: Splenectomy, impaired immune system.
Average death rate of other forms of meningitis is 10 - 30%; the rates of permanent damage vary.

Pro.: Chemoprophylaxis of Meningococci meningitis for persons who have been in close contact up to 10 days after contact with the patient: Rifampicin (dose for adults 2 x 600 mg/day orally for 2 days); adults can also receive Ciprofloxacin (500 mg/day orally); pregnant women receive a chemoprophylaxis with Ceftriaxon (dose: 250 mg i.m.). Epidemic Meningococci meningitis: search for asymptomatic carriers (throat swabs).

active immunisation:
- Meningococci-vaccine: the conjugated Men C-vaccine (e.g. Menjugat®) only protects against serotype C, but it provides a long immunity. The tetravalent polysaccharide-vaccine (e.g. Mencevax ACWY®) protects against serotypes A, C, W135 and Y (but not against the –
   in Germany prevalent - type B).
   Ind: - vaccinate all children in their 2nd year of age with a conjugated Men C-vaccine (earlier in the UK)
         - asplenic individuals and those with immunodeficiency
         - path lab staff etc.
         - travellers into high risk areas: „Meningococci-belt" (southern Sahara, Saudi-Arabia, India, Nepal, gulf countries, tropical South America etc.)
         - pupils/students intending to stay for a long time in countries where immunisation is highly recommended (e.g. Great Britain)
   SE: localised - and generalised reactions; very rarely allergic reactions
   Dose: A vaccination with a conjugated Men C-vaccine is recommended, providing long immunity. The 4-valent Polysaccharide-vaccine is given 6 months later (not any earlier than 2 years of age). This vaccine is compulsory for pilgrims travelling to Mecca (Hadj). Protection starts from day 10 and lasts 3 years.

   Children younger than 2 years receive the conjugated vaccine against serotype C, since only this vaccine can stimulate the immune system sufficiently.
- Pneumococci-vaccination
   Ind: 1. toddlers
        2. vaccination for everybody > 60 years
        3. high risk patients, e.g. those who underwent splenectomy
- Haemophilus influenzae B-vaccine protects infants and toddlers
- viral meningoencephalitis vaccine for travellers into high risk areas

SEXUALLY TRANSMITTED DISEASES - STD

Internet-Infos: *www.dstdg.de* (Deutsche STD-Gesellschaft)

Worldwide distribution of the following STD:
- Trichomoniasis
- thrush - Candida albicans
- Chlamydia
- Mycoplasma genitalium
- human immune deficiency virus (HIV)
- genital warts - human papilloma-virus (HPV)
- gonorrhoea
- genital herpes - HSV 1 + 2
- syphilis
- hepatitis B-viruses (HBV)

STD of mainly tropical/subtropical areas:
- genital ulcer (chancroid) - Haemophilus ducreyi
- lymphogranuloma venerum - Chlamydia trachomatis
- inguinal (lympho)granuloma - Calymmatobacterium granulomatis

SYPHILIS   [A53.9] Notifiable disease!

Def.: Chronic infectious disease, running in 3 stages; direct transmission, usually via sexual intercourse, rarely via blood transfusion; intrauterine fetal infection is possible.

PPh.: Treponema pallidum, fine helix-shaped Spirochetes, which -- if untreated - will persist inside the body. It is impossible to grow a culture in vitro.

Ep.: Incidence (2003): Western Europe 2/100.000/year; Russia 100 - 150/100.000/year! m : f = 2 : 1; increasing incidence in homosexual men

Inc: primary infection usually 14 - 24 days, sometimes 10 - 90 days

Sym.: A) congenital syphilis:
multisystem disease causing changes of skin and bones, affecting the inner organs, rarely Hutchinson' triad (misshaped teeth, sensorineural deafness, parenchymatous keratitis)

B) acquired (postnatal syphilis):

   1. early disease (up to 1 year after infection):

      • primary syphilis:
Painless, usually single, eroded and indurated ulcer (chancre) of varying size, red, oozing, highly infectious, usually on the external genitalia (rarely extragenital), inguinal lymphadenopathy, vanishes without any intervention after 5 weeks.

      • secondary syphilis
The stage of haematogenic and lymphatic spreading starts after 2-3 months leading to a large variety of symptoms, mainly cutanous with an infectious (in most cases) rash: round maculae, papules, broad lesions (condyloma lata) , hair loss etc.; oral mucosa: mucous patches, which also affect the lips and pharynx; iritis, hepatitis, generalised lymphadenopathy etc. Secondary syphilis can last up to 5 years with varying and temporarily absent symptoms; in 30 % it is self limiting.

   2. late syphilis:

      • tertiary syphilis
5 to 50 years after infection, 1/3 of the patients will progress to this stage that consists of rubber-like pus and necrotising tissue. Any tissue can be affected, leading to a false diagnosis like tumour, Tbc etc.
skin: cutanous gumma; tongue: glossitis gummosa; "gummae" of the bones, muscles, heart (endocardium), lungs, stomach, bowel, rectum, liver, brain etc., also the typical mesaortitis syphilitica → aortic aneurysm, leaking aortic valve

      • Neurosyphilis:
Vertebral column: Tabes dorsalis → demyelinization of the posterior columns with stabbing pain in the abdomen and legs, ataxia, loss of sensitivity and pain sensation
(→ plantar pressure ulcers), Argyll-Robertson-phenonmenon (miosis + absent pupil reflex to light; convergent pupil reaction remains intact).
brain: meningovascular neurosyphilis causing ischaemic stroke etc.; progressive paralysis, mental, intellectual and personality changes, causing dementia.

DD: consider multiple infections: gonorrhoea, Chlamydia trachomatis; HIV-infection and other sexually transmitted diseases (HIV-diagnostics!)

Di.: clinical features and laboratory tests:
- ▸ microscopic direct pathogen detection through dark-field microscopy (unreliable) or fluorescent microscopy of fluid from open lesions in primary syphilitic lesions or needle aspirate from affected lymphnodes. (a culture is not possible; PCR is not a routine test).
- ▸ Treponema pallidum antibody detection:
  1. screening test: TPHA= Treponema pallidum-haemagglutination test
  2. FTA-ABS-test = fluorescence-treponema-antibody-absorption test or immunoblot
  Both tests are specific and sensitive in the detection of Treponema pallidum antibodies, they become positive 3 - 4 weeks after the infection (= reactive) and they remain permanently positive even when the disease has been cured.
- ▸ assessment of the disease's activity:
  - • a positive 19S-IgM-FTA-ABS-test always requires treatment, this also applies to congenital syphilis
  - • VDRL = venereal disease research laboratory-test is a cardiolipin test, detecting lipoid antibodies, which develop during the course of a treponema infection, but they are not specific for syphilis (false positive in the case of phospholipid-antibody-syndrome, SLE, Lepra etc.). VDRL becomes positive after 4 - 6 weeks > 1 : 4 (= reactive); in most cases it will be negative a few months after successful therapy; very rarely a low titre can persist.

  If the time of infection is unknown, then investigation of CSF is indicated to exclude neurosyphilis!

Th.: primary/secondary syphilis:
Benzathinpenicillin G: 1 x 2,4 million IU = 1.800mg i.m. stat, divided into 2 injections at 2 gluteal locations
Penicillin allergy: Doxycyclin 2 x 100 mg oral for 14 days or Erythromycin 4 x 500mg for 14 days

Oral therapy is only possible if compliance is good

Late syphilis or unknown stage:
Benzathinpenicillin G: 2.4 million IU i.m. on days 1, 8, 15. Neurosyphilis requires higher doses.
Penicillin allergy: Doxycyclin: 2 x 100 mg oral or Erythromycin: 4 x 500 mg oral for 28 days
Treatment of the sexual partner: All sexual partners of the last 90 days before a diagnose was made (primary, secondary or early latent syphilis) will need treatment. HIV-positive patients require hospital treatment.

Herxheimer-reaction (systemic reaction at the start of therapy-mainly in elderly patients or advanced disease): fever, myalgia, headache, hypotension; it is believed to be due to the release of endotoxin-like substances when large numbers of Treponema are killed. Patients should be informed about these side effects.
treatment: acetylsalicylic acid, bed rest
Therapeutic success must be followed up clinically + serologically.

Pro.: Treatment of asymptomatic sexual contacts, avoid frequent change of sexual partners, use of condoms; screen all pregnant women

GONORRHOEA [A54.9]

Ep.: most frequent bacterial sexually transmitted disease, many undiagnosed cases

PPh.: Neisseria gonorrhoeae (N.G.): gram-negative diplococci, often found in leucocytes; worldwide increasing incidence of penicillinase-producing strains (PPNG)

Inf.: sexual: genital, rectal, pharyngeal

Inc: 2 - 8 days

Sym.: ca 25% of infected men and 50% of infected women asymptomatic carriers = source of infection!
- • women: urethritis, cervicitis sometimes with mucous-like/purulent discharge, bartholinitis
- • men: acute urethritis with dysuria (itching, burning) and purulent discharge. rectal infection: proctitis

Co.:
- • women: pelvic inflammatory disease (PID), endometritis, adnexitis, peritonitis, perihepatitis (Fitz Hugh-Curtis-syndrome), infertility
- • men: prostatitis, epidydimitis, infertility
- • disseminated gonococci infections: gonococci sepsis, endocarditis, meningitis
- • reactive arthritis (often monoarthritis of the knee)
- • newborns: purulent conjunctivitis → treatment: oral antibiotics (local treatment is not sufficient)

DD: non- gonorrhoic urethritis (NGU): mainly Chlamydia or Ureaplasma urealyticum

Di.: - microscopy + culture from fresh swabs (urethra, cervix; depending on the site of symptoms-throat, rectum); needs a special transport medium!
- nucleic acid amplification test (NAT): first catch urine or urethral swab for men, cervical swab for women

Th.: one-dose treatment with a Cephalosporin of the 2nd or 3rd generation (e.g. Cefixim 400 mg oral or Ceftriaxon 250 mg parenteral): mind the SE + CI!

Note: In South-East-Asia the resistance to Chinolones is > 50 %.

> *Note:*
> - check treatment success via culture after 1 week
> - screen for concomitant syphilis and HIV before and also 6 weeks after treatment!
> - consider multiple infections: non-gonorrhoic urethritis caused by Chlamydia trachomatis, syphilis, HIV etc.
> - always treat the sexual partners!

Pro.: treat infected sexual partners, avoid multiple sexual partners, use condoms

---

HIV-INFECTION [Z21] and AIDS [B24] (<u>a</u>cquired <u>i</u>mmune <u>d</u>eficiency <u>s</u>yndrome)

Notifiable infection!

Internet-Infos:

International:
www.unaids.org
www.aidsinfo.nih.gov
www.hiv.net
www.hivinsite.ucsf.edu

Germany:
www.kompetenznetz-hiv.de
www.daignet.de

PPh.: 2 types of the human immunodeficiency virus (HIV):
- HIV-1: most frequent type, found worldwide, 3 groups:
  group M (major) is worldwide the most frequently found group, with the <u>subtypes A to K</u>. In Europe and the USA HIV-1M:B is the predominant virus, in West Africa it is HIV-1M:A, in South Africa HIV-1M:C and in East Africa HIV-1M:A and HIV-1M:D.
  group N (rare, 5 cases in Cameroon)
  group O (very rare, West Africa: Cameroon)
- HIV-2 with 6 subtypes (A - F): mainly in West Africa, late worldwide

Double infections with two different types are possible. During an HIV-infection various virus mutants can develop. Virus recombination of 2 subtypes are increasingly observed, e.g. HIV-1M:A/B in Kaliningrad, HIV-1M:B/C in China, HIV-1M:A/E in Thailand, HIV-1M:A/G in Nigeria.
HIV is an RNA <u>retrovirus</u>, in the possession of the enzyme '<u>R</u>everse <u>T</u>ranscriptase', which transcribes virus-RNA into proviral DNA. HIV damages the immune system and the nervous system directly; the patient does develop antibodies, but they don't lead to an eradication of the virus.

Ep.: The first HIV infection was reported in Zaire in 1959. The transmission was probably one from monkeys (= Simian) immunodeficiency virus (SIV) to humans. After 1980 spread of the pandemic from Central Africa to the Caribbean (Haiti) and USA, from there global spread to Europe and into other regions. In sub-Saharan Africa men and women are equally infected (heterosexual transmission is the most common way); in USA/Europe so far predominantly homo- and bisexual men and injecting drug users have been infected. In homosexuals there has been a significant increase over the last few years. 2007 worldwide ca. 40 million infected people, of those > 95 % in poor countries, particularly in Africa (70 %) south of the Sahara. This is 1/3 of the population in the south part of Africa: AIDS is the most frequent cause of death in this area! It also makes up to ca. 15 % in South-East Asia. There is an explosive spread of HIV in East Europe!
AIDS is amongst the 5 most frequent causes of death worldwide (infectious diarrhoea, pneumonia, tuberculosis, AIDS, malaria).

Inf.: 1. Sexual: (%-numbers relate to all AIDS cases in <u>Europe</u>)
   Risk factors: frequently changing sexual partners, unsafe sex, infection on trips to endemic areas
   - homo- and bisexual men: ca. 50 %
   - heterosexual individuals: ca. 20 %, mostly women
2. parenteral:
   - i.v.-drug abuse (very high risk if needles are exchanged): < 10 %
   - treatment with blood (-products): since the introduction of HIV-antibody tests for blood and its products (late 1985) this is a very rare source of infection in the industrialized countries (higher risk in poor countries of the 3rd world)
   - accidental injuries of medical staff: very rarely – the risk to catch HIV is lower than catching hepatitis B
3. vertical transmission from mother to the fetus is possible from the 12th week of pregnancy, usually in the last trimester: In Europe < 1 % of all AIDS patients are children(much more in Africa)! Risk of transmission without therapy is 15 - 20 % (in developing countries it is much higher). Chemoprophylaxis, elective caesarean section and avoidance of breast feeding will reduce the risk to < 2 %.
The remaining HIV patients in Europe are from areas with a high prevalence (e.g. Africa) or the source of infection is unknown.

Inc: 1. Serological definition: gap between infection and the detection of serum antibodies: 1-3 months, occ. longer

2. Clinical definition: gap between infection and the onset of AIDS: this depends on nutrition, immune system and age: adults living in rich industrialized countries: 10 ± 2 years (perinatal infection only ca. 5 years); malnourished HIV-patients living in poor countries: the incubation period is shorter!

Pg.: • target cells of HIV-infections are those carrying the CD4-surface antigen: T-helper lymphocytes (CD4+), macrophages, monocytes, Langerhans' cells of the epidermis, parts of the microglia. Another cellular component or co-receptor is required to enable penetration of HIV-1 into the target cells. Macrophagotrop HIV-1-viruses possess the beta-chemokin receptor CCR5. T-lymphocytotrop HIV-1-viruses possess the alpha-chemokin receptor CXCR4 as a co-receptor. Individuals with an inherited homozygous genetic defect of the CCR-5 receptor may confer relative resistance to HIV infection.

• damage to the immune system: destruction of T-helper cells leads to a decreased number below 400/µl; this in turn results in a reduced T-helper cells/T-suppressor cells ratio < 1,2 (normal is ca. 2) → opportunistic infections, malignancies.
Note: according to the surface antigen we divide into:
T-helper cells: T4- lymphocytes (because they carry the CD4-antigen)
T-suppressor cells: T8-lymphocytes (because they carry the CD8-antigen)

• damage to the CNS: HIV-associated encephalopathy (HIVE) affects ca. 20 % of all patients; there are typical multinuclear cells, loss of myelin, brain atrophy. HIV-viruses replicate inside the CNS in the macrophages and in the microglia.

Sym.: **CDC-staging of HIV-infections:**
(CDC = Centers for Disease Control / USA, 1993)

	3 stages of clinical features		
	A	B	C
3 numbers of T-helper lympho-cytes (/µl)	asymptomatic or acute HIV-illness or LAS	symptomatic, but not A or C	AIDS-complications
1 > 500	A1	B1	C1
2 200 - 499	A2	B2	C2
3 < 200	A3	B3	C3

Individual staging is done according to the most advanced stage; stepping back is not possible; this unidirectional classification is unpractical for antiretroviral therapy, where reconstructions of the immune system are possible. Also the virus concentration has not been taken into consideration.

Note: Viraemia and hence infectiousness peak twice during the illness: in the very beginning (acute HIV-illness) and towards the end (terminal AIDS-illness).

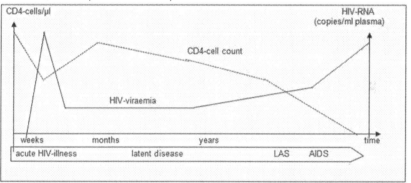

Category A:

▶ **acute HIV-illness (= acute retroviral syndrome):**
Between 1 - 6 weeks after exposure to HIV ca. 30% of those infected will develop a non-specific illness, similar to infectious mononucleosis with fever, lymphadenopathy, splenomegaly, tonsillitis, occ. rash, myalgia. Mononucleosis can be excluded by negative serology and a lymphopenic blood count. HIV-antibody test is usually negative at this stage.
HIV-antibodies usually become positive 1 - 3 months after infection. If they are still not detectable 6 months after a presumed exposure, an infection can usually be excluded.

▶ **asymptomatic infection (latent disease):**
virus replication inside lymphatic tissue; HIV-antibodies are positive 1 - 3 (-6) months after infection
Patients are clinically healthy but infectious carriers. This latent illness lasts about 10 years (it is shorter in infants and toddlers, and in malnourished, immune incompetent patients in poor countries).

▶ **persistent generalized lymphadenopathy (PGL) = lymphadenopathy-syndrome (LAS):**
Ca. 40 % of AIDS-patients will develop LAS:
- HIV-antibody test is positive
- generalized lymphadenopathy: persistently (> 3 months) swollen lymph nodes in at least 2 extra-inguinal areas
  → diagnosis: biopsy + histology
- absence of any unspecific symptoms
- 30 % will develop seborrhoic dermatitis.

Category B:
Note: the following are an indication for a progression of the HIV-infection:
- increased number of viruses
- decrease of T-helper cells

▶ **Non-AIDS-specific diseases:**
Diseases which are promoted by an immunodeficiency, but are not part of Category C:
- subfebrile temperature (< 38,5^{0}C) or chronic diarrhoea (> 1 month)
- idiopathic thrombocytopenic purpura
- inflammatory diseases of the female pelvis; cervical dysplasia or carcinoma in situ
- HIV-associated peripheral neuropathy (ca. 40 %)
- bacterial angiomatosis
- listeriosis
- oropharyngeal or vulvovaginal thrush
- herpes zoster, several dermatomes can be affected (intraocular complications)
- oral hairy leucoplakia (white, non-removable plaques on the side of the tongue), caused by Epstein-Barr-virus

Category C:

AIDS – specific diseases (AIDS-indicating diseases):

▶ **wasting-syndrome:**

Def.: involuntary weight loss of > 10% and chronic diarrhoea (> 30 days) or fever/ tiredness

When: CD4-cell count < 200/μl, in ca. 14 % of all untreated patients

Di.: ask for nutrition, check weight, exclude other (infectious, malignant, or endocrinological) diseases, which might be an explanation for the presenting symptoms, check testosterone levels (exclude hypogonadism)

Th.: involve a nutritionist, start highly active antiretroviral therapy (HAART), if necessary give an appetite stimulant, oral or parenteral supplements

Pro.: high mortality, which is much reduced when HAART is given

▶ **HIV-associated encephalopathy = HIVE:**

Def.: direct infection of the connective tissue (microglia) with consecutive destruction of the CNS

Ep.: untreated 15 - 20 %

When: CD4-cell count < 200/μl.

PPh.: HIV-virus

Sym.: subcortical, slowly progressive dementia; patients develop cognitive (impaired concentration, memory), motor (impaired gait, fine motor activity), emotional (depression) and rarely vegetative (incontinence) abnormalities

DD: opportunistic infections of the CNS, psychiatric diseases, dementia of other origin

Di.: MRI (diffuse brain atrophy, to exclude other diseases), analysis of CSF (minor barrier dysfunction, virus count, to exclude other diseases), psycho-mental tests (can be used for early diagnose).

Th.: highly active antiretroviral therapy (HAART) (suitable: AZT, 3TC, NVP, LPV)

Pro.: good if treated with HAART, often resulting deficiency

▶ **opportunistic infections, typical for AIDS:**
In 80% AIDS becomes manifest in opportunistic infections. Since the introduction of HAART opportunistic infections have been on the retreat. Hence, any numbers will refer to patients not receiving antiretroviral therapy.
Note: most opportunistic infections in the immunodeficient patient often present an atypical, complicated course and are difficult to treat (compared to immunocompetent patients). Diagnostics are difficult, since serological tests (antibodies) are not very convincing in AIDS-patients; there are often multiple infections, and a differentiation between asymptomatic colonisation and pathogen organisms can be very difficult.

- **Protozoa-infections:**

  - **cerebral toxoplasmosis**
    Sym.: fever, confusion, psychotic syndromes, headache, cerebral convulsions
    Di.: intracerebral abscess concentrating contrast medium, MRI
    Th.: e.g. Clindamycin (2400 mg/day) or Sulfadiazine (4 g/ day) + Pyrimethamine (100 mg/day) + Foline acid (3 tbl. per week), alternatively Atovaquone, after 4 - 8 weeks start maintenance therapy
    Primary prophylaxis if CD4-cell count < 200/µl: e.g. Cotrimoxazole (also provides protection against PcP).

  - **Cryptosporidiosis**, microsporidiosis, Isospora belli infection → watery diarrhoea, tenesmus
    Th.: HAART

- **fungal infections:**

  - **Pneumocystis jiroveci- (used to be: carinii) pneumonia (PcP):**
    Most frequent type of pneumonia! (85 % of all AIDS-patients, 50 % presenting manifest)
    Sym.: exercise triggered shortness of breath, dry cough, subfebrile temperatures
    Di.: mainly normal auscultation of the chest. CXR: bilateral infiltrations; hypoxia, raised LDH, CRP normal. Pathogen detection from sputum, bronchial-rinse, transbronchial pulmonary biopsy
    Th.: 1. choice: Cotrimoxazole; give additional steroids in case of hypoxia
    2. choice: Pentamidine infusion (SE: nephro-, hepato-, myelotoxicity; watch out for hypoglycaemia and hypotension!)
    Primary prophylaxis of PcP: give Cotrimoxazole when the T-helper cells are down ≤ 200/µl at the latest
    Secondary prophylaxis (to prevent relapse): Cotrimoxazole (protects against Pneumocystis + toxoplasmosis). 2nd choice: Pentamidine-inhalations

  - **Candida-oesophagitis:** retrosternal burning, dysphagia, impaired taste
    (→ oesophagoscopy + brush biopsy), possible bronchopulmonal infection and Candida-sepsis
    Th.: Amphotericin B locally only if the immune system is stable, give systemic Fluconazole if the immune system is impaired. Alternative treatment for candida sepsis or resistances: systemic Amphotericin B (may be combined with Flucytosin), Itraconazole, Caspofungin, Voriconazole; no primary prophylaxis

  - **Cryptococcosis**
    Sym.: pulmonal and extrapulmonal infection + meningo-encephalitis, headache is typical
    Th.: Amphotericin B + Flucytosin + Fluconazole; secondary prophylaxis using Fluconazole or Itraconazole; alternative: Voriconazole

  - **other fungal infections: aspergillosis, histoplasmosis**

- **Bacterial infections:**

  - **recurrent pneumonias** within a year

  - **atypical mycobacteriosis**
    30 % of AIDS-patients, especially Mycobacterium avium/ Mycobacterium intracellulare (MAI-strains), but also other atypical mycobacteria. This happens when CD4-cell count < 100/µl
    Sym.:fever,weight loss, abdominal pain, raised alkaline phosphatase, hepatosplenomegaly
    Di.: blood cultures, culture from drainage, culture from respiratory and gastrointestinal secretions (also unspecific results), PCR for MAI
    Th.: Ethambutol (1200 mg/day) + Clarithromycin (1000 mg/day) + Rifabutin (300 mg/day, beware interactions with antiretroviral drugs)

  - **tuberculosis:** the risk is ca. 10 % per year (Tbc makes up 30 % of all AIDS - deaths!)
    Sym.: mainly affecting the lower lobes without cavities, frequently atypical or severe (miliar) illness
    Di.: mainly by culture in various media, PCR; Mendel-Mantoux-test is often negative in the immunodeficient patient
    Th.: anti - tuberculosis combination therapy, beware simultaneous antiretroviral therapy, since interactions between Rifampicin and non-nucleoside reverse transcriptase-inhibitors and protease inhibitors are common (e.g. this could require a treatment change to Rifabutin)

  - **Salmonella sepsis**, mainly Salmonella typhimurium
    Th.: Ciprofloxacin, alternatively Ceftriaxon

- **virus infections:**

  - **Cytomegaly-virus infection (CMV)**
    Inc.: frequently, if CD4-cell count < 100/µl, causes blindness in up to 30 % of non treated patients. Initially active replication, followed by organ manifestation
    Sym.: gastrointestinal, retinal invasion, pneumonia, encephalitis
    Di.: fundoscopy, endoscopy, quantitative PCR from serum and biopsies
    Th.: if renal function is normal, give Valganciclovir (2 x 2 tbl/day) or Ganciclovir (2 x 5 mg/kg/day), alternatively Foscarnet (2 x 90 mg/kg/day), no primary prophylaxis

  - **Herpes zoster**
    Sym.: often several dermatomes are affected, prolonged illness, atypical varicella-like manifestations
    Th.: high dose Aciclovir

  - **Herpes simplex-infection** (particularly HSV-2) → genital herpes, ano-rectal, oropharyngeal infection (persistent and ulcering)

- **progressive multifocal leucencephalopathy (PML):** reactivation of a JC-virus-infection
Th.: antiretroviral therapy

▶ **malignancies, defining AIDS:**
in ca. 20 % certain malignancies lead to the diagnosis of 'AIDS'.

- **Kaposi-sarcoma** (pronounce: "Kaposhi")
    4 forms: 1. classical Kaposi-sarcoma
                2. African Kaposi-sarcoma
                3. Kaposi-sarcoma in transplant-patients receiving immune-suppressing drugs
                4. HIV-associated Kaposi-sarcoma
    Aet.: HHV-8 + co-factors
    The classical Kaposi-sarcoma is a very rare form of sarcoma, occurring usually in elderly men in the mediterranian in a localised form. In Africa and the Peloponnese it often presents with a vehement illness. The HIV-associated Kaposi-sarcoma is a generalised illness, in form of a multifocal tumour, and it affects mainly homosexual men.
    skin: purple or brown-blue maculae, plaques, tumour nodes, mainly in the cutanous lines and on the legs. Oral mucosa: blue-red palatal nodules
    GI-tract: polyp-like changes
    lymph nodes and other organs can be affected (e.g. lungs)
    DD: bacterial angiomatosis: red pinpoint sized papules and nodules in HIV-patients. PPh.: Bartonella henselae or B. quintana (Th.: e.g. Erythromycin or Doxycyclin)
    Th.: antiretroviral combined therapy (HAART)
    Local therapy: e.g. excision, laser therapy

- **Non-Hodgkin-lymphoma:** usually B-cell type (EBV-associated), mostly men, usually highly malignant; manifestation: mainly extra-nodal in stage IV
    Th.: usually CHOP-scheme, early combination with HAART

- **invasive cervix carcinoma:** most frequent malignancy in women; often it is the first AIDS-defining illness

- **CNS-lymphomas**

▶ **other diseases found in AIDS, e.g.**
- Hodgkin's disease: frequent illness, but so far it has not been defined as HIV-associated
- immune reconstitution inflammatory syndrome = IRIS: 1-3 months after a successful antiretroviral therapy, inflammatory illnesses will appear, these can be self limiting opportunistic infections
- visceral leishmaniosis (Kala Azar): mostly in the mediterranian

▶ **HIV-infections in children:**
Mostly perinatal infection; if searching for antibodies, HIV-infections of the newborn can only be excluded after 18 months; PCR enables an earlier diagnose.
Clinical features of congenital HIV-infection:
- premature babies
- dystrophy
- craniofacial dysmorphies
- CNS -damage: cortical atrophy + calcification of the basal ganglia with subsequent ataxia
- opportunistic infections (mostly Pneumocystis jiroveci (obsolete: carinii-Pneumonia (PcP); also: Haemophilus influenzae, Candida infections, CMV-infections, Herpes-virus-infections)
- chronic lymphoid interstitial pneumonia (LIP)

Course of illness:
1. fast course with typical clinical features within the first year of life (1/5 of all children)
2. slower course with an average incubation period of 4 - 5 years

Older children and adolescents show a similar course of illness as adults. Indication and form of antiretroviral treatment is chosen according to the consensus of the paediatric working group AIDS and the German Society for Paediatric Infectiology (DGPI).

DD: • immune deficiency of a different origin
        • idiopathic CD4 - lymphocytopenia = ICL: very rare immunodeficiency syndrome in which
        T-helper (CD4) - lymphocytes are < 300/μl without the presence of a HIV-infection

Di.:   history - clinical features - pathogen/antibody detection

HIV-serology:

- HIV-1 and HIV-2 antibody detection:
    Before a HIV-test is done, the patient's consent must be obtained. The CDC (Centres for Disease Control) in the USA recommends, that a HIV-antibody-test should be offered as a routine after doctor-patient-contact, in order to enable an earlier detection of HIV-infections.
    Considering the consequences of a positive HIV-antibody test result, a reactive (= positive) screening test (e.g. ELISA) must always be followed by a confirmation test (e.g. Westernblot), in order to exclude the very rare event

of a false positive result: e.g. in autoimmune diseases, pregnancy, other infectious diseases, after transfusions and after transplantations, after influenza vaccination etc.

In order to avoid a mix-up of samples, a second sample must always be sent. Only after 2 positive results should the patient be informed about the test result. Antibody detection is usually positive 6 weeks after the infection. A negative test result only excludes a HIV-infection, if there were no risks for an infection for 6 months before the test → reliable diagnostics via virus detection!

- detection of virus (or their components):
  - HIV-isolation (takes ca. 6 weeks, not suitable for routine diagnostics)
  - nucleic acid detection-test (NAT): ca. from day 11 onwards, PCR HIV-DNA or HIV-RNA can be found in lymphocytes, proving free viruses. A negative result of NAT does not exclude the presence of HIV, and it requires further investigations after a short time, if there is sufficient suspicion (e.g. for transfusions).

- virus quantification (e.g. via PCR): A unit of viruses is defined as a given number of virus equivalents/ml plasma or RNA-copies/ml plasma. Significance and indication:
  therapy- /therapy control/ prognosis: The target of a therapy is the reduction of HIV-replication below a level where detection is possible (20 or 50 copies/ml). A drop in HIV-RNA less than 1 log 10 after 4 weeks, or the absence of a drop below the detection level within maximal 6 months is regarded as an insufficient therapeutic success, and it is an indication, to exchange all drugs ASAP.

- counting T-helper lymphocytes (CD4 - cell count):
  The CD4-cell count provides information about the extend of the immune deficiency, and it is being used for staging within the CDC-classification. It is just as important for the assessment of a prognosis in untreated patients as the virus quantification. Counts of > 400 - 500/µl rarely mean a risk to catch AIDS-defining diseases. Counts of < 200/µl means a high risk to catch those diseases, and must be followed by immediate therapeutic measures including the prophylaxis against opportunistic infections. The CD4 - cell count is also used for therapy monitoring; a target is a rise of the count.

- determination of HIV-resistance: genotype tests

  ind: before onset of therapy (to exclude primary resistances) and before a change of the antiretroviral therapy due to inefficiency of a treatment or insufficient response; indicated in children and pregnant women

- check serum levels of antiretroviral medication if insufficient compliance or reduced absorption/enhanced metabolism is suspected.

  ind: if unusual/complex combinations are being used, lack of efficiency of HAART, suspected reduced absorption, side effects, hepatic impairment

Th.:   1. healthy lifestyle, eliminating any factors that might reduce resistance
       2. highly active antiretroviral therapy = HAART
       3. prophylaxis against and therapy of opportunistic infections and other complications
       4. psychosocial support

Antiretroviral therapy:

ingredient	generic name	Brand name	extras, important side effects	Group of ingredient
AZT	Zidovudin	Retrovir®	bone marrow depression	NRTI
d4T	Stavudin	Zerit®	neuropathy, lactate acidosis	= nucleoside reverse-transcriptase-inhibitor
3TC	Lamivudin	Epivir®	headache	= nucleoside analoga
ddI	Didanosin	Videx®	pancreatitis, lactate acidosis; to be taken on an empty stomach	dosis reduction in renal failure!
ABC	Abacavir	Ziagen®	hypersensitivity	
FTC	Emtricitabin	Emtriva®	as 3TC	
TDF	Tenofovir Disoproxil Fumarate	Viread®	diarrhoea, nausea, hypophosphataemia, dosis reduction in renal failure!	NtRTI = nucleotid-analoge reverse transcriptase-inhibitors
TDF/FTC/EFV		Atripla®		NtRTI + NNRTI
AZT/3TC	as above	Combivir®	as above	combinations of various NRTI/ NtRTI and NNRTI
AZT/3TC/ABC		Trizivir®		
3TC/ABC		Kivexa®		
FTC/TDF		Truvada®		
NVP	Nevirapin	Viramune®	allergy, hepatic impairment	NNRTI = non-nucleoside reverse-transcriptase-inhibitor
EFV	Efavirenz	Sustiva®	nightmares, depression	
SQV	Saquinavir	Invirase®	nausea, diarrhoea	PI = protease-inhibitors combination of 2 PI: RTV 100 - 200 mg/day can increase the plasma level of PI(PI-boosting)
NFV	Nelfinavir	Viracept®	nausea, diarrhoea	
IDV	Indinavir	Crixivan®	diarrhoea, nephrolithiasis	
RTV	Ritonavir	Norvir®	nausea, hyperlipidaemia	
DRV	Darunavir	Prezista®	Diarrhoea	
F-APV	Fosamprenavir	Telzir®	Amprenavir-prodrug nausea, diarrhoea	
LPV/RTV	Lopinavir/ Ritonavir	Kaletra®	diarrhoea, hyperlipidaemia	
ATV	Atazanavir	Reyataz®	Hyperbilirubinaemia	
TPV	Tipranavir	Aptivus®	rash, diarrhoea, GPT↑,intracranial haemorrhage	
T-20	Enfuvirtide	Fuzeon®	Reactions at injection site	fusion inhibitor

Indication for treatment:
therapy should be started when clinical features or chemical parameters start to show signs of immunodeficiency.

Recommendations for indication of therapy:

clinical features	CD4-count	HIV-virus count	recommendation for therapy
HIV-associated symptoms (CDC-staging B,C)	any count	any count	absolute indication for therapy based on the reports of randomised studies with clinical targets
asymptomatic patients (CDC-staging A)	< 200 /μl	any count	
	200-350 /μl	any count	generally advisable (surrogate marker studies)
	350-500 /μl	> 50.000-100.000 c/ml	acceptable (surrogate marker studies)
		< 50.000 c/ml	acceptable (various studies reveal different results, consult an expert)
	> 500 /μl	any count	
acute retroviral syndrome	any count	any count	acceptable (studies)

In order to prevent/postpone resistance, and in order to reduce the virus count as much as possible, a combined treatment with at least 3 antiretroviral medications is required. If the virus count is high, and if AIDS-defining diseases are present, a PI should be part of the treatment. The highly active antiretroviral therapy (HAART) usually consists of 2 NRTI and 1 NNRTI or a PI boosted with RTV. As a rule one of the drugs should easily pass the blood/brain barrier (AZT or d4T). NNRTI and PI have several interactions because they are metabolised via the Cytochrom-P-450-system. Hence the dosage and contraindications of each HAART must be evaluated individually for each patient. All retroviral drugs can have long term side effects like the lipodystrophy syndrome (LDS). This is a redistribution of fat (lipoatrophy of the face and limbs, build-up of intra-abdominal fat) and pathological glucose tolerance, as well as mixed hyperlipidaemia.

If there is simultaneous treatment with Ribaverin for hepatitis C, a combination with Didanosin, Stavudin or Zidovudin is contraindicated.

Patients who underwent a particularly fast immune reconstruction or those, who started HAART at very low T-helper cell count (< 50 CD4/µl), frequently suffer immune reconstitutional inflammatory syndrome (IRIS). This is a temporary deterioration of an illness or new onset of opportunistic infections, despite rising T-helper cells.

Basic combinations of antiretroviral therapy:

recommendation	nucleos(t)ide analoga		Combination drug
preferred combinations	TDF + FTC	+	EFV**
	ABC + 3TC		NVP***
	AZT + 3TC*		LPV + RTV
			FPV + RTV
alternatives	Ddl + 3TC or FTC	+	SQV + RTV
			ATV + RTV
			IDV+ RTV
			NFV
			ATV
	AZT + 3TC + ABC****		None
combinations which should be avoided	Monotherapy		
	dual combinations		
	T-20, TPV and DRV for primary therapy		
	IDV, LPV or SQV without RTV		
	AZT + d4T	+	any drug
	3TC + FTC		
	d4T + ddl		
	d4T + ddC		
	ddl + ddC		
	TDF + ddl		
	TDF + ABC		

*   The combination AZT/3TC is as effective as all the other recommended N(t)RTI-combinations, but side effects are much more frequent; there are signs that the rise of the CD4-cell count is not as high.
**   Do not use for pregnant women or those trying for a family.
***   Reduce dose in hepatic impairment, in men with > 400 CD4-cells/µl and in women with > 250 CD4-cells/µl.
****   In special situations; virologically not as effective for primary therapy as the recommended combinations.

HAART is a lifelong therapy, and it requires a very high patient compliance. A 95% compliance is needed, in order to achieve a target of more than 80 %; if the compliance is 70%, only 30% of patients will have a virus count below the detectable threshold after 6 months. Therapeutic monitoring is done by CD4-cell count and virus count. The other part of therapeutic monitoring consists of drug level checks and drug resistance tests.

Pro.: • general:
 - inform high risk groups and the whole population about the mode of infection and prophylaxis: HIV is mainly transmitted via sexual intercourse or contact with infected blood.
 - avoid frequent change of sexual partners and prostitution, use condoms
 - screen all blood donors for HIV (NAT)
 - minimise heterologous blood transfusions using:
  · autologous blood donation and transfusions for elective surgery
  · auto-transfusion ("recycling" of suction blood during operations)
 - take great care when handling blood: gloves, safe needles and instruments; if indicated-use a mask and glasses/goggles. Discard of any needles, syringes and sharp instruments safely.
 - patients should inform doctors/dentists about their HIV-infection before any diagnostic or surgical procedure.
 - disinfect the skin after any possible contact with any infected material; any mucosal contamination requires intensive rinsing with water. Accidental injuries have to be bled for at least one minute, disinfected for 10 minutes and reported to occupational health. HIV-antibody tests (+ HB- and HC-serology) immediately, after 6 weeks, 3, 6 and 12 months. An antiviral post exposure prophylaxis (PEP) –ideally performed within 2 hours after the injury- will reduce the risk of HIV-infection by > 80 %: triple combination immediately post exposition for 4 weeks: 2 nucleoside analoga + 1 protease inhibitor, e.g. AZT + 3TC (Combivir® 2 x 1 tbl.) + LPV/RTV (Kaletra® 2 x 2 cps.). The medication has to be available!

medium risk of a needle prick injury causing an inoculation of virus-positive blood leading to an infection	Hepatitis B-virus	Hepatitis C-virus	HIV
	up to 30 %	ca. 3 %	up to 1,5%

- the risk of HIV-transmission from an infected pregnant woman to her newborn can be reduced to < 1 % by the following means:
  1. antiretroviral therapy from the 32nd week onwards
  2. in the case of premature labour before 34 weeks: tocolysis
  3. elective caesarean section in week 36 before the onset of labour
  4. antiretroviral prophylaxis of the newborn for 6 weeks
  5. no breastfeeding

- active immunisation:
  the development of a vaccine is difficult due to the large variety of HIV-mutants. HIV has a low antigen potential, a strong antigen drift (changing antigen potential of the HIV-envelope gp41 and gp120) and "shedding" (discarding of surface glycoproteins).

Prg.: signs for a poor prognosis:
- high virus count (> 10.000 copies/ml) at the first investigation
- rising virus count, persistent virus count > 10.000 copies/ml
- drop of T-helper cell count
- rising β2-microglobulin-level
- progression of the clinical staging

long term non-progressors: asymptomatic and high CD4-cell count after > 10 years: maximal 5% of all infected patients (effective cytotoxic immune response towards infected CD4-cells).

A small number of HIV-infected children are able to eliminate the virus. In poor countries 50% of HIV-positive children die before their 2nd birthday.

HAART reduced the incidence of AIDS-defining diseases significantly, and the average CD4-cell count raises. A cure is not possible, but the prognosis has improved significantly: A 25 year old patient with a HIV-infection receiving optimal treatment, has an average life expectancy of another 35 years (Danish study). Sadly in third world countries most HIV/AIDS-patients have no access to HAART due to poverty and a lacking health system.

Prognosis of HIV-infections in case of optimal therapy (HAART): probability of an AIDS-defining disease or death within the next 3 years depending on the CD4 - cell count and virus count:

CD 4 cell count	virus count< 10^5 c/ml	virus count> 10^5 c/ml
0 - 49 /µl	16 %	20 %
50 - 99 /µl	12 %	16 %
100 - 199 /µl	9 %	12 %
200 - 349 /µl	5 %	6 %
≥ 350 /µl	3 %	4 %

SELECTED TROPICAL DISEASES

Internet-Infos:
Germany: www.fit-for-travel.de
            www.crm.de
            www.travelmed.de
            www.dtg.mwn.de (DTG)
            www.tropinst.med.uni-muenchen.de
International: www.cdc.gov (CDC)
            www.safetravel.ch
            www.who.int (WHO)

YELLOW FEVER   Notification of suspected disease, illness and death! [A95.9]

PPh.: The yellow fever virus is an RNA flavivirus; transmission mainly via various mosquitoes, possibly also via ticks. Monkeys and long-tailed monkeys of the tropical forests of Central-/South America + Africa are the main reservoir.

Ep.: yellow fever belt: tropical Africa (ca. 15° S - 15° N latitude – also in Kenya) and tropical Central- and South America (40° S - 20° N latitude); in Africa the incidence is rising! Watch out for the updated WHO-reports! Yellow fever does not exist in Asia. 2 ways (cycles) of transmission:
- jungle- or sylvanian yellow fever: the virus circulates between mosquitoes and primates; human beings will only be sporadically infected.
- city- or urban yellow fever: the virus circulates between mosquitoes and the non- immune population → epidemic.

Inf.: transmitters of south American yellow fever: various types of Haemagogus, of the African yellow fever: Aedes aegypti, A. africanus und A. simpsoni

Inc: 3 - 6 days

Sym.: 3 stages:
- initial stage (viraemic stage: 3 days): sudden onset of fever up to 40 °C associated with shivers, severe headache and myalgia, conjunctivitis, nausea, vomiting, sometimes relative bradycardia!
- remission: 3. or 4. Day – temperature drops, and the patient may be cured. In case the illness takes a more severe course, there will be another rise in temperature, this is when organ damage takes place.
- stage of hepato-renal damage:
  1. hepatitis + jaundice and vomiting
  2. nephritis + proteinuria
  3. haemorrhagic diathesis, mucosal haemorrhage, epistaxis, gastrointestinal haemorrhage (coffee ground vomit)

Lab.: - mild leukopenia, thrombocytopenia, lympho-/monocytosis
- transaminases, bilirubin ↑, INR ↑, proteinuria
- detection of virus-RNA in the blood (PCR) is the first choice method
- IgM-antibody detection is possible after a few days.

Co.: liver-/renal failure, meningoencephalitis, multi organ failure, coma

course: broad spectrum varying from mild (flu-like) symptoms to severe courses associated with a high mortality.

DD: hepatitis, malaria, rickettsiosis, Weil's disease, Dengue-fever, viral haemorrhagic fever (VHF): see Dengue-fever

Di.:
- symptoms (fever, jaundice, haemorrhagia) + tropical journeys of non-vaccinated patients + detection of virus-RNA
- autopsy: typical hepatic histology: non-continuing coagulation necrosis of the intermediate zone of the lobules, and Torres' inoculations in the cell nuclei + virus-detection (PCR)

Th.: strict isolation (quarantine) in mosquito-proof rooms (in cases of suspected disease, proven disease and non-vaccinated contacts for the length of incubation = 6 days)
· symptomatically: intensive care
· trial of antiviral therapy (e.g. Ribavirin)

Prg.: up to the age of 14 years the mortality is low (flu-like illness).
mortality is high in adults: up to 85 %.
once the disease has been survived, there is lifelong immunity.

Pro.: protect yourself from mosquito bites; vaccination with the attenuated 17 D-life vaccine; this is only available with WHO approved vaccination centres (the vaccine needs to be kept refrigerated). Since 1996 there have been 13 cases with severe complications (6 of them deaths).
Ind: compulsory in many countries (see WHO-information)
CI + SE must be considered (see vaccination table). Consider exemption of vaccination in certain cases, in case of an underlying contra-indication. The target countries are NOT obliged to accept these "exemption certificates"!
Dos: adults 0,5 ml s.c.

Note: protection starts 10 days post vaccination. Vaccination must not take place less than 10 days before travelling! Protection ceases after 10 years.

Do not vaccinate in the same session with other vaccinations (see vaccination table!).

DENGUE FEVER   Notifiable disease !   [A90]

PPh.: Dengue-virus=ARBO-virus of the genus Flavivirus; 4 serotypes: DENV 1-4 (no cross immunity); reservoir: humans and monkeys

Ep.: Worldwide distribution (30 °N - 40 °S), ca. 50 million cases per year worldwide. Dengue-haemorrhagic fever is a common cause of infant mortality in Asia. Ca. 150 notified cases/year in Germany – most frequently obtained during travel to Thailand. There is a high number of unreported cases (up to 10 x higher than the number of reported cases).
It is the most common tropical virus infection imported by tourists.
There is a significant rise of cases in Asia (particularly Thailand) and South America. The disease can also be found in the south of the USA, in the Caribbean (e.g. Cuba).

Inf.: It is a mosquito-borne infection (Aedes aegypti and Aedes albopictus): Virus transmission from infected humans (urban DF) or monkeys (rural-sylvatic DF); the viruses replicate in the female mosquito after ingestion of blood, and they are transmitted after another bite. Mosquitoes bite at daytime and at dawn/dusk!

Inc: 2 - 10 days

Sym.: The majority (> 90 %) of infections remains asymptomatic or like a flu-like infection.
< 10 % feverish illness :

1. stage:
- sudden onset of high fever (fever drop after 1 - 2 days) and severe arthralgia and myalgia of the spine, arms and legs ("break bone fever"); occ. relative bradycardia!
- chills
- headache (particularly retro orbital)
- often a bitter or metallic taste
2. stage: (after 4 - 5 days)
- another rise in fever
- measles-like rash (gross, confluent)
- lymphadenopathy
3. stage:
these symptoms cease after 5 - 6 days; recovery may take several weeks.

Co.: in endemic areas children often suffer severe illness, caused by sequential secondary infections of various Dengue-serotypes.
2 % develop a thrombocytopenic haemorrhagia = Dengue-haemorrhagic fever (DHF); 4 levels (level 4 = Dengue-haemorrhagic shock = DHS or Dengue shock syndrome = DSS).
Meningoencephalitis, temporary visual impairment (retinal haemorrhage).

DD: - malaria, abdominal typhus, Hantavirus-infections
- other tropical virus infections associated with fever and arthralgia (e.g. Chikungunya in Southeast Asia, Ross-River-Virus in Australia, West-Nile-Virus in Israel/Egypt)
- Viral haemorrhagic fevers (VHF): yellow fever, VHF in Africa (Ebola-, Marburg-, Lassa-fever, Crim - Congo, Rift - Valley - fever) and VHF in South America (Argentina-, Bolivia-, Venezuela-fever) are often fatal: even suspected cases are notifiable!

Di.: • history + clinical features: tropical journey, patient suffering from sudden biphasic fever, headache, arthralgia (sometimes the onset of the illness does not happen until the patient has returned home from his journey- despite the short incubation period!)
• Lab.:
- mild leucopenia; relative lymphocytosis, thrombocytopenia, occ. transaminases ↑
- detection of the virus, virus antigen or virus-RNA in the blood (within the first 24 - 48 hrs of the illness)
- detection of antibodies: IgM are a sign for an early infection (3 - 6 days; if the test is done too early, it may be negative!); they may be absent in the case of a previous infection with Dengue-fever. If the patient had been vaccinated against yellow fever or tick-borne encephalitis, there may be cross-reacting antibodies against Dengue!

Th.: symptomatic, fluid replacement, occ. intensive care

Note: Avoid Acetylsalicylic acid in a haemorrhagic patient (alternative: Paracetamol). In cases of DHS: Glucocorticosteroids and thrombocyte-transfusions

Prg.: Adults who have been infected for the first time have a good prognosis.
Infections only leave a short immunity of a few months, but no cross immunity against the 3 other Dengue-viruses. Second infections are usually more severe than the first ones: DHF, particularly in young children. 20% of untreated patients suffering from DHF/DSS will die. If the patient receives the best treatment, the death rate is 1 %.

Pro.: prophylaxis = protection from mosquito bites → apply mosquito repellent all day long (even at daytime).Sleep in air conditioned rooms (see also prophylaxis against malaria).

MALARIA [B54] Notifiable disease

Internet: *www.who.int/topics/malaria/en*

Ep.: Worldwide it is the second most common infectious disease (after tuberculosis); prevalence is up to 500 million, 90% in Africa; ca. 2 million children die every year from Malaria in Africa. In Germany there are 500 - 1.000 imported cases per year, 70 % of that is tropical Malaria (ca. 80 % of that from Africa, mainly from Ghana, Kenya, Nigeria, Gambia, Cameroon). 40% of the world's population in tropical/subtropical areas is exposed to Malaria!

PPh.: 4 species of Plasmodia
Transmission of malaria parasites:
- bite by female Anopheles mosquito in endemic areas
- rarely „introduced" (imported) malaria: „aircraft-, airport-, baggage"-malaria and malaria transmitted via blood

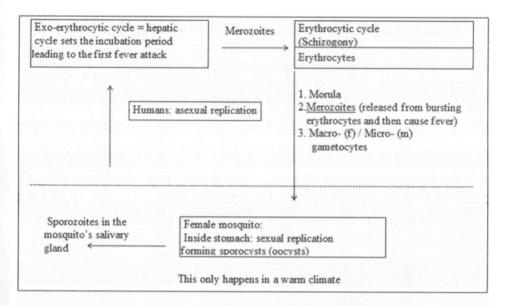

	Merozoites	
Exo-erythrocytic cycle = hepatic cycle sets the incubation period leading to the first fever attack	⟶	Erythrocytic cycle (Schizogony)
		Erythrocytes

Humans: asexual replication

1. Morula
2. Merozoites (released from bursting erythrocytes and then cause fever)
3. Macro- (f) / Micro- (m) gametocytes

Sporozoites in the mosquito's salivary gland ←

Female mosquito:
Inside stomach: sexual replication forming sporocysts (oocysts)

This only happens in a warm climate

Kulex-mosquito:
no Malaria transmitter

Anopheles-mosquito:
Malaria transmitter

Malaria type / pathogen	Incubation period (may be longer!)	pyrexial rhythm	Recrudescence = relapse
A) **Benign form:** (1/3 of cases)			
Malaria quartana: Pl. malariae	21 - 42 days	No fever for 2 days	No spontaneous cure
Malaria tertiana: Pl. vivax and ovale	10 – 21 days	No fever for 1 day **)	Spontaneous cure after max. 5 years
B) **Malignant form:** (2/3) Can be fatal after only a few days Malaria tropica: Pl. Falciparum	7 - 20 days (90 %) *) (10 % longer)	Irregular pyrexial rhythm (caused by absent synchronisation of the parasite replication)	up to 2 years(if not fatal beforehand)

*) a pyrexial illness with onset < 7 days in a malaria endemic area is unlikely to be malaria.

**) in case of M. Tertiana: daily paroxysms occur when parasite broods are undergoing schizogony on successive days (2 generations of parasites are replicating postponed by 24 hrs) (= Tertiana duplicata showing a quotidiana rhythm).
In benign forms of malaria there is synchronisation of the parasite replication causing regular recurrent paroxysms (fever attacks); the interval is determined by the period of schizogony. There are usually some days of an irregular rhythm before the typical paroxysm is established! Malignant malaria tropica always shows an irregular paroxysm!

SYMPTOMS and courses of disease	MISDIAGNOSIS (DD)
- fever, occ. chills M. tropica: sometimes just subfebrile temperatures! - headache/pain in arms and legs, occ. cough	flu-like infection, sepsis
- pain in the right upper abdomen - hepato-/splenomegaly - biliary malaria: jaundice	hepatic/ gall bladder diseases
gastrointestinal malaria: nausea, vomiting, diarrhoea	Gastroenteritis

- haemolytic anaemia (LDH↑,haptoglobin↓),haemolytic crisis
- thrombocytopenia (important for diagnose),occ. leucocytopenia
  Note: the more severe the illness, the more pronounced is the
  thrombocytopenia.
- hypoglycaemia
	haematological diseases

Co.:	M. tropica: cytoadherence of infected parasite erythrocytes can lead to impaired <u>microcirculation and</u> ischaemia of important organs: • cerebral malaria: drowsiness, confusion, coma	psychosis, meningitis
	• heart/lung: pulmonary oedema, shock	cardiac disease pneumonia
	• kidneys: Acute renal failure	renal diseases

Di.: 1. **Note:** Always consider malaria in a pyrexial patient during or up to 2 years after a tropical journey, and organise immediate diagnostics! The diagnose must be confirmed the same day! Even a proper malaria prophylaxis does not always prevent an infection! 90 % of all imported cases of malaria will occur within the first month after return from the journey.

2. Microscopic parasite detection:

Blood films and thick smears have to be obtained twice a day on 2 subsequent days, and they have to be assessed under the microscope immediately. If no parasites can be found straight away, the smear/film should be assessed for 30 minutes. A single negative blood film never excludes malaria! The distinction of the different types of parasites is more accurate in the blood film rather than in the thick smear, which is just an accumulation. The thick smear enables a concentration of otherwise just scanty parasites: 1 drop of capillary blood is applied onto a slide using the corner or a second slide, and it is then stirred and spread for 30 seconds until it has a diameter of 1 cm. The film should be thin enough, so that the letters of a newspaper still can be read through the smear. Let dry for 30 minutes, and then apply Giemsa-stain, let dry again and then examine under the microscope (commercial quick dyes may be used, e.g. Diff-Quick/manufacturer: Baxter).

Note: The fluorescence-microscopic QBC-method (<u>q</u>uantitative <u>b</u>uffy-<u>c</u>oat) is too expensive for small laboratory units.

Plasmodium vivax	young parasites		Schüffner' dots in erythrocytes (magnified)
Plasmodium malariae			Sometimes a dark stripe inside the erythrocytes
Plasmodium Falciparum	inside the erythrocytes		semilunar macrogametocyte

M. tertiana and quartana: the amount of infected erythrocytes (parasitaemia) is limited to 2%; M. Tropica: in the worst case all erythrocytes can be infected with Pl. falciparum.

Criteria for complicated malaria tropica in adults (one criterion is sufficient to diagnose complicated malaria):
• hyperparasitaemia (> 5 % of erythrocytes are infected with Plasmodia or > 100.000 Plasmodia/µl)
• lethargy, coma, epileptic fits
• severe anaemia (Hb < 5 g/dl)
• acute renal failure (output < 400 ml/24 hrs and/or
  creatinine > 265 µmol/l
• pulmonary oedema or ARDS
• hypoglycaemia (< 2mmol/l)
• shock
• spontaneous haemorrhagia
• acidosis (pH < 7,25; bicarbonate < 15 mmol/l)
• haemoglobinuria
• jaundice, bilirubin > 50 µmol/l

3. Molecular biological diagnostics:
   - detection of Pl. falciparum-histidin-rich proteine-2 (PfHRP-2) for the diagnose of M. tropica. Optimising of the tests lead to a reduced number of false negative results. This test is not suitable for a quick diagnose in the average tourist. It does not replace microscopy. This test can be positive for up to 24 hours after the disappearance of parasites from the blood (hence it is not suitable as a therapy follow up). Non-professionals should not perform this test, as it is often done incorrectly!
   - detection of Plasmodium-DNA via PCR (no routine test, result takes 24 hrs; expensive!)

4. detection of plasmodium antibodies: (IFAT = indirect immune fluorescence antigen test)
   It is not suitable for the diagnose of an acute case, since it is not positive until 6 - 10 days after the onset of symptoms (2 - 4 weeks later the titre has reached its peak). It can be used to proof an asymptomatic infection retrospectively or to measure the spread of the infection in any given population.

Th.: Even the slightest suspicion (pyrexial patient returning from a tropical country) justifies immediate admission to hospital, diagnosis and therapy. If the unit has insufficient experience with the treatment of malaria, telephone advice can be obtained in experienced centres for tropical medicine! Therapeutic efficiency must be checked by repeated investigations for parasitaemia!

A) Benign forms of malaria caused by Pl. vivax or Pl. ovale (M. tertiana) and Pl. malariae (M. quartana):
   Chloroquine
   Chloroquine has no effect on the exo-erythrocytic hepatic schizonts (= reservoir). In order to avoid relapses in M. tertiana (Pl. vivax, Pl. ovale), any course of Chloroquine has to be followed by Primaquine, which acts on hepatic schizonts and gametes. A G-6-PD-deficiency must be excluded beforehand (otherwise patients suffering from this disorder will suffer a massive haemolysis).

   Note: There are occasional cases of Chloroquine-resistance in Pl. vivax-infections in Southeast Asia/Pacific Ocean.

B) Malignant malaria caused by Pl. falciparum (M. tropica):
   Treating M. tropica is difficult due to increasing resistance. Hence therapeutic advice from a tropical unit should always be obtained!

Level of resistance	Parasite concentration and course of illness
S (= sensitivity)	Complete eradication of parasites from the, no relapse of parasitaemia - sym.: cure
R I-resistance	Temporary eradication of parasites, but followed by a relapse of parasitaemia after weeks or months. - sym.: initial cure, but then relapse of malaria
R II-resistance	Significant reduction of parasites, but no eradication. - sym.: ongoing malaria after a certain period of improvement
R III-resistance	No reduction of parasites - sym.: no clinical improvement

The WHO divided the malaria areas of the world into 3 zones (considering risk and resistance):
Zone A: areas without Chloroquine resistance or without prevalence of Pl. falciparum
Zone B: areas of Chloroquine resistance
Zone C: areas of high Chloroquine resistance or multiple resistance

I. **Therapy of uncomplicated M. tropica:**
   3 alternatives:
   Atovaquon + Proguanil (e.g. Malarone®) or Mefloquine (Lariam®) - Mefloquine resistance in SO-Asia), Artemether + Lumefantrine (Riamet®)

II. **Therapy of complicated M. tropica:**
   - Quinine: first choice
     initial i.v.-infusion of Quinine dihydrochloride, change to oral Quinine as soon as possible.
     Therapy for 10 days; daily control of parasitaemia
   - due to possible resistance Quinine must be combined with Doxycycline. Doxycycline is contraindicated in severe hepatic impairment (→ seek advice from a unit for tropical medicine).
   - supportive therapy:
     Input/output balance; correct electrolytes (be aware of pulmonary oedema caused by over-infusion → CVP-control!), control of renal function and blood sugar (beware hypoglycaemia!)
   - lower fever: tepid sponging, if needed give Paracetamol (no Aspirin because of the thrombocytopenia !)
   - if necessary: exchange transfusion for severest cases of M. tropica with high parasitaemia (> 20 % of erythrocytes infected with Plasmodia). There are no data available from any studies.

Prg.: Death rate of M. tropica in Germany is 1 - 2 % (untreated > 20 %). Early diagnosis and appropriate therapy can prevent fatalities!

Pro.: There is no 100% prophylaxis!

Always follow the most up to date recommendations when advising a patient (institutes for travel medicine/tropical medicine; Internet).
Insufficient medical prophylaxis is the main cause for deaths!

I. Exposure prophylaxis

Prevent mosquito bites: reduces the risk of infection 10 fold! Anopheles mosquitoes usually don't fly into air conditioned rooms, and they are active between sunset and dawn.

- mosquito-proof screen for windows and doors
- for non air-conditioned rooms Pyrethrum based insect sprays are advisable
- use a mosquito net for the bedroom
- wear light coloured protecting clothes with long sleeves/legs
- use an insect repellent
- in the countryside: stay inside at dusk and at night!

II. Chemoprophylaxis:

Chemoprophylaxis does not prevent an infection, but it suppresses the clinical manifestation. Always follow the most recent recommendations!

- if a person intends to stay in a tropical area for a long time, specific advice needs to be obtained.
- Short-term stays up to 6 weeks: Mefloquine (e.g. Lariam®) or Atovaquone/Proguanil (e.g. Malarone®) alternative: Doxycycline for areas of Mefloquine-resistance (in some countries off-label use)

III. emergency self treatment (stand-by-medication):

Journeys into areas of low risk (e.g. many big cities in Asia and South America → follow the most recent recommendations!) the following is advised: Just strict exposition prophylaxis + a suitable stand-by-medication for an emergency.

Medications for emergency treatment:

- Artemether + Lumefantrine (Riamet®) or
- Atovaquone + Proguanil (Malarone®) or
- Mefloquine (e.g. Lariam®)

Still the traveller is advised to seek medical help immediately after the onset of fever. Only if this is not available, he is advised to take the stand-by-medication, and then try to obtain a diagnosis as soon as possible.

Malaria prophylaxis in pregnancy:
Pregnant women should avoid travelling into endemic areas if at all possible.

Sensitive strains of Pl. falciparum and Pl. malariae will be eradicated by chemoprophylaxis; but Pl. ovale and Pl. vivax may cause delayed fever due to their persistency outside the erythrocytes!

Note: There is active research for an active immunisation (vaccine against malaria tropica). It can't prevent malaria infection, but it may reduce the number of severe cases by 50 %. It may be useful for the local population of endemic areas (not for tourists).

Medication for prophylaxis/therapy: (adult doses)

Always follow the manufacturer's advise for children's dosage; there have been overdoses with toxic side effects in children (e.g. after Chloroquine).

- Chloroquine:

Action mode: Chloroquine is a 4-Aminochinoline, it is a schizontocide.

SE: Occ. gastrointestinal symptoms, rarely allergic rash and photosensitivity, occ. haemolytic crisis in patients suffering from G-6-PD-deficiency; parenteral treatment may cause a drop in blood pressure.

Long-term use may rarely cause neuropathy, cardiomyopathy and damage to the eyes: reversible deposits in the cornea, irreversible retinopathy: after a total dose of 100 g base (= 6,5 years of 2 tbl./week) there is a risk of irreversible retinopathy. Annual eye checks are advised for those on long-term medication.

Note: Therapeutic width of Chloroquine is fairly narrow; the recommended dose must not be exceeded in any case.

Ind: Prophylaxis and therapy only of malaria quartana and tertiana

CI: G-6-PD-deficiency, retinopathy, chronic renal failure, myasthenia gravis. Relative CI in patients suffering from psoriasis, porphyria and epilepsy.

Dos: A) Therapy:

1 tablet Chloroquine à 250 mg contains 150 mg of the therapeutic effective Chloroquine base
- initial dose: 600 mg Chloroquine base = 4 tablets
- after 6 hours: 300 mg Chloroquine base = 2 tablets
- after 24 and 48 hours: 300 mg Chloroquine base = 2 tablets

B) Prophylaxis:

Adults up to 75 kg: 2 tablets (= 300 mg base) 1 x/week
> 80 kg 3 tablets (= 450 mg base) 1 x/week
start: 1 week before entering endemic area
stop: 4 weeks after leaving the area

- Primaquinee (in some countries not available)
  action: Primaquine is a 8-Aminochinoline, it is schizontocide and gametocide.
  SE: Haemolysis in those suffering from G-6-PD-deficiency, gastrointestinal side effects, rarely granulocytopenia.
  Ind: only to be used for the final treatment of M. tertiana.
  CI: G-6-PD-deficiency (must be excluded before the therapy is commenced!), children
  (→ Methaemoglobinaemia)
  Dos: 15 mg (= 1 tbl.)/day for 2 weeks. Needs to be repeated in a case of relapse.

- Mefloquine (Lariam®) - T50: 21 days!
  SE: · Gastrointestinal SE: nausea, vomiting, diarrhoea, raised transaminases
       · central nervous/psychotic SE (ca. 10 %): dizziness, anxiety, restlessness, confusion, depression, nightmares,
         hallucinations, convulsions, increase of an existing suicidal tendency
       · cardiac SE: extrasystoles, AV-block, bradycardia
       · dermatological SE: rash, hair loss
       · haematological SE: leucopenia, thrombocytopenia
  Ind: prophylaxis, therapy of uncomplicated M. tropica, emergency self medication of M. tropica
  CI: pregnancy (1st trimester; women of childbearing age must use sufficient contraception), breastfeeding, young
       children, renal failure, hepatic failure, epilepsy, psychiatric illness, driving/diving (reduced awareness),
       cardiomyopathy, concurrent medication which might prolong the QT-interval; beware patients with cardiac
       arrhythmia → consider drug interactions!
  Dos: A. therapy and stand by-medication:
           1. Dose: 750 mg = 3 tbl
           2. Dose: 500 mg = 2 tbl (6 - 8 hours after 1. dose)
           3. Dose only if bodyweight > 60 kg: 250 mg = 1 tbl (6 - 8 hours after 2nd dose)

        B. prophylaxis:
           1 x 1 tbl (= 250 mg)/week (always on the same day). Often the application at night is better tolerated.
           start: ideally 3 weeks before entering the endemic area (to build up a sufficient blood level, and to reveal any
           unwanted side effects in time).
           stop: 4 weeks after return

- Quinine:
  action: Quinine is a Chinoline derivate extracted from the bark of the cinchona tree; its main action is schizontocide.
  SE: headache, nausea, impaired vision, deafness, tinnitus, allergic reactions, changes in the blood count, haemolysis
       in G-6-PD-deficiency, very rarely haemolysis, arrhythmia, low blood pressure, hypoglycaemia(!), haemolytic-
       uraemic-syndrome
  CI: allergy to Quinine, G-6-PD-deficiency, tinnitus, optical nerve damage, myasthenia gravis; be careful in cases of
       underlying cardiac arrhythmia
  Ind: therapy of complicated M. tropica
  Dos: for complicated malaria initially i.v., change to oral ASAP:
       Daily dose: 20 - 25 mg/kg divided into 3 doses.
       length: 7 - 10 days (exact dose see leaflet)

- Atovaquone (250 mg) + Proguanil (100 mg) (Malarone®):
  SE: nausea, vomiting, abdominal pain, diarrhoea, headache, dizziness, tinnitus etc.
  CI: allergy, pregnancy, breast feeding etc.
  Ind: A) therapy and emergency self medication of uncomplicated Malaria tropica, better tolerated than Mefloquine
       B) prophylaxis of M. tropica in areas of multiple resistance and in cases where Mefloquine caused adverse effects.
          Only stays of < 28 days and weight ≥ 40 kg.
  Dos: A) therapy: 4 tbl as a single dose on 3 subsequent days
       B) prophylaxis: 1 tbl/day . start: 1 - 2 days before entering the endemic area. Stop: 7 days after return

- Artemether + Lumefantrine (Riamet®, Coartem®)
  Ind: therapy of uncomplicated M. tropica, emergency self medication for infections with Pl. falciparum or in cases
       where Mefloquine is contraindicated; beware SE + CI (read leaflet!).
  SE: gastrointestinal SE, headache, dizziness, neurological impairment, rise in transaminases, prolonged QT-interval
  interactions: always be aware of interactions with other medication!
  CI: e.g. if the following are given at the same time: Macrolides, tricyclic antidepressants, Azole-antifungals etc.
  Dos: 6 doses of 4 tbl at a time at 0, 8, 24, 36, 48, 60 hours

- Doxycycline:
  Ind: 1. therapy of complicated M. tropica combined with Quinine
       2. chemo-prophylaxis of M. tropica in areas of Mefloquine resistance (e.g. Cambodia, Vietnam, Myanmar [Burma]).
          In some countries it is used off-label.
  SE: beware SE (particularly photosensitivity) and CI (e.g. pregnancy, breastfeeding, children < 8 years). (further
       details: see chapter 'pneumonia')
  Dos: 100 mg/day

BILHARZIA- SCHISTOSOMIASIS [B65.9]

PPh.: Schistosoma (Bilharzia): 1-2 cm long trematode worms

Urogenital bilharzias: S. haematobium
Intestinal/hepatic schistosomiasis: S. mansoni, S. japonicum, S. intercalatum, S. mekongi

Lifecycle: infective larvae (cercariae) are released from freshwater snails (= intermediate host), penetrate intact human skin inside the water and may cause cercaria dermatitis.
Schistosomulum migrates along lymphatic vessels and the right heart to the lungs → sometimes this causes Katayama-syndrome (fever, cough, eosinophilia etc.) → cercariae arrive in the liver and develop into worms → mating (the female is enclosed inside a furrow (gynaecophoric canal) of the male's body: Schistosoma = „split body".)
Schistosoma haematobium settles in the bladder's blood vessels → eggs are excreted with the urine (S. haematobium eggs are equipped with a terminal spine; S. mansoni eggs have got a lateral spine).
Schistosoma mansoni travel into the mesenteric blood vessels → eggs are secreted in stool.
Eggs deposited in fresh water hatch, releasing the miracidium → ingestion by snails → development of cercariae.

Ep.: tropical/subtropical areas, worldwide ca. 400 million infected people, ca. 90 % are found in Africa, south of the Sahara.
Germany: ca. 50 - 100 notified imported cases/year (50 % are asylum seekers from Africa). Travellers may become infected during adventurous journeys and long term stays in endemic areas, when they ignore the warning not to swim in fresh water.

Inf.: Percutaneous penetration during a swim in fresh water

Inc: 2 - 7 weeks up to general malaise, 4 - 12 weeks to organ stage

Sym.: General: occ. Cercarial dermatitis, Katayama-syndrome, fever, eosinophilia
Organ manifestation: bladder – haematuria
intestinal – abdominal pain, colitis, melaena

Co.: • increased risk of bladder carcinoma, colon carcinoma and hepatic carcinoma
• portal hypertension (eggs obstructing the portal vein)
• focal epilepsy seen in infections with S. japonicum

Di.: tropical journey + detection of eggs (urine, stool, tissue), detection of antibodies

Th.: Praziquantel (Biltricide®): cure > 80 %
Therapy control: therapy is successful when no further eggs are being excreted in urine or stool

Pro.: avoid swimming or bathing in contaminated waters. Don't drink contaminated water; eradication of snails, don't discard urine/faeces into any waters
Note: in central Europe and North America there are cercariae which are non-pathogen for humans (e.g. duck cercariae); they can cause harmless (swimmers itch) after swimming in fresh waters.

IX. ANNEX – INFECTIOUS DISEASES

DIFFERENTIAL DIAGNOSIS „FEVER" [R50.0]

Def.: Physiological temperature curve shows a diurnal rhythm, with the valley during the second half of the night, and the peak in the afternoon: axillary up to 37,0°C - oral / sublingual and forehead up to 37,2°C - rectal and auricular up to 37,6 °C. Frontal and oral readings are the same, but vasoconstriction can lead to false low readings. Temperature rises by ca. 0,5°C post ovulation, this also happens after physical exertion. Clothing and outside temperature also have an influence on the result. Children show much higher, faster and more frequent fluctuations in temperature. Normal range also varies between individuals. Once physiological limits have been passed, it is called fever.

Types of fever and common causes (readings auricular or rectal):

1. <u>Subfebrile</u> (< 38,5 °C rectal)
   - infections, pyelonephritis, endocarditis lenta, tuberculosis
   - hyperthyroidism (fever can be can be very high in thyreotoxic crisis !)
   - fever caused by medication ("drug fever")
   - Hodgkin-/non Hodgkin-lymphoma, tumours of the GI tract

   > **Note:** Any sign of inflammation can be caused by a tumour!

   - hypernephroma, relapsing pulmonary embolism, collagenosis, vasculitis
   - <u>postoperative fever</u>

2. <u>Febrile</u> (> 38,5 °C):
	a) <u>Continua</u> (diurnal fluctuations up to 1 °C)
	b) <u>Relapsing</u> (diurnal fluctuations 1 - 2 °C)
	c) <u>Intermittent</u> (high diurnal fluctuations > 2°C)
	d) <u>Septic fever:</u> Intermittent high phases of fever with or without shivers
Continua:	bacterial infection (this is rarely seen today because of early start of antibiotic therapy).
Two peaks:	a) complication after bacterial infection
	b) virus infection
Oscillating:	brucellosis, Hodgkin disease
Malaria:	(Subicterus, hepatic pain, trips to tropical areas!)
	Malaria quartana: **2** days no fever!
	Malaria tertiana: **1** days no fever!
	Malaria tropica (or mixed infection): irregular fever

fever and white blood count:
a) leucocytosis: bacterial infection
b) leucopenia: virus infection, typhoid (!), brucellosis, raised peripheral consumption of granulocytes; therapy with cytostatics or immune suppressants.

Diagnostic tips:

- obtain a temperature reading, if in doubt: measure temperature in 3 different ways: axillary < buccal < rectal; buccal reading lies between the other two (patients usually don't know this) → this excludes any malingerers with fake fever.

- history:
  - journeys? (malaria, typhoid, amoebiasis, Dengue fever, tropical virus infections etc.)
  - any contact to (sick) animals? contact to infectious patients?
  - what medication was taken <u>before</u> the onset of fever? ("drug fever")

- culture/serology, if history and symptoms give a hint to the origin and type of infection (urine culture, sputum culture etc.)

- arrange repeated blood cultures (at least 2 - 3 samples each on two subsequent days) in fever of unknown origin

- "thick film" in fever after tropical holidays

- basal TSH, CRP/ESR, rheumatic serology (RF, ANA etc.)

  Note: untargeted serological and immunological random tests are rarely useful.

- organ screening (sonography of abdomen, heart and thyroid gland, CXR, transoesophageal echocardiography (endocarditis?), gastroenterological diagnostic, TFT, gynaecological/urological examination)

(fever of unknown origin = FUO):

1. <u>FUO in neutropenic patients</u> (neutrophile granulocytes < 1.000/µl) <u>during/after cytostatic therapy</u> is seen in ca. 75 % of all patients. In 50 % the origin remains unknown. An infection has to be assumed unless the opposite has been proven. In most cases the pathogens are Staphylococci, Streptococci or gram-negative bacteria (Pseudomonas aeruginosa, E. coli, Klebsiella); fungi. After the initial basic diagnostics (aerobic/anaerobic blood cultures, urine culture, CXR, abdominal ultrasound, oral and ano-genital inspection, removal of intravenous catheter etc.) treatment with broad spectrum antibiotic should be initiated. The sooner the onset of therapy, the better the result!

Recommended therapy:
e.g. Piperacillin/Tazobactam Aminoglycoside or third generation-Cephalosporin Aminoglycoside. If there is no improvement after 72 hours: change to antibacterial antimykotic therapy, e.g. Carbapenem (Imipenem, Meropenem) Glycopeptide (Teicoplanin, Vancomycin) Amphotericin B. Pulmonary infection requires Amphotericin B in the first trial of therapy, since in ca. 30 % there will be a fungal infection.

2. FUO without neutropenia: abscess, endocarditis, tuberculosis, HIV-infection and opportunistic infections

3. Nosocomial FUO: fever > 38 °C in inpatients, who showed no sign of infection on admission.
   Cause: Infected intravasal catheter, UTI, pneumonia, DVT/pulmonary embolism

   3 common causes of postoperative fever:
   - wound infections
   - nosocomial pneumonia (ICU !) and UTI (urinary catheter !)
   - DVT and thrombo-embolism

4. FUO caused by malignancies, collagenosis, medication (drug fever) etc.

In up to 25 % the cause of FUO cannot be found.

Tips - patients without neutropenia and without threatening symptoms:
Observe patient for 2 - 3 days; check temperature (fever can be faked in „Munchhausen"-syndrome) and detailed diagnostics.

- provide sufficient fluid intake !
  insensible loss (= invisible fluid loss via skin/lungs):
  - normal temperature ca. 1,0 ℓ/24 hours
  - rule of thumb: each 1°C > 37°C requires additional 0,5 - 1,0 ℓ water
  Consider these minimal volumes when balancing fluid, and add to the visible excretions of renal tract, GI tract and wounds/fistula/tubes.

- stop all non life-saving medication (drug fever ?).

- if symptomatic lowering of fever is necessary, then it should be done slowly and regularly, to avoid circulatory stress (tepid sponging or Paracetamol).

- no therapeutic means, which could mask the diagnose, e.g.
  - no antibiotics before bacteriological investigation
  - no contrast medium before thyroid diagnostics, if there is the slightest suspicion for hyperthyroidism.
  - biopsy to diagnose collagenosis/vasculitis is pointless after corticosteroid-therapy for several weeks.

- negative serological investigations for antibodies in early disease and in immune deficiency/suppression do not exclude the disease!

- antibiotics given for unknown infection have to cover a broad spectrum. This can be changed later after culture/sensitivity.

HEREDITARY FEVER SYNDROMES
(Autoinflammatory Disorders, Hereditary periodic fever syndromes)

<u>Def.:</u> Hereditary diseases associated with periodic fever flare ups. These diseases are fairly rare, but one should be aware of them in view of differential diagnose. The mutations cause a disturbed cytokine balance. Leading symptoms are: relapsing fever spells and facultative serositis, pleuritis, pericarditis, peritonitis, arthralgia and rash; occ.hepatosplenomegaly.

Name	familial mediterranean fever (FMF) (most common)	hyper-IgD-syndrome (HIDS)	TNF-receptor-associated periodical fever (TRAPS)	Muckle-Wells-syndrome (MWS)	familial cold urticaria (FCU)	cyclic neutropenia (CN)
manifestation	< 10	< 1	< 20	variable	< 1	< 5
period in days	1 – 3	3 – 7	> 7	days	days-weeks	4 - 5
interval	weeks-months	4 - 8 weeks	months	variable	Exposition	20 days
symptoms	polyserositis peritonitis pleuritis	lymphadenopathy conjunctivitis abdominal pain	oedema myalgia abdominal pain	deafness arthralgia	cold intolerance conjunctivitis	aphthous stomatitis
arthritis	monoarthritis	polyarthritis	rare	Synovitis	painful arthritis	none
skin	erysipeloid-like (mainly on legs)	maculopapulous plaques	painful erythematous plaques	urticaria	Urticaria	cutanous infections
complications	amyloidosis	not known	amyloidosis	amyloidosis deafness	Amyloidosis	septicaemia
laboratory	Ca5a-inhibititor MVK	IgD / (IgA) (↑)	type1-TNF ↓	none	None	neutropenia
hereditary path	autosomal-recessive	autosomal-recessive	autosomal-dominant	autosomal-dominant	autosomal-dominant	autosomal-dominant
gene location	16p13	12q24	12p13	1q44	1q44	19p13.3
protein	Pyrin / Marenostrin	Mevalonatkinase	type1-TNFR	Cryopyrin	Cryopyrin	--
therapy	Colchicin	None	Cortisone Etanercept	Kineret	Kineret	G-CSF

<u>Other rare hereditary fever syndromes:</u>
PFAPA (periodic fever, aphthous stomatitis, pharyngitis, adenitis)
CINCA (chronic infantile neurological and arthritis syndrome)

Notifiable infectious diseases §§ 6/7 (Germany)

The aim of the act for protection from infectious diseases is to prevent transmission of infectious diseases in humans. An important part of it is the duty to notify any cases. It enables the health authorities to obtain immediate information about the outbreak of infections. We differentiate between:

- Notification by name (infectious disease) - immediately, within 24 hours to the health authority. Duty to report is given in the case of suspicion already! (§ 6)
- Notification by name (detection of pathogens) - immediately, within 24 hours to the health authority. Duty to report is given when pathogens have been detected, indicating an acute infection. (§ 7)
- Anonymous notification, when pathogens have been detected; report encoded cases to the Robert-Koch-institute within 2 weeks. (§ 7)

A full notification consists up to 15 details, hence pre-designed forms are being used. Apart from the written notification, it is advisable to inform the heath authorities immediately in urgent cases, in order to accelerate any preventative epidemic means (e.g. to find any contact persons for early antibiotic prophylaxis).

Failure to do so, or a delay, is considered an infringement, or if causing more cases of disease – even a crime. Hence one must not rely on any oral notification. Only omit your own report, when there is written proof of notification by another doctor.

Staff entitled to report according to § 6:
- diagnosing doctor (hospital: additionally the consultant/ clinic director)
- leading personnel of pathological-anatomical institutes
- qualified nursing staff; directors of nursing/ residential homes, camps, prisons and other institutions, non-medical practitioners, pilots, ship captains, vets (rabies, if animal was in contact with humans!).

Staff entitled to report according to § 7:
- Leading personnel of laboratories and pathological-anatomical institutes
- Doctors organising laboratory investigations.

One must not rely on other people reporting the case. You are only allowed to refrain from reporting if you have proof that someone else already has done it!

Important

- Hepatitis: Notifiable by name: suspicion, disease and death of an acute virus hepatitis. Notification of any proven acute hepatitis A - E-infection and any detection of hepatitis C-Virus, if a chronic infection is not known.
- Tuberculosis: Notifiable by name: disease and death of a tuberculosis (even if there was no bacteriological detection!), and if the patient refuses to undergo treatment for pulmonary tuberculosis. Any detection of acid proof rods and the result of sensitivity has to be reported.
- Rabies: Suspicion, disease and death; also any contact to an animal suspicious of rabies (injury or touch) has to be reported by name.
- Dangerous diseases/ pathogens, not mentioned in the act:
  Notification by name, if a high risk for the public is suspected, and any pathogens are suspected.
- Accumulation of nosocomial infections: Anonymous notification, if an epidemic is likely or suspected.
- In some countries stipulations vary, and there may be more notifiable diseases (e.g. borreliosis, pertussis).
- Post-vaccination injury: Notifiable by name, as long as it is not an ordinary side effect due to vaccination.

Disease	Notification disease § 6	Notification pathogen § 7
Adenovirus-conjunctivitis		+
Botulism	+	+
Brucellosis		+
Cholera	+	+
Diphtheria	+	+
Echinococcosis		(+)
Infectious enteritis caused by		
- Adenovirus	✳	
- Astrovirus	✳	
- Campylobacter	✳	+
- Coronavirus	✳	
- Cryptosporidium parvum	✳	+
- Entamoeba histolytica	✳	
- Escherichia coli (all pathogenic forms)	✳	+
- Giardia lamblia	✳	+
- Norovirus	✳	+
- Rotavirus	✳	+
- Salmonella	✳	+
- Yersinia enterocolitica	✳	+
- other forms incl. microbiotic food poisoning	✳	
Enteropathic haemolytic uraemic syndrome (HUS)	+	+
Typhus fever		+
viral encephalitis		+
Haemophilus influenzae (direct detection in blood or CSF)		+
HIV		(+)
Human spongiform encephalopathy	+	+
Influenza (avian flu)		+
Beware: Avian Influenza	+	
Louse fever		+
Legionellosis		+
Lepra		+
Leptospirosis		+
Listeriosis (detection in CSF, blood and in other usually sterile substrates and swabs in newborns)		+
Syphilis		(+)
Malaria		(+)
Measles	+	+
Meningococcal-meningitis or –sepsis	+	+
Anthrax	+	+
Ornithosis		+
Paratyphus A, B and C	+	+
Plague	+	+
Poliomyelitis (any non-spastic paralysis, unless posttraumatic)	+	+
Psittacosis		+
Q-fever		+
Rubella (only congenital)		(+)
Shigella dysentery	✳	+
Rabies	+	+
Toxoplasmosis (only congenital)		(+)
Trichinosis		+
Tuberculosis	+	+
Tularaemia		+
Abdominal typhoid	+	+
Virus associated haemorrhagic fever (e.g. Ebola, yellow fever, Hanta, Lassa, Marburg)	+	+
Virus hepatitis A / B / C / D / E	+	+
Virus hepatitis :all other forms	+	

+ = name patient

(+) = anonymous directly to an das Robert-Koch-Institute   ✳ Report if the patient handles food (professionally) (§ 42) or if there are 2 or more similar cases and suspected epidemic

Recommended initial therapy in family practice in bacterial infections in adults
(Paul-Ehrlich-Society for chemotherapy)

Diagnosis	Most common pathogens	First choice	Alternatives
Pharyngeal and respiratory infectious diseases			
Acute bronchitis	Usually viral: no antibiotics rarely: Pneumococci Haemophilus influenzae Moraxella catarrhalis Chlamydia pneumoniae	---- Cephalosporine Gr. 2/3 Aminopenicillin BLI (Beta-Lactamase-Inhibitor) Makrolides	Fluorchinolone Gr. 3/4 Ketolide Doxycycline
Acute exacerbation of COPD = AECB		Refer to chapter COPD	
Pneumonia (community acquired) = CAP		Refer to chapter pneumonia	
Tonsillitis, pharyngitis Erysipelia	A-Streptococci	Phenoxymethylpenicillin Cephalosporine Gr. 2/3	Macrolides
Lyme-Borreliosis	Borrelia burgdorferi	Amoxicillin Cephalosporine Gr. 2/3 Doxycycline	Macrolides
GI-infections			
Acute enteritis	Salmonella Campylobacter jejuni Yersinia Shigella	Ciprofloxacin	Aminopenicillins TMP/Sulfonamide Macrolides only for Campylobacter
Note: Infections caused by Salmonella, Campylobacter or Yersinia should only be treated in special cases.			
Duodenal / gastric ulcer MALT-lymphoma	Helicobacter pylori	Amoxicillin Clarithromycin Proton pump inhibitors	Clarithromycin Metronidazole Proton pump inhibitors
Diverticulitis	Escherichia coli Enterococci Bacteriodes fragilis	Amoxicillin BLI Amoxicillin Metronidazole Fluorchinolone group 2/3 Metronidazole or Clindamycin	
Infections of the biliary system			
Cholangitis Cholecystitis	Escherichia coli Enterococci Klebsiella Anaerobic and aerobic Streptococci rarely Clostridium perfringens 1-3%	Ciprofloxacin Aminopenicillins BLI	
Note: Cholelithiasis requires endoscopic or surgical therapy! Endoscopic investigation requires prophylaxis using Ciprofloxacin			
UTI			
Acute uncomplicated cystitis of sexually active women	Escherichia coli 75 - 85 % Proteus mirabilis 10 - 15 % Staphylococci 5 - 15 % Other pathogens are rare	Fluorchinolones group 1 / 2 Trimethoprim (with or without Sulfonamide)	Aminopenicillins (first choice in pregnancy)
Note: For typical symptoms (acute dysuria) and leucocyturia short therapy (up to 3 days). Repeat urine after 1 to 2 weeks.			
Acute uncomplicated pyelonephritis	Escherichia coli 70 - 85 % Proteus mirabilis 10 - 18 % Rarely other pathogens	Ciprofloxacin Trimethoprim (with or without Sulfonamide)	Aminopenicillins BLI
Note: typical symptoms (loin pain, fever) and leucocyturia you may commence therapy (7 - 14 days) without microbiological tests. Atypical course or relapse requires lab investigations.			
Complicated UTI	Escherichia coli 30 - 50 % Proteus mirabilis 10 - 15 % other Enterobacteria 10 - 20 % Pseudomonas aeruginosa 5 - 10 % Enterococci 10 - 20 % Staphylococci 10 - 20 %	after tests	
Note: Therapy for at least 7 - 10 days or more. Multiple resistance of many pathogens requires culture and sensitivity; exception (e.g. fever) :start therapy after taking a urine sample for culture/sensitivity, using a broad spectrum antibiotic; consult a urologist.			
Genital infections			
Syphilis	Treponema pallidum	Benzathinpenicillin i.m.	Doxycycline if allergic to Penicillin
Gonorrhoea	Neisseria gonorrhoeae	Cephalosporin i.v. (group 2/3)	
Unspecific urethritis	Chlamydia trachomatis	Doxycycline	Macrolides
Note: Genital infections require a treatment of the sex partner !			

Antibiotic groups
(with permission of Prof. Dr. med. F. Vogel, Hofheim)
(o = oral / p = parenteral)

Group	Example	p / o	Spectrum
PENICILLINS			
Penicillins	Benzylpenicillin (= Penicillin G) Phenoxymethyl-Penicillin	p o	effective against Streptococci incl. Pneumococci
Aminopenicillins	Amoxicillin Ampicillin	o p / o	- Penicillin-spectrum - effective against Enterococci and some gram-negative non- Betalactamase producing pathogens - not effective against Staphylococci and anaerobic Betalactamase producing pathogens
Aminopenicillins / Betalactamase-inhibitors	Amoxillin / Clavulanic acid Ampicillin / Sulbactam	p / o p / o	- Penicillin-spectrum - effective against Enterococci and some gram-negative Betalactamase-producing pathogens
Acylaminopenicillins	Azlocillin Mezlocillin Piperacillin	p p p	- effective against gram-positive pathogens incl. Enterococci - not effective against Betalactamase producing Staphylococci - effective against gram-negative non-Betalactamase producing pathogens - varying efficiency against Pseudomonas
Acylaminopenicillins / Betalactamase-inhibitors	Piperacillin / Tazobactam Piperacillin / Sulbactam	p p	- effective against gram-positive pathogens incl. Enterococci - effective against some gram-negative Betalactamase-producing pathogens - effective against Pseudomonas
Isoxazolylpenicillins	Dicloxacillin Flucloxacillin Oxacillin	o p / o p / o	effective against gram-positive Betalactamase-producing pathogens (Staphylococci-Penicillins)
CEPHALOSPORINS			
group 1	Cefazolin Cefalexin Cefadroxil Cefaclor	p o o o	- effective against gram-positive and some gram-negative bacteria - stable against Penicillase from Staphylococci - unstable against Betalactamase from gram-negative bacteria
group 2	Cefuroxim Cefotiam Cefamandol Loracarbef	p / o p p o	- highly effective against gram-positive and gram-negative bacteria - stable against Penicillase from Staphylococci and most types of Betalactamase from gram-negative bacteria
group 3a	Cefotaxim Ceftriaxon Ceftibuten Cefixim	p / o p o o	- significantly more effective than group 1 and 2 against gram-negative bacteria - stable against many types of Betalactamase of gram-negative bacteria - less effective against some gram-positive bacteria - not effective against Enterococci, slightly effective against Staphylococci
group 3b	Ceftazidim Cefepim Cefpirom	p p p	Same spectrum as Cephalosporins group 3a, additionally effective against Pseudomonas

Group	example	p / o	Spectrum
CARBAPENEMES			
	Imipenem / Cilastatin	p	broad spectrum (gram-positive and gram-
	Meropenem	p	negative pathogens incl.
	Ertapenem	p	anaerobics
GLYCOPEPTIDES			
	Vancomycin	p	- effective against Streptococci incl.
	Teicoplanin	p	Enterococci
			- effective against Staphylococci incl. MRSA
FLUORCHINOLONES			
group 1	Norfloxacin	o	- indication: mainly UTI
			- effective against gram-negative bacteria
group 2	Ofloxacin	p / o	- partly systemic application, many indications
	Ciprofloxacin	p / o	- more effective than group 1 against gram-
	Fleroxacin		negative pathogens, partly effective against Pseudomonas
			- limited efficiency against Pneumococci, Staphylococci and „atypical" Pneumonia (Chlamydia, Mycoplasma, Legionella)
group 3	Levofloxacin	p / o	- very effective against gram-negative and gram-
	Sparfloxacin	p	positive pathogens incl. Pneumococci, Staphylococci, Streptococci
			- very effective against „atypical" Pneumonia (Chlamydia, Mycoplasma, Legionella)
group 4	Clinafloxacin	p / o	similar antibacterial spectrum as group 3, but
	Gatifloxacin	o	improved effectiveness against anaerobics
	Moxifloxacin	o	
MACROLIDES			
older Macrolides	Erythromycin	o	- effective against „atypical" Pneumonia (Chlamydia, Mycoplasma, Legionella)
			- effective against Streptococci incl. Pneumococci
			- no sufficient efficiency against Haemophilus influenzae
newer Macrolides	Azithromycin	o	spectrum is similar to the one of older
	Clarithromycin	o	Macrolides, but improved efficiency against
	Roxithromycin	o	Haemophilus influenzae
	Telithromycin		-Telithromycin is also efficient against Erythromycin-resistant Pneumococci
AMINOGLYCOSIDES			
	Amikacin	p	- effective against Enterobacteria
	Gentamicin	p	- effective against Pseudomonas (particularly
	Tobramycin	p	Tobramycin)
TETRACYCLINES			
	Doxycycline	p / o	- effective against „atypical" Pneumonia (Chlamydia, Mycoplasma, Legionella), increasing resistance in Pneumococci
TRIMETHOPRIM			
Trimethoprim with or without Sulfonamide	Trimethoprim is as effective as Co-trimoxazol (the latter may cause side effects due to the Sulfonamide)	o	- effective against various gram-positive and gram-negative bacteria - effective in purulent bronchitis, traveller's diarrhoea, Pneumocystis jiroveci. Resistant in some UTIs

IMPORTANT VACCINATIONS FOR ADULTS note most recent STIKO-
recommendations

(**Note:** These are recommended immunisations in **Germany;** different immunisation programs may apply to the UK and other countries)

	TETANUS (S)	DIPHTHERIA (S)	POLIOMYELITIS (S)	INFLUENZA (S / I)	OTHER VACCINATIONS
Type of vaccination	active adsorbed vaccine from tetanus toxoid treated with formalin	Active adsorbed vaccine using aluminium-formalin-toxoid	active: IPV= parenteral adsorbed vaccine (Salk)	Active adsorbed vaccine (according to WHO recommended antigen combination)	**measles-, mumps-, rubella (MMR) Pneumococci-, Meningococcal-vaccination**
Performance of basic immunisation in adults (children: refer to manufacturer's guide!)	2 inj. i.m. à 0,5 ml 4 weeks apart, 3.inj. after 6 - 12 months	basic immunisation: 3 inj. (0 - 1 - 6 months); after age 6 years: reduced toxoid dose **d** = at least 2 IE toxoid	2 injections 2 - 6 months apart	annual vaccination, ideally in autumn (before onset of an Influenza epidemic)	Refer to chapters! **rabies-vaccination (I):** Ind: longer stay in countries with high risk (e.g. India), professions with high risk
Protection period (Interval for booster)	10 years post 3. injection (not earlier)	10 years booster ideally with Td-vaccine	10 years; booster for adults, who travel or live in high risk areas.	1 year	Dose: HDC- (e.g. Rabivac®) or PCEC-vaccine (Rabipur®) 3 doses on days 0, 7, 21 (28), occ. 4. dose
Complications	very rarely diseases of the peripheral nervous-systems			very rarely Guillain-Barré-syndrome; thrombocytopenia, vasculitis	after 1 year, booster after 5 years; Avoid contact with animals
Special contraindications		Thrombocytopenia	no life vaccine for patients suffering from immune deficiency	Allergy to eggs (allergic reactions!)	
Epidemiology	worldwide use combination with Diphtheria-toxoid-vaccination (Td-vaccination)	since 1990 epidemics in Russia, Ukraine, Baltic countries, spreading into other countries	Worldwide target of the WHO: eradication in 2005	Worldwide every 2-3 years influenza A epidemics, every 10-40 years pandemics with millions of deaths; peak in winter	
Remarks	after exposition in non-immune individuals perform simultaneous prophylaxis using tetanus-hyperimmunoglobuline/ active vaccination	Tetanus booster using Td- or Td/IPV-vaccine. Immunisation should be obtained by all means before travelling to Eastern Europe!	OPV = oral life vaccine (Sabin) is not being used anymore due to the risk of a vaccine-associated poliomyelitis (VAP) (1 : 4,5 Mio.)	individuals > 60 and chronic ill patients should also receive vaccination against Pneumococci!	

S = Standard vaccination for all humans; I = Indication vaccination for certain individual at risk; T = travel vaccination

Local- and generalized reactions after vaccinations: 1. Local: reddening, swelling, pain in injection area, localized lymphadenopathy

2. Generalized: tiredness, fever, flulike reaction, rarely allergic reaction

general contraindications against active vaccination:
1. Acute (febrile) illness (wait 2 weeks)
2. known SE/adverse reactions, allergy against an ingredient of the vaccine (egg/preservative)
3. no i.m. injections for those on Warfarin/Cumarins (→ s.c. injections not contraindicated)
4. life vaccines are contraindicated in pregnancy and (depending on immunity) in immune deficiencies (no contraindication for adsorbed vaccines).

	HEPATITIS A (I T)	HEPATITIS B (I T)	TYPHOID (T)	YELLOWFEVER (T)	Viral meningo-encephalitis (I T)
type of vaccination	Active adsorbed vaccine: Al-OH-Adj.-vaccine Liposomal-vaccine	Active adsorbed vaccine with HBs-Ag children: general vacc. adults: indication vacc.	Active: 2 types: • oral life vaccine • parenteral adsorbed vaccine	Active life vaccine with attenuated virus YF-strain "17D" from chicken embryos	Active adsorbed vaccine; Formalin inactivated Vaccine
performance of basic immunisation in adults (children: refer to manufacturer's guide!)	Single vaccine 2 doses i.m. (deltoid) 6 months apart; antibody testing (anti-HAV) may be advisable	3 doses (deltoid): 2. Dose after 2 months, 3.dose after 6 months Adults only: test for antibodies before vaccination (anti-HBc)	1 capsule on days 1,3,5 one hour before meal or 1 parenteral dose	1 injection à 0,5 ml s.c.	3 inj. à 0,5 ml i.m.; 2. inj. after 1 - 3 months 3. inj. after 9 -12 months
protection period (Interval for booster)	10 years	10 years Check immunity (anti-HBs) in exposed professionals, dialysis patients, immune deficient patients	oral vaccine: 1 year parenteral vaccine: 3 years	10 years	3 years
complications	doubtful association with neurological diseases (few cases)		gastrointestinal symptoms	anaphylaxis when allergic to eggs; very rarely encephalitis, occ. deaths	very rarely diseases of the peripheral nervous system
special contraindications		allergy to formalde-hyde and mercury salts	immunisation may be insufficient when the patient suffers from diarrhoea	allergy to eggs; acute liver- and kidney diseases etc.	allergy to eggs and preservatives, neuropathies
epidemiology	worldwide; particularly when hygiene is poor; young people after travelling to southern countries	worldwide; increased risk within high risk groups (refer to 'hepatitis')	tropical and subtropical countries	refer to WHO-report	Russia, Baltic countries, Eastern Europe, Bavaria, Baden-Württemberg, Carinthia, Balkan etc.
remarks	Exposure of individuals, for whom a Hepatitis A might be dangerous, may require simultaneous application of immunoglobuline (5 ml i.m) Use combined vaccine hepatitis A/B.	Post exposure: active Passive simultaneous pro-phylaxis with Hepatitis B-immunoglobuline within 6 hours if at all possible! The vaccination also protects against hepatitis D.	vaccination does not protect against Paratyphus; patients must not take any antibiotic or Resochin just before, during or immediately after the oral vaccination!	vaccinations are only available at certain centres defined by the WHO. validity: from 10 days post vaccination; protection lasts 10 years	Passive vaccination: Immunoglobulines (up to max. 4 days post exposure), not for children < 14

Minimal gap between 2 vaccinations:
(and the other way round)

• no gap needed between 2 adsorbed vaccines and adsorbed-/life vaccine

• between 2 life vaccines with attenuated pathogens: either simultaneous vaccination, otherwise 2. vaccination after 4 weeks
• Immunoglobulines: → Parenteral life vaccination: 3 months
  Parenteral life vaccination → Immunoglobulines: 2 weeks
• adsorbed vaccine ↔ Immunoglobulines: no gap.

Basic immunisation: always follow the advice regarding minimal intervals; there are no maximum intervals: *Each vaccination counts!*

X.   SOMATOFORM DISORDERS [F45]
Somatisation disorder [F45.0]
Somatoform autonomic functional disorder [F45.3]

Def.: The term "Somatoform disorder" was introduced into the official classifying system in 1980. Somatoform means, that existing physical symptoms copy somatic diseases, but physical investigations of the organs are all normal. Symptoms can occur in any organ system. In somatoform autonomous functional disorder patients allocate these symptoms to organs which are mainly innervated by the autonomous nervous system.

Ep.: Somatoform disorders are fairly common: they contribute up to 15 % of all patients in general practice, and > 10 % on medical wards.

Aet.: Trigger factors can be found in the biological, psychological and social area. Genetic, perception, personality, childhood, history of abuse, and interpersonal relationships, triggering organic diseases, trauma, stress, pressure and going through a rough period can all contribute.

Sym.: Patients are usually convinced that their symptoms are of organic origin. They last for at least 2 years, but they may recur, they can be multiple and change frequently. Somatoform autonomous functional disorder is not defined by any length of time. Symptoms show a certain preference of the gastrointestinal, cardiovascular and respiratory system. Symptoms like dry mouth, hot or cold sweats, blushing, paraesthesias, restlessness, tremor, palpitations, dizziness and additional subjective symptoms relate to a certain organ or system; also feeling of bloatedness, flatulence, epigastric burning, chest pain (sharp or pressure), chest tightness and shortness of breath.
List of somatoform autonomous functional disorder (phenomenological definitions):
- cardiovascular system F45.30 – associated term: **cardiac neurosis**
- upper GI tract F45.31 - associated term: **dyspepsia**
- lower GI tract F45.32 - associated term: **irritable bowel syndrome (IBS)**
- respiratory system F45.33 - associated term **hyperventilation syndrome**

DD: Disassociation of organic diseases can be very difficult. The disorder can be accompanied by depression and anxiety of various intensity. In hypochondriac disorders the patient rather concentrates on the presence of a severe disease.

Di.: Definitions according to ICD-10 - somatoform disorders:
   • repeated physical symptoms; no sufficient organic cause can be found.
   • persistent request for medical investigations despite repeated negative results and reassurance by the doctor, that there is no organic cause for the symptoms.
   • Patient rejects the idea of a psychological cause.
   • the situation is equally disappointing for doctor and patient.

Th.:   1. General management:
   Ideally one should favour a cognitive-educative approach. Explain the connection between symptoms and investigation results, aiming to achieve a certain patient understanding of what a psychosomatic disorder is. Mentioning any minor out-of-range results should be avoided. Then attempt a gradual introduction to get the patient used to the idea of a somatoform disorder. Nicotine, alcohol and caffeine should be reduced or avoided because of their negative influence on the vegetative nervous system.

   2. Psychotherapy and pharmacotherapy
      ▶ Somatoform autonomous functional disorder of the cardiovascular system:
         • Cognitive behavioural therapy (CBT) for cardiac neurosis und psychogenic chest pain
         • Betablocker for hyperkinetic cardiac syndrome
      ▶ Somatoform autonomous functional disorder of the respiratory system:
         Hyperventilation syndrome: Biofeedback training and progressive muscle relaxation; pharmacologically: Metoprolol and Clomipramin (tricyclic antidepressant). Otherwise follow the guidelines as for 'anxiety'.
      ▶ Somatoform autonomous functional disorder of the lower GI-tract:
         CBT, cognitive-behavioural treatments using relaxation and stress management, psychodynamic psychotherapy, conflict centred talks, progressive Jacobson muscle relaxation and hypnosis.

(For further information refer to the according chapters of organic diseases.)

XI. BULLYING AT WORK AND ILLNESS

Def.: <u>Conflict related communication</u> at work; the defeated person is being systematically harassed (directly or indirectly) by one or several other people over a long period (for at least 6 months) with the aim to exclude him/her from the community.

Ep.: In Germany ca. 3 % of all employees are affected; rates in other European countries are similar. 2/3 of victims are women, more frequently in the public sector, in the health- and social sector; in ca. 50 % the bullies are superiors.

Aet.: <u>Mode of disease:</u> victim of harassment, accusing others, sometimes internal causes.
<u>cause within a group:</u> insufficient influence, poor information, limited communication, role conflicts, unclear hierarchy, lacking respect for each other, lacking social respect and support, inconsistent tasks/orders, excessive demands and social stress; risk to lose job, pressure at work
<u>causes found with the bullies:</u> non-competent leadership and diligent hard working colleagues, risk to lose one's status, feeling of inferiority.
<u>Causes found with the victim:</u> lack in confidence, poor social skills, fighting against injustice, very diligent, rigidity, passivity and helplessness, lacking distance, extremely dedicated.

<u>bullying strategies:</u> competence withdrawal, allocating pointless tasks, social isolation, attacking the person's privacy, verbal aggression e.g. threats, spreading rumour etc.

Sym.: Mainly <u>depressive symptoms</u>, often moderate depressive episode or assimilation disorder; headaches, insomnia, lack of concentration, fear of failure, also <u>somatisation disorder, anxiety</u>, tinnitus, sometimes post traumatic stress disorder; obvious association with bullying or chronic conflicts, frequently psychosomatic or psychiatric <u>comorbidity</u>, suicide.

Di.: unclear symptoms, particularly if associated with depressive mood, unclear anxiety and somatisation disorder: always consider the possibility of bullying or other problems at work.

Th.: Behaviour analysis, organigram, therapeutic course: gain distance - learn to understand, decide and behave appropriately; construction of a meaningful work- and life perspective. Inpatient therapy/ rehabilitation in a special clinic specialised in bullying therapy. The latter only makes sense when there is a need for rehabilitation, when the patient shows commitment, and when the predicted prognosis is good.

<u>Removing the patient from work</u> is most important for organisational problems and to release the patient from his/her duties.
Involve the employee representative committee, ombudsmen, bullying officers, mediation, occupational doctor, solicitor.

XII. ALCOHOLISM

__Def.:__ • **dangerous alcohol intake**: daily alcohol intake > 30 g/d (m) and. > 20 g/d (f).
• **damaging abuse** [F10.1]:
  Alcohol intake, causing physical, psychological or social damage, without signs of any dependence.

• **Alcohol dependence** [F10.2]: ≥ 3 criteria must be fulfilled:
  1. A strong desire or obsession to drink alcohol („craving").
  2. Loss of control: inability to control alcohol intake!
  3. physical withdrawal symptom, when alcohol intake is stopped or reduced
  4. Tolerance: larger amounts of alcohol are needed to achieve the desired effect.
  5. negligence of hobbies and interests in favour of alcohol intake
  6. persisting alcohol intake despite proven damage

also:
- drinking at inappropriate times: before driving, at work → loss of driving licence, problems at work
- drinking without consideration of social consequences → marriage-/relationship-/family problems; isolation

Typology of alcoholism (Jellinek):
There are various types of drinkers:
• Alpha-drinker = conflict- and relief drinker
• Beta-drinker = occasional drinker
  Many people are conflict- and occasional drinkers, without the presence of alcohol dependence; drinkers of these two groups are at risk of developing alcohol dependence.

Alcohol dependence is present in the following 3 types:
• Gamma-drinker = Alcoholic, who lost control over his drinking behaviour.
• Delta-drinker = level-drinker: Must keep a certain blood level, to be psychologically stable. Withdrawal symptoms when alcohol intake is stopped.
• Epsilon-drinker = episodic or quarterly drinker with periodical binge drinking.

__Ep.:__ Ca. 3 % of the german population is alcohol dependent; in ca. 5 % there is damaging use (alcohol abuse); in eastern european countries the prevalence is up to 5 x higher; m : f = ca. 3 : 1 (high number of unreported cases in women). Peak in the 3rd – 5th decade.
Mortality due to alcohol in Germany: 30 men (10 women) per 100.000/year (most common cause of death: liver cirrhosis). In Russia up to 40 % of all men die from alcohol abuse.
Children of alcoholics have a 4 fold risk to develop alcoholism.

__Aet.:__ • Primary alcohol dependence (80 %): congenital personality, social surroundings and stress
• Secondary alcohol dependence due to underlying psychiatric disease

__Pg.:__ Humans who have congenitally too few intracranial Dopamin-D2-receptors, are supposed to be at higher risk of developing addictions.

__Sym.:__ 1. Alcohol dependence

2. consequences of alcoholism:

A) Neuropsychiatric disorders:

• Acute alcohol intoxication [F10.0]:
  > Alcohol tolerance varies significantly. Alcohol concentration in blood > 5 ‰ is usually lethal.
    Sym.: impaired behaviour (e.g. loss of inhibition), neurological impairment (e.g. loss of coordination and slurred speech), amnesic gaps (blackouts), impaired consciousness ranging from somnolence to coma; death due to aspiration, bolus death, respiratory depression, body temperature too low etc.
  > Pathologic intoxication (rare): can occur after small amounts alcohol (depends on individual disposition).
    Sym.: impaired behaviour (often aggression), impaired consciousness and amnesia
    Th.: monitor vital functions (risk of aspiration), high dependency unit
• Alcohol withdrawal syndrome [F10.3]:
Typical signs of physical dependence; occurs after interruption of the regular alcohol intake.

2 types:

1. withdrawal syndrome without delirium [F10.3]:
   onset ca. 10 hours after alcohol was stopped, peak after 1 - 2 days.
   Sym.:- GI- symptoms (e.g. nausea, vomiting, diarrhoea)
         - cardiovascular symptoms: tachycardia, hypertension
         - Vegetative symptoms: insomnia, sweating, mydriasis, facial flush, fever
         - Neurological symptoms: fine tremor, slurred speech, often epileptic fits
         - Psychiatric symptoms: agitation, fear, irritability, depressions

2. withdrawal syndrome with delirium (Syn: Delirium tremens = Alcohol delirium)[F10.4]:
   onset 2 - 3 days after alcohol was stopped.

Most severe form of alcohol withdrawal syndrome; may occur during an intense drinking phase in form of a continuity delirium. It may have been triggered by medical or surgical illness and its associated interruption of alcohol intake. This is a life threatening condition, with a mortality rate of up to 20 % if untreated, if treated ca. 2 %.

Syn.: as 1, additionally:
- disorientation to place and time
- optic and acoustic hallucinations (creepy crawlies, mice, birds etc.)
- severe psychomotoric restlessness causing danger to others and themselves
- urge to keep themselves busy, fiddling, looking for things

Th.: ICU:
- monitoring of circulation, respiration, water-, electrolyte- and glucose balance (risk of hypoglycaemia!), monitoring of CK (risk of rhabdomyolysis)
- underlying cardio-pulmonary diseases: Diazepam (also prevents convulsions)
  SE: low effect in complete delirium compared to Clomethiazol; little improvement of arterial hypertension and tremor. In case of hypertension add Clonidin (central sympathicolysis). For hallucinations use e.g. Haloperidol.
- No cardiopulmonary diseases: Clomethiazol (Distraneurin®) or Clonidin

Note: Clomethiazol-vials are not available anymore due to respiratory SE (respiratory depression, bronchial hypersecretion)! Oral use of Clomethiazol may cause dependence, hence therapy should not last longer than 2 weeks as an inpatient. Clomethiazol therapy in the community is contraindicated. Ethanol is not a therapeutic drug for withdrawal syndrome.

- Addition of Vitamin B1 (Thiamine) 100 mg/day as a prophylaxis against Wernicke-encephalopathy
- Alcoholic ketoacidosis: Glucose infusion

- Epileptic fits (common in withdrawal syndrome) → prophylaxis using Diazepam, Carbamazepine

- Wernicke-Korsakoff-syndrome = Wernicke-encephalopathy (amnestic syndrome):
  Onset usually acute, occ. after delirium.
  PPh.: damage to paraventricular cerebral areas
  Pg.: Vitamin B1-deficiency due to malnutrition of alcoholics
  Sym.: Triad: 1. Reduced consciousness and confusion, 2. Ocular muscle palsy, 3. Ataxia
        Impaired memory, orientation and concentration, confabulation
  Th.: Vitamin B1 = Thiamine (50 mg/day parenteral) alcohol abstinence. Glucose infusions without vitamin B1 can cause a deterioration, hence all alcoholics receiving Glucose should receive vitamin B1 beforehand!
  Prg.: mortality ca. 10 %.

- Atrophic cerebral changes:
  Inc.: common, 50 % of all alcoholics; 10 % of all cases of dementia due to alcohol
  Sym.: impaired concentration and memory, fine motor function, change in personality, final state: dementia
  Di.: CT, MRI: enlarged intracranial ventricles and cerebral sulci
  Prg.: only partially reversible in the early stage.

- Dementia syndrome in vitamin B2- (= Niacin) deficiency (Pellagra)

- Polyneuropathy: 20 % of all alcoholics show signs of symptomatic PN: Distal- and lower limb sensomotoric disorder. Good prognosis if the patient is abstinent.

DD: Polyneuropathy (PNP):
  1. Acquired: Most frequently alcoholism and diabetes mellitus; also hepatic diseases, uraemia, porphyria, medication (Vincristin, Paclitaxel, Platinderivates, Interferon, antiretroviral drugs etc.), toxins, rarely inflammatory diseases (e.g. Guillain-Barré-syndrome etc.)
  2. occ. hereditary PNP

- Alcohol psychosis: jealousy, depressive syndromes, phobias, hallucinations, paranoid disorders, suicide risk!

- Cerebellum atrophy:
  Inc.: rarely, 1 % of all alcoholics
  Sym.: ataxia, nystagmus, dysarthry etc. → MRI
  Th.: Vitamin B1 (Thiamin) alcohol abstinence
  Prg.: poor

- Central pontine myelinolysis (rare):
  cause: too rapid correction of hyponatraemia!
  Sym.: Para-, quadriplegia, pseudobulbar paralysis, drowsiness etc.

- Alcoholic myopathy (= alcohol myopathy): Here up to 50 % of all alcoholics
  - rarely acute necrotising form with rhabdomyolysis and risk of acute renal failure
  - Subacute painful myopathy (occ. hypokalaemia and CK ↑)
  - Chronic painful alcohol myopathy and muscle atrophy

B) Further consequences of alcoholism

- **GI tract:**
  - (severely) neglected dental health
  - Reflux oesophagitis and increased risk of Barret-oesophagus and oesophagus carcinoma.
  - Acute gastritis, occ. gastric haemorrhage (erosive gastritis)
  - Mallory-Weiss-syndrome (mucosal tears of the oesophago-cardial zone, triggered by vomiting→ bleeding)
  - Bolus death (obstruction of the pharynx/larynx by pieces of meat, common in alcohol intoxication)
  - Impaired intestinal absorption, malnutrition

- **Liver:** There are two alcohol metabolizing enzyme systems:
  - · Alcoholdehydrogenase (ADH)
  - · Microsomal ethanol-oxidizing system (MEOS)

  ADH/MEOS transforms ethanol into acetaldehyde. This is metabolized by acetaldehyde-dehydrogenase (via acetate) to CO_2 and water.
  If alcohol intake is increased, the activity of MEOS rises significantly, whilst ADH-activity remains stable. Acetaldehyde is hepatotoxic.
  Individual toxic hepatic alcohol limit varies, depending on underlying diseases (particularly chronic hepatitis), malnutrition, gender (Alcoholdehydrogenase capacity is much lower in women): Toxic limit for men: ca. 40 g ethanol daily, for women only ca. 20 g !
  - Alcoholic fatty liver (90 %; γ-GT raised, normal transaminases)
  - Alcoholic steatohepatitis = ASH = fatty liver hepatitis (50 %; γ -GT and transaminases raised)
  - Alcoholic liver cirrhosis (25 %) and its complications incl. primary hepatic cell carcinoma
  Zieve-syndrome: Jaundice, haemolytic anaemia and hyperlipoproteinaemia associated with alcohol toxic hepatic damage. – refer to chapter 'liver'!

- **Pancreas:** Acute pancreatitis, chronic-calcifying pancreatitis

- **cardiovascular:**
  - Alcohol toxic cardiac arrhythmias (holiday-heart-syndrome): Paroxysmal atrial fibrillation and other (supraventricular) arrhythmias after alcohol abuse
  - Alcoholtoxic dilatative cardiomyopathy (1 %)
  - Arterial hypertension!
  - Increased risk of stroke if alcohol intake is> 30 g/day
    Alcohol intake and CAD: The connection between alcohol intake and mortality is U-shaped: Total mortality in moderate alcohol intake is reduced up to 40 %, but rises significantly when intake is high. HDL-cholesterol increases, LDL-cholesterol drops. Flavonoids (e.g. Catechin) in red wine are of some additional benefit (protects LDL-cholesterol against oxidation) → „French paradox": In France the mortality from CHD is fairly low despite high fat intake (explanation: red wine intake!).

- **metabolism:**
  - Hypertriglyceridaemia, hyperuricaemia
  - Hypoglycaemia (high risk in alcohol intoxication, particularly in additional Diabetes mellitus)
  - Porphyria cutanea tarda
  - Folic acid deficiency and hyperchrome anaemia → folic acid substitution

- **Immune system:** immune deficiency, frequent infections (e.g. pneumonia, tuberculosis)

- **Endocrine disorders:** men: loss of libido and erectile dysfunction (testosterone↓), women: Oligo- or amenorrhoea (oestrogen ↓), Pseudo-Cushing-syndrome

- **Fetal alcohol syndrome (alcohol embryopathy):**
  Ep.: Ca. 40 % of all alcohol dependant pregnant women; most frequent cause of a mental handicap!
  1. Pre- and postnatal growth retardation
  2. Dysfunctional CNS (virtually any neurological symptom, developmental retardation, intellectual damage/disorder)
  3. Two of the three following characteristic craniofacial dysmorphies:
     - microcephalus
     - narrow eyes (ptosis)
     - thin upper lip, not well developed philtrum, flat face

C) Increased risk of cancer:
high risk: oral cavity, pharynx, larynx, oesophagus
moderate risk: liver
minor risk: breasts, colon/rectum

D) Almost all alcoholics are also cigarette addicts with all its associated late damages, particularly CAD and cancer of the oropharynx, larynx, oesophagus; tobacco-alcohol-amblyopia etc.

E) Psychosocial consequences of alcoholism:
Partner-/family conflicts (alcohol disease = family disease), problems at work (many days off sick, loss of job), increased incidence of accidents and violence (~ 25 % of all work- and road traffic accidents happen after drinking!), financial and legal problems (e.g. repeated loss of driving licence because of drink driving).

Lab.:
- γ-GT ↑ (<u>DD:</u> other hepatic diseases, cholestasis)
- MCV ↑ (<u>DD:</u> Megaloblastic anaemia caused by vitamin B12- or folic acid deficiency)
- CDT ↑ (Carbohydrate Deficient-Transferrin): Sensitivity in women is often insufficient
- Additional lab tests, concerning any complications

Di.:
- <u>History:</u> ask for alcohol intake (patients report is unreliable), ask relatives/friends.
- <u>You can use a standardised questionnare</u>
- <u>Symptoms:</u> seek/ask for alcohol associated problems/longterm damage

<u>To diagnose alcohol disease, at least 3 of the following 8 criteria must be fulfilled:</u>
1. Strong desire or obsession to drink alcohol
2. Reduced ability to drink moderate amounts
3. Drinking alcohol with the target to limit or avoid withdrawal symptoms
4. Physical withdrawal symptoms, e.g. tremor, restlessness
5. Proven tolerance (larger and larger amounts are needed to achieve the same effect)
6. The amount of intake is set by the increasing demand.
7. Loss of interest , replaced by drinking
8. Ongoing alcohol intake despite obvious presence of damage

The <u>CAGE-test</u> has been proven useful:
Alcoholism can be assumed when 2 or more of the following 4 questions are answered with "yes":

<u>C</u>ut Down: Have you (without success) tried to reduce alcohol intake?
<u>A</u>nnoyed: Were you annoyed when others criticised your drinking behaviour?
<u>G</u>uilty: Do you feel guilty because of your drinking?
<u>E</u>ye Opener: Do you use alcohol to "get started" in the morning?

Th.: Alcohol dependence is a chronic disease, and cannot be cured. Apart from Acamprosat there is no medical therapy for alcohol dependence (***Beware:*** dependence may change to tablets!). Contact persons must not tolerate the addiction, otherwise they will become "co-alcoholics". The disease can only come to a halt by lifelong abstinence.

1. <u>Contact- or motivation phase:</u> Clarification of the situation; the patient must be motivated to acknowledge the disease and must be committed to treatment. If this stipulation is not fulfilled, therapy is not possible.
   In practice <u>short intervention</u> in 3 steps: Empathic listening - objective information - cooperative problem solution.

2. <u>Detoxification phase:</u> if after the start of abstinence significant withdrawal symptoms and delirium occur, the patient has to be admitted to a special clinic for 1 to 2 weeks.

3. <u>Weaning off phase:</u> In the community and – if necessary – as an inpatient treatment lasting several months. Target: The patient must learn to tackle daily problems without alcohol. <u>Regular (weekly) attendance of self help groups is the base of therapy</u>, e.g. Anonymous Alcoholics (AA); the marital partner can be involved.
   A <u>supporting</u> therapy with Acamprosat can reduce the risk of relapse in a small number of patients.
   <u>Mode of action:</u>   Reduction of craving for alcohol by normalising the cerebral glutaminergic system.
   <u>SE:</u>   occ. diarrhoea etc.
   <u>Dose:</u> Campral® 6 - 12 months (2 x 3 capsules/day)

4. <u>Follow Up:</u> In the community, lasting years, sometimes lifelong. Regular attendance of self help groups. Rebuild social contacts. Occupational rehabilitation.

Prg.: Without therapy life expectance is reduced by 15 years, and prognosis is poor. Most common cause of death: Suicide (15 % of all alcoholics!), accidents, cardiac diseases, cancer, liver cirrhosis etc. If therapy is resolute, <u>committed</u> patients can be rehabilitated in ca. 70% (e.g. FORD-method)!

XIII. POVERTY AND DISEASE

Def.:   • **Absolute poverty (a threat to physical existence, not enough to live)**
        • **Relative poverty is income related:**
        - living on income support
        - 50% or less of the average national income

Ep.:   Increase of poverty in Germany (relative poverty ca. 15 %), Europe, but particularly in the third world and in the previous communist countries. In Europe there is a <u>west-east-</u> and a noticeable <u>north-south-gradient</u>, regarding wealth, health, medical and social care.
        <u>Affected groups:</u>

- migrants (legal and illegal)
- unemployed people
- homeless people
- Families with many children and single parents

- the children of those affected
- elderly, retired people
- Illegal prostitutes
- alcoholics, drug addicted

Interaction between poverty and health:
Poor health increases the risk of long term unemployment, poverty and the other way round. Poverty sometimes means homelessness.

Examples for increased risk of disease in poverty:
- increased infant- and mother mortality
- CHD (MI: 2-3 fold increased risk)
- stroke (2-3 fold increased risk)
- increased disposition of infections, more cases of tuberculosis
- Dermatological diseases (common)
- Parasitic diseases (lice, scabies)
- Psychiatric diseases (depressions, more suicides)

- cancer
- hepatic diseases
- dental diseases (!)
- alcohol disease
- more accidents
- reduced life expectance

Health damaging behaviour: Heavy smoking and drinking (often starting in childhood), unhealthy nutrition; medication dependence (higher in women), alcohol- and drug addiction (higher in men).

Violence is frequently associated with poverty. Women are frequently victims of physical violence and sexual exploitation. In former times women rather indulged violence than homelessness. Since 2000 in Germany violent men can be removed from the common flat with a police order.

Medical support:
Unapproachable services are a common problem, because often the stipulations cannot be met (e.g. no night shelter if the patient is not free of parasites, proof of place of residence, no dogs or children allowed). Medical and social care often does not reach poor people. The classical "patient goes to doctor" has to be replenished with "doctor goes to patient". Preventative medicine (e.g. vaccinations) and hence therapeutic offers must be easily available and nearby. They also must include offers for women with children (shelter homes and emergency shelter, community drug clinics, social information service, debt advice, mobile medical service, jobseeker offices, schools, kindergardens).

Practical tips for handling patients:
- always consider the patient's life circumstances (holistic approach)!
- can the patient read written information? (increasing number of illiterate people)
- exact history, use simple language
- ask for nicotine abuse, alcohol intake/ alcohol damage
- ask for nutrition habits
- ask for vaccination, fill any vaccination gaps. Offer influenza- and Pneumococci vaccination!
- Beware: Medication with addictive potential – consider any interactions between alcohol and medication (e.g. neuroleptics, antiepileptics, antihistamines, sedatives etc.)
  Medication compliance may be insufficient and short course-antibiotics may be preferred.
- Scabies and head lice! – Examine the naked patient if a dermatological disease is suspected!
- Parasitary diseases: treat all family members as well.
- check and repair teeth.
- inform the hospital social services about any inpatients; involve social services before discharging the patient (home circumstances, care, homeless people). Is there a family doctor or other easily available medical care (e.g. for drug addicted/homeless patients) the discharge letter can be sent to?!

XIV. MEDICAL REPORTS

Any doctor is obliged to act as a medical expert, as long as the requested report refers to his/her medical knowledge.

- civil law
- social law
- criminal law

arbitration committee
expert commission

Medical reports are a piece of evidence; the medical expert is a "court assistant". The expert must not interview any witness or party on his own initiative and must not try to clear up the facts.

science

client/agent

expert-report

law

person to be examined

Medical reports are either a free text or pre-designed forms to be filled in.

Types of reports:
1. Reports for statutory pensions; to define pensions due to inability to work.
2. Reports for statutory accident insurance: accidents at work (including the journey to and from work)
3. Reports to determine a compensation settlement or to define a reduced ability to work
4. Reports to determine the grade of a handicap
5. Reports for private insurance

General foundations by law:
The person to be examined has the duty to cooperate; the question is, how much one can ask of him?
The expert is not supposed to exceed the limits of the request without the court's approval.
The report must be checked immediately after being received; they can only be challenged on grounds of bias, when the report hasn't been completed within a certain time (ca. 6 months) or when the request is not within the experts expertise and competence.
The report must be written by the expert himself and on request of the person to be examined (no delegation).
The expert has to pay a fine when he has not produced the report within a set time after several warnings.
The report has to be written "to the best of the expert's knowledge". There may be a detention fine for a false report.

Report request:
Only provide an answer to the asked questions: "an expert is not a judge". The expert must declare any assistants, and replenish his signature with an appendix.

Foundation for a report:
It is a medical document. There is a punishment for "favour certificates"; good medical note keeping (including absent symptoms) is the base for any report and certificate.

Note: It is not the diagnosis which matters, but the question how much this affects the patient's capability.

Special expert report terms:
- subsequent damage
- handicap
- reduced ability to work
- Inability to work
- helplessness
- significant changes of circumstances
- temporary annuity
- report fee

XV. OCCUPATIONAL DISEASES

Def.: Occupational diseases are clearly defined by the law and published in a list, or diseases fulfilling the criteria to be added to this list due to new scientific knowledge.

Currently there are 67 listed occupational diseases; however considering a certain clause, patients can claim for compensation for several other diseases. The medical diagnosis "disease associated with occupation" is not the same as the term "occupational disease".

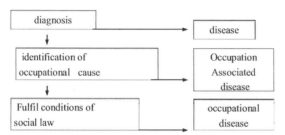

Every doctor or dentist is obliged to inform the authority in charge when he suspects an occupational disease. The subsequent investigations are expensive and time consuming. Hence only justified suspicions should be reported.

The recognition of an occupational disease requires 3 full evidences (almost certainly) and two likelihood evidences (> 50 % = in favour of the facts of the case). Full evidences are: The question if the occupation was covered by an accident insurance, the proof of a damaging influence and a medical diagnosis. Simple likelihood has to prove the liability based causality (the damaging influence is in a causal relationship with the insured occupation) and the liability filling causality (th damaging influence is likely to have caused the disease). The occupational influence must be either the sole cause (monocausal) or a significant partial cause (multicausal) for the disease or the significant deterioration of an underlying disease [e.g. hepatitis infection in medical personnel, smoking and Asbestos; baker's asthma in underlying seasonal asthma].

Causality in occupational disease law:

    a) full evidence:
                                            1. Insured occupation
                                            2. Damaging influence
                                            3. Medical diagnose
    b) likelihood evidence:
                                              1. Liability based causality
                                            2. Liability filling causality

> *Note:* The discussion about likelihood evidence only makes sense when full evidence is present!

Report: In the medical part the doctor must define the diagnosis and give an answer to the question, if the occupational influence was significant for the development of the disease or the deterioration of an underlying disease. All other evidence has to be provided by the insurance.

Reduced ability to work is the difference of the ability to work on the job market before and after the damaging event (accident or exposure to toxin). Any restrictions to private life don't play any role. A reduced ability to work of>= 20 % will qualify for a monthly pension as a compensation, according to the annual income. This pension is totally unrelated to any other income of the insured person.

XVI. PHYSICAL EXERCISE AND HEALTH

Def.: Moderate physical training is defined as regular and individual physical exercise with the aim to promote, remain or re-install health.

Only ca. 20 % of the adult population exercises sufficiently. 80 % don't exercise enough. Increasing age is associated with reduced exercise. Even children and adolescents exercise less these days.

Interactions between physical exercise and health:
Sufficient physical activity has a positive effect on:
• cognition
• mood
• cardiopulmonary system
• glucose- and lipid metabolism
• immune system
• musculoskeletal system

Ongoing lack of exercise causes a variety of diseases, often seen in industrialised countries. This includes:
• cardiovascular diseases, particularly CHD and arterial hypertension
• obesity
• metabolic syndrome and type 2 diabetes mellitus
• disorder of lipid metabolism
• degenerative diseases of the musculoskeletal system

Prevention, therapy and rehabilitation using physical therapy
Physical activity and weight loss have a positive effect on the metabolic syndrome (preventative and therapeutically)! This requires a general change of lifestyle and qualitative change of nutritional behaviour. Frequently medication can be reduced or even stopped!
Studies have shown a reduction in risk regarding cardiovascular mortality and total mortality up to 50 % after regular physical training. The same applies to rehabilitation-programmes and cardiac rehabilitation of patients suffering from CHD or those who survived a MI.

Practical tips
Before starting any health sport activity, the patient should undergo a (sports) medical examination to determine their exercise tolerance. Medical and orthopaedic aspects should be considered. The most recent recommendations are: for primary prevention 3-4 days per week 45-60 minutes exercise of moderate intensity. Ideal health sports are: walking, cycling, swimming, cross country skiing and inline-skating. The intensity of training should be performed at 50-70% of the maximal capacity guided by the heart rate. If possible modified power training for the large muscles should accompany this basic training. Secondary prevention (e.g. community cardiac groups): Physical therapy under medical supervision.

Risks of physical activities
Overuse of musculoskeletal structures, particularly in obese patients with a pre-damaged musculoskeletal system. MI risk is high for those suffering from so far undiagnosed CHD (who didn't undergo a medical examination and exercising vigorously being totally untrained).

Internet-Infos: *www.labtestsonline.de*

There is a grey line between heath and illness. Reference intervals have been obtained from healthy people, considering certain aspects that might influence the result.

General

■ **Certain factors** have an influence on the measurement level. There are fixed (longterm) and variable (shortterm) factors.

fixed	variable	
gender	nutrition	height
age	alcohol intake	diurnal rhythm
genetic factors	physical activity	medication
	amount of muscle tissue	etc.

Tips for phlebotomy: Phlebotomy should be performed at the same time every day before the administration of morning medication, with the patient lying on his back. The tourniquet should only be slightly tightened. The ideal time is 7.00 - 9.00. Last food intake should have taken place 12 hrs ago, the last alcohol 24 hrs ago. If blood is taken from a venflon, then the first bit should be discarded. Last food intake and the time of blood sample always has to be taken into consideration when assessing a lab test result.

factors height, torniquet and physical exercise: The change from a lying to a standing position moves fluid from the intravascular to the interstitial space; within one hour the blood volume drops by ca. 8 %. Cell concentration, proteins and protein bound molecules raise; e.g. Hb, erythrocyte count, Hct, leucocyte count, thrombocytes, total protein, protein subfractions, enzymes, cholesterol and calcium. Low molecular particles such as sodium, potassium, urea and creatinine are distributed in both compartments, and hence they don't change.
Prolonged application of a torniquet or „pumping" with the hand may cause pseudo-hyperkalaemia. A torniquet < 2 min doesn't cause any significant changes. Patients who are prone to oedema are an exception.
Strenuous physical exercise also leads o concentration of the blood. Longterm intensive physical exercise increases the muscle enzyme CK. There is a moderate increase of GOT and LDH. Plasma half life of CK is 15 hrs.

Time and diurnal rhythm: There are variation due to diurnal rhythm in some measurements and in some hormones, like Cortisol and Somatotropin they are quite significant. Urine electrolyte excretion depends highly on the diurnal rhythm; hence 24 hrs collection is necessary.

Nutrition: A light, low fat breakfast doesn't have a major influence on the levels of most blood(serum)- tests. There may be a minor rise in phosphate, bilirubin, GPT, potassium, triglycerides and glucose.
Urine tests depend much more on nutrition. In healthy individuals sodium, potassium, magnesium, chloride and phosphate depends on the intake of those ions, urea depends on protein intake, and urate on purines.

Artificial haemolysis: causes: needle too small, aspiration too fast, leaving blood over night no matter if room temperature or in the refrigerator. Haemolysis is visible in serum or plasma from a Hb-concentration 20 mg/dl onwards. Artificial haemolysis increases the concentration of those things, which come in larger concentration inside the erythrocytes than in plasma. (multiplying factors in brackets): LDH (160), GOT (40), potassium (23) and GPT (7). Increased potassium levels can be found when the sample hadn't been proceeded for a long period of time – even without haemolysis.
Lab results stating "sample haemolysed" indicate falsely increased LDH-, GOT- and potassium-levels.

Pregnancy: Plasma volume is raised by 50% (from 2.600 to 3.900 ml). This is only partially compensated by an increased total amount of erythrocytes (ca.20%) → Pregnancy induced hydraemia associated with a drop in Hb by 10 g/dl is possible!
Only the red blood count is affected. All other plasma components are subject to other regulation mechanisms; the concentration of some components increases as the pregnancy advances.

Drug interactions: e.g. thromboplastin time (Quick) → refer to "anticoagulants".

■ **sample handling**

Serum: leave at room temperature for 30 - 60 minutes and let the sample clot, then separate using a jelly and put into a centrifuge. Heparinised blood must be mixed very well and must be in the centrifuge within 60 minutes.

Erythrocytes lose potassium when kept too long (even worse when kept in the fridge), leading to false high potassium levels. At room temperature blood glucose is metabolised, causing false low glucose levels (pseudohypoglycaemia).
Serum and heparinised plasma may be kept at 4-8°C for 3 days, without any significant changes in most basic components (< 10%). When the samples are kept any longer, there is a marked drop particularly in enzyme activities. Bilirubin concentration is reduced by light exposure, particularly sunlight.

Citrate blood (1 : 10) for coagulation tests: When the torniquet is left too long, there will be contamination with cellular/interstitial fluid, leading to premature activation. This leads to changes of the thromboplastin time/Quick.
The ratio of blood and anticoagulant (9 parts blood + 1 part 0,11 Mol/l sodium citrate) must be exact (sample bottle must be completely full). Blood and anticoagulant must be immediately mixed. Centrifugation should take place within 2 hrs, clotting tests within 4 hrs. aPTT is the most sensitive test: if kept too long, readings will be too high. Clotting bottles must never be kept in the refrigerator; citrate blood and plasma should be kept at room temperature.
Thromboplastin time, fibrinogen, AT and D-Dimer results will be correct in most cases even after 4 -10 hrs correct (but not aPTT).

Citrate blood (1 : 5) can be kept at room temperature for 3-4 hrs for ESR.

EDTA-blood for haematological tests and PCR-diagnostics: Shake the bottle gently in order to achieve an even distribution of the EDTA to obtain correct haematology and PCR-diagnostics.

Blood film: should be constructed within, and fixated the same day.

Thrombocyte count must be performed within 2 - 4 hrs. Thrombocyte count drop varies considerably after phlebotomy.

Erythrocytes, Hb, Hct and leucocytes results are accurate even after 3 days in the fridge.

acid-base-balance/blood gases: arterial heparinised blood is the gold standard. Arterialised capillary blood is more practical; the reference intervals are also valid for this type of blood. Blood circulation is increased in an area of skin by applying a vasodilating cream; then the blood is taken with a heparinised glass capillary from the ear lobe or from the finger tip; in infants it can be taken from the heel without squeezing and deep puncture. The capillary must be completely filled. Then a wire is inserted to mix blood and Heparin. The capillary is then closed on either end.

Urine tests: Early morning urine and 24 hrs-urine are preferable. Tests should take place within 4 hrs. Microscopy and culture: There is contamination from cells, microorganisms and/or mucus. For that reason the first part of the urine should be discarded, and only the next part should be used (= mid stream urine =MSU) .
If porphyrine metabolism is investigated, the urine must be protected from light in a brown vessel from the point of collection onwards, and it must be kept in the refrigerator. Catecholamines investigations require 10 ml 10 % HCl in the collection bottle.

The following list just contains the most common haematology and biochemistry measurements; some of them only show therapy aims in high risk patients.

abbreviations:

AB	= arterial blood
B	= blood
CB	= capillary blood
L	= CSF
P	= plasma
S	= serum
ST	= stool
U	= urine
VB	= venous blood
ZB	= citrate blood

Decimal units:

Factor	prefix	Symbol
10^{-1}	Deci-	d
10^{-2}	Centi-	c
10^{-3}	Milli-	m
10^{-6}	Micro-	μ
10^{-9}	Nano-	n
10^{-12}	Pico-	p
10^{-15}	Femto-	f
10^{-18}	Atto-	a

Note: Apart from S-electropheresis heparinised plasma is equally acceptable as serum and has the same reference interval.

The following reference intervals are for adults (children's references will be indicated). The arrows indicate that levels are often outside those intervals (reduced/increased).

Test	unit	reference interval male | female	multiplication factor deviation
Inflammation marker			
ZB-ESR (ESR)			
1 hrs			
up to 50 years	mm/hrs	up to 15| up to 20	pregnancy ↑
> 50 years	mm/hrs	up to 20 | up to 30	
S-C-reactive protein (CRP)	mg/l	< 5	
S-Interleukin 6	µg/ml	up to 10	
S-Procalcitonin	µg/l	up to 0,5	
Haematology tests			
B-haemoglobin	g/dl	13,0 - 17 | 12 - 16	x 0,62 = mmol/l children ↓/pregnancy ↓
	mmol/l	8,1 - 10,5 | 7,4 - 9,9	
B-haematocrit	%	42 - 50 | 38 – 44	children ↓/pregnancy ↓
B-erythrocytes	Mill./µl	4,3 - 5,6 | 4,0 - 5,4	children ↓/pregnancy ↓
B-Ery-diameter	µm	6, 8 - 7,3	
B-Ery-diameter-deviation (+/-)	µm	0,6 - 0,9	
Erythrocytes indices:			
B-Ery-MCV (mean corpuscular volume)	fl	85 – 98	alcoholism ↑
B-Ery-MCH (mean corpuscular haemoglobin = HBE)	pg	28 – 34	
B-Ery-MCHC (mean corpuscular haemoglobin concentration)	g/dl	31 – 37	
B-Reticulocytes	/1.000 Erys	3 – 18	
B-Thrombocytes	1.000/µl	140 – 345	
B-Leucocytes			
2 - 3 years	/µl	6.000 - 17.000	
4 - 12 years	/µl	5.000 - 13.000	
adults	/µl	3.800 - 10.500	pregnancy ↑ / physical exercise ↑
Blood film			
Neutrophiles	%	0 – 5	
Neutrophiles (end cells)	%	30 – 80	young children ↓
Eosinophiles	%	0 – 6	
Basophiles	%	0 – 2	
Lymphocytes	%	15 – 50	young children ↑
Monocytes	%	1 – 12	
B-Neutrophiles (Granulocytes)	/µl	1.830 - 7.250	
B-Lymphocytes (total)	/µl	1.500 - 4.000	toddlers (up to 3 years) up to 10.500
B-Eosinophiles	/µl	80 – 360	
B-Basophiles	/µl	20 – 80	
B-Monocytes	/µl	90 – 600	
Lymphocytes status			
T-cells (CD3)			
1. year	/µl	1.700 - 3.600	
2 - 6 years	/µl	1.800 - 3.000	
7 - 17 years	/µl	1.400 - 2.000	
> 18 years	/µl	900 - 2.300	
B-cells (CD19)			

Test	unit	reference interval male | female	multiplication factor deviation
1. year	/µl	500 - 1.500	
2 - 6 years	/µl	700 - 1.300	
7 - 17 years	/µl	300 - 500	
>18 years	/µl	105 - 620	
T4(helper-)cells (CD4+)			
1. year	/µl	1.700 - 2.800	
2 - 6 years	/µl	1.000 - 1.800	
7 - 17 years	/µl	700 - 1.100	
> 18 years	/µl	435 - 1.600	
T8(Suppressor)-cells (CD8+)			
1. year	/µl	800 - 1.200	
2 - 6 years	/µl	800 - 1.500	
7 - 17 years	/µl	600 - 900	
ab18 years	/µl	285 - 1.300	
T4/T8-Quotient =			
CD4/CD8-Ratio			
1. year		1,5 - 2,9	
2 - 6 years		1,0 - 1,6	
7 - 17 years		1,1 - 1,4	
> 18 years		0,6 - 2,8	
NK-cells (CD16/56+)			
1. year	/µl	300 - 700	
2 - 6 years	/µl	200 - 600	
7 - 17 years	/µl	200 - 300	
> 18 years	/µl	200 - 400	

special anaemia diagnostics

iron metabolism

Test	unit	reference interval male | female	multiplication factor deviation
S-iron	µg/dl	50 - 160 | 50 - 150	x 0,179 = µmol/l
	µmol/l	9 - 29 | 9 - 27	
S-Ferritin			
2 - 17 years	µg/l	7 - 142	
18 - 45 years	µg/l	10 - 220 | 6 - 70	
> 46 years	µg/l	15 - 400 | 18 - 120	
S-Transferrin	g/l	2,0 - 3,6	pregnancy ↑ oral contraceptives ↑
Transferrin-saturation	%	16 - 45	
S-soluble transferrin receptor (sTfR)	mg/l	0,83 - 1,76	depends on methods used
Vitamins: S-folic acid	nmol/l	7 - 36	
S-Vitamin B12	pmol/l	150 - 800	

Osmotic Erythrocytes resistance

Test	unit	reference interval	multiplication factor deviation
Haemolysis (start)	NaCl g/dl	0,42 - 0,46	pregnancy ↑ (= reduced
Haemolysis total	NaCl g/dl	0,30 - 0,34	osmotic resistance)

Haemoglobin tests

Test	unit	reference interval	
Hb-A$_2$-quantitative	% of total-Hb	up to 3	
Hb-F-quantitative	% of total-Hb	up to 0,5	
VB-Co-Hb	% of total-Hb	up to 2	
VB-Met-Hb	% of total-Hb	up to 1	
Ery-glucose-6-P-DH	U/g Hb	4,6 - 13,5	

Coagulation

Test	unit	reference interval male | female	multiplication factor deviation
Bleeding time (in vitro)	min	6 - 9	(depending on method used),
P-aPTT	sec	20 – 35	pregnancy ↓
P-TPZ (Quick)	%	≥ 70	pregnancy ↑
- INR (target)			INR = International Normalized Ratio
- normal (no therapy)		ca. 1,0	
- simple risk		2,0 – 3,0	
- increased risk		up to 4,0	
P-Thrombin time	sec	14 – 20	(depending on method used)
P-Fibrinogen	mg/dl	160 – 400	x 0,03 = µmol/l,
	µmol/l	4,8 - 12,0	pregnancy ↑
P-Antithrombin III	%	70 - 120	
P-D-Dimer	µg/ml	< 0,5	pregnancy ↑
P-Protein C	%	70 - 140	depends on methods used
P-Protein S,	%	70 - 140	depends on methods used
P-Protein S-activity	%	60 - 130	depends on methods used
Serum osmolality	mosmol/kg	280 - 296	

Serum electrolytes

Test	unit	reference interval male | female	multiplication factor deviation
S-sodium	mmol/l	children 130-145	
		adults 135 - 145	
S-potassium	mmol/l	children 3,2 -5,4	
		adults 3,6 - 5,0	
S-Calcium (total)	mmol/l	2,2 – 2,6	pregnancy ↓
S-Calcium (ionised)	mmol/l	1,1 – 1,3	
S-Magnesium	mmol/l	0,75 – 1,05	
S-Chloride	mmol/l	97 - 108	
S-Phosphate	mmol/l	children 1,1- 2,0	
		adults 0,84 -1,45	

acid-base-balance

Test	unit	reference interval male | female	multiplication factor deviation
AB-pH	--	7,37 – 7,45	
AB-PCO$_2$	mm Hg	35 - 46 | 32 - 43	pregnancy ↓
	kPA	4,69-6,16 | 4,29 - 5,76	x 0,134 = kPA
AB-PO$_2$	mm Hg	72 - 107	depends on age: 102 - 0,33 x age in years
	kPA	9,65 - 14,34	x 0,134 = kPA •
AB-O$_2$-saturation	%	94 - 98	depends on age
AB-bicarbonate	mmol/l	22 - 26	pregnancy ↓
AB-base excess	mmol/l	- 2 up to + 2	
P-Lactate	mmol/l	0,6 – 2,4	

Renal Function Tests

Test	unit	reference interval male | female	multiplication factor deviation
S-urea	mg/dl	12 - 50	X 0,1665 = mmol
	mmol/l	2,0 – 8,3	thirst, nutrition rich in protein ↑
S-creatinine	mg/dl	up to 1,1 |up to 0,9	x 88,4 = µmol/l
	µmol/l	44 - 97 | 44 - 80	(reference interval varies between laboratories) muscle waste ↓ / old people ↓

Test	unit	reference interval male | female	multiplication factor deviation
creatinine clearance 24 hrs	ml/min	≥ 110 | ≥ 95	up to age 30 years; after that -10 for each further decade marker to estimate GFR
S-Cystatin C	mg/l	0,50 - 0,96	

hepatic metabolism

Test	unit	reference interval male | female	multiplication factor deviation
S-Bilirubin total	mg/dl	up to 1,1	x 17,104 = µmol/l
	µmol/l	up to 19	fasting ↑
S-Bilirubin direct	mg/dl	up to 0,3	
	µmol/l	up to 5	
S-copper	µg/dl	79 - 131 | 74 - 122	x 0,157 = µmol/l
	µmol/l	12,4 – 20,6 | 11,6 - 19,2	oral contraceptives ↑ pregnancy ↑
S-Coeruloplasmin	g/l	0,2 - 0,6	oral contraceptives ↑
	µmol/l	0,94 - 3,75	
P-Ammonium	µg/dl	up to 94 | up to 82	x 0,588 = µmol/l
	µmol/l	up to 55,3| up to 48,2	

Enzyme activities at 37 °C[1]

Test	unit	reference interval male | female	multiplication factor deviation
S-GOT = AS(A)T[2] without Pp	U/l	up to 38 | up to 32	
with Pp		< 50 | < 35	
S-GPT = ALT[2] without Pp	U/l	up to 41 | up to 31	
with Pp		< 50 | < 35	
S-γ-GT[2]	U/l	< 60 | < 40	
S-AP[2] 1 – 12 years	U/l	up to 300	
13 – 17 years		up to 390 | up to 190	
adults		40 – 130 | 35 - 105	obese women ↑
S-GLDH	U/l	< 7 | < 5	
S-CHE children, adults	kU/l	5,3 – 12,9	women: > 40 years
w: not pregnant		4,3 – 11,3	
w: pregnant		3,7 – 9,1	also seen in patients taking contraceptives
S-LDH[2]	U/l	< 250	children: up to 300
S-HBDH	U/l	72 – 182	
S-Pancreas-Amylase	U/l	28 – 100	
S-α-Amylase	U/l	< 110	
S-Lipase	U/l	13 – 60	
S-Elastase 1	ng/ml	up to 2	
S-CK[2]	U/l	up to 190 | up to 170	physical exercise ↑
S-CK-MB	U/l	< 25	Up to 6 % of total-CK

[1] current reference intervals; in the following years there may be minor changes.
[2] IFCC-methods (International Federation of Clinical Chemistry); Pp = Pyridoxalphosphate

cardio specific proteins

Test	unit	reference interval	multiplication factor deviation
S-Troponin T	ng/ml	< 0,1	Use EDTA-blood for fast test!
P-BNP	pg/ml	Up to 100	Refer to "Cardiac failure"
P-NT-pro BNP	pg/ml	Up to 266	Depending on age

Serum proteins (based on new reference compound CRM 470)

Test	unit	reference interval	multiplication factor deviation
S-Protein total	g/l	66 - 83	pregnancy ↓
S-Albumin	g/l	35 - 52	pregnancy ↓
S-Albumin	%	54 - 65	

Test	unit	reference interval male | female	multiplication factor deviation
S-α₁-Globulin	%	2 - 5	
S-α₂-Globulin	%	7 - 13	
S-β-Globulin	%	8 - 15	
S-γ-Globulin	%	11 - 22	
S-Coeruloplasmin	g/l	0,2 - 0,6	oral contraceptives ↑ pregnancy ↑
S-Transferrin	g/l	2,0 - 3,6	oral contraceptives ↑ pregnancy ↑
S-Haptoglobin	g/l	0,3 - 2,0	
S-IgA	g/l	0,7 - 4,0	children ↓
S-IgE	µg/l	12 - 240	
	kU/l	5 - 100	
S-IgG	g/l	7 - 16	children ↓
S-IgM	g/l	0,4 - 2,3	children ↓
S-α₁-Antitrypsin	g/l	0,9 - 2,0	oral contraceptives ↑ pregnancy ↑
S-C₃-Komplement	g/l	0,9 - 1,8	
S-C₄-Komplement	g/l	0,1 - 0,4	
S-α₂-Macroglobulin	g/l	1,3 - 3,0	pregnancy ↑
S-β₂-Microglobulin	mg/l	up to 2,4 (> 60 years up to 3,0)	

Lipid metabolism / urate / homocysteine

Test	unit	reference interval male | female	multiplication factor deviation
S-urate	mg/dl	up to 6,4　|　up to 6,0	x 59,485 = µmol/l
	µmol/l	up to 381 | up to357	
S-triglycerides	mg/dl	< 150	x 0,0114 = mmol/l
	mmol/l	< 1,7	
S-cholesterol	mg/dl	< 200	x 0,0259 = mmol/l
	mmol/l	< 5,2	

Test	unit	Ia	Ib	II< 160	targets:
S-LDL-cholesterol (bad cholesterol)	mg/dl mmol/l	< 130 < 4,1	< 100 < 3,4	< 2,6	I　Primary prevention a) without risk factors b) with risk factors für Atherosclerosis II　Secondary prevention in CHD/Atherosclerosis
S-HDL-cholesterol (good cholesterol)	mg/dl mmol/l	> 40 > 1,0	> 40 > 1,0	> 40 > 1,0	Sometimes the lower limit is set as 35 mg/dl .
S-Lipoprotein(a) = Lp(a)	mg/dl	< 30			Risk limit

glucose metabolism

Test	unit	reference interval	multiplication factor deviation
P-glucose (fasting)	mg/dl	up to 100	normal (x 0,0555 = mmol/l)
	mmol/l	< 5,6	normal
	mg/l	≥ 126	diabetes mellitus
	mmol/l	≥ 6,9	diabetes mellitus

OGTT administering75 g glucose (equivalent):

Test	unit	reference interval	multiplication factor deviation
CB-glucose level at 2 hrs	mg/dl	< 140	normal
	mmol/l	< 7,8	normal
	mg/dl	140 - 199	IGT = impaired glucose tolerance
	mmol/l	7,8 - 11,0	
	mg/dl	≥ 200	diabetes mellitus
	mmol/l	≥ 11,1	diabetes mellitus
Ery-HBA₁c	%	< 6,5	use EDTA-blood

381

Test	unit	reference interval male | female	multiplication factor deviation
Porphyrine (beware drug interaction)			
U-D-Aminolaevulinic acid	µg/l	1.000 - 4.500	x 0,00763 = µmol/l
U-Porphobilinogen	µg/l	400 - 1.200	x 0,00442 = µmol/l
U-Porphyrine total	µg/l	40 - 150	x 1,2 = nmol/l
Digestion- and resorption tests			
ST-fatty acids/24 hrs	g	7	
ST-Elastase	µg/g	> 200	
Xylose-tolerance test (administering 25 g)	g/5 hrs	> 4	
U-Xylose excretion			
CSF investigations			
L-Leukocytes count	/µl	0 - 5	
L-Protein total	mg/dl	12 - 50	
L-glucose	mg/dl	49 - 74	x 0,0555 = mmol/l
Q(L/S)-glucose		> 0,5	
Tumour markers			
S-α-Fetoprotein (AFP)	U/l	up to 8,5	pregnancy ⇑
S-CEA	µg/l	up to 5	smokers ↑
S-CA 19-9	U/ml	up to 37	depending on method
S-CA72-4	U/ml	up to 6	
S-PSA (Prostate-Ag) (in men)	ng/ml	up to 49 years up to 2,0 50 - 59 years up to 3,0 60 - 69 years up to 4,0 < 0,1 post prostatectomy	post PR examination ↑ ratio free PSA: total-PSA → levels < 0,15 are suspicious for prostate-carcinoma (4 - 10 ng/ml total-PSA)
S-CA 15-3	U/ml	< 28	
S-CA 125	U/ml	< 35	

Note: *Only PSA (prostate carcinoma) and AFP (hepatocellular carcinoma)are suitable for tumour screening. We have to consider total levels and also a rise in tumour markers even within normal reference intervals!*

Medication plasma levels (therapeutic concentration)			
S-Digitoxin	µg/l	10 - 25	
S-Digoxin	µg/l	0,5 – 0,8	
S-Theophyllin	mg/l	10 - 20	

Rheuma serology tests			
S-Rheuma factor Latex	IU/ml	up to 20	
S-Rheuma-Waaler-Rose	IU/ml	up to 10	
S-Antistreptolysin titre	IU/ml	up to 200	
(further: refer to chapter Rheumatology)			

Hormones: FT3, FT4, TSH → refer to chapter "Thyroid gland"

Urine tests

Test	unit	reference interval male | female	multiplication factor deviation
U-specific gravity	g/l	1.012 - 1.022	
U-osmolality	mosmol/kg	855 - 1.335	
U-pH	PH	4,8 - 7,6	Vegetarian nutrition ↑

Quantitative tests

(Electrolytes, urate and urea depend very much on nutrition)

Test	unit	reference interval	multiplication factor deviation
U-sodium	mmol/24 hrs	90 - 300	children ↓
U-potassium	mmol/24 hrs	25 - 105	
U-Calcium	mmol/24 hrs	2,0 -8,0 | 1,5 - 6,5	
U-Magnesium	mmol/24 hrs	2,0 - 8,0 | 1,5 - 7,0	
U-Chloride	mmol/24 hrs	80 - 270	
U-Phosphate	mmol/24 hrs	4 - 36	
U-urate	g/24 hrs	0,3 -0,8 | 0,3 - 0,7	x 59,485 = μmol/l
U-urea	g/24 hrs	13 - 33	x 0,1665 = mmol/l
U-Protein	mg/24 hrs	< 150	
U-Albumin	mg/24 hrs	< 30	
U-glucose (24 hrs)	mg/dl	up to 20	x 0,0555 = mmol/l
U-α_1-Microglobulin	mg/24 hrs	up to 13,3	
U-β_2-Microglobulin	mg/l	up to 0,3	
U-Amylase	U/l	up to 530	

Urine microscopy

Test	unit	reference interval	
U-Erythrocytes	/μl	up to 5	
U-Leukocytes	/μl	up to 10	

ESR = ERYTHROCYTE SEDIMENTATION RATE

Unspecific screening test for suspected inflammatory disease and its course.

Methods (Westergren): mix 3,8 % Na-Citrate with venous blood (1 : 5 → 1,6 ml blood + 0,4 ml Na-Citrate solution) and put into a vertical 200 mm pipette.

reference interval at 1 hr (in mm): m up to 15, f up to 20

Patients > 50 years: m up to 20, f up to 30.
Reading ESR at 1 hr is sufficient; the 2 hrs-reading doesn't provide any additional information. Slightly raised ESR may be normal in older people.

ESR depends on the following factors:
• Erythrocytes: macrocytosis, anaemia: ESR ↑
• Poikilocytosis, Polycythaemia: ESR ↓
• Plasma proteins: increased acute-phase-proteins (Fibrinogen, α_2-Macroglobulin etc.), Immunoglobulins, Immunocomplexes): ESR ↑
• source of errors:
  performance of ESR:
  Pipette is wet, pipette is not completely vertical (even when just tilted by 10° will double the ESR), moving the pipette whilst reading the result or beforehand, changing amount of added citrate (too little citrate: ESR ↓, too much citrate: ESR ↑), clouded citrate solution, changing temperature (sun, heating; cold temperature reduces, high temperature accelerates), inappropriate mixture of citrate and blood; citrate-blood-mix must be discarded after 5 hrs (inhibits ESR up to 20 %).

Physiological ESR-changes:
• Pre-menstrual and taking oral contraceptives cause minor rise in ESR.
• From 4th Pregnancy week ESR-rise; highest levels during the first postnatal week (up to 55 mm/hrs); Cause: Fibrinogen ↑

Iatrogenic ESR-changes:
Dextran infusions increase ESR (up to 75 mm/hrs)

Diseases associated with raised ESR:
Inflammations (infectious, non- infectious), subacute thyroiditis de Quervain, neoplasm (mainly those with metastases), autoimmune diseases (e.g. giant cell arteritis: Polymyalgia rheumatica and Arteriitis temporalis Horton), nephrotic syndrome, haematological diseases (leukaemia, anaemia, haemolysis caused by antibodies), multiple myeloma, M. Waldenström etc.

Further tests when ESR is much increased:
1. history and patient examination, laboratory screening: blood count, urine, serum electropheresis, creatinine, hepatic enzymes etc.
2. search for any inflammatory focus, tumours, autoimmune diseases (US, CXR, GI investigation, rheumatology serology, Coombs-test); refer to gynaecology, urology, ENT etc.

In 5 % of all raised ESR there is no cause to be found.

C-REACTIVE PROTEIN (CRP)

CRP is a classic unspecific "Acute-Phase-Protein", produced by inflammations and tumours. Interleukin-6 and other cytokines induce CRP-production in the liver. CRP is equally useful for diagnosis as ESR. But pregnancy and changes in erythrocytes do not influence the CRP. CRP got a short plasma-half-life of 24 hrs; compared to ESR the is an immediate rise at the beginning of the disease, and it also returns much quicker to normal (ca. 2 weeks after the disease is over) (ESR: ca. 4 weeks). Hence for the diagnostics of acute infections CRP is superior to ESR. Acute uncomplicated virus infections are not associated with a rise of CRP; bacterial infections increase CRP!
reference interval: < 5 mg/l

INDEX

HEROLD's INTERNAL MEDICINE – Vol. 1
ISBN: 978-1-4467-6367-4
Lulu-ID: 10003820
http://www.lulu.com/
€ 23.50

- Evidence based medicine
- Haematology
- Cardiology
- Pulmonology
- Gastroenterology (part 1)

HEROLD's INTERNAL MEDICINE – The E-Book
ISBN: 978-1-4466-4798-1
Lulu-ID: 9545211
http://www.lulu.com/
€ 35.00

Made in the USA
Lexington, KY
30 December 2011